Lecture Notes in Computer Sci

Commenced Publication in 1973
Founding and Former Series Editors:
Gerhard Goos, Juris Hartmanis, and Jan van Leeuwen

T0238823

Editorial Board

Serge Vaudenay (Ed.)

Progress in Cryptology – AFRICACRYPT 2008

First International Conference on Cryptology in Africa
Casablanca, Morocco, June 11-14, 2008
Proceedings

 Springer

Volume Editor

Serge Vaudenay
Ecole Polytechnique Fédérale de Lausanne (EPFL), I&C, ISC, LASEC
Station 14 - Building INF, 1015 Lausanne, Switzerland
E-mail: serge.vaudenay@epfl.ch

Library of Congress Control Number: 2008926815

CR Subject Classification (1998): E.3, F.2.1-2, G.2.1, D.4.6, K.6.5, C.2, J.1

LNCS Sublibrary: SL 4 – Security and Cryptology

ISSN   0302-9743
ISBN-10   3-540-68159-0 Springer Berlin Heidelberg New York
ISBN-13   978-3-540-68159-5 Springer Berlin Heidelberg New York

Springer is a part of Springer Science+Business Media

springer.com

© Springer-Verlag Berlin Heidelberg 2008
Printed in Germany

Typesetting: Camera-ready by author, data conversion by Scientific Publishing Services, Chennai, India
Printed on acid-free paper   SPIN: 12271184   06/3180   5 4 3 2 1 0

Preface

The AFRICACRYPT 2008 conference was held during June 11–14, 2008 in Casablanca, Morocco. Upon the initiative of the organizers from the Ecole normale supérieure in Casablanca, this event was the first international research conference in Africa dedicated to cryptography.

The conference was honored by the presence of the invited speakers Bruce Schneier, Jacques Stern, and Alexander W. Dent who gave talks entitled "The Psychology of Security" "Modern Cryptography: A Historical Perspective" and "A Brief History of Provably-Secure Public-Key Encryption", respectively. These proceedings include papers by Bruce Schneier and by Alexander Dent.

The conference received 82 submissions on November 24, 2007. They went through a careful doubly anonymous review process. This was run by the iChair software written by Thomas Baignères and Matthieu Finiasz. Every paper received at least three review reports. After this period, 25 papers were accepted on February 12, 2008. Authors then had the opportunity to update their papers until March 13, 2008. The present proceedings include all the revised papers.

At the end of the review process, the paper entitled "An Authentication Protocol with Encrypted Biometric Data" written by Julien Bringer and Hervé Chabanne was elected to receive the Africacrypt 2008 Best Paper Award.

I had the privilege to chair the Program Committee. I would like to thank all committee members for their tough work on the submissions, as well as all external reviewers for their support. I also thank my assistant Thomas Baignères for maintaining the server and helping me to run the software. I thank the invited speakers, the authors of the best paper, the authors of all submissions. They all contributed to the success of the conference.

Finally, I heartily thank the General Chair Abdelhak Azhari, Chair of the AMC (Moroccan Association for Cryptography), as well as his team for having organized this wonderful conference, and especially Abderrahmane Nitaj with whom I had a very good interaction. I hope their efforts will contribute to the successful development of academic research in cryptology in Africa.

June 2008   Serge Vaudenay

Africacrypt 2008

June 11–14, 2008, Casablanca, Morocco

General Chair

Abdelhak Azhari, AMC Casablanca, Morocco

Program Chair

Serge Vaudenay, EPFL Lausanne, Switzerland

Program Committee

Tom Berson	Anagram, USA
Alex Biryukov	University of Luxembourg, Luxembourg
Xavier Boyen	Voltage Inc., USA
Anne Canteaut	INRIA, France
Jean-Marc Couveignes	Toulouse University, France
Mohamed El Marraki	Faculty of Science Rabat, Morrocco
Steven Galbraith	Royal Holloway University of London, UK
Helena Handschuh	Spansion, France
Tetsu Iwata	Nagoya University, Japan
Pascal Junod	Nagravision, Switzerland
Tanja Lange	TU Eindhoven, The Netherlands
Arjen Lenstra	EPFL, Switzerland, and Alcatel-Lucent Bell Laboratories, USA
Javier Lopez	University of Malaga, Spain
Stefan Lucks	Bauhaus University Weimar, Germany
Mitsuru Matsui	Mitsubishi Electric Corp., Japan
Alexander May	Bochum University, Germany
Atsuko Miyaji	JAIST, Japan
David Molnar	Berkeley University, USA
Refik Molva	Eurecom, France
Jean Monnerat	UCSD, USA
David Naccache	ENS, France
Raphael Phan	EPFL, Switzerland
Josef Pieprzyk	Macquarie University, Australia
Bart Preneel	K.U. Leuven, Belgium
Jean-Jacques Quisquater	UCL, Belgium
C Pandu Rangan	Indian Institute of Technology, Madras, India
Vincent Rijmen	Graz University of Technology, Austria and K.U. Leuven, Belgium

Rei Safavi-Naini   University of Calgary, Canada
Louis Salvail   University of Aarhus, Denmark
Ali Aydın Selçuk   Bilkent University, Turkey
Serge Vaudenay   EPFL, Switzerland
Michael Wiener   Cryptographic Clarity, Canada
Amr Youssef   Concordia University, Canada

External Reviewers

Murat Ak   Marcelo Kaihara   Kenny Paterson
Cristina Alcaraz   Said Kalkan   Deike Priemuth-Schmid
Elena Andreeva   Alexandre Karlov   Mohammed-Reza
                                                         Reyhanitabar
Nuttapong Attrapadung   Kamer Kaya   Maike Ritzenhofen
Jean-Philippe Aumasson   Dmitry Khovratovich   Kazuo Sakiyama
Roberto Avanzi   Jongsung Kim   Takashi Satoh
John Aycock   Varad Kirtane   Christian Schaffner
Daniel J. Bernstein   Ulrich Kühn   Berry Schoenmakers
Emmanuel Bresson   Gregor Leander   Nicolas Sendrier
Christophe de Cannière   Jesang Lee   Siamak F. Shahandashti
Chris Charnes   Kerstin Lemke-Rust   Andrew Shallue
Kim-Kwang Raymond Choo   Marco Macchetti   Abdullatif Shikfa
Véronique Cortier   Fréderic Magniez   Masaaki Shirase
Ivan Damgård   Alexander Maximov   Igor Shparlinski
Ed Dawson   Yusuke Naito   Michal Sramka
Hüseyin Demirci   Ivica Nikolic   Kannan Srinathan
Alex Dent   Marine Minier   Dirk Stegemann
Gerardo Fernandez   David Mireles   Damien Stehlé
Ewan Fleischmann   Pradeep Mishra   Daisuke Suzuki
Fabien Galand   Michael Naehrig   Emin Tatli
Elisa Gorla   Adam O'Neill   Frederik Vercauteren
Michael Gorski   Hiroyuki Okazaki   Damien Vergnaud
Aline Gouget   Katsuyuki Okeya   Dai Watanabe
Safuat Hamdy   Melek Önen   Christopher Wolf
Shoichi Hirose   Dag Arne Osvik   Juerg Wullschleger
Toshiyuki Isshiki   Raphael Overbeck   Go Yamamoto
Shaoquan Jiang   Dan Page
Ellen Jochemsz   Sylvain Pasini

Table of Contents

Public-Key Cryptography

Pseudorandomness

Analysis of Stream Ciphers

Hash Functions

Broadcast Encryption

Invited Talk

Implementation

Improving Integral Attacks Against Rijndael-256 Up to 9 Rounds

Samuel Galice and Marine Minier

CITI / INSA-Lyon
F-69621 Villeurbanne
{samuel.galice,marine.minier}@insa-lyon.fr

Abstract. Rijndael is a block cipher designed by V. Rijmen and J. Daemen and it was chosen in its 128-bit block version as AES by the NIST in October 2000. Three key lengths - 128, 192 or 256 bits - are allowed. In the original contribution describing Rijndael [4], two other versions have been described: Rijndael-256 and Rijndael-192 that respectively use plaintext blocks of length 256 bits and 192 bits under the same key lengths and that have been discarded by the NIST. This paper presents an efficient distinguisher between 4 inner rounds of Rijndael-256 and a random permutation of the blocks space, by exploiting the existence of semi-bijective and Integral properties induced by the cipher. We then present three attacks based upon the 4 rounds distinguisher against 7, 8 and 9 rounds versions of Rijndael-256 using the extensions proposed by N. ferguson et al. in [6]. The best cryptanalysis presented here works against 9 rounds of Rijndael-256 under a 192-bit key and requires $2^{128} - 2^{119}$ chosen plaintexts and 2^{188} encryptions.

Keywords: block cipher, cryptanalysis, integral attacks, Rijndael-256.

1 Introduction

Rijndael [4] is an SPN block cipher designed by Vincent Rijmen and Joan Daemen. It has been chosen as the new advanced encryption standard by the NIST [7] with a 128-bit block size and a variable key length k, which can be set to 128, 192 or 256 bits. It is a variant of the Square block cipher, due to the same authors [3]. In its full version, the block length b is also variable and is equal to 128, 192 or 256 bits as detailed in [5] and in [10]. We respectively called those versions Rijndael-b. The recommended Nr number of rounds is determined by b and k, and varies between 10 and 14.

Many cryptanalyses have been proposed against Rijndael for the different block sizes and more particularly against the AES. The first attack against all the versions of Rijndael-b is due to the algorithm designers themselves and is based upon the integral (or saturation) property ([3], [4], [12]) that allows to efficiently distinguish 3 Rijndael inner rounds from a random permutation. This attack has been improved by Ferguson et al. in [6] allowing to cryptanalyse an 8 rounds version of Rijndael-b with a complexity equal to 2^{204} trial encryptions and $2^{128} - 2^{119}$ plaintexts and a 9 rounds version using a related-key attack.

S. Vaudenay (Ed.): AFRICACRYPT 2008, LNCS 5023, pp. 1–15, 2008.

In [13], S. Lucks presented an other improvement of the Square Attack using a particular weakness of the key schedule against a 7 rounds version of Rijndael-b where 2^{194} executions are required for a number of chosen plaintexts equal to 2^{32}. H. Gilbert and M. Minier in [8] also presented an attack against a 7 rounds version of Rijndael-b (known under the name of "Bottleneck Attack") using a stronger property on three inner rounds than the one used in the Square Attack in order to mount an attack against a 7 rounds version of Rijndael requiring 2^{144} cipher executions with 2^{32} chosen plaintexts.

Many other attacks ([2], [14]) have been exhibited against the AES using algebraic techniques exploiting the low algebraic degree of the AES S-box. Other attacks that use related keys and rectangle cryptanalysis have been proposed in [9] and in [11]. But none of these attacks exploits new intrinsic structure of the transformations used in Rijndael-b.

This paper describes an efficient distinguisher between 4 Rijndael-256 inner rounds and a random permutation based upon a particular integral (or saturation) property due to a slow diffusion, presents the resulting 7 rounds attacks on Rijndael-256 which are substantially faster than an exhaustive key search for all the key lengths and the corresponding 8 and 9 rounds extension of the previous attacks for $k = 192$ and $k = 256$.

This paper is organized as follows: Section 2 provides a brief outline of Rijndael-b. Section 3 recalls the original Integral property on three inner rounds, investigates the new four rounds property and describes the resulting distinguisher for 4 inner rounds. Section 4 presents 7, 8 and 9 rounds attacks based on the 4 rounds distinguisher of Section 3. Section 5 concludes this paper.

2   A Brief Outline of Rijndael-b

Rijndael-b is a symmetric block cipher that uses a parallel and byte-oriented structure. The key length is variable and equal to 128, 192 or 256 bits whereas the block length is equal to 128, 192 or 256 bits. The current block at the input of the round r is represented by a $4 \times (b/32)$ matrix of bytes $A^{(r)}$. We give its representation for $b = 256$:

$$A^{(r)} = \begin{array}{|c|c|c|c|c|c|c|c|}
\hline
a_{0,0}^{(r)} & a_{0,1}^{(r)} & a_{0,2}^{(r)} & a_{0,3}^{(r)} & a_{0,4}^{(r)} & a_{0,5}^{(r)} & a_{0,6}^{(r)} & a_{0,7}^{(r)} \\
\hline
a_{1,0}^{(r)} & a_{1,1}^{(r)} & a_{1,2}^{(r)} & a_{1,3}^{(r)} & a_{1,4}^{(r)} & a_{1,5}^{(r)} & a_{1,6}^{(r)} & a_{1,7}^{(r)} \\
\hline
a_{2,0}^{(r)} & a_{2,1}^{(r)} & a_{2,2}^{(r)} & a_{2,3}^{(r)} & a_{2,4}^{(r)} & a_{2,5}^{(r)} & a_{2,6}^{(r)} & a_{2,7}^{(r)} \\
\hline
a_{3,0}^{(r)} & a_{3,1}^{(r)} & a_{3,2}^{(r)} & a_{3,3}^{(r)} & a_{3,4}^{(r)} & a_{3,5}^{(r)} & a_{3,6}^{(r)} & a_{3,7}^{(r)} \\
\hline
\end{array}$$

The key schedule derives $Nr + 1$ b-bits round keys K_0 to K_{Nr} from the master key K of variable length.

The round function, repeated $Nr - 1$ times, involves four elementary mappings, all linear except the first one:

- SubBytes: a bytewise transformation that applies on each byte of the current block an 8-bit to 8-bit non linear S-box (that we call S) composed of the inversion in the Galois Field $GF(256)$ and of an affine transformation.

- ShiftRows: a linear mapping that rotates on the left all the rows of the current matrix (0 for the first row, 1 for the second, 3 for the third and 4 for the fourth in the case of Rijndael-256 as described in [4]).
- MixColumns: another linear mapping represented by a 4×4 matrix chosen for its good properties of diffusion (see [5]). Each column of the input matrix is multiplied by the MixColumns matrix M in the Galois Field $GF(256)$ that provides the corresponding column of the output matrix. We denote by $M_{i,j}$ for i and j from 0 to 3, the coefficients of the MixColumns matrix.
- AddRoundKey: a simple x-or operation between the current block and the subkey of the round r denoted by K_r. We denote by $K_r^{(i,j)}$ the byte of K_r at position (i, j).

Those $Nr - 1$ rounds are surrounded at the top by an initial key addition with the subkey K_0 and at the bottom by a final transformation composed by a call to the round function where the MixColumns operation is omitted.

3   The Integral Properties

We describe in this section the three inner rounds property named Integral property explained in the original proposal [4] and the new four rounds property of Rijndael-256.

3.1   The Integral Property of Rijndael-b

This particular property studied in [12] was first used to attack the Square block cipher [3] and holds for the three size of blocks (128, 192 or 256 bits) of the initial version of Rijndael-b. As previously mentioned, we denote by $A^{(r)}$ the input of the round r.

Let us define the set Λ which contains 256 plaintext blocks (i.e. 256 matrices of bytes of size $4 \times (b/32)$) all distinct. Two blocks belong to the same set Λ if they are equal everywhere except on a particular predefined byte (called the active byte). This active byte takes all possible values between 0 and 255:

$$\forall A^{(1)}, A'^{(1)} \in \Lambda : \begin{cases} a_{i,j}^{(1)} \neq a_{i,j}'^{(1)} \text{ for a given } i \text{ and a given } j \\ a_{i,j}^{(1)} = a_{i,j}'^{(1)} \text{ elsewhere} \end{cases}$$

for $0 \leq i \leq 3$ and $0 \leq j \leq (b/32)$.

The Λ set contains then one active byte whereas the other bytes are passive. Notice that this definition could be generalized to several active bytes as we will see in the next subsections. In all the cases, the transformations SubBytes and AddRoundKey transform a set Λ into another set Λ with the positions of the active bytes unchanged (see [1] and [12] for more details).

Now, if we look at the semi-bijective properties of the internal transformations of Rijndael-b on three rounds - especially the ones of the ShiftRows and of the MixColumns operations -, we could observe the following results (as shown in figure 1):

- The MixColumns of the first round transforms the active byte of the Λ set into a complete column of active bytes.
- The ShiftRows of the second round diffuses this column on four distinct columns whereas the MixColumns converts this to four columns of only active bytes. This stays a Λ set until the input of MixColumns of the third round.
- Until the input of the MixColumns of the third round, the implied transformations constitute a bijection. Then, since the bytes of this Λ set, range over all possible values and are balanced over this set, we have if we denote by $M^{(3)}$ the active blocks belonging to the Λ set at the input of the third MixColumns operation:

$$\bigoplus_{A^{(4)}=MC(M^{(3)}),M^{(3)}\in\Lambda} a_{k,l}^{(4)} = \bigoplus_{M^{(3)}\in\Lambda} \left(2m_{k,l}^{(3)} \oplus 3m_{k+1,l}^{(3)} \oplus m_{k+2,l}^{(3)} \oplus m_{k+3,l}^{(3)}\right)$$
$$= 0 \qquad\qquad (1)$$

where MC represents the MixColumns of the third round and k and l taking all possible values.

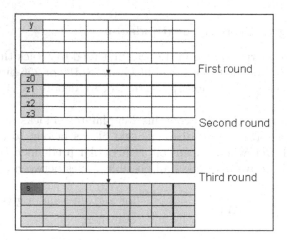

Fig. 1. The three rounds integral property in the case of Rijndael-256: $\bigoplus_{y\in\Lambda} s = 0$

Then, we can easily deduce that each byte at the input of the fourth round is balanced in order to construct an efficient distinguisher between three Rijndael-b inner rounds and a random permutation testing if equality (1) occurs and requiring 256 plaintexts belonging to a same Λ set. Notice also that this property holds for all the possible positions of the active byte.

3.2   An Improvement of the Saturation Property for 4 Rounds of Rijndael-256

In the case of Rijndael-256, we describe in this section a particular and stronger property on three rounds of Rijndael-256 and how to extend this property to four rounds for a particular Λ set with three active bytes.

A Stronger Three Rounds Property. Suppose now that the same Λ set than the previous one with one active byte, say y at byte position (i, j), is defined. Let us see how this set crosses three Rijndael-256 inner rounds:

- The MixColumns of the first round transforms the active byte of the Λ set into a complete column of active bytes (called $z_0, \cdots, z_3$ in figure 2).
- The ShiftRows of the second round diffuses this column on four among eight distinct columns whereas the MixColumns converts this to four columns among eight of only active bytes.
- The ShiftRows of the third round diffuses those four columns into the eight columns of the current block but the $(j+2 \mod 8)$-th and the $(j+6 \mod 8)$-th columns only depend on one byte each, say $a^{(2)}_{i+2 \mod 4, j}$ and $a^{(2)}_{i+1 \mod 4, j}$. Using the notations of figure 2, we could say that at the end of the third round, the third and the seventh columns only depend respectively on the byte $a^{(2)}_{2,0} = z_2$ and $a^{(2)}_{1,0} = z_1$; thus, the bytes of those two columns bijectively depend on the y value and each of those two columns represent a Λ set.

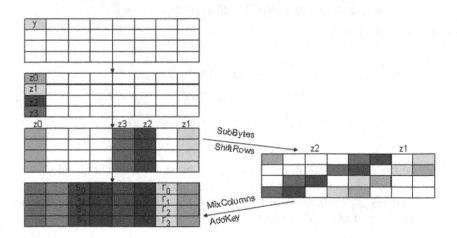

Fig. 2. The three rounds property in the case of Rijndael-256: the bytes $s_0, \cdots, s_3$ only depend on z_2 and the bytes $r_0, \cdots, r_3$ only depend on z_1

So, we have demonstrated that two particular columns stay two different Λ sets at the output of the third round.

How to Exploit this Property? Because two complete Λ sets remain for two particular columns at the end of the third round, we want to find a way to exploit this particular property on four rounds of Rijndael-256 by adding one round at the end of the three previous rounds. We always consider here that the input of the four rounds is a Λ set with one active byte y.

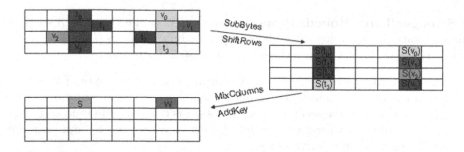

Fig. 3. The fourth round added after the three previous ones

Using the notation of figure 3, we could write the output bytes s and w at the end of the fourth round according to the output bytes of the third round, noticing that t_0, v_3, v_0 and t_3 belongs to two Λ sets:

$$s = 2 \cdot S(t_0) \oplus 3 \cdot S(t_1) \oplus S(t_2) \oplus S(t_3) \oplus K_4^{(0,2)}$$

$$w = 2 \cdot S(v_0) \oplus 3 \cdot S(v_1) \oplus S(v_2) \oplus S(v_3) \oplus K_4^{(0,6)}$$

More formally, we obtain:

$$a_{0,j+2 \bmod 8}^{(5)} = 2S(a_{0,j+2 \bmod 8}^{(4)}) \oplus 3S(a_{1,j+3 \bmod 8}^{(4)})$$
$$\oplus S(a_{2,j+5 \bmod 4}^{(4)}) \oplus S(a_{3,j+6 \bmod 4}^{(4)}) \oplus K_4^{(0,2)} \qquad (2)$$
$$a_{0,j+6 \bmod 8}^{(5)} = 2S(a_{0,j+6 \bmod 8}^{(4)}) \oplus 3S(a_{1,j+7 \bmod 8}^{(4)})$$
$$\oplus S(a_{2,j+1 \bmod 8}^{(4)}) \oplus S(a_{3,j+2 \bmod 8}^{(4)}) \oplus K_4^{(0,6)}$$

If we use the notations of figure 3 and if we consider as in the previous subsection that the input of the four rounds is a Λ set with one active byte, say y, then we have: $\bigoplus_{y \in \Lambda} 2S(t_0) = 0$ and $\bigoplus_{y \in \Lambda} S(t_3) = 0$ because t_0 and t_3 belongs to the same Λ set at the end of the third round. Thus, we could write:

$$\bigoplus_{y \in \Lambda} s = \bigoplus_{y \in \Lambda} (2S(t_0) \oplus 3S(t_1) \oplus S(t_2) \oplus S(t_3)) \qquad (3)$$

$$= 2 \bigoplus_{y \in \Lambda} S(t_0) \oplus 3 \bigoplus_{y \in \Lambda} S(t_1) \oplus \bigoplus_{y \in \Lambda} S(t_2) \oplus \bigoplus_{y \in \Lambda} S(t_3)$$

$$= 0 \oplus 3 \bigoplus_{y \in \Lambda} S(t_1) \oplus \bigoplus_{y \in \Lambda} S(t_2) \oplus 0$$

The same property holds for w.

So, we want to find a way to obtain

$$3 \bigoplus_{y \in \Lambda} S(t_1) \oplus \bigoplus_{y \in \Lambda} S(t_2) = 0 \tag{4}$$

considering that t_1 depends on z_1 and on z_3 and that t_2 depends on z_2 and on z_0. Thus, more input blocks are required to satisfy this equality. A good solution to produce such equality is to take $(256)^2 \Lambda$ sets to completely saturated the values of z_1 and z_3 for t_1 and of z_0 and z_2 for t_2. To produce a such number of plaintexts, let us define the following Λ set with three active bytes - say y, n, p as denoted in figure 4 - at the positions (i, j), $(i + 1 \mod 4, j)$ and $(i + 2 \mod 4, j)$. More formally, we could write this new Λ set as follows:

$$\forall A^{(1)}, A'^{(1)} \in \Lambda : \begin{cases} a_{i,j}^{(1)} \neq a_{i,j}'^{(1)}, a_{i+1 \bmod 4,j}^{(1)} \neq a_{i+1 \bmod 4,j}'^{(1)} \text{ and} \\ a_{i+2 \bmod 4,j}^{(1)} \neq a_{i+2 \bmod 4,j}'^{(1)} \\ \text{for a given } i \text{ and a given } j \\ a_{i,j}^{(1)} = a_{i,j}'^{(1)} \text{ elsewhere} \end{cases}$$

Using such a Λ set with 2^{24} elements generated from three different active bytes (say y, n and p) belonging to a same input column, at the end of the fourth round, equality (4) is verified and we then could write using equality (3), $\bigoplus_{y,n,p \in \Lambda} s = 0$ and $\bigoplus_{y,n,p \in \Lambda} w = 0$. More formally, we have:

$$\bigoplus_{y,n,p \in \Lambda} a_{i,j+2 \bmod 8}^{(5)} = 0 \tag{5}$$

$$\bigoplus_{y,n,p \in \Lambda} a_{i,j+6 \bmod 8}^{(5)} = 0 \tag{6}$$

for all $i \in \{0..3\}$.

We performed some computer experiments which confirm the existence of those properties for arbitrarily chosen key values. The complete property is represented on figure 4. Notice also that when only two bytes - say y and n - at position (i, j) and $(i + 1 \mod 4, j)$ are saturated, the corresponding properties are less strong and could be written as partial sums:

$$\bigoplus_{y,n \in \Lambda} a_{0,j+2 \bmod 8}^{(5)} = \bigoplus_{y,n \in \Lambda} a_{3,j+2 \bmod 8}^{(5)}$$

$$\bigoplus_{y,n \in \Lambda} a_{0,j+6 \bmod 8}^{(5)} = \bigoplus_{y,n \in \Lambda} a_{3,j+6 \bmod 8}^{(5)}$$

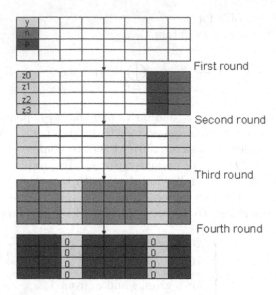

Fig. 4. The four Rijndael-256 rounds property

3.3   The 4 Rounds Distinguisher

Then, we can easily use equality (5) or equality (6) at the input of the fifth round in order to construct an efficient distinguisher between four Rijndael-256 inner rounds and a random permutation testing if equality (5) or (6) occurs and requiring 2^{24} plaintexts belonging to a same Λ set with three active bytes at positions (i, j), $(i + 1 \mod 4, j)$ and $(i + 2 \mod 4, j)$.

The existence of such property for Rijndael-256 is not really surprising even if it has never been observed before. This property is due to a slower diffusion in Rijndael-256 than in Rijndael-128 (the AES) and in Rijndael-192. Note also that this particular property doe not work for the AES and Rijndael-192: there is no particular output byte after the third round that only depends on one particular byte of the corresponding input and we do not find such a property for the AES and Rijndael-192.

4   The Proposed Attacks

We could use the properties previously described on four Rijndael-256 inner rounds to mount elementary attacks against 7, 8 and 9 rounds versions of Rijndael-256. To attack 7 rounds of Rijndael-256, we use first the extension by one round at the beginning proposed in [6] and the partial sums technique described in [6] with equality (5) to add two rounds at the end of our four rounds distinguisher. To extend this 7 rounds attack by one round at the end and/or by one round at the beginning, we directly apply the techniques proposed in [6] and the weakness of the Rijndael key-schedule proposed in [13].

4.1 The 7 Rounds Attack

Extension at the Beginning. Usually and as done in [4] and in [13], to extend the distinguisher that use equality (1) by one round at the beginning, the authors first choose a set of 2^{32} plaintexts that results in a Λ set at the output of the first round with a single active byte (see figure 5). This set is such that one column of bytes at the input of the first MixColumns range over all possible values and all other bytes are constant. Then, under an assumption of the four well-positioned key bytes of K_0, a set of 256 plaintexts that result in a Λ set at the input of the second round is selected from the 2^{32} available plaintexts.

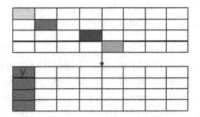

Fig. 5. The extension by one round at the beginning

Instead of guessing four bytes of the first subkey K_0, the authors of [6] simply use all the 2^{32} plaintexts that represents in our case 2^8 Λ sets with three active bytes (2^8 groups of 2^{24} encryptions that vary only in three bytes of $A^{(1)}$). Then, for some partial guesses of the key bytes at the end of the cipher, do a partial decryption to a single byte of $A^{(5)}$, sum this value over all the 2^{32} encryptions and check for a zero result.

This first improvement save a factor 2^{32} corresponding with 4 exhaustive key bytes of K_0 compared to the attack proposed in [4] using always 2^{32} plaintexts/ciphertexts.

Note also, as done in [10], that we could see this extension as a distinguisher with one more round implying a Λ set with 4 active bytes that results after the first round into an other Λ set with a complete column of active bytes.

Partial Sums Technique. Using the method of [6], we could use the equality (5) to attack a 7 rounds version of Rijndael-256 by adding one round at the beginning using the technique previously described and adding two rounds at the end using the two rounds extension proposed in [6] and described in figure 6. We describe here the original attack and then directly apply it in our case.

This extension works in the original paper on a 6 rounds version of Rijndael and looks at a particular byte of $A^{(5)}$ that verifies (1) and how it relates to the ciphertext. First, the authors rewrite the cipher slightly by putting the AddRoundKey before the MixColumns in round 5. Instead of applying MixColumns and then adding K_5, they first add in K_5', which is a linear combination of four

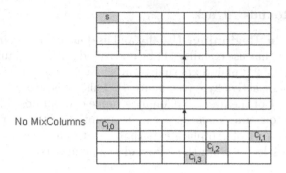

Fig. 6. The extension by two rounds at the end proposed in [6], considering that the last round does not contain a MixColumns operation

bytes of K_5, and then apply MixColumns. Under this assumption, it is easy to see that any byte of $A^{(5)}$ depends on the ciphertext, on four bytes of K_6 and one byte of K_5' considering that the sixth round is the last one and does not contain a MixColumns operation. Then, only the five key bytes of the two last rounds remain unknowns.

Moreover, the authors improve the complexity of their attack using a technique called "partial sums" to sequentially decipher the two last rounds (the last not containing the MixColumns operation) according the values of the five unknown key bytes. To use the three rounds distinguisher given by equation (1), they compute from the i-th ciphertext c_i:

$$\sum_i S^{-1} \left[S_0 \left[c_{i,0} \oplus k_0 \right] \oplus S_1 \left[c_{i,1} \oplus k_1 \right] \oplus S_2 \left[c_{i,2} \oplus k_2 \right] \oplus S_3 \left[c_{i,3} \oplus k_3 \right] \oplus k_4 \right] \quad (7)$$

where S_0, S_1, S_2, S_3 represent the inverse of the S-box S multiplied by a component of InvMixColumns, $c_{i,j}$ the byte number j of c_i; $k_0, \cdots, k_3$ the four bytes of K_6 and k_4 the implied byte of K_5'.

To improve the general complexity of the attack, they associate the following partial sums to each ciphertext c:

$$x_k := \sum_{j=0}^{k} S_j \left[c_j \oplus k_j \right]$$

for k from 0 to 3. They use the transformation $(c_0, c_1, c_2, c_3) \rightarrow (x_k, c_{k+1}, \cdots, c_3)$ to sequentially determine the different values of k_k and to share the global computation into 4 steps of key bytes search (always testing if equation (1 happens) with 2^{48} operations for each one corresponding with 2^{50} S-box lookups for each set of 2^{32} ciphertexts, corresponding with 2^{24} particular Λ sets of plaintexts with one active byte (see [6] for the details of the complexities). To discard false alarms (i.e. bad keys that pass the test), they need to repeat this process on 6 different Λ sets. Then, the general complexity of the partial sums attacks against a 6 rounds version of Rijndael-b is about 2^{44} encryptions (considering that 2^8 S-box applications are roughly equivalent with one trial encryption) using $6 \cdot 2^{32}$ plaintexts.

The Corresponding 7 Rounds Attacks. Applying the first extension at the beginning and the partial sums technique, we could directly mount an attack against 7 rounds of Rijndael-256 using the equality (5) and the corresponding four Rijndael-256 rounds distinguisher: we test if equality (5) holds for $A^{(6)}$ by summing on the 2^{32} values of the partial decryptions corresponding with the 2^{32} plaintexts that represent in our case 2^8 Λ sets with three active bytes. Then, we exploit the partial sums technique on the four corresponding bytes of K_7 and the implied byte of K_6'. For a set of 2^{32} ciphertexts, the cost of the four steps of the deciphering process is exactly the same than in the previous attack and is about 2^{50} S-box lookups. We need to repeat the process using around 6 different sets of 2^{32} ciphertexts to detect false alarms as in [6]. Then, the total number of S-box lookups is 2^{52} corresponding with 2^{44} encryptions, always considering that 2^8 S-box applications is roughly equivalent with one trial encryption.

4.2 The 8 Rounds Attack

The Naive Approach. As done in [6] and in [13], we could directly improve the previous 7 rounds attack by adding one round at the end. To express a single byte of $A^{(6)}$ in the key and the ciphertext, we could extend equation (5) to three levels at the end with 16 ciphertexts bytes and 21 key bytes. However, the partial sums technique is only helpful during the last part of the computation.

For a 192-bit master key, we first guess the required 112 bits of the last round 256-bit subkey. The two last bytes of this subkey required for the computations could be directly deduced from the other 112 bits due to the weakness of the key schedule described in [13]: if we know some bytes of the subkey at position i and $i-1$, we directly deduce those at position $i - Nk$ with $Nk = 6$ for a 192-bit master key. (Note that this property is only true for some particular positions of the byte $a_{i,j}^{(6)}$, for example if $i = 0$ and $j = 2$.) Thus after guessing the 14 required bytes of this subkey, we could directly use the partial sums technique requiring about 2^{50} S-box lookups. Thus, the total cost for a structure of 2^{32} ciphertexts is about 2^{162} S-box lookups. As noticed in [6], we need to process three structures of 2^{32} ciphertexts before we start eliminating guesses for the last round key, so the overall cost of this 8-rounds attack is on the order of 2^{164} S-box lookups or about 2^{156} trial encryptions.

For a 256-bit master key, the alignment in the key schedule is different and guessing the eighth round subkey does not give any information about round keys of round 7 and of round 6. So, we could not improve the general complexity of the attack. Working in a similar fashion as before, we first guess the 128 bits of the last round key and using the partial sums technique, compute the four bytes of K_7 and the byte of K_6' for each of the 2^{32} ciphertexts belonging to a same structure. Thus, the complete cost of this attack is about 2^{178} S-box lookups for one structure. As noticed in [6], we need to process five structures of 2^{32} ciphertexts before we start eliminating guesses for the last round key, so the overall cost of this 8-rounds attack is on the order of 2^{180} S-box lookups or about 2^{172} trial encryptions.

The Herd Technique. In [6], the authors develop a technique to improve their 6 rounds attack by adding one round at the beginning. This new attack require naively the entire codebook of 2^{128} known plaintexts that could be divided into 2^{96} packs of 2^{32} plaintexts/ciphertexts that represent 2^{24} Λ sets with one active byte after this first round. But this property could not be directly exploited because in this case even the wrong keys pass the test at the end of the fifth round since equality (1) holds on for the 2^{120} Λ sets.

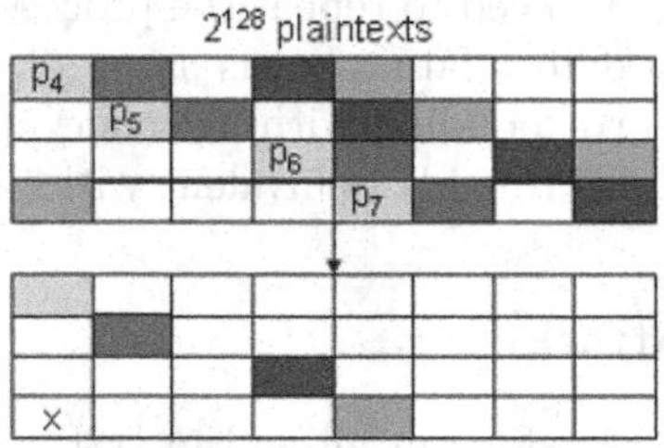

Fig. 7. The herd technique: adding one more round at the beginning

Instead, they use a particular byte at the end of the first round, say $a_{a,b}^{(2)}$ different from the four bytes of the Λ set with a fixed value x (see figure 7). With $a_{a,b}^{(2)} = x$, they obtain a set of 2^{120} possible encryptions composed of 2^{88} packs, where each pack contains 2^{24} groups of Λ sets. They call this structure with 2^{120} elements a *herd*. If they sum up equality (1) on a herd, then the property is only preserved for the correct key.

Thus, they notice that this particular byte $a_{a,b}^{(2)}$ depends on only four bytes of plaintext, say $(p_4, \cdots, p_7)$ and on four bytes of the key K_0. As done for the partial sums technique, they could share the key exhaustive search on the four key bytes of K_0 required to entirely determine the value of $a_{a,b}^{(1)}$ in a three-phase attack using 2^{64} counters m_y for the first phase, 2^{32} counters n_z for the second whereas the third phase filters information for key guesses.

The attack works as follows: in the first phase, the counter m_y is incremented at bit level according the 64-bit value $y = (c_0, \cdots, c_3, p_4, \cdots, p_7)$; in the second phase, the four bytes of K_0 are guessed to compute $a_{a,b}^{(2)}$ and to share the counters into herds; then select a single herd and update n_z by adding $z = (c_0, \cdots, c_3)$ for each y that is in the good herd; in the third phase, guess the five key bytes of K_7 and of K_6' to decrypt each z to a single byte of $A^{(6)}$, sum this byte over all the 2^{32} values of z (with multiplicities) and check for zero. This last phase must be repeated for each initial guess of the four bytes of K_0.

The first phase requires about 2^{120} trial encryptions and the rest of the attack has a negligible complexity compared to it (see [6] for some details about the attack complexity). Then, the total complexity of this attack is 2^{120} trial encryptions and 2^{64} bits of memory using 2^{128} chosen plaintexts. The authors provide another improvement of their attack remarking that the four plaintext bytes $(p_4, \cdots, p_7)$ and the four guessed key bytes of K_0 define four bytes of $A^{(1)}$. So they can create 2^{24} smaller herds with 2^{104} elements by fixing three more bytes of $A^{(1)}$ to reduce the plaintext requirements to $2^{128} - 2^{119}$ texts.

So, we could directly apply this attack to an 8 rounds version of Rijndael-256 using the particular equality (5) by adding two rounds at the beginning and two rounds at the end using $2^{128} - 2^{119}$ plaintexts that will be separated into herds during the second phase of the attack. However, in the previous case, they consider that all the codebook is known due to the huge amount of plaintexts required. So, they do not take into account the ciphering process. This is not our case and we first need to cipher $2^{128} - 2^{119}$ chosen plaintexts among the 2^{256} possible values with four active columns that lead to 2^{24} herds with 2^{104} elements at the end of the first round. Then, the complexity of the attack itself is the same but the total cost is dominated by the $2^{128} - 2^{119}$ trial encryptions. Notice that the same problem remains for Rijndael-192.

4.3   The 9 Rounds Attack

As done in [6], we could use the herd technique (with 2^{23} undamaged herds) combined with the partial sums technique to mount an attack against a 9 rounds version of Rijndael-256. In this case, we guess four bytes of K_0 and the 21 subkey bytes - 16 bytes of K_9, 4 bytes of K_8 and one byte of K'_7 - required to add three rounds at the end of the 4 rounds distinguisher. We always consider that this distinguisher is extended with one round at the beginning summing on sets with 2^{32} elements. The attack then works as follows: first, construct 2^{23} undamaged herds of 2^{104} elements using $2^{128} - 2^{119}$ plaintexts; guess the four key bytes of K_0 to determine a particular herd; then apply the partial sums technique to this set to compute each x_k and to obtain a single byte of $A^{(7)}$ depending on 16 bytes of the ciphertext and 21 subkey bytes; then use the fact that summing the 2^{104} values on a single byte of $A^{(7)}$ will yield zero (from equality (5)) for the good key. The required storage is about 2^{104} bits and the total complexity of this attack is about $2^{32} \cdot 2^{170} = 2^{202}$ trial encryptions for one herd and a 256-bit key (see [6] for the details of the complexity of the attack). We need to test four herds before discarding the first bad keys and at least 26 herds to get exactly the good key (with a decreasing complexity). Then, the total complexity of this attack is about 2^{204} trial encryptions.

This attack could only work for a 256-bit key. However, in the case of a 192-bit key, using the weakness of the key-schedule described in section 4.2, we know that we could preserve 2 bytes of the exhaustive search of K_9 that are directly determined by the 14 others. Then, we could save a 2^{16} factor from the previous attack and we obtain a complexity of about $2^{204-16} = 2^{188}$ trial encryptions for the same number of plaintexts and the same required storage.

5   Conclusion

In this paper, we have presented a new particular property on four rounds of Rijndael-256 that relies on semi-bijective properties of internal foldings. Then we have built the best known attack against a 9 rounds version of Rijndael-256 requiring for a 192-bit keys 2^{188} trial encryptions with $2^{128} - 2^{119}$ plaintexts. We have summed up in table 1 all known results concerning the attacks against Rijndael-b.

In [6], the authors also present a related key attack against a 9 rounds version of the AES. Moreover, in [9] and in [11], two related key rectangle attacks have been proposed against the AES under keys of length 192 and 256 bits. We do not find a way to extend the attacks that use related keys against Rijndael-256. The main problem in this case comes from the higher number of 32-bit key words that must be generated to construct 256-bit subkeys: we do not find a key pattern that sufficiently preserves an integral property.

Table 1. Summary of Attacks on Rijndael-b - CP: Chosen plaintexts, RK: Related-key

Cipher	nb rounds	Key size	Data	Time Complexity	source
AES	6	(all)	2^{32} CP	2^{72}	[4] (Integral)
	7	(all)	$2^{128} - 2^{119}$ CP	2^{120}	[6] (Part. Sum)
	8	(192)	$2^{128} - 2^{119}$ CP	2^{188}	[6] (Part. Sum)
	8	(256)	$2^{128} - 2^{119}$ CP	2^{204}	[6] (Part. Sum)
	9	(256)	2^{85} RK-CP	2^{224}	[6] (Related-key)
	10	(192)	2^{125} RK-CP	2^{182}	[11] (Rectangle)
Rijndael-192	6	(all)	2^{32} CP	2^{72}	[4] (Integral)
	7	(all)	$2^{128} - 2^{119}$ CP	$2^{128} - 2^{119}$	[6] (Part. Sum)
	8	(192)	$2^{128} - 2^{119}$ CP	2^{188}	[6] (Part. Sum)
	8	(256)	$2^{128} - 2^{119}$ CP	2^{204}	[6] (Part. Sum)
Rijndael-256	6	(all)	2^{32} CP	2^{72}	[4] (Integral)
	7	(all)	$2^{128} - 2^{119}$ CP	$2^{128} - 2^{119}$	[6] (Part. Sum)
	7	(all)	6×2^{32} CP	2^{44}	this paper
	8	(all)	$2^{128} - 2^{119}$ CP	$2^{128} - 2^{119}$	this paper
	9	(192)	$2^{128} - 2^{119}$ CP	2^{188}	this paper
	9	(256)	$2^{128} - 2^{119}$ CP	2^{204}	this paper

References

1. Biryukov, A., Shamir, A.: Structural cryptanalysis of SASAS. In: Pfitzmann, B. (ed.) EUROCRYPT 2001. LNCS, vol. 2045, pp. 394–405. Springer, Heidelberg (2001)
2. Courtois, N., Pieprzyk, J.: Cryptanalysis of block ciphers with overdefined systems of equations. In: Zheng, Y. (ed.) ASIACRYPT 2002. LNCS, vol. 2501, pp. 267–287. Springer, Heidelberg (2002)
3. Daemen, J., Knudsen, L.R., Rijmen, V.: The block cipher Square. In: Biham, E. (ed.) FSE 1997. LNCS, vol. 1267, pp. 149–165. Springer, Heidelberg (1997)

4. Daemen, J., Rijmen, V.: AES proposal: Rijndael. In: The First Advanced Encryption Standard Candidate Conference, N.I.S.T (1998)
5. Daemen, J., Rijmen, V.: The Design of Rijndael. Springer, Heidelberg (2002)
6. Ferguson, N., Kelsey, J., Lucks, S., Schneier, B., Stay, M., Wagner, D., Whiting, D.: Improved cryptanalysis of Rijndael. In: Schneier, B. (ed.) FSE 2000. LNCS, vol. 1978, pp. 213–230. Springer, Heidelberg (2001)
7. FIPS 197. Advanced Encryption Standard. Federal Information Processing Standards Publication 197, U.S. Department of Commerce/N.I.S.T (2001)
8. Gilbert, H., Minier, M.: A collision attack on 7 rounds of Rijndael. In: AES Candidate Conference, pp. 230–241 (2000)
9. Hong, S., Kim, J., Lee, S., Preneel, B.: Related-key rectangle attacks on reduced versions of SHACAL-1 and AES-192. In: Gilbert, H., Handschuh, H. (eds.) FSE 2005. LNCS, vol. 3557, pp. 368–383. Springer, Heidelberg (2005)
10. Nakahara Jr, J., de Freitas, D.S., Phan, R.C.-W.: New multiset attacks on Rijndael with large blocks. In: Dawson, E., Vaudenay, S. (eds.) Mycrypt 2005. LNCS, vol. 3715, pp. 277–295. Springer, Heidelberg (2005)
11. Kim, J., Hong, S., Preneel, B.: Related-key rectangle attacks on reduced AES-192 and AES-256. In: Biryukov, A. (ed.) FSE 2007. LNCS, vol. 4593, pp. 225–241. Springer, Heidelberg (2007)
12. Knudsen, L.R., Wagner, D.: Integral cryptanalysis. In: Daemen, J., Rijmen, V. (eds.) FSE 2002. LNCS, vol. 2365, pp. 112–127. Springer, Heidelberg (2002)
13. Lucks, S.: Attacking seven rounds of Rijndael under 192-bit and 256-bit keys. In: AES Candidate Conference, pp. 215–229 (2000)
14. Murphy, S., Robshaw, M.J.B.: Essential algebraic structure within the AES. In: Yung, M. (ed.) CRYPTO 2002. LNCS, vol. 2442, pp. 1–16. Springer, Heidelberg (2002)

Implementation of the AES-128
on Virtex-5 FPGAs

Philippe Bulens[1,*], François-Xavier Standaert[1,**], Jean-Jacques Quisquater[1],
Pascal Pellegrin[2], and Gaël Rouvroy[2]

[1] UCL Crypto Group, Place du Levant, 3, B-1348 Louvain-La-Neuve, Belgium
[2] IntoPIX s.a., Place de l'Université, 16, B-1348 Louvain-La-Neuve, Belgium

Abstract. This paper presents an updated implementation of the Advanced Encryption Standard (AES) on the recent Xilinx Virtex-5 FPGAs. We show how a modified slice structure in these reconfigurable hardware devices results in significant improvement of the design efficiency. In particular, a single substitution box of the AES can fit in 8 FPGA slices. We combine these technological changes with a sound intertwining of the round and key round functionalities in order to produce encryption and decryption architectures that perfectly fit with the Digital Cinema Initiative specifications. More generally, our implementations are convenient for any application requiring Gbps-range throughput.

1 Introduction

Reprogrammable hardware devices are highly attractive options for the implementation of encryption algorithms. During the selection process of the AES [1], an important criterion was the efficiency of the cipher in different platforms, including FPGAs. Since 2001, various implementations have consequently been proposed, exploring the different possible design tradeoffs ranging from the highest throughput to the smallest area [2]. Each of those implementations usually focuses on a particular understanding of "efficiency". Furthermore, every time a new hardware platform is introduced, a new implementation is to be made in order to comply with and take advantage of its specificities.

Therefore, this paper aims to provide an update on the performances of the AES, taking the new Xilinx's Virtex-5 FPGAs as evaluation devices. Our results show how the modified slice structure (*i.e.* the 6-input Look-Up-Tables combined with multiplexors) allows an efficient implementation of the AES substitution box (S-box). We include these technological advances in a state-of-the art architecture for an encryption module. The resulting IP core typically complies with the Digital Cinema Initiative specifications [3]: the presented encryption and decryption designs can be used to decrypt the incoming compressed data stream in a digital cinema server and re-encrypt the uncompressed data between the server and the projector. More generally, it is convenient for any application requiring

---

* Supported by Walloon Region, Belgium / First Europe Program.
** Postdoctoral Researcher of the Belgian Fund for Scientific Research.

S. Vaudenay (Ed.): AFRICACRYPT 2008, LNCS 5023, pp. 16–26, 2008.

Gbps-range throughput and compares positively with most recently published FPGA implementations of the AES. Although the improvements described in this paper straightforwardly derive from technological advances, we believe they are of interest for the cryptographic community in order to keep state-of-the-art implementation results available and detailed in the public literature.

The rest of the paper is structured as follows. Section 2 briefly reminds how the AES cipher processes the data. Then, we discuss the specificities of our target platform before fully describing the architecture developed in Section 4. The implementation results are summarized in Section 5, together with some selected results from the literature. Finally, our conclusions are in Section 6.

2   Cipher Description

The AES is a substitution permutation network (SPN) allowing the encryption/decryption of data by blocks of 128-bits and supporting key lengths of 128, 192 and 256 bits. In the following, we focus on the 128-bit key version. Its internal state, usually represented as a 4×4 matrix of bytes, is updated by iterating through the round structure (10, 12 or 14 times according to the key size). The round is described as four different byte-oriented transformations.

First, **SubBytes** introduces the non-linearity by taking, for each byte, the modular inverse in $GF(2^8)$ and then applying an affine transformation. Instead of computing distinctly these two steps, the full transformation is achieved by passing each byte through an S-box (Figure 1). Then **ShiftRows** modifies the state. It simply consists of a circular left shift of the state's rows by 0, 1, 2 and 3 bytes respectively (Figure 2). Third, **MixColumns** applies a linear transformation to the state's columns (Figure 3). Each of them is regarded as a polynomial and is multiplied by a fixed polynomial $c(x) = 3 \cdot x^3 + x^2 + x + 2 \pmod{x^4 + 1}$. Finally, the **AddRoundKey** transform mixes the key with the state. As each subkey has the same size as the state, the combination is performed by a simple bitwise XOR between subkey bytes and their corresponding state bytes (Figure 4). A first key addition is performed before entering the first round, and the last round omits the **MixColumns** transformation.

Prior to the en/de-cryption process, the subkeys have to be generated.

The key schedule takes the main key K_0 and expand it as shown in Fig. 5 for the case of a 128-bit key, where **SubWord** applies the S-box to the 32-bit input word, **RotWord** rotates the word one byte to the left and $RC(i)$ is an 8-bit constant associated to each round i [1].

---

[1] In encryption mode, this can easily be performed "on-the-fly", *i.e.* in parallel to the rounds execution in order to get the subkey at the exact time it is needed. In decryption mode, the round keys generally have to be derived prior to the decipher. Solutions allowing "on-the-fly" derivation of the decryption subkeys require specific features (*i.e.* knowledge of a decryption key that corresponds to the last encryption subkeys) and hardware overhead. Therefore, these are not considered in this paper.

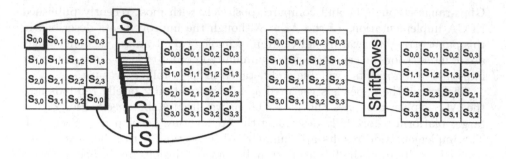

Fig. 1. SubBytes Transformation   **Fig. 2. ShiftRows** Transformation

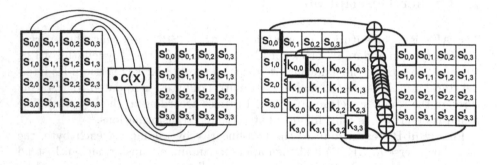

Fig. 3. MixColumns Transformation.   **Fig. 4. AddRoundKey** Transform

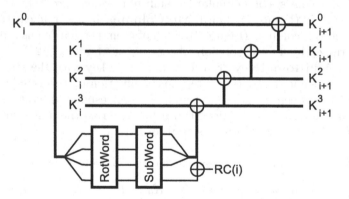

Fig. 5. AES 128-bit Key Expansion Round

Let us finally mention that the decryption process slightly differs from the encryption one. In order to decipher data blocks, the inverse transformations have to be applied to the state. These operations are respectively called **Inv-***Transform*, except for the **AddRounKey** as it is its own inverse, and are applied in the same order as described above. This may result in a different performance of the AES in encryption and decryption. Regarding the key schedule, the operations remain the same as for encryption but the subkeys have to be introduced in reverse order. For more details, we refer to [1].

3   Target Platform

The chosen platform for implementation is a Xilinx Virtex-5 FPGA. Nowadays, such devices embed programmable logic blocks, RAM memories and multipliers[2]. Compared to the early reconfigurable devices, these features allow recent FPGAs to provide interesting solutions for a wide range of applications. In this description, the focus will be set on logic elements as other resources will not be used. In Xilinx FPGAs, the slice is the logic unit that is used to evaluate a design's area requirement. The Virtex-5 platform exhibits two kinds of slices. Each of these contains 4 Look-Up Tables (LUTs), 4 flips-flops and some additional gates. These elements, defining the "basic" slice (sliceL), provide the user with logic, arithmetic and ROM functions. In addition, another version of the slice (sliceM) adds the capability to be configured as a distributed RAM or as a shift register. Those enhanced slices represent about 25% of the total amount of available slices. Figure 6 shows the difference between sliceLs and sliceMs.

4   Architecture

A lot of architectures for the AES have been presented in the open literature. Each of those target a specific application and accordingly, various tradeoffs of pipelining, unrolling, datapath width and S-boxes designs have been presented. These contributions do generally agree that the most critical part in an AES design is the S-box. To our knowledge, three different methods have been exploited in order to achieve efficient implementations:

Logic. In this first proposal, one 256×8-bit S-box is required for each byte of the state. Implementing this as a large multiplexor on platforms where LUTs provide 4-to-1 computation and by taking advantage of special FPGA configurations (*i.e.* MuxF5s and MuxF6s), led the authors of [4] to consume 144 LUTs for one S-box. In the case of a full length datapath (128-bit), 2304 LUTs (144×16) are required to perform the whole substitution. This method results in a logic depth of 2 slices. Those two levels can be advantageously pipelined to prevent frequency reduction.

Algorithmic. Another approach to compute **SubBytes** is to directly implement the multiplicative inverse and affine transforms. In order to make it more efficient, Rijmen suggested in [5] to move computations from $GF(2^8)$ to the composite field $GF((2^4)^2)$. The main advantage relies on the reduced size of the inversion table: $2^4 \times 4$ instead of $2^8 \times 8$. However, some logic is required to implement transformations to and from such composite field. This type of implementation has been exploited in [6,7] for example.

---

[2] Additionally, certain devices also embed microprocessors (PowerPC).

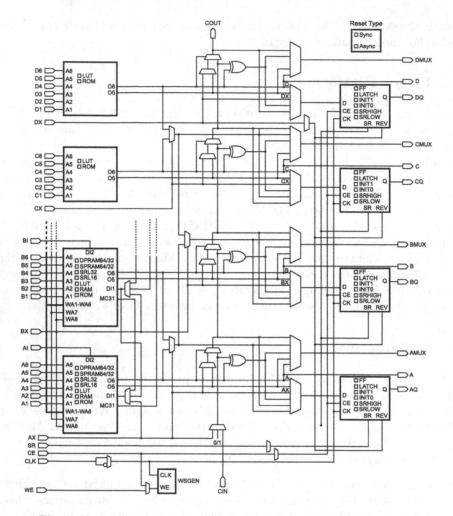

Fig. 6. Virtex-5 mixed view of top-half sliceL and bottom-half sliceM

RAM. Embedded BlockRAMs on FPGA platforms can also be used to store the S-boxes. Such an approach achieves high-throughput performances in [8]. If enough of these memories are available, another idea is to combine **SubByte** and **MixColumns** in a single memory table, as sometimes proposed in software implementations. Examples of such designs are in [9,10]. Depending on the size and availability of RAM blocks, these solutions may be convenient in practice for recent FPGA devices.

In our context, the choice is straightforward. Due to the technology evolution, a 256×8-bit S-box fits in 32 LUTs. Using a similar approach as in [4], four LUTs make four 6-to-1 tables from which the correct output is chosen thanks to the F7MUXs and F8MUX of the slice. It allows packing a 256×1-bit table in four LUTs, that is a single slice. This solution has the significant advantage of both reducing the area requirements in LUTs and performing the S-box computation in a single clock cycle, thus reducing the latency.

The architecture developed for 128-bit key encryption is in Figure 7. It has a 128-bit datapath for both data and keys. The state of the cipher iterates over a single round structure. The **ShiftRows** operation is not shown on the figure below as it simply consists in routing resources.

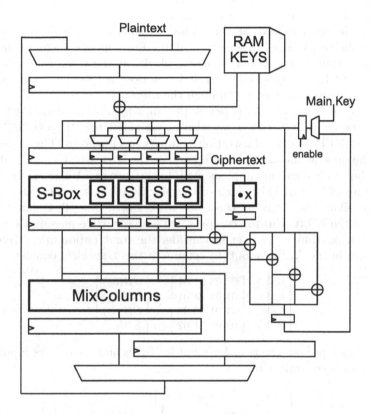

Fig. 7. AES Encryption Architecture

As far as the key expansion is concerned and when dealing with large amount of data, like in the Digital Cinema context, computing all subkeys prior to en/decryption seems a better alternative than an "on-the-fly" key schedule. Indeed, the overhead due to a master key change quickly vanishes as the number of messages using this same key increases. It also allows us to reduce the area requirements of the complete design. As the key schedule and the encryption module use the same S-boxes, these are shared in our architecture. Multiplexors allow the S-box inputs to change between the state and the subkey. These multiplexors do not increase the implementation cost as they are packed in LUTs together with the key addition. The remaining of the subkeys'computation proceeds as explained in Section 2. Each subkey is written in a RAM configured sliceM.

They are brought back when needed for en/de-cryption. We note that the key schedule must be performed before decryption takes place anyway, as the subkeys have to be applied in reverse order. Also, the S-Boxes sharing does not hold for the decryption architecture. Indeed, the S-boxes implementing **SubBytes** for the key schedule can not be reused for deciphering. This is because decryption involves the **InvSubBytes** operation that do not yield the same S-boxes.

This architecture perfectly suits the needs of ECB and Counter modes of operations. It could also be tuned to handle the CBC mode, at the cost of a reduced efficiency. Indeed, as the plaintext block is to be XORed with the previous ciphertext before being encrypted, the four pipeline stages of the round do not allow encryption of four plaintext blocks at the same time. Note that in the case of CBC decryption, this restriction does not hold as this additional XOR is performed after looping through the round.

Although the focus of this paper is the implementation of an AES en/de-cryption module on a Virtex-5, we also investigated this IP core in Virtex-4 and Spartan-3 FPGAs, for illustration/comparisons purposes. The architecture remains the same as the one presented here for the Virtex-5. The only difference relies on the way S-boxes are implemented. In the case of a Virtex-4, the S-boxes are made up of BlockRAMs. Each of the blockRAM has a datapath width of 32-bit that allows the output of the S-box to be stored times 0, 1, 2 and 3. That is, a part of the **MixColumns** computation is made while passing through the RAMs. To make things clear, let us consider the combination of SubBytes and MixColumns in the AES. An output column of this transform equals:

$$
\begin{bmatrix} b_0 \\ b_1 \\ b_2 \\ b_3 \end{bmatrix} = \begin{bmatrix} 02 & 03 & 01 & 01 \\ 01 & 02 & 03 & 01 \\ 01 & 01 & 02 & 03 \\ 03 & 01 & 01 & 02 \end{bmatrix} \times \begin{bmatrix} SB(a_0) \\ SB(a_1) \\ SB(a_2) \\ SB(a_3) \end{bmatrix},
$$

where the b_i's represent the transform output bytes and the a_i's its input bytes. The b_i vector is equivalent to:

$$
\begin{bmatrix} 02 \\ 01 \\ 01 \\ 03 \end{bmatrix} \times [SB(a_0)] \ \oplus \ \begin{bmatrix} 03 \\ 02 \\ 01 \\ 01 \end{bmatrix} \times [SB(a_1)] \ \oplus \ \begin{bmatrix} 01 \\ 03 \\ 02 \\ 01 \end{bmatrix} \times [SB(a_2)] \ \oplus \ \begin{bmatrix} 01 \\ 01 \\ 03 \\ 02 \end{bmatrix} \times [SB(a_3)]
$$

Therefore, if we define four tables as:

$$
T_0(a) = \begin{bmatrix} 02 \times SB(a) \\ SB(a) \\ SB(a) \\ 03 \times SB(a) \end{bmatrix}, \ T_1(a) = \begin{bmatrix} 03 \times SB(a) \\ 02 \times SB(a) \\ SB(a) \\ SB(a) \end{bmatrix},
$$

$$
T_2(a) = \begin{bmatrix} SB(a) \\ 03 \times SB(a) \\ 02 \times SB(a) \\ SB(a) \end{bmatrix}, \ T_3(a) = \begin{bmatrix} SB(a) \\ SB(a) \\ 03 \times SB(a) \\ 02 \times SB(a) \end{bmatrix},
$$

the combination of SubBytes and MixColumns equals:

$$\begin{bmatrix} b_0 \\ b_1 \\ b_2 \\ b_3 \end{bmatrix} = T_0(a) \oplus T_1(a) \oplus T_2(a) \oplus T_3(a)$$

In our Virtex-4 implementation, these T tables are stored in RAM and all what is left to complete the **MixColumns** transform is a single level of logic handling the XOR of the four bytes. On Spartan-3 devices, the situation is different since the BlockRAMs do not provide a dedicated latch at their output. Reproducing the behavior of Virtex-4 requires 24-bits to be stored using the slice flip-flops which consumes much more area. Since XORing the table outputs without using the slice flip flops causes a reduction of the work frequency, the most efficient solution is to implement **MixColumns** and **SubBytes** independently.

5   Results

The AES designs were described using VHDL. Synthesis and Place & Route were achieved on Xilinx ISE 9.1i. The selected devices are Xilinx's Virtex-5, Virtex-4 and Spartan-3. Table 1 summarizes the results achieved for both the encryption and decryption (Enc/Dec) modules. Moreover, some previous results are summarized in Table 2. As it is generally true for any comparison of hardware performances, those results have to be taken with care since they relate to different FPGA devices. In the Virtex-5 FPGAs, a slice is made up of 4 LUTs instead of 2 for previous Xilinx devices. In order to allow fair(er) comparison, it then makes sense to double the figures as if a slice was 2 LUTs. This is taken into account into the parenthesis of Table 1. Compared to previous devices, the benefit of Virtex-5 is easily underlined. It corresponds either to the removal of the blockRAMs from the design on the Virtex-4 or a 50% slice reduction from a full logic design on Spartan-3 FPGAs. This strongly emphasized the advantage of technology evolution shifting from 4 to 6 input bits LUTs.

Table 1. Implementation Results: encryption/decryption designs

Device	Slices	BRAM	Freq. (MHz)	Thr. (Gbps)	Thr. / Area (Mbps/slice)
Virtex-5	400 / 550 (800 / 1100)	0	350	4.1	10.2 / 7.4
Virtex-4*	700 / 1220	8	250	2.9	4.1 / 2.3 *
Spartan-3	1800 / 2150	0	150	1.7	0.9 / 0.8

Additional insights on our implementation results can be obtained by looking at Table 2. Namely, the proposed architectures range among the efficient ones found out in the literature. Again, these observations have to be considered as general intuitions rather than fair comparisons since they consider different FPGA technologies: more recent FPGAs have higher work frequencies and thus throughput. In addition, the hardware efficiency (*i.e.* throughput/area ratio) of

Table 2. Previous Implementations

Device	Datapath	Slices	BRAM	Freq. (MHz)	Thr. (Gbps)	Thr./Area (Mbps/slice)
Spartan-2 [11]*	8	124	2	–	0.002	0.02*
Virtex-2 [10]*	32	146	3	123	0.358	2.45*
Virtex-E [12]*	128	542	10	119	1.45	2.67*
Virtex-E [4]	128	2257	0	169	2.0	0.88
Virtex-2 [13]*	128	387	10	110.16	1.41	3.64*
Virtex-2 [13]	128	1780	0	77.91	1.0	0.56
Virtex-4 [14]	128	18400	0	140	17.9	0.97
Virtex-5 [15]	128	349	0	350	4.1	11.67

the *-marked implementations is not meaningful since they consumes FPGA RAM blocks. Finally, the hardware cost can only be compared if the respective implementation efficiencies (*e.g.* measured with the throughput/area ratio) are somewhat comparable. As a matter of fact, it is always possible to reduce the implementation cost, by considering smaller datapaths (*e.g.* [11] uses an 8-bit datapath for the AES, [10] uses a 32-bit datapath, all the others use 128-bit architectures) at the cost of a reduced throughput.

In the case of Helion Technology's implementation [15], the comparison is more interesting since it relates to the same Virtex-5 platform as ours. At first sight, their Fast AES Encryption core seems to consume less area than the proposed architecture. However, the gap can be reduced if we assume that their core uses an "on-the-fly" key schedule. In such a case, the distributed RAM used to store subkeys is to be removed from our presented design (along with its control logic) which allows to earn at least 32 slices. This makes both designs very close. As a matter of fact, the differences between these cores mainly relate to different optimization goals. Our was to design encryption and decryption IPs exploiting a very similar architecture with a key scheduling algorithm executed prior to the encryption/decryption process. We note that not using the "on-the-fly" key scheduling for encryption makes sense for power consumption reasons. If the implementation context does not require frequent key changes, there is no need to re-compute these keys for every plaintext.

6   Conclusion

This paper reports implementation results of the AES algorithm on the new Virtex-5 devices. It exhibits the (straightforward but significant) benefits that can be drawn from the technology evolution within recent FPGAs. In particular it is shown how the AES substitution box perfectly suits the new Virtex-5 slice structure using 6-bit LUTs. This enables reducing the cost of a single S-box from 144 down to 32 LUTs ! Compared to 4 input bit LUTs-based designs, this advantage roughly corresponds to either the removal of blocks of embedded RAM memory or a slice count reduction of 50%, depending on the design choices.

The proposed architectures range among the most efficient ones published in the open literature. Their reasonable implementation cost make them a suitable solution for a wide range of application requiring Gbps-range throughput, including digital cinema and network encryption.

References

1. National Institute of Standards and Technology. Advanced Encryption Standard (AES). Federal Information Processing Standards Publications – FIPS 197 (November 2001), http://csrc.nist.gov/publications/fips/fips197/fips-197.pdf
2. Järvinen, K., Tommiska, M., Skyttä, J.: Comparative survey of high-performance cryptographic algorithm implementations on FPGAs. In: IEE Proceedings on Information Security, vol. 152, pp. 3–12 (October 2005)
3. Digital Cinema Initiative. DCI System Specification, V1.0 (July 20, 2005)
4. Standaert, F.-X., Rouvroy, G., Quisquater, J.-J., Legat, J.-D.: Efficient Implementation of Rijndael in Reconfigurable Hardware: Improvements and Design Trade-offs. In: D.Walter, C., Koç, Ç.K., Paar, C. (eds.) CHES 2003. LNCS, vol. 2779, pp. 334–350. Springer, Heidelberg (2003)
5. Rijmen, V.: Efficient implementation of the Rijndael S-box, www.citeseer.ist.psu.edu/rijmen00efficient.html
6. Rudra, A., Dubey, P.K., Jutla, C.S., Kumar, V., Rao, J.R., Rohatgi, P.: Efficient rijndael encryption implementation with composite field arithmetic. In: Koç, Ç.K., Naccache, D., Paar, C. (eds.) CHES 2001. LNCS, vol. 2162, pp. 171–184. Springer, Heidelberg (2001)
7. Hodjat, A., Verbauwhede, I.: A 21.54 Gbits/s Fully Pipelined AES coprocessor on FPGA. In: 12th Annual IEEE Symposium on Field Programmable Custom Computing Machine, FCCM 2004, April 2004, pp. 308–309 (2004)
8. Saggese, G.P., Mazzeo, A., Mazzocca, N., Strollo, A.G.M.: An FPGA-based Performance Analysis of the Unrolling, Tiling and Pipelining of the AES Algorithm. In: Y. K. Cheung, P., Constantinides, G.A. (eds.) FPL 2003. LNCS, vol. 2778, pp. 292–302. Springer, Heidelberg (2003)
9. McLoone, M., McCanny, J.V.: Rijndeal FPGA Implementation Utilizing Look-Ups Tables. In: IEEE Workshop on Signal Processing Systems — SIPS 2001, September 2001, pp. 349–360 (2001)
10. Rouvroy, G., Standaert, F.-X., Quisquater, J.-J., Legat, J.-D.: Compact and Efficient Encryption/Decryption Module for FPGA Implementation of the AES Rijndael Well Suited for Small Embedded Applications. In: International Symposium on Information Technology: Coding and Computing, ITCC 2004, April 2004, pp. 583–587. IEEE Computer Society Press, Los Alamitos (2004)
11. Good, M., Benaissa, M.: AES on FPGA from the Fastest to the Smallest. In: Rao, J.R., Sunar, B. (eds.) CHES 2005. LNCS, vol. 3659, pp. 427–440. Springer, Heidelberg (2005)
12. Standaert, F.-X., Rouvroy, G., Quisquater, J.-J., Legat, J.-D.: A Methodology to Implement Block Ciphers in Reconfigurable Hardware and its Application to Fast and Compact AES RIJNDAEL. In: 11th ACM International Symposium on Field-Programmable Gate Arrays — FPGA 2003, February 2003, pp. 216–224 (2003)
13. Zambreno, J., Nguyen, D., Choudary, A.: Exploring Area/Delay Tradeoffs in an AES FPGA Implementation. In: Becker, J., Platzner, M., Vernalde, S. (eds.) Field-Programmable Logic and Applications, August–September 2004, pp. 575–585. Springer, Heidelberg (2004)

14. Lemsitzer, S., Wolkerstorfer, J., Felbert, N., Braendli, M.: Multi-gigabit GCM-AES Architecture Optimized for FPGAs. In: Paillier, P., Verbauwhede, I. (eds.) CHES 2007. LNCS, vol. 4727, pp. 227–238. Springer, Heidelberg (2007)
15. Helion, http://www.heliontech.com
16. Canright, D.: A Very Compact S-box for AES. In: Rao, J.R., Sunar, B. (eds.) CHES 2005. LNCS, vol. 3659, pp. 441–456. Springer, Heidelberg (2005)
17. Chodowiec, P., Gaj, K.: Very Compact FPGA Implementation of the AES Algorithm. In: Walter, C.D., Koç, Ç.K., Paar, C. (eds.) CHES 2003. LNCS, vol. 2779, pp. 319–333. Springer, Heidelberg (2003)
18. Feldhofer, M., Dominikus, S., Wolkerstorfer, J.: Strong Authentication for RFID Systems Using the AES Algorithms. In: Joye, M., Quisquater, J.-J. (eds.) CHES 2004. LNCS, vol. 3156, pp. 357–370. Springer, Heidelberg (2004)
19. IP Cores, http://ipcores.com/index.htm
20. McLoone, M., McCanny, J.V.: High Performance Single-Chip FPGA Rijndael Algorithm Implementation. In: Koç, Ç.K., Naccache, D., Paar, C. (eds.) CHES 2001. LNCS, vol. 2162, pp. 65–76. Springer, Heidelberg (2001)
21. McLoone, M., McCanny, J.V.: Single-Chip FPGA Implementation of the Advanced Encryption Standard Algorithm. In: Brebner, G., Woods, R. (eds.) FPL 2001. LNCS, vol. 2147, pp. 152–161. Springer, Heidelberg (2001)
22. Satoh, A., Morioka, S.: Unified Hardware Architecture for the 128-bit Block Ciphers AES and Camellia. In: Walter, C.D., Koç, Ç.K., Paar, C. (eds.) CHES 2003. LNCS, vol. 2779, pp. 304–318. Springer, Heidelberg (2003)
23. Saqib, N.A., Rodríquez-Hendríquez, F., Díaz-Pérez, A.: AES Algorithm Implementation–An Efficient Approach for Sequential and Pipeline Architectures. In: 4th Mexican International Computer Science — ENC 2003, September 2003, pp. 126–130 (2003)

Weaknesses in a Recent Ultra-Lightweight RFID Authentication Protocol

Paolo D'Arco and Alfredo De Santis

Dipartimento di Informatica ed Applicazioni
Università degli Studi di Salerno, 84084, Fisciano (SA), Italy
{paodar, ads}@dia.unisa.it

Abstract. In this paper we show weaknesses in **SASI**, a new Ultra-Lightweight RFID Authentication Protocol, designed for providing **S**trong **A**uthentication and **S**trong **I**ntegrity. We identify three attacks, namely, a *de-synchronisation* attack, through which an adversary can break the synchronisation between the RFID Reader and the Tag, an *identity disclosure* attack, through which an adversary can compute the identity of the Tag, and a *full disclosure* attack, which enables an adversary to retrieve all secret data stored in the Tag. The attacks are effective and efficient.

1 Introduction

RFID Technology. Radio Frequency Identification (RFID, for short) is a rapidly growing technology enabling automatic objects identification[1]. Each object is labeled with a tiny integrated circuit equipped with a radio antenna, called *Tag*, whose *information content* can be received by another device, called *Reader*, without physical contact, at a distance of several meters.

RFID tags can perform computations. They are usually divided in *passive* tags and in *active* tags. The first ones do not have a power source. They receive energy for computation from the readers and can perform very simple operations. The second are powered by small batteries and are capable of performing more significant and computational heavy operations.

An important security concern associated to the RFID technology is the *privacy* of the tag content. Indeed, it is pretty much easy for anybody with technical skills to set up a device for reading the tag content. Nevertheless, to preserve user privacy, only authorised RFID readers should be enabled to access the tag content. An authentication protocol, which grants access to the tag content only to a legitimate reader, is therefore required.

Based on the computational cost and the operations supported on tags, authentication protocols can be divided in classes. Using the terminology of [2], the

---

[1] Ari Juels [4] has recently pointed out that the RFID technology ...*In essence ... is a form of computer vision... RFID has an advantage over even the most acute eyes and brain: it is in fact a form of X-ray vision...RFID is poised to become one of the sensory organs of our computing networks.*

S. Vaudenay (Ed.): AFRICACRYPT 2008, LNCS 5023, pp. 27–39, 2008.

full-fledged class refers to protocols demanding support on tags for conventional cryptographic functions like symmetric encryption, hashing, or even public key cryptography. The *simple* class refers to protocols requiring random number generation and hashing. The *lightweight* class refers to protocols which require random number generation and simple checksum functions. The *Ultra-Lightweight* class refers to protocols which only involve simple bitwise operations, like *and*, *or*, exclusive *or*, and modular addition.

An overview of the applications of RFID and of the main security issues can be found in [5]. Moreover, we refer the reader to [1] for references to the full body of research papers dealing with RFID technology and its challenges.

A few lightweight and ultra-lightweight authentication protocols have appeared in the literature during the last two years. For example, a series of ultra-lightweight authentication protocols involving only bitwise operations and modular addition have been proposed in [8,9,10]. Unfortunately, the vulnerabilities of these protocols have been showed in [7,6,3].

Our Contribution. We focus our attention on a new ultra-lightweight authentication protocol, recently proposed in [2], to provide strong authentication and strong integrity data protection. We identify three attacks, namely, a *de-synchronisation* attack, through which an adversary can break the synchronisation between the RFID Reader and the Tag, an *identity disclosure* attack, through which an adversary can compute the identity of the Tag, and a *full disclosure* attack, which enables an adversary to retrieve all secret data stored in the Tag. The attacks are effective and efficient.

2   The Authentication Protocol

Let us focus on the protocol proposed by Chien in [2]. Three entities are involved: a Tag, a Reader and a Backend Server. The channel between the Reader and the Backend Server is assumed to be secure, but the channel between the Reader and the Tag is susceptible to all the possible attacks.

Each Tag has a *static identifier*, ID, a *pseudonym*, IDS, and *two keys*, K_1 and K_2. All of them are 96-bit strings X. A string is represented as a sequence $X[95] \ldots X[0]$, from the most significant bit to the least significant bit. The pseudonym and the keys are shared with the Backend Server which, for each Tag with static identifier ID, stores in a table the tuple (IDS, K_1, K_2). After each successfull execution of the authentication protocol, the Tag and the Backend Server *update* such values.

The authentication protocol is a four-round protocol. To simplify the description we do not introduce explicitly the Backend Server and will say that the Reader performs some computations. However, the Reader just forwards the values received from the Tag to the Backend Server and gets back the output of the computation the Backend Server performs.

The computations involve the following operations: $\oplus$ (bitwise exclusive or), $\vee$ (bitwise or), $+ \bmod 2^{96}$, and $Rot(x, y)$, where x and y are two 96-bit values, and the $Rot(\cdot, \cdot)$ operator shifts to the left in a cyclic way x by y positions.

Let us look at Fig. 1. The Reader starts the authentication protocol by sending an **Hello** message to the Tag. The Tag replies with the pseudonym **IDS**. Then, the Reader chooses, uniformly at random, two 96-bit random values n_1 and n_2, computes

$$\mathbf{A} = \mathbf{IDS} \oplus K_1 \oplus n_1$$
$$\mathbf{B} = (\mathbf{IDS} \vee K_2) + n_2$$
$$\overline{K_1} = Rot(K_1 \oplus n_2, K_1)$$
$$\overline{K_2} = Rot(K_2 \oplus n_1, K_2)$$
$$\mathbf{C} = (K_1 \oplus \overline{K_2}) + (\overline{K_1} \oplus K_2),$$

and sends to the Tag $\mathbf{A}\|\mathbf{B}\|\mathbf{C}$, the concatenation of $\mathbf{A}, \mathbf{B}$ and $\mathbf{C}$. The Tag, upon receiving $\mathbf{A}\|\mathbf{B}\|\mathbf{C}$, extract n_1 from $\mathbf{A}$, n_2 from $\mathbf{B}$, computes it own values

$$\overline{K_1} = Rot(K_1 \oplus n_2, K_1)$$
$$\overline{K_2} = Rot(K_2 \oplus n_1, K_2)$$
$$\widetilde{C} = (K_1 \oplus \overline{K_2}) + (\overline{K_1} \oplus K_2),$$

and verifies whether $\widetilde{C} = \mathbf{C}$. If the equality holds, i.e., the computed value is equal to the received value, then the Tag computes and sends to the Reader the value

$$\mathbf{D} = (\overline{K_2} + ID) \oplus ((K_1 \oplus K_2) \vee \overline{K_1})$$

and updates its pseudonym and secret keys. Similarly, the Reader, once $\mathbf{D}$ has been received, computes his own value

$$\widetilde{D} = (\overline{K_2} + ID) \oplus ((K_1 \oplus K_2) \vee \overline{K_1}),$$

checks whether $\widetilde{D} = \mathbf{D}$, and if the equality holds, updates the pseudonym and the keys shared with the Tag.

The pseudonym and the keys are updated has follows: The Reader sets

$$IDS = (IDS + ID) \oplus (n_2 \oplus \overline{K_1})$$
$$K_1 = \overline{K_1}, \quad K_2 = \overline{K_2}$$

while the Tag sets

$$IDS_{old} = IDS, \qquad K_{1,old} = K_1, \quad K_{2,old} = K_2,$$
$$IDS = (IDS_{old} + ID) \oplus (n_2 \oplus \overline{K_1}), \qquad K_1 = \overline{K_1}, \quad K_2 = \overline{K_2}.$$

The Tag stores two tuples (IDS, K_1, K_2) and $(IDS_{old}, K_{1,old}, K_{2,old})$ because it might happen that the Tag updates the pseudonym and the keys, while the Server does not. Such an event for example might occur if a simple communication fault does not permit the Reader to get the value $\mathbf{D}$, sent by the Tag during the 4-th round of the authentication protocol. The old tuple is used as follows: any time the Reader gets **IDS** from the Tag, the Reader/Backend Server looks for a tuple (IDS, K_1, K_2). If no entry is found, the Reader sends another **Hello** message to the Tag and the Tag replies with **IDS_old**. Hence, even if the Reader has not updated the tuple, the authentication protocol can be run by using the old one, i.e., $(IDS_{old}, K_{1,old}, K_{2,old})$. Of course, if no match is found also at the second trial, the protocol fails.

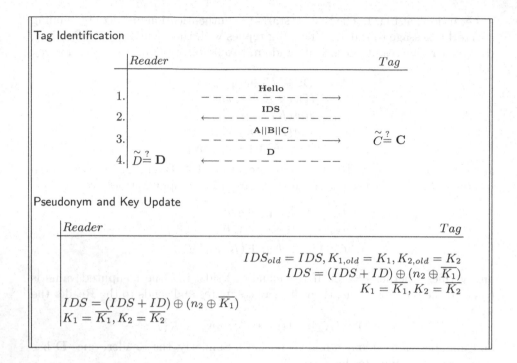

Fig. 1. SASI: Identification - Pseudonym and Key Update

3   De-synchronisation

In this section we propose a de-synchronisation attack.

Let Adv be an adversary who controls the channel between the Reader and the Tag. Adv might simply look at and store the messages exchanged between the parties before forwarding them correctly. Such a behaviour models *passive* attacks. As well as, Adv might intercept/delay/inject/modify messages as he likes. Such a behaviour models *active* attacks.

In our attack, Adv, in a first stage just looks at an honest execution of the authentication protocol and stores the messages **Hello**, **IDS**, **A||B||C** and **D** the parties send to each other.

Then, in a second stage, Adv interacts with the Tag. Roughly speaking, Adv resets the Tag to the state in which the Tag was at the time of the interaction with the Reader and, then, by using the transcript of the execution of the authentication protocol, induces the Tag to accept a new sequence $\mathbf{A'}||\mathbf{B'}||\mathbf{C'}$. Such a sequence drives the Tag to overwrite the new **IDS**, computed at the end of the honest execution. If Adv succeeds, then Tag and Reader are de-synchonised. The new sequence $\mathbf{A'}||\mathbf{B'}||\mathbf{C'}$ is constructed by computing $\mathbf{C'}$ as a modification of $\mathbf{C}$ and by looking for an appropriately chosen $\mathbf{A'}$, obtained by flipping a single bit of $\mathbf{A}$. The value $\mathbf{B}$ stays the same. The attack if described in Fig. 2.

Adv's Computation.

1. Let $\mathbf{C}' = \mathbf{C} + 2^0$ and set $j = 0$.
2. Sends the **Hello** message to the Tag and gets back the (new) **IDS**, computed *at the end* of the execution Reader-Tag of the authentication protocol. Indeed, the Tag has updated $(\mathbf{IDS}, K_1, K_2)$.
3. Sends again the **Hello** message to the Tag and gets back the (old) **IDS**, the one used *during* the execution Reader-Tag of the authentication protocol.
4. Computes $\mathbf{A}'$ by flipping the j-th bit of $\mathbf{A}$ and sends to the Tag $\mathbf{A}'||\mathbf{B}||\mathbf{C}'$.
5. If the Tag accepts and replies with $\mathbf{D}'$, the attack has succeeded and *Adv* terminates. Otherwise, if $j < 96$ then sets $j = j + 1$ and repeats from step 2., else *Adv* sets $\mathbf{C}' = \mathbf{C} - 2^0$ and $j = 0$ and repeats from step 2.

Fig. 2. SASI: De-synchronisation Attack

As we will show in a while, on average, after 48.5 trials, the Tag accepts a message $\mathbf{A}'||\mathbf{B}||\mathbf{C}'$. Hence, the Tag updates the pseudonym and the keys (IDS, K_1, K_2), while the old tuple $(IDS_{old}, K_{1,old}, K_{2,old})$, used in the interaction with *Adv* stays the same. At this point, Reader and Tag have been de-synchronized. The tuple held by the Reader has been overwritten in the tag's memory. Hence, they do not share a tuple anymore!

Why does the attack work? Notice that, by definition

$$\mathbf{A} = \mathbf{IDS} \oplus K_1 \oplus n_1.$$

By flipping a bit in $\mathbf{A}$, *Adv* implicitly flips a bit of n_1. On the other hand, n_1 is used to compute

$$\overline{K_2} = Rot(K_2 \oplus n_1, K_2).$$

Hence, by flipping a bit of n_1, *Adv* flips a bit of $\overline{K_2}$, but he *does not know* which one, since K_2 is unknown. Moreover, it is easy to see that, given two different positions, i and j, in n_1 by flipping the corresponding bits, the bits flipped in $\overline{K_2}$ lie in two different positions i' and j'. In other words, any time *Adv* flips a bit in n_1, he flips a bit in a different position of $\overline{K_2}$. Since,

$$\mathbf{C} = (K_1 \oplus \overline{K_2}) + (\overline{K_1} \oplus K_2),$$

and the only element *Adv* can partially control is $\overline{K_2}$, *Adv* changes the value of $\mathbf{A}$ as long as he does change *the first bit* of $\overline{K_2}$ and gets $\mathbf{C}'$.

Notice that, *Adv* does not know a-priori if, by changing the first bit of $\overline{K_2}$, gets $\mathbf{C} + 2^0$ or $\mathbf{C} - 2^0$. Therefore, the attack, in order to find a sequence $\mathbf{A}'||\mathbf{B}||\mathbf{C}'$, accepted by the Tag, need to be applied once by setting $\mathbf{C}' = \mathbf{C} + 2^0$ and, if going through all possible $\mathbf{A}'$ no sequence is found, one more time by setting $\mathbf{C}' = \mathbf{C} - 2^0$ and trying again. Eventually, the Tag will accept a sequence and reply with $\mathbf{D}'$. Since $\mathbf{A}$ is a uniformly distributed random 96-bit string, Adv has to try on average $\frac{192+1}{2} = 96.5$ times.

However, the above de-synchronization attack can be improved. Indeed, notice that, if Adv computes $\mathbf{C}' = \mathbf{C} + 2^{95} = \mathbf{C} - 2^{95}$ instead of the $\mathbf{C}' = \mathbf{C} + 2^0$ and $\mathbf{C}' = \mathbf{C} - 2^0$, the average number of trials can be reduced. The key observation is that

$$2^{95} \bmod 2^{96} = -2^{95} \bmod 2^{96}.$$

Hence, Adv has to try on average $\frac{96+1}{2} = 48.5$ times to find a suitable sequence $\mathbf{A}'||\mathbf{B}||\mathbf{C}'$ for the tag.

Another important observation, which will play a key role in the next section, is the following: once a sequence $\mathbf{A}'||\mathbf{B}||\mathbf{C}'$ which the Tag accepts is found, Adv discovers the value used in the rotation $Rot(K_2 \oplus n_1, K_2)$ to compute $\overline{K}_2$. Indeed, if the sequence $\mathbf{A}'||\mathbf{B}||\mathbf{C}'$ has been computed by flipping the i-th bit of $\mathbf{A}$, then the value of the rotation is exactly $96 - i$.

4   Identity Disclosure

In this section, by building on the above de-synchronisation attack, we show how Adv can compute the static ID stored in the Tag.

We start by noticing that, the same argument used before by modifying $\mathbf{A}$ and $\mathbf{C}$, in order to find a sequence the Tag accepts, can be applied to *every* position, i.e., we can compute a sequence $\mathbf{A}'||\mathbf{B}||\mathbf{C}'$ by working on *any one* of $\mathbf{C}[95], \ldots, \mathbf{C}[0]$. Indeed, all Adv has to do when working on the i-th position, with $i \in \{0, \ldots, 95\}$, is to set $\mathbf{C}' = \mathbf{C} \pm 2^i$, and then look for an $\mathbf{A}'$ such that the sequence $\mathbf{A}'||\mathbf{B}||\mathbf{C}'$ is accepted.

We also know that the attack is more efficient, i.e., it requires a small number of trials, if applied to position 95, and that, once a sequence $\mathbf{A}'||\mathbf{B}||\mathbf{C}'$ is found, Adv also knows the amount $y = y(z)$ of the rotation $Rot(x, z)$ used in the protocol to get $\overline{K}_2$. Therefore, once Adv has received a reply $\mathbf{D}'$ from the Tag, then Adv can compute a new sequence $\mathbf{A}''||\mathbf{B}||\mathbf{C}''$ by working on any position of $\mathbf{C}$ in 1.5 trials on average. Indeed, Adv knows *exactly* in which position of $\mathbf{A}$ he has to flip a bit in order to add or subtract 2^i to $\mathbf{C}$. Therefore, Adv just needs to check if, by flipping a certain bit in $\mathbf{A}$, he gets $\mathbf{C}' = \mathbf{C} + 2^i$ or $\mathbf{C}' = \mathbf{C} - 2^i$. Let us represent the static identifier ID as $ID[95] \ldots ID[0]$. The key idea in the attack is *to collect pairs of values* $\mathbf{D}, \mathbf{D}'$, sent from the Tag to Adv as replies to forged sequences $\mathbf{A}'||\mathbf{B}||\mathbf{C}'$, and to analyse the *differences* given by $\mathbf{D} \oplus \mathbf{D}'$. As we will show in a while, the differences give to Adv information about the ID and some other values used in the computation both by the Reader and the Tag. Notice that the attack described in this section does not enable Adv to compute the MSB of the ID, and gives two possible candidate values for the ID (and for $\overline{K}_2$).

We will proceed as follows: we first describe an identity disclosure attack which works only in a *special case*. Then, we show how to turn the general case to the special case through a pre-processing stage. The identity disclosure attack for the special case is given in Fig. 3.

Adv's Computation.

1. Let **IDS, A∥B∥C, D** be the transcript of an honest execution *Adv* looks at.
2. Apply the desynchronisation attack, described in Fig. 2, in order to compute the amount $y = y(z)$ of the rotation $Rot(x, z)$. (Notice that, w.r.t. the description given in Fig. 2, for the aforementioned efficiency reasons, in step 1. compute $\mathbf{C} = \mathbf{C} + 2^{95}$ instead of $\mathbf{C} = \mathbf{C} + 2^0$.)
3. Let $i = 0$.
4. Using the knowledge of y, compute a sequence $\mathbf{A}^i\|\mathbf{B}\|\mathbf{C}^i$, where $\mathbf{A}^i$ is obtained by flipping in **A** the bit $\mathbf{A}[(i + y) \bmod 96]$ and the values $\mathbf{C}^i$ is either $\mathbf{C} + 2^i$ or $\mathbf{C} - 2^i$.
5. Send one of the sequence $\mathbf{A}^i\|\mathbf{B}\|\mathbf{C}^i$ to the Tag. If it is not accepted, then send the second one. One of them will be accepted.
6. The Tag sends back to *Adv*, as a reply to one of $\mathbf{A}^i\|\mathbf{B}\|\mathbf{C}^i$, a value, say $\mathbf{D}^i$.
7. From $\mathbf{D}^i$ and **D**, the value the Tag sends to the Reader during the honest execution of the authentication protocol *Adv* looks at, *Adv* computes the *i*-th bit of the static identifier as

$$ID[i] = \mathbf{D}[i + 1] \oplus \mathbf{D}^i[i + 1]. \tag{1}$$

8. If $i < 95$, set $i = i + 1$ and repeat from step 4.

Fig. 3. SASI: Identity disclosure attack

When (and why) does the attack of Fig. 3 work? Notice that, by definition

$$\mathbf{D} = (\overline{K_2} + ID) \oplus ((K_1 \oplus K_2) \vee \overline{K_1})$$

and, hence, denoting with $\overline{K_2^i}$ the value of $\overline{K_2}$ obtained when the Tag accepts $\mathbf{A}^i\|\mathbf{B}\|\mathbf{C}^i$, it holds that

$$\mathbf{D}^i = (\overline{K_2^i} + ID) \oplus ((K_1 \oplus K_2) \vee \overline{K_1}).$$

Therefore

$$\mathbf{D} \oplus \mathbf{D}^i = (\overline{K_2} + ID) \oplus (\overline{K_2^i} + ID).$$

Let us look at the *i*-th position. It holds that

$$\mathbf{D}[i] \oplus \mathbf{D}^i[i] = (\overline{K_2}[i] + ID[i] + c_i) \oplus (\overline{K_2^i}[i] + ID[i] + c_i),$$

where c_i denotes the carry from the sum of the bits of the previous positions. Let us assume that $c_i = 0$. The bits $\overline{K_2}[i]$ and $\overline{K_2^i}[i]$ are one the complement of the other. Hence, either if $ID[i] = 0$ or $ID[i] = 1$, it holds that $\mathbf{D}[i] \oplus \mathbf{D}^i[i] = 1$. On the other hand, if $ID[i] = 0$, it holds that $\mathbf{D}[i + 1] \oplus \mathbf{D}^i[i + 1] = 0$, since $\overline{K}_2[i + 1] = \overline{K}_2^i[i + 1]$; while, if $ID[i] = 1$, then either $(\overline{K}_2[i] + ID[i])$ or $(\overline{K}_2^i[i] + ID[i])$ produces *a carry* to the next position in the computation of $\mathbf{D}$ or of $\mathbf{D}^i$, respectively. Therefore, $\mathbf{D}[i + 1] \neq \mathbf{D}^i[i + 1]$. Hence, equation (1) holds.

By giving a closer look at the identity disclosure attack presented in Fig. 3, it comes out that it works *surely* for the LSB of the ID, i.e., to compute $ID[0]$. Indeed, in the first position there is *no carry* from the sum operation of the bits of previous positions. In other words, in

$$(\overline{K}_2[0] + ID[0]) \oplus (\overline{K}_2^1[0] + ID[0])$$

only one of the sums provides a carry to the next position if $ID[0] = 1$, since $\overline{K}_2^1[0]$ is the complement of $\overline{K}_2[0]$. However, in general, we need to consider

$$(\overline{K}_2[i] + ID[i] + c_i) \oplus (\overline{K}_2^i[i] + ID[i] + c_i).$$

where c_i might be different from 0. Notice that, for example, if $c_i = 1$ and $ID[i] = 0$, by computing $ID[i]$ through equation (1) we draw a wrong conclusion!
    Hence, we have two possibilities: either we find a strategy to keep track of the carries generated and propagated during the computation of $\mathbf{D}$ and $\mathbf{D}^i$ by $(\overline{K}_2 + ID)$ and $(\overline{K}_2^i + ID)$ or we identify a method *to reduce the general case to the special case*, i.e., one where *almost no carry* is generated. We succeeded in pursuing the second approach.

First of all, notice that the bit-string P obtained by computing $\mathbf{D} \oplus \mathbf{D}^0$ has the form $P = 0^{96-r}1^r$, where $r \geq 1$. Moreover, the substring of $r - 2$ bits equal to 1, from the second position to the $(r - 1)$-th, tells us that, for $i = 2, \ldots, r - 1$, $\overline{K}_2[i] \neq ID[i]$. On the other hand, the last 1 tells us that $\overline{K}_2[r] = ID[r]$. Indeed, let us look at $\mathbf{D} \oplus \mathbf{D}^0$:

$$
\begin{array}{llllll}
\mathbf{D} \ldots & (\overline{K}_2[r] + ID[r]) & (\overline{K}_2[r-1] + ID[r-1]) & \ldots & (\overline{K}_2[1] + ID[1]) & (\overline{K}_2[0] + ID[0]) \\
\oplus \ldots & \oplus & \oplus & \ldots & \oplus & \oplus \\
\mathbf{D}^0 \ldots & (\overline{K}_2^1[r] + ID[r]) & (\overline{K}_2^1[r-1] + ID[r-1]) & \ldots & (\overline{K}_2^1[1] + ID[1]) & (\overline{K}_2^1[0] + ID[0]) \\
\hline
P = 0 \ldots & 0 & 1 & \ldots & 1 & 1
\end{array}
$$

    Assume that $ID[0] = 1$. Since $\overline{K}_2^1[0] = 1 - \overline{K}_2[0]$, then one of the two strings $\mathbf{D}$ or $\mathbf{D}^0$ *generates* a carry c_1 to the next position. Then, in $\mathbf{D}$ and $\mathbf{D}^0$, for $j = 1, \ldots, 95$, it holds that $\overline{K}_2[j] = \overline{K}_2^1[j]$. Therefore, we can conclude that the *only reason* to get a sequence of 1's in P is that $\overline{K}_2[j] \neq ID[j]$ as long as $j < r$. In such a case, the carry generated in position 0 is *propagated* until position $r-1$.

On the other end, it is easy to check that, in position $r - 1$, it holds that $\overline{K}_2[r-1] = ID[r-1]$. Indeed, either if $\overline{K}_2[r-1] = ID[r-1] = 0$ or if $\overline{K}_2[r-1] = ID[r-1] = 1$, the bit $P[r-1] = 1$ and the bit $P[r] = 0$.

We are now ready to describe the whole attack. It works in two steps:

- Pre-processing. Modify the string $\overline{K}_2$, in order to get an *all* 1's string P, by using the procedure described in Fig. 4.
- Identity Disclosure. Apply the identity disclosure attack given in Fig. 3.

Indeed, we have control over $\overline{K}_2$ and we can efficiently get a pair $(\overline{\mathbf{D}}, \overline{\mathbf{D}}^0)$ such that $\overline{P} = \overline{\mathbf{D}} \oplus \overline{\mathbf{D}}^0$ is all 1's. We proceed as follows: assume that Adv, after sending $\mathbf{A}^0||\mathbf{B}||\mathbf{C}^0$ gets back $\mathbf{D}^0$ such that[2] $P = \mathbf{D} \oplus \mathbf{D}^0 = 0^{96-r}1^r$, with $r > 1$. Then, let us look at Fig. 4.

Adv's Computation.

1. Constructs and sends to the Tag a new sequence $\mathbf{A_r}||\mathbf{B}||\mathbf{C_r}$, modifying $\mathbf{A}^0$ and $\mathbf{C}^0$, in order to flip the r-th bit of $\overline{K}_2$.
2. The Tag replies with a value $\mathbf{D_r}$.
3. Then, Adv sends to the Tag a new sequence $\mathbf{A_r}'||\mathbf{B}||\mathbf{C_r}'$, constructed from $\mathbf{A_r}||\mathbf{B}||\mathbf{C_r}$, in order to flip the first bit of the new $\overline{K}_2$, i.e., $\overline{K}_2 = 0$.
4. The tag replies with $\mathbf{D_r}'$.
5. Adv computes $P = \mathbf{D_r} \oplus \mathbf{D_r}' = 0^{96-t}1^t$, where $t > r$. If $t = 96$, then Adv has finished; otherwise, Adv repeats the procedure working on the t-th bit, that is, setting $r = t$ and $\mathbf{A}^0||\mathbf{B}||\mathbf{C}^0 = \mathbf{A_r}||\mathbf{B}||\mathbf{C_r}$.

Fig. 4. Pre-processing for the identity disclosure attack

Notice that step 1., as we have seen, takes on average 1.5 trials and aims at *extending* the *all* 1's substring in P. Step 3. also takes on average 1.5 trials.

Once Adv has a pair $(\overline{\mathbf{D}}, \overline{\mathbf{D}}^0)$ such that $\overline{P} = \overline{\mathbf{D}} \oplus \overline{\mathbf{D}}^0 = 1^{96}$, it mounts the identity disclosure attack described in Fig. 3, starting from step 3. and setting $i = 1$.

More precisely, let $\mathbf{A_s}||\mathbf{B}||\mathbf{C_s}$ the sequence which gave rise to $\overline{\mathbf{D}}$. The transcript $\mathbf{Hello}, \mathbf{IDS}, \mathbf{A_s}||\mathbf{B}||\mathbf{C_s}, \overline{\mathbf{D}}$ plays the role of the transcript of the honest

---

[2] In the special case, in which the string we get is $0^{95}1$, it means that $ID[0] = 0$. Hence, we fix this bit of ID and apply again the procedure starting from the 2nd position of the strings and going on, as long as we do not get, a string P with the form $P = 0^{96-j-r}1^r0^j$, where $j > 0$ and $r > 1$. To simplify the analysis and simplify the description, w.l.o.g., let us assume that $ID[0] = 1$ and $P = \mathbf{D} \oplus \mathbf{D}^0 = 0^{96-r}1^r$, with $r > 1$.

execution in step 1. of Fig. 3. Then, Adv, for $i = 1, \ldots, 94$, waits for $\overline{\mathbf{D}}^i$ from the tag as a reply to a sequence $\mathbf{A}^i || \mathbf{B} || \mathbf{C}^i$, constructed from $\mathbf{A_s} || \mathbf{B} || \mathbf{C_s}$ by working on the i-th position. Once received $\overline{\mathbf{D}}^i$, he computes the i-th bit of the ID using equation (1), i.e., $ID[i] = \overline{\mathbf{D}}[i+1] + \overline{\mathbf{D}}^i[i+1]$.

At this point, notice that Adv computes *two possible* ID values. According to our assumption $ID[0] = 1$ in both. However, depending on the value of $\overline{K}_2[0]$, Adv gets two values for the ID, which are one the complement of the other up to the LSB and the MSB. More precisely, let us assume that $\overline{K}_2[0] = 0$. Through the pre-processing stage and the identity disclosure attack, Adv computes ID_1. Notice that he also computes $\overline{K}_{2,1}$. Indeed, he knows that, for $j = 1, \ldots, 94$, it holds that $ID[j] \neq \overline{K}[j]$.

   On the other hand, if $\overline{K}_2[0] = 1$, it is easy to see that the bits in position 0 either of $\overline{K}_2 + ID$ or of $\overline{K}_2^1 + ID$, *generate* a carry which is propagated until position 95. Hence, the second pair $(ID_2, \overline{K}_{2,2})$ is obtained, for $j = 1, \ldots, 94$, by flipping the bits of ID_1 and $\overline{K}_{2,1}$ one by one, i.e., $ID_2[j] = 1 - ID_1[j]$, and $\overline{K}_{2,1} = 1 - \overline{K}_{2,1}$.

   We stress that the above method does not enable Adv to compute the MSB of the possible ID and $\overline{K}_2$. Therefore, Adv has to guess such bits.

Notice that, we can also recover the string $\overline{K}_2$, used by the Reader to update the tuple (IDS, K_1, K_2) to the new tuple $(IDS_{New}, \overline{K}_1, \overline{K}_2)$. Indeed, we have modified such a string in the preprocessing stage of our attack, in order to get a new string $\overline{K}_2$, such that $ID[i] \neq \overline{K}_2[i]$, for $i = 2, \ldots, 94$. Such a condition has been obtained by changing some bits of the former $\overline{K}_2$. However, we know *in which positions* the bits have been flipped. Hence, from the two strings $\overline{K}_{2,1}$ and $\overline{K}_{2,2}$, obtained at the end of the attack, we can compute the two possible $\overline{K}_2$, by *reversing* the flipped bits in $\overline{K}_{2,1}$ and $\overline{K}_{2,2}$.

Remark. The attack is effective and efficient. It requires on average, 48.5 interactions with the Tag to find out the amount y of the rotation, $\ell \cdot (1.5 + 1.5)$ interactions for the pre-processing stage, and $95 \cdot 1.5$ interactions to compute the first 95 bits of the ID. The parameter ℓ represents the number of positions in which $\overline{K}_2$ and ID are equal. Since the strings are uniformly distributed, it holds that, on average, $\ell = (96+1)/2$. Therefore, the whole identity disclosure attack requires, on average, $48.5 + 144 + 48.5 = 240$ interactions.

5   Full Disclosure

In this section we show how Adv can efficiently extract *all secret data*.

   Assume that we have a black box procedure, let us say $\mathsf{BB}(\mathbf{IDS}, \mathbf{A} || \mathbf{B} || \mathbf{C}, \mathbf{D})$, to recover the ID of the tag and the string $\overline{K}_2$. Then, Adv can extract all secret data from the tag as described in Fig. 5:

   The Reader, at the end of the first execution of the authentication protocol, once received the value $\mathbf{D}$, updates the old tuple (IDS, K_1, K_2) to the new tuple $(IDS_{New}, \overline{K}_1, \overline{K}_2)$.

Then, Adv computes the ID and $\overline{K}_2$, re-sets the Tag, and eavesdrops a second execution of the authentication protocol.

It is immediate to see that, at this point, Adv has enough information to compute everything. Indeed:

Adv's Computation

1. **Eavesdropping stage.** Looks at an execution of the authentication protocol and stores **Hello, IDS, A||B||C, D**.
2. **Identity disclosure stage.** Applies BB(**IDS, A||B||C, D**) and gets ID and $\overline{K}_2$.
3. **Re-set stage.** Re-sets the Tag by sending again the sequence **A||B||C**, engaging an instance of the authentication protocol
4. **Eavesdropping stage.** Looks at a new execution of the authentication protocol and stores **Hello, IDS$_\mathbf{New}$, A$_{New}$||B$_{New}$||C$_{New}$, D$_{New}$**.
5. **Secret Data Extraction.** Computes K_1, K_2, n_1, n_2 of the current state.

Fig. 5. Full disclosure attack

- From $\mathbf{B}_{New}$, since knows K_2 (the former $\overline{K}_2$) and **IDS**$_{New}$, gets n_2, i.e., he computes

$$n_2 = \mathbf{B}_{New} - (\mathbf{IDS}_{New} \vee K_2).$$

- From **IDS**$_{New} = (\mathbf{IDS} + ID) \oplus (n_2 \oplus \overline{K}_1)$, since knows **IDS**, ID and n_2, computes $\overline{K}_1$, i.e.,

$$\overline{K}_1 = \mathbf{IDS}_{New} \oplus (\mathbf{IDS} + ID) \oplus n_2.$$

Notice that the current K_1 is equal to the former $\overline{K}_1$, used to compute **IDS**$_{New}$.

- From $\mathbf{A}_{New}$, since knows **IDS**$_{New}$ and K_1, computes n_1, i.e.,

$$n_1 = \mathbf{A}_{New} \oplus \mathbf{IDS}_{New} \oplus K_1.$$

Remark. The above attack requires a black box procedure which *uniquely* computes ID and $\overline{K}_2$. However, the ambiguities present in the method we have described, can be solved in several ways. Either by directly "testing" the computed secrets, by interacting with the Tag, or by using the available values of $\mathbf{C}$ and $\mathbf{D}$, for testing which ones are the correct hypothesis. The complexity of the full disclosure attack is the *same* complexity of the identity disclosure attack, up to some simple computation. All the secret data involved in the authentication protocol can be efficiently retrieved. Moreover, it is easy to check that, Adv, once computed the value K_1, K_2, n_1, n_2 of the current state, by using the transcript of previous executions of the authentication protocol, can also compute the secret data therein used.

6   Conclusions

We have showed that **SASI** [2], a new ultra-lightweight authentication protocol, proposed to provide strong authentication and strong integrity protection for RFID tag presents vulnerabilities.

We have described three attacks. A *de-synchronisation* attack, through which an adversary can break the synchronisation between the RFID Reader and the Tag. An *identity disclosure* attack, through which an adversary can compute the identity of the Tag. A *full disclosure* attack, which enables an adversary to retrieve all secret data stored in the Tag. The attacks are effective and efficient.

During the writing of the camera-ready version of this paper, we have find out that other researchers [11] have proposed two de-synchronisation attacks to **SASI**.

The recent history shows that all ultra-lightweight authentication protocols proposed have been broken through efficient attacks relatively soon after they have been published. Almost all of them are designed in order to provide confusion and diffusion of the output values. Informal security arguments are used to support the merits of the obtained protocols. The above paper confirms that sound security arguments should be used to support design strategies.

Acknowledgement

We would like to thank an anonymous referee for helpful comments.

References

1. Avoine, G.: Bibliography on Security and Privacy in RFID Systems, Massachusetts Institute of Technology, Cambridge, Massachusetts, USA (last update in Jun 2007), Available online at: http://lasecwww.epfl.ch/~gavoine/rfid/
2. Chien, H.: SASI: A new Ultralightweight RFID Authentication Protocol Providing Strong Authentication and Strong Integrity. IEEE Transactions on Dependable and Secure Computing 4(4), 337–340 (2007)
3. Chien, H., Hwang, C.: Security of ultra-lightweight RFID authentication protocols and its improvements. ACM SIGOPS Operating Systems Review 41(4), 83–86 (2007)
4. Juels, A.: The Vision of Secure RFID. Proceedings of the IEEE 95(8), 1507–1508 (2007)
5. Juels, A., Pappu, R., Garfinkel, S.: RFID Privacy: An Overview of Problems and Proposed Solutions. IEEE Security and Privacy 3(3), 34–43 (2005)
6. Li, T., Deng, R.: Vulnerability Analysis of EMAP-An Efficient RFID Mutual Authentication Protocol. In: Proc. of the The Second International Conference on Availability, Reliability and Security, pp. 238–245 (2007)
7. Li, T., Wang, G.: Security Analysis of Two Ultra-Lightweight RFID Authentication Protocols. In: Proc. of the 22-nd IFIP SEC 2007 (May 2007)
8. Peris-Lopez, P., Hernandez-Castro, J.C., Estevez-Tapiador, J.M., Ribagorda, A.: LMAP: A Real Lightweight Mutual Authentication Protocol for Low-cost RFID tags. In: Proc. of the Second Workshop RFID Security, July11-14, Graz University of Technology (2006)

9. Peris-Lopez, P., Hernandez-Castro, J.C., Estevez-Tapiador, J.M., Ribagorda, A.: EMAP: An Efficient Mutual-Authentication Protocol for Low-Cost RFID Tags. In: Meersman, R., Tari, Z., Herrero, P. (eds.) OTM 2006 Workshops. LNCS, vol. 4277, pp. 352–361. Springer, Heidelberg (2006)
10. Peris-Lopez, P., Hernandez-Castro, J.C., Estevez-Tapiador, J.M., Ribagorda, A.: M^2AP: A Minimalist Mutual-Authentication Protocol for Low-Cost RFID Tags. In: Ma, J., Jin, H., Yang, L.T., Tsai, J.J.-P. (eds.) UIC 2006. LNCS, vol. 4159, pp. 912–923. Springer, Heidelberg (2006)
11. Sun, H., Ting, W., Wang, K.: On the Security of Chien's Ultralightweight RFID Authentication Protocol, eprint archieve, report 83 (February 25, 2008)

Differential Cryptanalysis of Reduced-Round PRESENT*

Meiqin Wang

Key Laboratory of Cryptologic Technology and Information Security, Ministry of
Education, Shandong University,
Jinan, 250100, China
mqwang@sdu.edu.cn

Abstract. PRESENT is proposed by A.Bogdanov et al. in CHES 2007
for extremely constrained environments such as RFID tags and sensor
networks. In this paper, we present the differential characteristics for
r-round($5 \leq r \leq 15$), then give the differential cryptanalysis on reduced-
round variants of PRESENT. We attack 16-round PRESENT using 2^{64}
chosen plaintexts, 2^{32} 6-bit counters, and 2^{64} memory accesses.

1 Introduction

RFID systems and sensor networks have been aggressively deployed in a variety
of applications, but their further pervasive usage is mainly limited by lots of
security and privacy concerns. As RFID tags and sensor networks are low cost
with limited resources, the present cryptographic primitives can not be feasible.
So the security primitives suitable for these environments must be designed.

PRESENT is an Ultra-Lightweight block cipher proposed by A.Bogdanov,
L.R.Knudsen and G.Leander et al.[3] and has implementation requirements similar to many compact stream ciphers. Compared to other current block ciphers
for low-cost implementation requirements such as TEA[12,13], MCRYPTON[7],
HIGHT[5], SEA[11] and CGEN[9], PRESENT has the lowest implementation
costs.

PRESENT is a 31-round SP-network with block length 64 bits and 80 bits or
128 bits key length. Serpent[1] and DES have excellent performance in hardware,
so the design of PRESENT makes use of the characteristics of the two block
ciphers. The non-linear substitution layer S-box of PRESENT is similar to that
of Serpent and the linear permutation layer pLayer of PRESENT is similar to
that of DES.

Differential cryptanalysis, proposed by Biham and Shamir[4], is one of the
most general cryptanalytic techniques. Although the original PRESENT proposal provided theoretical upper bounds for the highest probability characteristics of 25-round PRESENT[3], the proposal did not give the concrete differential
cryptanalysis.

* Supported by National Natural Science Foundation of China Key Project
No.90604036, National Outstanding Young Scientist No.60525201 and 973 Program
No.2007CB807902.

In this paper, we consider actual differential attack against reduced-round PRESENT. First we give some differential characteristics for PRESENT. 14-round differential characteristics occur with the probability of 2^{-62} and 15-round differential characteristics occur with the probability of 2^{-66}. However, the original PRESENT proposal can provided theoretical upper bounds for the characteristics of 15-round PRESENT with the highest probability 2^{-60}. Second, we attack 16-round PRESENT with 14-round differential characteristics using 2^{64} chosen plaintexts, 2^{32} 6-bit counters, and 2^{64} memory accesses.

The paper is organized as follows. Section 2 introduces the description of PRESENT. In Section 3, we give some notations used in this paper. In Section 4, we present the best differential characteristics we found for PRESENT, and give the differential attack on 16-round PRESENT-80. Section 5 concludes this paper.

2 Description of PRESENT

2.1 The Encryption Process

PRESENT is a 31-round Ultra-Lightweight block cipher. The block length is 64-bit. PRESENT uses only one 4-bit S-box S which is applied 16 times in parallel in each round. The cipher is described in Figure 1. As in Serpent, there are three stages involved in PRESENT. The first stage is addRoundKey described as follows,

$$b_j \rightarrow b_j \bigoplus k^i{}_j$$

where $b_j, 0 \le j \le 63$ is the current state and $k^i{}_j, 1 \le i \le 32, 0 \le j \le 63$ is the $j - th$ subkey bit of round key K_i.

The second stage is sBoxLayer which consists of 16 parallel versions of the 4-bit to 4-bit S-box, which is given in Table 1.

Table 1. Table of S-box

x	0	1	2	3	4	5	6	7	8	9	A	B	C	D	E	F
S[x]	C	5	6	B	9	0	A	D	3	E	F	8	4	7	1	2

The third stage is the bit permutation pLayer, which is given by Table 2. From pLayer, bit i of stage is moved to bit position $P(i)$.

2.2 The Key Schedule

PRESENT's key schedule can take key sizes of 80 bits or 128 bits. We will cryptanalyze 80 bits version, so we will only give the schedule algorithm for 80 bits version.

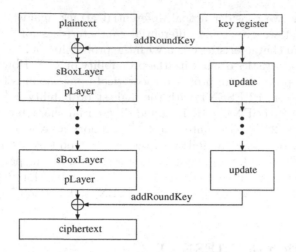

Fig. 1. 31-round PRESENT Encryption Algorithm

Table 2. Table of pLayer

i	0	1	2	3	4	5	6	7	8	9	10	11	12	13	14	15
$P(i)$	0	16	32	48	1	17	33	49	2	18	34	50	3	19	35	51
i	16	17	18	19	20	21	22	23	24	25	26	27	28	29	30	31
$P(i)$	4	20	36	52	5	21	37	53	6	22	38	54	7	23	39	55
i	32	33	34	35	36	37	38	39	40	41	42	43	44	45	46	47
$P(i)$	8	24	40	56	9	25	41	57	10	26	42	58	11	27	43	59
i	48	49	50	51	52	53	54	55	56	57	58	59	60	61	62	63
$P(i)$	12	28	44	60	13	29	45	61	14	30	46	62	15	31	47	63

Firstly, the 80-bit key will be stored in a key register K denoted as $K = k_{79}k_{78}\ldots k_0$. In round j, PRESENT firstly extracts 64-bit subkeys K_j in the following ways,

$$K_j = \kappa_{63}\kappa_{62}\ldots\kappa_0 = k_{79}k_{78}\ldots k_{16}$$

Then it updates key register $K = k_{79}k_{78}\ldots k_0$ as follows,

$$[k_{79}k_{78}\ldots k_1 k_0] = [k_{18}k_{17}\ldots k_{20}k_{19}]$$

$$[k_{79}k_{78}k_{77}k_{76}] = S[k_{79}k_{78}k_{77}k_{76}]$$

$$[k_{19}k_{18}k_{17}k_{16}k_{15}] = [k_{19}k_{18}k_{17}k_{16}k_{15}] \oplus round\_counter$$

3   Some Notations

In the remainder of the paper, we use $X = x_0, x_1, \ldots, x_{15}$ to denote the intermediate difference in each step. $x_0, x_1, \ldots, x_{15}$ are 16 nibble differences and x_0 is the least significant nibble difference. We denote K_i as the subkey for the $i-th$ round.

4   Differential Characteristics for PRESENT

Firstly, we give the XORs differential distribution of S-box in Table 3. From the XOR's distribution table for S-box, one bit input difference will cause at least two bits output difference, which will cause two active S-boxes in the next round. Then each of the two active S-boxes will have at least two bits output difference, which will cause at least four active S-boxes in the next round.

4.1   Searching for Differential Characteristics

The differential cryptanalysis of DES[4] makes use of 2-round iterative characteristics to form 13-round differential characteristics. Knudsen has searched the better iterative characteristics for DES[6], which is an efficient method to find the differential characteristics for more rounds. We have searched for the differential characteristic in the following way:

  – We searched the iterative characteristics from 2-round to 7-round, which are more advantage than the 2-round iterative characteristic given in [3]. As the maximum probability in the differential distribution table for PRESENT S-box is 2^{-2}, we only consider the maximum number of active S-boxes from

Table 3. Differential Distribution Table of S-box

	0_x	1_x	2_x	3_x	4_x	5_x	6_x	7_x	8_x	9_x	A_x	B_x	C_x	D_x	E_x	F_x
0_x	16	0	0	0	0	0	0	0	0	0	0	0	0	0	0	0
1_x	0	0	0	4	0	0	0	4	0	4	0	0	0	4	0	0
2_x	0	0	0	2	0	4	2	0	0	0	2	0	2	2	2	0
3_x	0	2	0	2	2	0	4	2	0	0	2	2	0	0	0	0
4_x	0	0	0	0	0	4	2	2	0	2	2	0	2	0	2	0
5_x	0	2	0	0	2	0	0	0	0	2	2	2	4	2	0	0
6_x	0	0	2	0	0	0	2	0	2	0	0	4	2	0	0	4
7_x	0	4	2	0	0	0	2	0	2	0	0	0	2	0	0	4
8_x	0	0	0	2	0	0	0	2	0	2	0	4	0	2	0	4
9_x	0	0	2	0	4	0	2	0	2	0	0	0	2	0	4	0
A_x	0	0	2	2	0	4	0	0	2	0	2	0	0	2	2	0
B_x	0	2	0	0	2	0	0	0	4	2	2	2	0	2	0	0
C_x	0	0	2	0	0	4	0	2	2	2	2	0	0	0	2	0
D_x	0	2	4	2	2	0	0	2	0	0	2	2	0	0	0	0
E_x	0	0	2	2	0	0	2	2	2	2	0	0	2	2	0	0
F_x	0	4	0	0	4	0	0	0	0	0	0	0	0	0	4	4

2-round to 4-round are 4, 7 and 9 respectively. The possible distribution of the number of active S-boxes in them is listed in Table 4. As a result, only 4-round iterative characteristics with the probability 2^{-18} have been found, one of which is given in Table 5.

- We searched the best differential characteristics from 5-round to 10-round which are more advantage than the characteristics based on 4-round iterative characteristics we have found.
- Based on 4-round iterative characteristic, we have found the differential characteristics from 11-round to 15-round.

Table 4. Possible Distribution of Active S-box for Iterative Characteristics

Rounds	2	3	4
Possible Distribution of Active S-box	2-2	2-2-2	2-2-2-2
		3-2-2	3-2-2-2
		2-3-2	2-3-2-2
		2-2-3	2-2-3-2
			2-2-2-3

Table 5. 4-round Iterative Differential of PRESENT

Rounds		Differences	Pr
I		$x_0 = 4, x_3 = 4$	1
R1	S	$x_0 = 5, x_3 = 5$	$\frac{1}{2^4}$
R1	P	$x_0 = 9, x_8 = 9$	1
R2	S	$x_0 = 4, x_8 = 4$	$\frac{1}{2^4}$
R2	P	$x_8 = 1, x_{10} = 1$	1
R3	S	$x_8 = 9, x_{10} = 9$	$\frac{1}{2^4}$
R3	P	$x_2 = 5, x_{14} = 5$	1
R4	S	$x_2 = 1, x_{14} = 1$	$\frac{1}{2^6}$
R4	P	$x_0 = 4, x_3 = 4$	1

The differential characteristics we found in the way are given in Table 6. It is noted that the number of active S-boxes in each round are all 2.

We have found 24 14-round differential characteristics with the probability 2^{-62} with different input differences but the same output difference. All the characteristics have the same differences after the $8 - th$ round, and all have the same probability 2^{-62}. Table 7 gives one of the 14-round characteristics we have found.

Table 6. Probability of the Best Characteristics We Found

Rounds	Differential Probability	Number of Active S-box
5	2^{-20}	10
6	2^{-24}	12
7	2^{-28}	14
8	2^{-32}	16
9	2^{-36}	18
10	2^{-42}	20
11	2^{-46}	22
12	2^{-52}	24
13	2^{-56}	26
14	2^{-62}	28
15	2^{-66}	30

4.2   Attacking 16-Round PRESENT

We will attack 16-round PRESENT using the 14-round differential characteristics with probability of 2^{-62}.

All of the 24 differential characteristics we found have 2 active S-boxes in the first round located in position 0,1,2,12,13 and 14, so the S-boxes in position from 3 to 11 and 15 are all non-active. This attack requires 2^{40} structures of 2^{24} chosen plaintexts each. In each structure, all the inputs to the 14 non-active S-boxes in the first round can be random value, while 8 bits of input to 2 active S-boxes take 2^8 possible values. In all structures, there are $2^{40} \cdot 2^{16} \cdot 2^7 = 2^{63}$ pairs for each possible characteristics. Each characteristic has the probability 2^{-62}, so the number of right pairs is $2^{63} * 2^{-62} * 24 = 48$ satisfying any one characteristic. For each structure, the number of possible pairs is $(2^{24})^2/2 = 2^{47}$, thus we have $2^{47} \cdot 2^{40} = 2^{87}$ pairs of plaintext to be considered in total.

According to the output difference of 14-round differential characteristics, there are two active S-boxes in round-15 which are x_0 and x_8 whose input difference is 9 and output difference will be 2, 4, 6, 8, 12 or 14. The least significant bit of their output difference must be zero, so at most 6 bits are non-zero for the output difference of S-boxes in round 15. After the pLayer of round 15, the maximum number of active S-boxes for round 16 is 6 and the active S-boxes will be $x_4, x_6, x_8, x_{10}, x_{12}$ and x_{14}, and the minimum number of active S-boxes for round 16 is 2.

For each structure, each pair satisfying one of the characteristics should have 10 non-active S-boxes in round 16, so the wrong pairs should be discarded. Thus about $2^{47} * 2^{-40} = 2^7$ candidates for right pairs remain from each structure.

Table 7. The 14-round Differential of PRESENT

Rounds		Differences	Pr
I		$x_2 = 7, x_{14} = 7$	
R1	S	$x_2 = 1, x_{14} = 1$	$\frac{1}{2^4}$
R1	P	$x_0 = 4, x_3 = 4$	1
R2	S	$x_0 = 5, x_3 = 5$	$\frac{1}{2^4}$
R2	P	$x_0 = 9, x_8 = 9$	1
R3	S	$x_0 = 4, x_8 = 4$	$\frac{1}{2^4}$
R3	P	$x_8 = 1, x_{10} = 1$	1
R4	S	$x_8 = 9, x_{10} = 9$	$\frac{1}{2^4}$
R4	P	$x_2 = 5, x_{14} = 5$	1
R5	S	$x_2 = 1, x_{14} = 1$	$\frac{1}{2^6}$
R5	P	$x_0 = 4, x_3 = 4$	1
R6	S	$x_0 = 5, x_3 = 5$	$\frac{1}{2^4}$
R6	P	$x_0 = 9, x_8 = 9$	1
R7	S	$x_0 = 4, x_8 = 4$	$\frac{1}{2^4}$
R7	P	$x_8 = 1, x_{10} = 1$	1
R8	S	$x_8 = 9, x_{10} = 9$	$\frac{1}{2^4}$
R8	P	$x_2 = 5, x_{14} = 5$	1
R9	S	$x_2 = 1, x_{14} = 1$	$\frac{1}{2^6}$
R9	P	$x_0 = 4, x_3 = 4$	1
R10	S	$x_0 = 5, x_3 = 5$	$\frac{1}{2^4}$
R10	P	$x_0 = 9, x_8 = 9$	1
R11	S	$x_0 = 4, x_8 = 4$	$\frac{1}{2^4}$
R11	P	$x_8 = 1, x_{10} = 1$	1
R12	S	$x_8 = 9, x_{10} = 9$	$\frac{1}{2^4}$
R12	P	$x_2 = 5, x_{14} = 5$	1
R13	S	$x_2 = 1, x_{14} = 1$	$\frac{1}{2^6}$
R13	P	$x_0 = 4, x_3 = 4$	1
R14	S	$x_0 = 5, x_3 = 5$	$\frac{1}{2^4}$
R14	P	$x_0 = 9, x_8 = 9$	1

Among 16 S-boxes in round 16, 10 S-boxes must be non-active, 2 S-boxed must be active and 4 S-boxes can be active or non-active. If it is active, the input difference must be 1, and the output difference will be 3, 7, 9 or 13. Discarding any pair with a wrong output difference using the above filter should keep only a fraction of $\frac{5}{16}^6 = 2^{-10.07}$. So only about $2^7 * 2^{-10.07} = 2^{-3.07}$ pairs remain for each structure.

For each structure, we check if the remaining pairs satisfy one of the 24 possible plaintext differences corresponding to 24 characteristics. As there are about 2^{24}

possible input differences, only a fraction of about $2^{-24} * 20 = 2^{-19.68}$ of the pairs remain. So the expected number of remaining pairs in all the 2^{40} structures is $2^{40} * 2^{-3.07} * 2^{-19.68} = 2^{17.25}$.

Only the ciphertext bits corresponding to active S-boxes in the last two round need to be decrypt, so 8 bits of round subkey K_{16} and 24 bits of round subkey K_{17} will be involved during decrypt from round 16 to round 14. After deriving the subkey K_{16} and K_{17} from the master key K, the 24-bit subkey from K_{17} above are independent from the 8-bit subkey from K_{16}, so the total number of subkey bits involved in the decryption from round 16 to round 14 is 32.

For each remaining pair, we guess the 24-bit subkey of K_{17} and 8-bit subkey of K_{16} in round 16, and decrypt the remaining ciphertext pairs from round 16 to round 14. According to the differential distribution table of S-box for PRESENT, given the input difference and output difference, there will be at most 4 pairs occurrences, so the average count per counted pair of the subkey nibble corresponding to one active S-box will be 4. According to the number of active S-boxes in round 16 denoted as $t(2 \leq t \leq 6)$ for the remaining ciphertext pairs, we will consider 5 cases according to the value of t:

- If $t = 2$, only $2^{17.25} * 2^{-16} = 2^{1.25}$ pairs of ciphertext satisfy the condition of 2 active S-boxes, so the total counted times of subkeys are about $2^{1.25} * 4^4 = 2^{9.25}$ for the remaining pairs.
- If $t = 3$, about $2^{17.25} * (2^{-12} - 2^{-16}) = 2^{5.16}$ pairs of ciphertext satisfy the condition of 3 active S-boxes, so the total counted times of subkeys are about $2^{5.16} * 4^5 = 2^{15.16}$ for the remaining pairs.
- If $t = 4$, about $2^{17.25} * (2^{-8} - 2^{-12}) = 2^{9.16}$ pairs of ciphertext satisfy the condition of 4 active S-boxes, so the total counted times of subkeys are about $2^{9.16} * 4^6 = 2^{21.16}$ for the remaining pairs.
- If $t = 5$, about $2^{17.25} * (2^{-4} - 2^{-8}) = 2^{13.16}$ pairs of ciphertext satisfy the condition of 5 active S-boxes, so the total counted times of subkeys are about $2^{13.16} * 4^7 = 2^{27.16}$ for the remaining pairs.
- If $t = 6$, the remained pairs satisfying the condition of 6 active S-box will be $2^{17.25} * (1 - 2^{-4}) = 2^{17.16}$, so the total counted times of subkeys are about $2^{17.16} * 4^8 = 2^{33.16}$ for the remaining pairs.

The total counted times of the subkeys are $2^{33.16} + 2^{27.16} + 2^{21.16} + 2^{15.16} + 2^{9.25} = 2^{33.18}$, so the wrong subkey hits are average about $2^{33.18}/2^{32} = 2^{1.18} = 2.27$ times, but the right subkey is counted for the right pairs about 48 times, so it can be easily identified. In total, we retrieve 32 subkey bits using at most $2^{33.18}$ 2-round PRESENT encryptions and 2^{32} 6-bit counters.

By exhaustively searching the remaining 48 bits master key, we can find out 80-bit master key. In this step, the time complexity is 2^{48} 16-round PRESENT encryptions.

In order to reduce the time of analysis we perform the follow algorithm [11]:

1. For each structure:
   (a) Insert all the ciphertext into the hash table according to the 40-bit ciphertext'bit of the non-active S-boxes in the last round.

(b) For each entry with collision(a pair of ciphertext with equal 40-bit values) check whether the plaintexts'difference(in round 1) is one of the 24 characteristics's input difference.

(c) If a pair passes the above test, check whether the difference(in the 24 bits) can be caused by the output difference of the characteristics.

(d) For each possible subkey of K_{17}, we decrypt the last round to obtain the output difference of 2 two active S-boxes for round 15 , and check whether the difference(in the 8 bits) can be caused by the output difference of the characteristics. If a pair passes the above test, add 1 to the counter related to 24 bits of K_{17} and 8 bits of K_{16}.

2. Collect all the subkeys whose counter has at least 48 hits. With the high probability the correct subkey is in this list.

3. Exhaustive searching the remaining 48-bit master key, we can obtain the whole 80-bit master key.

4.3   Complexity Estimations

In step (a), the time complexity is 2^{24} memory accesses. In step (b), about 2^7 pairs remain through the filter of step (a), so the time complexity is 2^8 memory accesses. The time complexity of step (c) (d) (e) and step 2 can be ignored for the fewer remaining pairs for each structure. In all, the time in step 1 is 2^{64} memory accesses.

In step 3, the time complexity is about 2^{48} 16-round PRESENT encryptions.

According to the relationship between the memory accesses and the encryption time in [4], 2^{64} memory accesses is the main term in the implementing time, so the time complexity is about 2^{64} memory accesses.

In our attack, the ratio of signal to noise is as follows:

$$S/N = \frac{p*2^k}{\alpha*\beta} = \frac{2^{-62}*2^{32}}{2^{33.18-17.25}*2^{17.25-67.32}} = 17.63$$

The success probability is as follows:

$$Ps = \int_{-\frac{\sqrt{\mu S_N}-\Phi^{-1}(1-2^{-a})}{\sqrt{S_N+1}}}^{\infty} \Phi(x)dx = 0.999999939$$

where $a = 32$ is the number of subkey bits involved in the decryption and μ is the number of right pairs which can be obtained

$$\mu = pN = 2^{-62} * 2^{63} * 24 = 48$$

In all, our attack needs encrypt the whole plaintext space. The time complexity is 2^{64} memory accesses. The memory requirements are about 2^{32} 6-bit counters and 2^{24} cells for hash table. We can obtain the right key with the probability 0.999999939.

5   Summary

In this paper, we give the differential cryptanalysis on reduced-round variants of PRESENT. We attack 16-round PRESENT using 2^{64} chosen plaintexts, 2^{32} 6-bit counters and 2^{24} hash cells, the time complexity in our attack is about 2^{64} memory accesses. Our attack requires the encryption of the whole plaintext space.

Acknowledgments

We wish to thank Martin Albrecht and reviewers for their suggestions to revise the paper.

References

1. Anderson, R.J., Biham, E., Knudsen, L.R.: Serpent: A Proposal for the Advanced Encryption Standard. Available at, http://www.cs.technion.ac.il/biham/Reports/Serpent
2. Biham, E., Dunkelman, O., Keller, N.: The Rectangle Attack - Rectangling the Serpent. In: Pfitzmann, B. (ed.) EUROCRYPT 2001. LNCS, vol. 2045, pp. 340–357. Springer, Heidelberg (2001)
3. Bogdanov, A., Knudsen, L.R., Leander, G., Paar, C., Poschmann, A., Robshaw, M.J.B., Seurin, Y., Vikkelsoe, C.: PRESENT: An Ultra-Lightweight Block Cipher. In: Paillier, P., Verbauwhede, I. (eds.) CHES 2007. LNCS, vol. 4727, pp. 450–466. Springer, Heidelberg (2007)
4. Biham, E., Shamir, A.: Differential Cryptanalysis of DES-like Cryptosystems. Journal of Cryptology 4(1), 3–72 (1991)
5. Hong, D., Sung, J., Hong, S., Lim, J., Lee, S., Koo, B.-S., Lee, C., Chang, D., Lee, J., Jeong, K., Kim, H., Kim, J., Chee, S.: HIGHT: A New Block Cipher Suitable for Low-Resource Device. In: Goubin, L., Matsui, M. (eds.) CHES 2006. LNCS, vol. 4249, pp. 46–59. Springer, Heidelberg (2006)
6. Knudsen, L.R.: Iterative Characteristics of DES and s2-DES. In: Brickell, E.F. (ed.) CRYPTO 1992. LNCS, vol. 740, pp. 497–511. Springer, Heidelberg (1993)
7. Lim, C., Korkishko, T.: mCrypton - A Lightweight Block Cipher for Security of Low-cost RFID Tags and Sensors. In: Song, J.-S., Kwon, T., Yung, M. (eds.) WISA 2005. LNCS, vol. 3786, pp. 243–258. Springer, Heidelberg (2006)
8. NIST, A Request for Candidate Algorithm Nominations for the AES, http://www.nist.gov/aes/
9. Robshaw, M.J.B.: Searching for Compact Algorithms: cgen. In: Nguyên, P.Q. (ed.) VIETCRYPT 2006. LNCS, vol. 4341, pp. 37–49. Springer, Heidelberg (2006)
10. Selcuk, A.A., Bicak, A.: On Probability of Success in Linear and Differential Cryptanalysis. In: Cimato, S., Galdi, C., Persiano, G. (eds.) SCN 2002. LNCS, vol. 2576, pp. 174–185. Springer, Heidelberg (2003)
11. Standaert, F.-X., Piret, G., Gershenfeld, N., Quisquater, J.-J.: SEA: A Scalable Encryption Algorithm for Small Embedded Applications. In: Domingo-Ferrer, J., Posegga, J., Schreckling, D. (eds.) CARDIS 2006. LNCS, vol. 3928, pp. 222–236. Springer, Heidelberg (2006)
12. Wheeler, D., Needham, R.: TEA, a Tiny Encryption Algorithm. In: Preneel, B. (ed.) FSE 1994. LNCS, vol. 1008, pp. 363–366. Springer, Heidelberg (1995)
13. Wheeler, D., Needham, R.: TEA extensions (October 1997) (Also Correction to XTEA, October 1998), Available via: www.ftp.cl.cam.ac.uk/ftp/users/djw3/

The Psychology of Security

Bruce Schneier

BT Counterpane, 1600 Memorex Drive, Suite 200, Santa Clara, CA 95050
bruce.schneier@bt.com

1   Introduction

Security is both a feeling and a reality. And they're not the same.

The reality of security is mathematical, based on the probability of different risks and the effectiveness of different countermeasures. We can calculate how secure your home is from burglary, based on such factors as the crime rate in the neighborhood you live in and your door-locking habits. We can calculate how likely it is for you to be murdered, either on the streets by a stranger or in your home by a family member. Or how likely you are to be the victim of identity theft. Given a large enough set of statistics on criminal acts, it's not even hard; insurance companies do it all the time.

We can also calculate how much more secure a burglar alarm will make your home, or how well a credit freeze will protect you from identity theft. Again, given enough data, it's easy.

But security is also a feeling, based not on probabilities and mathematical calculations, but on your psychological reactions to both risks and countermeasures. You might feel terribly afraid of terrorism, or you might feel like it's not something worth worrying about. You might feel safer when you see people taking their shoes off at airport metal detectors, or you might not. You might feel that you're at high risk of burglary, medium risk of murder, and low risk of identity theft. And your neighbor, in the exact same situation, might feel that he's at high risk of identity theft, medium risk of burglary, and low risk of murder.

Or, more generally, you can be secure even though you don't feel secure. And you can feel secure even though you're not. The feeling and reality of security are certainly related to each other, but they're just as certainly not the same as each other. We'd probably be better off if we had two different words for them.

This essay is my initial attempt to explore the feeling of security: where it comes from, how it works, and why it diverges from the reality of security.

Four fields of research—two very closely related—can help illuminate this issue. The first is behavioral economics, sometimes called behavioral finance. Behavioral economics looks at human biases—emotional, social, and cognitive—and how they affect economic decisions. The second is the psychology of decision-making, and more specifically bounded rationality, which examines how we make decisions. Neither is directly related to security, but both look at the concept of risk: behavioral economics more in relation to economic risk, and the psychology of decision-making more generally in terms of security risks. But both fields go a long way to explain the divergence between the feeling and the reality of security and, more importantly, where that divergence comes from.

S. Vaudenay (Ed.): AFRICACRYPT 2008, LNCS 5023, pp. 50–79, 2008.

There is also direct research into the psychology of risk. Psychologists have studied risk perception, trying to figure out when we exaggerate risks and when we downplay them.

A fourth relevant field of research is neuroscience. The psychology of security is intimately tied to how we think: both intellectually and emotionally. Over the millennia, our brains have developed complex mechanisms to deal with threats. Understanding how our brains work, and how they fail, is critical to understanding the feeling of security.

These fields have a lot to teach practitioners of security, whether they're designers of computer security products or implementers of national security policy. And if this paper seems haphazard, it's because I am just starting to scratch the surface of the enormous body of research that's out there. In some ways I feel like a magpie, and that much of this essay is me saying: "Look at this! Isn't it fascinating? Now look at this other thing! Isn't that amazing, too?" Somewhere amidst all of this, there are threads that tie it together, lessons we can learn (other than "people are weird"), and ways we can design security systems that take the feeling of security into account rather than ignoring it.

2   The Trade-Off of Security

Security is a trade-off. This is something I have written about extensively, and is a notion critical to understanding the psychology of security. There's no such thing as absolute security, and any gain in security always involves some sort of trade-off.

Security costs money, but it also costs in time, convenience, capabilities, liberties, and so on. Whether it's trading some additional home security against the inconvenience of having to carry a key around in your pocket and stick it into a door every time you want to get into your house, or trading additional security from a particular kind of airplane terrorism against the time and expense of searching every passenger, all security is a trade-off.

I remember in the weeks after 9/11, a reporter asked me: "How can we prevent this from ever happening again?" "That's easy," I said, "simply ground all the aircraft."

It's such a far-fetched trade-off that we as a society will never make it. But in the hours after those terrorist attacks, it's exactly what we did. When we didn't know the magnitude of the attacks or the extent of the plot, grounding every airplane was a perfectly reasonable trade-off to make. And even now, years later, I don't hear anyone second-guessing that decision.

It makes no sense to just look at security in terms of effectiveness. "Is this effective against the threat?" is the wrong question to ask. You need to ask: "Is it a good trade-off?" Bulletproof vests work well, and are very effective at stopping bullets. But for most of us, living in lawful and relatively safe industrialized countries, wearing one is not a good trade-off. The additional security isn't worth it: isn't worth the cost, discomfort, or unfashionableness. Move to another part of the world, and you might make a different trade-off.

We make security trade-offs, large and small, every day. We make them when we decide to lock our doors in the morning, when we choose our driving route, and when we decide whether we're going to pay for something via check, credit card, or cash. They're often not the only factor in a decision, but they're a contributing factor. And most of the time, we don't even realize it. We make security trade-offs intuitively.

These intuitive choices are central to life on this planet. Every living thing makes security trade-offs, mostly as a species—evolving this way instead of that way—but also as individuals. Imagine a rabbit sitting in a field, eating clover. Suddenly, he spies a fox. He's going to make a security trade-off: should I stay or should I flee? The rabbits that are good at making these trade-offs are going to live to reproduce, while the rabbits that are bad at it are either going to get eaten or starve. This means that, as a successful species on the planet, humans should be really good at making security trade-offs.

And yet, at the same time we seem hopelessly bad at it. We get it wrong all the time. We exaggerate some risks while minimizing others. We exaggerate some costs while minimizing others. Even simple trade-offs we get wrong, wrong, wrong—again and again. A Vulcan studying human security behavior would call us completely illogical.

The truth is that we're not bad at making security trade-offs. We are very well adapted to dealing with the security environment endemic to hominids living in small family groups on the highland plains of East Africa. It's just that the environment of New York in 2007 is different from Kenya circa 100,000 BC. And so our feeling of security diverges from the reality of security, and we get things wrong.

There are several specific aspects of the security trade-off that can go wrong. For example:

1. The severity of the risk.
2. The probability of the risk.
3. The magnitude of the costs.
4. How effective the countermeasure is at mitigating the risk.
5. How well disparate risks and costs can be compared.

The more your perception diverges from reality in any of these five aspects, the more your perceived trade-off won't match the actual trade-off. If you think that the risk is greater than it really is, you're going to overspend on mitigating that risk. If you think the risk is real but only affects other people—for whatever reason—you're going to underspend. If you overestimate the costs of a countermeasure, you're less likely to apply it when you should, and if you overestimate how effective a countermeasure is, you're more likely to apply it when you shouldn't. If you incorrectly evaluate the trade-off, you won't accurately balance the costs and benefits.

A lot of this can be chalked up to simple ignorance. If you think the murder rate in your town is one-tenth of what it really is, for example, then you're going to make bad security trade-offs. But I'm more interested in divergences between perception and reality that *can't* be explained that easily. Why is it that, even if someone knows that automobiles kill 40,000 people each year in the U.S. alone, and airplanes kill only hundreds worldwide, he is more afraid of airplanes than automobiles? Why is it that, when food poisoning kills 5,000 people every year and 9/11 terrorists killed 2,973 people in one non-repeated incident, we are spending tens of billions of dollars per year (not even counting the wars in Iraq and Afghanistan) on terrorism defense while the entire budget for the Food and Drug Administration in 2007 is only $1.9 billion?

It's my contention that these irrational trade-offs can be explained by psychology. That something inherent in how our brains work makes us more likely to be afraid of flying than of driving, and more likely to want to spend money, time, and other resources mitigating the risks of terrorism than those of food poisoning. And moreover, that these seeming irrationalities have a good evolutionary reason for existing: they've served our species well in the past. Understanding what they are, why they exist, and why they're failing us now is critical to understanding how we make security decisions. It's critical to understanding why, as a successful species on the planet, we make so many bad security trade-offs.

3   Conventional Wisdom about Risk

Most of the time, when the perception of security doesn't match the reality of security, it's because the perception of the risk doesn't match the reality of the risk. We worry about the wrong things: paying too much attention to minor risks and not enough attention to major ones. We don't correctly assess the magnitude of different risks. A lot of this can be chalked up to bad information or bad mathematics, but there are some general pathologies that come up over and over again.

In *Beyond Fear,* I listed five:

- People exaggerate spectacular but rare risks and downplay common risks.
- People have trouble estimating risks for anything not exactly like their normal situation.
- Personified risks are perceived to be greater than anonymous risks.
- People underestimate risks they willingly take and overestimate risks in situations they can't control.
- Last, people overestimate risks that are being talked about and remain an object of public scrutiny.[1]

David Ropeik and George Gray have a longer list in their book *Risk: A Practical Guide for Deciding What's Really Safe and What's Really Dangerous in the World Around You*:

- Most people are more afraid of risks that are new than those they've lived with for a while. In the summer of 1999, New Yorkers were extremely afraid of West Nile virus, a mosquito-borne infection that had never been seen in the United States. By the summer of 2001, though the virus continued to show up and make a few people sick, the fear had abated. The risk was still there, but New Yorkers had lived with it for a while. Their familiarity with it helped them see it differently.

- Most people are less afraid of risks that are natural than those that are human-made. Many people are more afraid of radiation from nuclear waste, or cell phones, than they are of radiation from the sun, a far greater risk.
- Most people are less afraid of a risk they choose to take than of a risk imposed on them. Smokers are less afraid of smoking than they are of asbestos and other indoor air pollution in their workplace, which is something over which they have little choice.
- Most people are less afraid of risks if the risk also confers some benefits they want. People risk injury or death in an earthquake by living in San Francisco or Los Angeles because they like those areas, or they can find work there.
- Most people are more afraid of risks that can kill them in particularly awful ways, like being eaten by a shark, than they are of the risk of dying in less awful ways, like heart disease—the leading killer in America.
- Most people are less afraid of a risk they feel they have some control over, like driving, and more afraid of a risk they don't control, like flying, or sitting in the passenger seat while somebody else drives.
- Most people are less afraid of risks that come from places, people, corporations, or governments they trust, and more afraid if the risk comes from a source they don't trust. Imagine being offered two glasses of clear liquid. You have to drink one. One comes from Oprah Winfrey. The other comes from a chemical company. Most people would choose Oprah's, even though they have no facts at all about what's in either glass.
- We are more afraid of risks that we are more aware of and less afraid of risks that we are less aware of. In the fall of 2001, awareness of terrorism was so high that fear was rampant, while fear of street crime and global climate change and other risks was low, not because those risks were gone, but because awareness was down.
- We are much more afraid of risks when uncertainty is high, and less afraid when we know more, which explains why we meet many new technologies with high initial concern.
- Adults are much more afraid of risks to their children than risks to themselves. Most people are more afraid of asbestos in their kids' school than asbestos in their own workplace.
- You will generally be more afraid of a risk that could directly affect you than a risk that threatens others. U.S. citizens were less afraid of terrorism before September 11, 2001, because up till then the Americans who had been the targets of terrorist attacks were almost always overseas. But suddenly on September 11, the risk became personal. When that happens, fear goes up, even though the statistical reality of the risk may still be very low.[2]

Others make these and similar points, which are summarized in Table 1.[3,4,5,6]

When you look over the list in Table 1, the most remarkable thing is how reasonable so many of them seem. This makes sense for two reasons. One, our perceptions of risk are deeply ingrained in our brains, the result of millions of years of evolution. And two, our perceptions of risk are generally pretty good, and are what have kept us alive and reproducing during those millions of years of evolution.

Table 1. Conventional Wisdom About People and Risk Perception

People exaggerate risks that are:	People downplay risks that are:
Spectacular	Pedestrian
Rare	Common
Personified	Anonymous
Beyond their control, or externally imposed	More under their control, or taken willingly
Talked about	Not discussed
Intentional or man-made	Natural
Immediate	Long-term or diffuse
Sudden	Evolving slowly over time
Affecting them personally	Affecting others
New and unfamiliar	Familiar
Uncertain	Well understood
Directed against their children	Directed towards themselves
Morally offensive	Morally desirable
Entirely without redeeming features	Associated with some ancillary benefit
Not like their current situation	Like their current situation

When our risk perceptions fail today, it's because of new situations that have occurred at a faster rate than evolution: situations that exist in the world of 2007, but didn't in the world of 100,000 BC. Like a squirrel whose predator-evasion techniques fail when confronted with a car, or a passenger pigeon who finds that evolution prepared him to survive the hawk but not the shotgun, our innate capabilities to deal with risk can fail when confronted with such things as modern human society, technology, and the media. And, even worse, they can be made to fail by others—politicians, marketers, and so on—who exploit our natural failures for their gain.

To understand all of this, we first need to understand the brain.

4 Risk and the Brain

The human brain is a fascinating organ, but an absolute mess. Because it has evolved over millions of years, there are all sorts of processes jumbled together rather than logically organized. Some of the processes are optimized for only certain kinds of situations, while others don't work as well as they could. And there's some duplication of effort, and even some conflicting brain processes.

Assessing and reacting to risk is one of the most important things a living creature has to deal with, and there's a very primitive part of the brain that has that job. It's the amygdala, and it sits right above the brainstem, in what's called the medial temporal lobe. The amygdala is responsible for processing base emotions that come from sensory inputs, like anger, avoidance, defensiveness, and fear. It's an old part of the brain, and seems to have originated in early fishes. When an animal—lizard, bird, mammal, even you—sees, hears, or feels something that's a potential danger, the amygdala is what reacts immediately. It's what causes adrenaline and other hormones to be pumped into your bloodstream, triggering the fight-or-flight response, causing increased heart rate and beat force, increased muscle tension, and sweaty palms.

This kind of thing works great if you're a lizard or a lion. Fast reaction is what you're looking for; the faster you can notice threats and either run away from them or fight back, the more likely you are to live to reproduce.

But the world is actually more complicated than that. Some scary things are not really as risky as they seem, and others are better handled by staying in the scary situation to set up a more advantageous future response. This means that there's an evolutionary advantage to being able to hold off the reflexive fight-or-flight response while you work out a more sophisticated analysis of the situation and your options for dealing with it.

We humans have a completely different pathway to deal with *analyzing* risk. It's the neocortex, a more advanced part of the brain that developed very recently, evolutionarily speaking, and only appears in mammals. It's intelligent and analytic. It can reason. It can make more nuanced trade-offs. It's also much slower.

So here's the first fundamental problem: we have two systems for reacting to risk—a primitive intuitive system and a more advanced analytic system—and they're operating in parallel. And it's hard for the neocortex to contradict the amygdala.

In his book *Mind Wide Open,* Steven Johnson relates an incident when he and his wife lived in an apartment and a large window blew in during a storm. He was standing right beside it at the time and heard the whistling of the wind just before the window blew. He was lucky—a foot to the side and he would have been dead—but the sound has never left him:

> But ever since that June storm, a new fear has entered the mix for me: the sound of wind whistling through a window. I know now that our window blew in because it had been installed improperly.... I am entirely convinced that the window we have now is installed correctly, and I trust our superintendent when he says that it is designed to withstand hurricane-force winds. In the five years since that June, we have weathered dozens of storms that produced gusts comparable to the one that blew it in, and the window has performed flawlessly.
>
> I know all these facts—and yet when the wind kicks up, and I hear that whistling sound, I can feel my adrenaline levels rise.... Part of my brain—the part that feels most *me*-like, the part that has opinions about the world and decides how to act on those opinions in a rational way—knows that the windows are safe.... But another part of my brain wants to barricade myself in the bathroom all over again.[7]

There's a good reason evolution has wired our brains this way. If you're a higher-order primate living in the jungle and you're attacked by a lion, it makes sense that you develop a lifelong fear of lions, or at least fear lions more than another animal you haven't personally been attacked by. From a risk/reward perspective, it's a good trade-off for the brain to make, and—if you think about it—it's really no different than your body developing antibodies against, say, chicken pox based on a single exposure. In both cases, your body is saying: "This happened once, and therefore it's likely to happen again. And when it does, I'll be ready." In a world where the threats are limited—where there are only a few diseases and predators that happen to affect the small patch of earth occupied by your particular tribe—it works.

Unfortunately, the brain's fear system doesn't scale the same way the body's immune system does. While the body can develop antibodies for hundreds of diseases, and those antibodies can float around in the bloodstream waiting for a second attack by the same disease, it's harder for the brain to deal with a multitude of lifelong fears.

All this is about the amygdala. The second fundamental problem is that because the analytic system in the neocortex is so new, it still has a lot of rough edges evolutionarily speaking. Psychologist Daniel Gilbert has a great quotation that explains this:

> The brain is a beautifully engineered get-out-of-the-way machine that constantly scans the environment for things out of whose way it should right now get. That's what brains did for several hundred million years— and then, just a few million years ago, the mammalian brain learned a new trick: to predict the timing and location of dangers before they actually happened.
>
> Our ability to duck that which is not yet coming is one of the brain's most stunning innovations, and we wouldn't have dental floss or 401(k) plans without it. But this innovation is in the early stages of development. The application that allows us to respond to visible baseballs is ancient and reliable, but the add-on utility that allows us to respond to threats that loom in an unseen future is still in beta testing.[8]

A lot of what I write in the following sections are examples of these newer parts of the brain getting things wrong.

And it's not just risks. People are not computers. We don't evaluate security trade-offs mathematically, by examining the relative probabilities of different events. Instead, we have shortcuts, rules of thumb, stereotypes, and biases— generally known as "heuristics." These heuristics affect how we think about risks, how we evaluate the probability of future events, how we consider costs, and how we make trade-offs. We have ways of generating close-to-optimal answers quickly with limited cognitive capabilities. Don Norman's wonderful essay, "Being Analog," provides a great background for all this.[9]

Daniel Kahneman, who won a Nobel Prize in Economics for some of this work, talks about humans having two separate cognitive systems: one that intuits and one that reasons:

The operations of System 1 are typically fast, automatic, effortless, associative, implicit (not available to introspection), and often emotionally charged; they are also governed by habit and therefore difficult to control or modify. The operations of System 2 are slower, serial, effortful, more likely to be consciously monitored and deliberately controlled; they are also relatively flexible and potentially rule governed.[10]

When you read about the heuristics I describe below, you can find evolutionary reasons for why they exist. And most of them are still very useful.[11] The problem is that they can fail us, especially in the context of a modern society. Our social and technological evolution has vastly outpaced our evolution as a species, and our brains are stuck with heuristics that are better suited to living in primitive and small family groups.

And when those heuristics fail, our feeling of security diverges from the reality of security.

5 Risk Heuristics

The first, and most common, area that can cause the feeling of security to diverge from the reality of security is the perception of risk. Security is a trade-off, and if we get the severity of the risk wrong, we're going to get the trade-off wrong. We can do this both ways, of course. We can underestimate some risks, like the risk of automobile accidents. Or we can overestimate some risks, like the risk of a stranger sneaking into our home at night and kidnapping our child. How we get the risk wrong—when we overestimate and when we underestimate—is governed by a few specific brain heuristics.

5.1 Prospect Theory

Here's an experiment that illustrates a particular pair of heuristics.[12] Subjects were divided into two groups. One group was given the choice of these two alternatives:

- Alternative A: A sure gain of $500.
- Alternative B: A 50% chance of gaining $1,000.

The other group was given the choice of:

- Alternative C: A sure loss of $500.
- Alternative D: A 50% chance of losing $1,000.

These two trade-offs aren't the same, but they're very similar. And traditional economics predicts that the difference doesn't make a difference.

Traditional economics is based on something called "utility theory," which predicts that people make trade-offs based on a straightforward calculation of relative gains and losses. Alternatives A and B have the same expected utility: +$500. And alternatives C and D have the same expected utility: -$500. Utility theory predicts that people choose alternatives A and C with the same probability and alternatives B and D with the same probability. Basically, some people prefer sure things and others prefer to take chances. The fact that one is gains and the other is losses doesn't affect the mathematics, and therefore shouldn't affect the results.

But experimental results contradict this. When faced with a gain, most people (84%) chose Alternative A (the sure gain) of $500 over Alternative B (the risky gain). But when faced with a loss, most people (70%) chose Alternative D (the risky loss) over Alternative C (the sure loss).

The authors of this study explained this difference by developing something called "prospect theory." Unlike utility theory, prospect theory recognizes that people have subjective values for gains and losses. In fact, humans have evolved a pair of heuristics that they apply in these sorts of trade-offs. The first is that a sure gain is better than a chance at a greater gain. ("A bird in the hand is better than two in the bush.") And the second is that a sure loss is worse than a chance at a greater loss. Of course, these are not rigid rules—given a choice between a sure $100 and a 50% chance at $1,000,000, only a fool would take the $100—but all things being equal, they do affect how we make trade-offs.

Evolutionarily, presumably it is a better survival strategy to—all other things being equal, of course—accept small gains rather than risking them for larger ones, and risk larger losses rather than accepting smaller losses. Lions chase young or wounded wildebeest because the investment needed to kill them is lower. Mature and healthy prey would probably be more nutritious, but there's a risk of missing lunch entirely if it gets away. And a small meal will tide the lion over until another day. Getting through today is more important than the possibility of having food tomorrow.

Similarly, it is evolutionarily better to risk a larger loss than to accept a smaller loss. Because animals tend to live on the razor's edge between starvation and reproduction, any loss of food—whether small or large—can be equally bad. That is, both can result in death. If that's true, the best option is to risk everything for the chance at no loss at all.

These two heuristics are so powerful that they can lead to logically inconsistent results. Another experiment, the Asian disease problem, illustrates that.[13] In this experiment, subjects were asked to imagine a disease outbreak that is expected to kill 600 people, and then to choose between two alternative treatment programs. Then, the subjects were divided into two groups. One group was asked to choose between these two programs for the 600 people:

- Program A: "200 people will be saved."
- Program B: "There is a one-third probability that 600 people will be saved, and a two-thirds probability that no people will be saved."

The second group of subjects were asked to choose between these two programs:

- Program C: "400 people will die."
- Program D: "There is a one-third probability that nobody will die, and a two-thirds probability that 600 people will die."

Like the previous experiment, programs A and B have the same expected utility: 200 people saved and 400 dead, A being a sure thing and B being a risk. Same with Programs C and D. But if you read the two pairs of choices carefully, you'll notice that—unlike the previous experiment—they are exactly the same. A equals C, and B equals D. All that's different is that in the first pair they're presented in terms of a gain (lives saved), while in the second pair they're presented in terms of a loss (people dying).

Yet most people (72%) choose A over B, and most people (78%) choose D over C. People make very different trade-offs if something is presented as a gain than if something is presented as a loss.

Behavioral economists and psychologists call this a "framing effect": peoples' choices are affected by how a trade-off is framed. Frame the choice as a gain, and people will tend to be risk averse. But frame the choice as a loss, and people will tend to be risk seeking.

We'll see other framing effects later on.

Another way of explaining these results is that people tend to attach a greater value to changes closer to their current state than they do to changes further away from their current state. Go back to the first pair of trade-offs I discussed. In the first one, a gain from $0 to $500 is worth more than a gain from $500 to $1,000, so it doesn't make sense to risk the first $500 for an even chance at a second $500. Similarly, in the second trade-off, more value is lost from $0 to -$500 than from -$500 to -$1,000, so it makes sense for someone to accept an even chance at losing $1,000 in an attempt to avoid losing $500. Because gains and losses closer to one's current state are worth more than gains and losses further away, people tend to be risk averse when it comes to gains, but risk seeking when it comes to losses.

Of course, our brains don't do the math. Instead, we simply use the mental shortcut.

There are other effects of these heuristics as well. People are not only risk averse when it comes to gains and risk seeking when it comes to losses; people also value something more when it is considered as something that can be lost, as opposed to when it is considered as a potential gain. Generally, the difference is a factor of 2 to 2.5.[14]

This is called the "endowment effect," and has been directly demonstrated in many experiments. In one,[15] half of a group of subjects were given a mug. Then, those who got a mug were asked the price at which they were willing to sell it, and those who didn't get a mug were asked what price they were willing to offer for one. Utility theory predicts that both prices will be about the same, but in fact, the median selling price was over twice the median offer.

In another experiment,[16] subjects were given either a pen or a mug with a college logo, both of roughly equal value. (If you read enough of these studies, you'll quickly notice two things. One, college students are the most common test subject. And two, any necessary props are most commonly purchased from a college bookstore.) Then the subjects were offered the opportunity to exchange the item they received for the other. If the subjects' preferences had nothing to do with the item they received, the fraction of subjects keeping a mug should equal the fraction of subjects exchanging a pen for a mug, and the fraction of subjects keeping a pen should equal the fraction of subjects exchanging a mug for a pen. In fact, most people kept the item they received; only 22% of subjects traded.

And, in general, most people will reject an even-chance gamble (50% of winning, and 50% of losing) unless the possible win is at least twice the size of the possible loss.[17]

What does prospect theory mean for security trade-offs? While I haven't found any research that explicitly examines if people make security trade-offs in the same way they make economic trade-offs, it seems reasonable to me that they do at least in part. Given that, prospect theory implies two things. First, it means that people are going to trade off more for security that lets them keep something they've become accustomed to—a lifestyle, a level of security, some functionality in a product or service—than they were willing to risk to get it in the first place. Second, when considering security gains, people are more likely to accept an incremental gain than a chance at a larger gain; but when considering security losses, they're more likely to risk a larger loss than accept than accept the certainty of a small one.

5.2   Other Biases That Affect Risk

We have other heuristics and biases about risks. One common one is called "optimism bias": we tend to believe that we'll do better than most others engaged in the same activity. This bias is why we think car accidents happen only to other people, and why we can at the same time engage in risky behavior while driving and yet complain about others doing the same thing. It's why we can ignore network security risks while at the same time reading about other companies that have been breached. It's why we think we can get by where others failed.

Basically, animals have evolved to underestimate loss. Because those who experience the loss tend not to survive, those of us remaining have an evolved experience that losses *don't* happen and that it's okay to take risks. In fact, some have theorized that people have a "risk thermostat," and seek an optimal level of risk regardless of outside circumstances.[18] By that analysis, if something comes along to reduce risk—seat belt laws, for example—people will compensate by driving more recklessly.

And it's not just that we don't think bad things can happen to us, we—all things being equal—believe that good outcomes are more probable than bad outcomes. This bias has been repeatedly illustrated in all sorts of experiments, but I think this one is particularly simple and elegant.[19]

Subjects were shown cards, one after another, with either a cartoon happy face or a cartoon frowning face. The cards were random, and the subjects simply had to guess which face was on the next card before it was turned over.

For half the subjects, the deck consisted of 70% happy faces and 30% frowning faces. Subjects faced with this deck were very accurate in guessing the face type; they were correct 68% of the time. The other half was tested with a deck consisting of 30% happy faces and 70% frowning faces. These subjects were much less accurate with their guesses, only predicting the face type 58% of the time. Subjects' preference for happy faces reduced their accuracy.

In a more realistic experiment,[20] students at Cook College were asked "Compared to other Cook students—the same sex as you—what do you think are the chances that the following events will happen to you?" They were given a list of 18 positive and 24 negative events, like getting a good job after graduation, developing a drinking problem, and so on. Overall, they considered themselves 15% more likely than others to experience positive events, and 20% less likely than others to experience negative events.

The literature also discusses a "control bias," where people are more likely to accept risks if they feel they have some control over them. To me, this is simply a manifestation of the optimism bias, and not a separate bias.

Another bias is the "affect heuristic," which basically says that an automatic affective valuation—I've seen it called "the emotional core of an attitude"—is the basis for many judgments and behaviors about it. For example, a study of people's reactions to 37 different public causes showed a very strong correlation between 1) the importance of the issues, 2) support for political solutions, 3) the size of the donation that subjects were willing to make, and 4) the moral satisfaction associated with those donations.[21] The emotional reaction was a good indicator of all of these different decisions.

With regard to security, the affect heuristic says that an overall good feeling toward a situation leads to a lower risk perception, and an overall bad feeling leads to a higher risk perception. This seems to explain why people tend to underestimate risks for actions that also have some ancillary benefit—smoking, skydiving, and such—but also has some weirder effects.

In one experiment,[22] subjects were shown either a happy face, a frowning face, or a neutral face, and then a random Chinese ideograph. Subjects tended to prefer ideographs they saw after the happy face, even though the face was flashed for only ten milliseconds and they had no conscious memory of seeing it. That's the affect heuristic in action.

Another bias is that we are especially tuned to risks involving people. Daniel Gilbert again:[23]

> We are social mammals whose brains are highly specialized for thinking about others. Understanding what others are up to—what they know and want, what they are doing and planning—has been so crucial to the survival of our species that our brains have developed an obsession with all things human. We think about people and their intentions; talk about them; look for and remember them.

In one experiment,[24] subjects were presented data about different risks occurring in state parks: risks from people, like purse snatching and vandalism, and natural-world risks, like cars hitting deer on the roads. Then, the subjects were asked which risk warranted more attention from state park officials.

Rationally, the risk that causes the most harm warrants the most attention, but people uniformly rated risks from other people as more serious than risks from deer. Even if the data indicated that the risks from deer were greater than the risks from other people, the people-based risks were judged to be more serious. It wasn't until the researchers presented the damage from deer as enormously higher than the risks from other people that subjects decided it deserved more attention.

People are also especially attuned to risks involving their children. This also makes evolutionary sense. There are basically two security strategies life forms have for propagating their genes. The first, and simplest, is to produce a lot of offspring and hope that some of them survive. Lobsters, for example, can lay 10,000 to 20,000 eggs at a time. Only ten to twenty of the hatchlings live to be four weeks old, but that's enough. The other strategy is to produce only a few offspring, and lavish attention on them. That's what humans do, and it's what allows our species to take such a long time to reach maturity. (Lobsters, on the other hand, grow up quickly.) But it also means that we are particularly attuned to threats to our children, children in general, and even other small and cute creatures.[25]

There is a lot of research on people and their risk biases. Psychologist Paul Slovic seems to have made a career studying them.[26] But most of the research is anecdotal, and sometimes the results seem to contradict each other. I would be interested in seeing not only studies about particular heuristics and when they come into play, but how people deal with instances of contradictory heuristics. Also, I would be very interested in research into how these heuristics affect behavior in the context of a strong fear reaction: basically, when these heuristics can override the amygdala and when they can't.

6 Probability Heuristics

The second area that can contribute to bad security trade-offs is probability. If we get the probability wrong, we get the trade-off wrong.

Generally, we as a species are not very good at dealing with large numbers. An enormous amount has been written about this, by John Paulos[27] and others. The saying goes "1, 2, 3, many," but evolutionarily it makes some amount of sense. Small numbers matter much more than large numbers. Whether there's one mango or ten mangos is an important distinction, but whether there are 1,000 or 5,000 matters less—it's a lot of mangos, either way. The same sort of thing happens with probabilities as well. We're good at 1 in 2 vs. 1 in 4 vs. 1 in 8, but we're much less good at 1 in 10,000 vs. 1 in 100,000. It's the same joke: "half the time, one quarter of the time, one eighth of the time, almost never." And whether whatever you're measuring occurs one time out of ten thousand or one time out of ten million, it's really just the same: almost never.

Additionally, there are heuristics associated with probabilities. These aren't specific to risk, but contribute to bad evaluations of risk. And it turns out that our brains' ability to quickly assess probability runs into all sorts of problems.

6.1   The Availability Heuristic

The "availability heuristic" is very broad, and goes a long way toward explaining how people deal with risk and trade-offs. Basically, the availability heuristic means that people "assess the frequency of a class or the probability of an event by the ease with which instances or occurrences can be brought to mind." [28] In other words, in any decision-making process, easily remembered (available) data are given greater weight than hard-to-remember data.

In general, the availability heuristic is a good mental shortcut. All things being equal, common events are easier to remember than uncommon ones. So it makes sense to use availability to estimate frequency and probability. But like all heuristics, there are areas where the heuristic breaks down and leads to biases. There are reasons other than occurrence that make some things more available. Events that have taken place recently are more available than others. Events that are more emotional are more available than others. Events that are more vivid are more available than others. And so on.

There's nothing new about the availability heuristic and its effects on security. I wrote about it in *Beyond Fear*,[29] although not by that name. Sociology professor Barry Glassner devoted most of a book to explaining how it affects our risk perception.[30] Every book on the psychology of decision making discusses it.

In one simple experiment,[31] subjects were asked this question:

– In a typical sample of text in the English language, is it more likely that a word starts with the letter K or that K is its third letter (not counting words with less than three letters)?

Nearly 70% of people said that there were more words that started with K, even though there are nearly twice as many words with K in the third position as there are words that start with K. But since words that start with K are easier to generate in one's mind, people overestimate their relative frequency.

In another, more real-world, experiment,[32] subjects were divided into two groups. One group was asked to spend a period of time imagining its college football team doing well during the upcoming season, and the other group was asked to imagine its college football team doing poorly. Then, both groups were asked questions about the team's actual prospects. Of the subjects who had imagined the team doing well, 63% predicted an excellent season. Of the subjects who had imagined the team doing poorly, only 40% did so.

The same researcher performed another experiment before the 1976 presidential election. Subjects asked to imagine Carter winning were more likely to predict that he would win, and subjects asked to imagine Ford winning were more likely to believe he would win. This kind of experiment has also been replicated several times, and uniformly demonstrates that considering a particular outcome in one's imagination makes it appear more likely later.

The vividness of memories is another aspect of the availability heuristic that has been studied. People's decisions are more affected by vivid information than by pallid, abstract, or statistical information.

Here's just one of many experiments that demonstrates this.[33] In the first part of the experiment, subjects read about a court case involving drunk driving. The defendant had run a stop sign while driving home from a party and collided with a garbage truck. No blood alcohol test had been done, and there was only circumstantial evidence to go on. The defendant was arguing that he was not drunk.

After reading a description of the case and the defendant, subjects were divided into two groups and given eighteen individual pieces of evidence to read: nine written by the prosecution about why the defendant was guilty, and nine written by the defense about why the defendant was innocent. Subjects in the first group were given prosecution evidence written in a pallid style and defense evidence written in a vivid style, while subjects in the second group were given the reverse.

For example, here is a pallid and vivid version of the same piece of prosecution evidence:

- On his way out the door, Sanders [the defendant] staggers against a serving table, knocking a bowl to the floor.
- On his way out the door, Sanders staggered against a serving table, knocking a bowl of guacamole dip to the floor and splattering guacamole on the white shag carpet.

And here's a pallid and vivid pair for the defense:

- The owner of the garbage truck admitted under cross-examination that his garbage truck is difficult to see at night because it is grey in color.
- The owner of the garbage truck admitted under cross-examination that his garbage truck is difficult to see at night because it is grey in color. The owner said his trucks are grey "because it hides the dirt," and he said, "What do you want, I should paint them pink?"

After all of this, the subjects were asked about the defendant's drunkenness level, his guilt, and what verdict the jury should reach.

The results were interesting. The vivid vs. pallid arguments had no significant effect on the subject's judgment immediately after reading them, but when they were asked again about the case 48 hours later—they were asked to make their judgments as though they "were deciding the case now for the first time"—they were more swayed by the vivid arguments. Subjects who read vivid defense arguments and pallid prosecution arguments were much more likely to judge the defendant innocent, and subjects who read the vivid prosecution arguments and pallid defense arguments were much more likely to judge him guilty.

The moral here is that people will be persuaded more by a vivid, personal story than they will by bland statistics and facts, possibly solely due to the fact that they remember vivid arguments better.

Another experiment[34] divided subjects into two groups, who then read about a fictional disease called "Hyposcenia-B." Subjects in the first group read about a disease with concrete and easy-to-imagine symptoms: muscle aches, low energy level, and frequent headaches. Subjects in the second group read about a disease with abstract and difficult-to-imagine symptoms: a vague sense of disorientation, a malfunctioning nervous system, and an inflamed liver.

Then each group was divided in half again. Half of each half was the control group: they simply read one of the two descriptions and were asked how likely they were to contract the disease in the future. The other half of each half was the experimental group: they read one of the two descriptions "with an eye toward imagining a three-week period during which they contracted and experienced the symptoms of the disease," and then wrote a detailed description of how they thought they would feel during those three weeks. And then they were asked whether they thought they would contract the disease.

The idea here was to test whether the ease or difficulty of imagining something affected the availability heuristic. The results showed that those in the control group—who read either the easy-to-imagine or difficult-to-imagine symptoms, showed no difference. But those who were asked to imagine the easy-to-imagine symptoms thought they were more likely to contract the disease than the control group, and those who were asked to imagine the difficult-to-imagine symptoms thought they were less likely to contract the disease than the control group. The researchers concluded that imagining an outcome alone is not enough to make it appear more likely; it has to be something easy to imagine. And, in fact, an outcome that is difficult to imagine may actually appear to be less likely.

Additionally, a memory might be particularly vivid precisely because it's extreme, and therefore unlikely to occur. In one experiment,[35] researchers asked some commuters on a train platform to remember and describe "the worst time you missed your train" and other commuters to remember and describe "any time you missed your train." The incidents described by both groups were equally awful, demonstrating that the most extreme example of a class of things tends to come to mind when thinking about the class.

More generally, this kind of thing is related to something called "probability neglect": the tendency of people to ignore probabilities in instances where there is a high emotional content.[36] Security risks certainly fall into this category, and our current obsession with terrorism risks at the expense of more common risks is an example.

The availability heuristic also explains hindsight bias. Events that have actually occurred are, almost by definition, easier to imagine than events that have not, so people retroactively overestimate the probability of those events. Think of "Monday morning quarterbacking," exemplified both in sports and in national policy. "He should have seen that coming" becomes easy for someone to believe.

The best way I've seen this all described is by Scott Plous:

> In very general terms: (1) the more *available* an event is, the more frequent or probable it will seem; (2) the more *vivid* a piece of information is, the more easily recalled and convincing it will be; and (3) the more *salient* something is, the more likely it will be to appear causal.[37]

Here's one experiment that demonstrates this bias with respect to salience.[38] Groups of six observers watched a two-man conversation from different vantage points: either seated behind one of the men talking or sitting on the sidelines between the two men talking. Subjects facing one or the other conversants tended to rate that person as more influential in the conversation: setting the tone, determining what kind of information was exchanged, and causing the other person to respond as he did. Subjects on the sidelines tended to rate both conversants as equally influential.

As I said at the beginning of this section, most of the time the availability heuristic is a good mental shortcut. But in modern society, we get a lot of sensory input from the media. That screws up availability, vividness, and salience, and means that heuristics that are based on our senses start to fail. When people were living in primitive tribes, if the idea of getting eaten by a saber-toothed tiger was more available than the idea of getting trampled by a mammoth, it was reasonable to believe that—for the people in the particular place they happened to be living—it was more likely they'd get eaten by a saber-toothed tiger than get trampled by a mammoth. But now that we get our information from television, newspapers, and the Internet, that's not necessarily the case. What we read about, what becomes vivid to us, might be something rare and spectacular. It might be something fictional: a movie or a television show. It might be a marketing message, either commercial or political. And remember, visual media are more vivid than print media. The availability heuristic is less reliable, because the vivid memories we're drawing upon aren't relevant to our real situation. And even worse, people tend not to remember *where* they heard something—they just remember the content. So even if, at the time they're exposed to a message, they don't find the source credible, eventually their memory of the source of the information degrades and they're just left with the message itself.

We in the security industry are used to the effects of the availability heuristic. It contributes to the "risk du jour" mentality we so often see in people. It explains why people tend to overestimate rare risks and underestimate common ones.[39] It explains why we spend so much effort defending against what the bad guys did last time, and ignore what new things they could do next time. It explains why we're worried about risks that are in the news at the expense of risks that are not, or rare risks that come with personal and emotional stories at the expense of risks that are so common they are only presented in the form of statistics.

It explains most of the entries in Table 1.

6.2 Representativeness

"Representativeness" is a heuristic by which we assume the probability that an example belongs to a particular class is based on how well that example represents the class. On the face of it, this seems like a reasonable heuristic. But it can lead to erroneous results if you're not careful.

The concept is a bit tricky, but here's an experiment makes this bias crystal clear.[40] Subjects were given the following description of a woman named Linda:

Linda is 31 years old, single, outspoken, and very bright. She majored in philosophy. As a student, she was deeply concerned with issues of discrimination and social justice, and also participated in antinuclear demonstrations.

Then the subjects were given a list of eight statements describing her present employment and activities. Most were decoys ("Linda is an elementary school teacher," "Linda is a psychiatric social worker," and so on), but two were critical: number 6 ("Linda is a bank teller," and number 8 ("Linda is a bank teller and is active in the feminist movement"). Half of the subjects were asked to rank the eight outcomes by the similarity of Linda to the typical person described by the statement, while others were asked to rank the eight outcomes by probability.

Of the first group of subjects, 85% responded that Linda more resembled a stereotypical feminist bank teller more than a bank teller. This makes sense. But of the second group of subjects, 89% of thought Linda was more likely to be a feminist bank teller than a bank teller. Mathematically, of course, this is ridiculous. It is impossible for the second alternative to be more likely than the first; the second is a subset of the first.

As the researchers explain: "As the amount of detail in a scenario increases, its probability can only decrease steadily, but its representativeness and hence its apparent likelihood may increase. The reliance on representativeness, we believe, is a primary reason for the unwarranted appeal of detailed scenarios and the illusory sense of insight that such constructions often provide."[41]

Doesn't this sound like how so many people resonate with movie-plot threats—overly specific threat scenarios—at the expense of broader risks?

In another experiment,[42] two groups of subjects were shown short personality descriptions of several people. The descriptions were designed to be stereotypical for either engineers or lawyers. Here's a sample description of a stereotypical engineer:

Tom W. is of high intelligence, although lacking in true creativity. He has a need for order and clarity, and for neat and tidy systems in which every detail finds its appropriate place. His writing is rather dull and mechanical, occasionally enlivened by somewhat corny puns and flashes of imagination of the sci-fi type. He has a strong drive for competence. He seems to have little feel and little sympathy for other people and does not enjoy interacting with others. Self-centered, he nonetheless has a deep moral sense.

Then, the subjects were asked to give a probability that each description belonged to an engineer rather than a lawyer. One group of subjects was told this about the population from which the descriptions were sampled:

– Condition A: The population consisted of 70 engineers and 30 lawyers.

The second group of subjects was told this about the population:

– Condition B: The population consisted of 30 engineers and 70 lawyers.

Statistically, the probability that a particular description belongs to an engineer rather than a lawyer should be much higher under Condition A than Condition B. However, subjects judged the assignments to be the same in either case. They were basing their judgments solely on the stereotypical personality characteristics of engineers and lawyers, and ignoring the relative probabilities of the two categories.

Interestingly, when subjects were not given any personality description at all and simply asked for the probability that a random individual was an engineer, they answered correctly: 70% under Condition A and 30% under Condition B. But when they were given a neutral personality description, one that didn't trigger either stereotype, they assigned the description to an engineer 50% of the time under both Conditions A and B.

And here's a third experiment. Subjects (college students) were given a survey which included these two questions: "How happy are you with your life in general?" and "How many dates did you have last month?" When asked in this order, there was no correlation between the answers. But when asked in the reverse order—when the survey reminded the subjects of how good (or bad) their love life was before asking them about their life in general—there was a 66% correlation.[43]

Representativeness also explains the base rate fallacy, where people forget that if a particular characteristic is extremely rare, even an accurate test for that characteristic will show false alarms far more often than it will correctly identify the characteristic. Security people run into this heuristic whenever someone tries to sell such things as face scanning, profiling, or data mining as effective ways to find terrorists.

And lastly, representativeness explains the "law of small numbers," where people assume that long-term probabilities also hold in the short run. This is, of course, not true: if the results of three successive coin flips are tails, the odds of heads on the fourth flip are not more than 50%. The coin is not "due" to flip heads. Yet experiments have demonstrated this fallacy in sports betting again and again.[44]

7   Cost Heuristics

Humans have all sorts of pathologies involving costs, and this isn't the place to discuss them all. But there are a few specific heuristics I want to summarize, because if we can't evaluate costs right—either monetary costs or more abstract costs—we're not going to make good security trade-offs.

7.1   Mental Accounting

Mental accounting is the process by which people categorize different costs.[45] People don't simply think of costs as costs; it's much more complicated than that.

Here are the illogical results of two experiments.[46]

In the first, subjects were asked to answer one of these two questions:

- Trade-off 1: Imagine that you have decided to see a play where the admission is $10 per ticket. As you enter the theater you discover that you have lost a $10 bill. Would you still pay $10 for a ticket to the play?
- Trade-off 2: Imagine that you have decided to see a play where the admission is $10 per ticket. As you enter the theater you discover that you have lost the ticket. The seat is not marked and the ticket cannot be recovered. Would you pay $10 for another ticket?

The results of the trade-off are exactly the same. In either case, you can either see the play and have $20 less in your pocket, or not see the play and have $10 less in your pocket. But people don't see these trade-offs as the same. Faced with Trade-off 1, 88% of subjects said they would buy the ticket anyway. But faced with Trade-off 2, only 46% said they would buy a second ticket. The researchers concluded that there is some sort of mental accounting going on, and the two different $10 expenses are coming out of different mental accounts.

The second experiment was similar. Subjects were asked:

- Imagine that you are about to purchase a jacket for $125, and a calculator for $15. The calculator salesman informs you that the calculator you wish to buy is on sale for $10 at the other branch of the store, located 20 minutes' drive away. Would you make the trip to the other store?
- Imagine that you are about to purchase a jacket for $15, and a calculator for $125. The calculator salesman informs you that the calculator you wish to buy is on sale for $120 at the other branch of the store, located 20 minutes' drive away. Would you make the trip to the other store?

Ignore your amazement at the idea of spending $125 on a calculator; it's an old experiment. These two questions are basically the same: would you drive 20 minutes to save $5? But while 68% of subjects would make the drive to save $5 off the $15 calculator, only 29% would make the drive to save $5 off the $125 calculator.

There's a lot more to mental accounting.[47] In one experiment,[48] subjects were asked to imagine themselves lying on the beach on a hot day and how good a cold bottle of their favorite beer would feel. They were to imagine that a friend with them was going up to make a phone call—this was in 1985, before cell phones—and offered to buy them that favorite brand of beer if they gave the friend the money. What was the most the subject was willing to pay for the beer?

Subjects were divided into two groups. In the first group, the friend offered to buy the beer from a fancy resort hotel. In the second group, the friend offered to buy the beer from a run-down grocery store. From a purely economic viewpoint, that should make no difference. The value of one's favorite brand of beer on a hot summer's day has nothing to do with where it was purchased from. (In economic terms, the consumption experience is the same.) But people were willing to pay $2.65 on average for the beer from a fancy resort, but only $1.50 on average from the run-down grocery store.

The experimenters concluded that people have reference prices in their heads, and that these prices depend on circumstance. And because the reference price was different in the different scenarios, people were willing to pay different amounts. This leads to sub-optimal results. As Thayer writes, "The thirsty beer-drinker who would pay $4 for a beer from a resort but only $2 from a grocery store will miss out on some pleasant drinking when faced with a grocery store charging $2.50."

Researchers have documented all sorts of mental accounting heuristics. Small costs are often not "booked," so people more easily spend money on things like a morning coffee. This is why advertisers often describe large annual costs as "only a few dollars a day." People segregate frivolous money from serious money, so it's easier for them to spend the $100 they won in a football pool than a $100 tax refund. And people have different mental budgets. In one experiment that illustrates this,[49] two groups of subjects were asked if they were willing to buy tickets to a play. The first group was told to imagine that they had spent $50 earlier in the week on tickets to a basketball game, while the second group was told to imagine that they had received a $50 parking ticket earlier in the week. Those who had spent $50 on the basketball game (out of the same mental budget) were significantly less likely to buy the play tickets than those who spent $50 paying a parking ticket (out of a different mental budget).

One interesting mental accounting effect can be seen at race tracks.[50] Bettors tend to shift their bets away from favorites and towards long shots at the end of the day. This has been explained by the fact that the average bettor is behind by the end of the day—pari-mutuel betting means that the average bet is a loss—and a long shot can put a bettor ahead for the day. There's a "day's bets" mental account, and bettors don't want to close it in the red.

The effect of mental accounting on security trade-offs isn't clear, but I'm certain we have a mental account for "safety" or "security," and that money spent from that account feels different than money spent from another account. I'll even wager we have a similar mental accounting model for non-fungible costs such as risk: risks from one account don't compare easily with risks from another. That is, we are willing to accept considerable risks in our leisure account—skydiving, knife juggling, whatever—when we wouldn't even consider them if they were charged against a different account.

7.2   Time Discounting

"Time discounting" is the term used to describe the human tendency to discount future costs and benefits. It makes economic sense; a cost paid in a year is not the same as a cost paid today, because that money could be invested and earn interest during the year. Similarly, a benefit accrued in a year is worth less than a benefit accrued today.

Way back in 1937, economist Paul Samuelson proposed a discounted-utility model to explain this all. Basically, something is worth more today than it is in the future. It's worth more to you to have a house today than it is to get it in ten years, because you'll have ten more years' enjoyment of the house. Money is worth more today than it is years from now; that's why a bank is willing to pay you to store it with them.

The discounted utility model assumes that things are discounted according to some rate. There's a mathematical formula for calculating which is worth more—$100 today or $120 in twelve months—based on interest rates. Today, for example, the discount rate is 6.25%, meaning that $100 today is worth the same as $106.25 in twelve months. But of course, people are much more complicated than that.

There is, for example, a magnitude effect: smaller amounts are discounted more than larger ones. In one experiment,[51] subjects were asked to choose between an amount of money today or a greater amount in a year. The results would make any banker shake his head in wonder. People didn't care whether they received $15 today or $60 in twelve months. At the same time, they were indifferent to receiving $250 today or $350 in twelve months, and $3,000 today or $4,000 in twelve months. If you do the math, that implies a discount rate of 139%, 34%, and 29%—all held simultaneously by subjects, depending on the initial dollar amount.

This holds true for losses as well,[52] although gains are discounted more than losses. In other words, someone might be indifferent to $250 today or $350 in twelve months, but would much prefer a $250 penalty today to a $350 penalty in twelve months. Notice how time discounting interacts with prospect theory here.

Also, preferences between different delayed rewards can flip, depending on the time between the decision and the two rewards. Someone might prefer $100 today to $110 tomorrow, but also prefer $110 in 31 days to $100 in thirty days.

Framing effects show up in time discounting, too. You can frame something either as an acceleration or a delay from a base reference point, and that makes a big difference. In one experiment,[53] subjects who expected to receive a VCR in twelve months would pay an average of $54 to receive it immediately, but subjects who expected to receive the VCR immediately demanded an average $126 discount to delay receipt for a year. This holds true for losses as well: people demand more to expedite payments than they would pay to delay them.[54]

Reading through the literature, it sometimes seems that discounted utility theory is full of nuances, complications, and contradictions. Time discounting is more pronounced in young people, people who are in emotional states – fear is certainly an example of this – and people who are distracted. But clearly there is some mental discounting going on; it's just not anywhere near linear, and not easily formularized.

8   Heuristics That Affect Decisions

And finally, there are biases and heuristics that affect trade-offs. Like many other heuristics we've discussed, they're general, and not specific to security. But they're still important.

First, some more framing effects.

Most of us have anecdotes about what psychologists call the "context effect": preferences among a set of options depend on what other options are in the set. This has been confirmed in all sorts of experiments—remember the experiment about what people were willing to pay for a cold beer on a hot beach—and most of us have anecdotal confirmation of this heuristic.

For example, people have a tendency to choose options that dominate other options, or compromise options that lie between other options. If you want your boss to approve your $1M security budget, you'll have a much better chance of getting that approval if you give him a choice among three security plans—with budgets of $500K, $1M, and $2M, respectively—than you will if you give him a choice among three plans with budgets of $250K, $500K, and $1M.

The rule of thumb makes sense: avoid extremes. It fails, however, when there's an intelligence on the other end, manipulating the set of choices so that a particular one doesn't seem extreme.

"Choice bracketing" is another common heuristic. In other words: choose a variety. Basically, people tend to choose a more diverse set of goods when the decision is bracketed more broadly than they do when it is bracketed more narrowly. For example,[55] in one experiment students were asked to choose among one of six different snacks that they would receive at the beginning of the next three weekly classes. One group had to choose the three weekly snacks in advance, while the other group chose at the beginning of each class session. Of the group that chose in advance, 64% chose a different snack each week, but only 9% of the group that chose each week did the same.

The narrow interpretation of this experiment is that we overestimate the value of variety. Looking ahead three weeks, a variety of snacks seems like a good idea, but when we get to the actual time to enjoy those snacks, we choose the snack we like. But there's a broader interpretation as well, one borne out by similar experiments and directly applicable to risk taking: when faced with repeated risk decisions, evaluating them as a group makes them feel less risky than evaluating them one at a time. Back to finance, someone who rejects a particular gamble as being too risky might accept multiple identical gambles.

Again, the results of a trade-off depend on the context of the trade-off.

It gets even weirder. Psychologists have identified an "anchoring effect," whereby decisions are affected by random information cognitively nearby. In one experiment[56], subjects were shown the spin of a wheel whose numbers ranged from 0 and 100, and asked to guess whether the number of African nations in the UN was greater or less than that randomly generated number. Then, they were asked to guess the exact number of African nations in the UN.

Even though the spin of the wheel was random, and the subjects knew it, their final guess was strongly influenced by it. That is, subjects who happened to spin a higher random number guessed higher than subjects with a lower random number.

Psychologists have theorized that the subjects anchored on the number in front of them, mentally adjusting it for what they thought was true. Of course, because this was just a guess, many people didn't adjust sufficiently. As strange as it might seem, other experiments have confirmed this effect.

And if you're not completely despairing yet, here's another experiment that will push you over the edge.[57] In it, subjects were asked one of these two questions:

- Question 1: Should divorce in this country be easier to obtain, more difficult to obtain, or stay as it is now?
- Question 2: Should divorce in this country be easier to obtain, stay as it is now, or be more difficult to obtain?

In response to the first question, 23% of the subjects chose easier divorce laws, 36% chose more difficult divorce laws, and 41% said that the status quo was fine. In response to the second question, 26% chose easier divorce laws, 46% chose more difficult divorce laws, and 29% chose the status quo. Yes, the order in which the alternatives are listed affects the results.

There are lots of results along these lines, including the order of candidates on a ballot.

Another heuristic that affects security trade-offs is the "confirmation bias." People are more likely to notice evidence that supports a previously held position than evidence that discredits it. Even worse, people who support position A sometimes mistakenly believe that anti-A evidence actually supports that position. There are a lot of experiments that confirm this basic bias and explore its complexities.

If there's one moral here, it's that individual preferences are not based on predefined models that can be cleanly represented in the sort of indifference curves you read about in microeconomics textbooks; but instead, are poorly defined, highly malleable, and strongly dependent on the context in which they are elicited. Heuristics and biases matter. A lot.

This all relates to security because it demonstrates that we are not adept at making rational security trade-offs, especially in the context of a lot of ancillary information designed to persuade us one way or another.

9   Making Sense of the Perception of Security

We started out by teasing apart the security trade-off, and listing five areas where perception can diverge from reality:

1. The severity of the risk.
2. The probability of the risk.
3. The magnitude of the costs.
4. How effective the countermeasure is at mitigating the risk.
5. The trade-off itself.

Sometimes in all the areas, and all the time in area 4, we can explain this divergence as a consequence of not having enough information. But sometimes we have all the information and *still* make bad security trade-offs. My aim was to give you a glimpse of the complicated brain systems that make these trade-offs, and how they can go wrong.

Of course, we can make bad trade-offs in anything: predicting what snack we'd prefer next week or not being willing to pay enough for a beer on a hot day. But security trade-offs are particularly vulnerable to these biases because they are so critical to our survival. Long before our evolutionary ancestors had the brain capacity to consider future snack preferences or a fair price for a cold beer, they were dodging predators and forging social ties with others of their species. Our brain heuristics for dealing with security are old and well-worn, and our amygdalas are even older.

What's new from an evolutionary perspective is large-scale human society, and the new security trade-offs that come with it. In the past I have singled out technology and the media as two aspects of modern society that make it particularly difficult to make good security trade-offs—technology by hiding detailed complexity so that we don't have the right information about risks, and the media by producing such available, vivid, and salient sensory input—but the issue is really broader than that. The neocortex, the part of our brain that has to make security trade-offs, is, in the words of Daniel Gilbert, "still in beta testing."

I have just started exploring the relevant literature in behavioral economics, the psychology of decision making, the psychology of risk, and neuroscience. Undoubtedly there is a lot of research out there for me still to discover, and more fascinatingly counterintuitive experiments that illuminate our brain heuristics and biases. But already I understand much more clearly why we get security trade-offs so wrong so often.

When I started reading about the psychology of security, I quickly realized that this research can be used both for good and for evil. The good way to use this research is to figure out how humans' feelings of security can better match the reality of security. In other words, how do we get people to recognize that they need to question their default behavior? Giving them more information seems not to be the answer; we're already drowning in information, and these heuristics are not based on a lack of information. Perhaps by understanding how our brains processes risk, and the heuristics and biases we use to think about security, we can learn how to override our natural tendencies and make better security trade-offs. Perhaps we can learn how not to be taken in by security theater, and how to convince others not to be taken in by the same.

The evil way is to focus on the feeling of security at the expense of the reality. In his book *Influence*,[58] Robert Cialdini makes the point that people can't analyze every decision fully; it's just not possible: people need heuristics to get through life. Cialdini discusses how to take advantage of that; an unscrupulous person, corporation, or government can similarly take advantage of the heuristics and biases we have about risk and security. Concepts of prospect theory, framing, availability, representativeness, affect, and others are key issues in marketing and politics. They're applied generally, but in today's world they're more and more applied to security. Someone could use this research to simply make people *feel* more secure, rather than to actually make them more secure.

After all my reading and writing, I believe my good way of using the research is unrealistic, and the evil way is unacceptable. But I also see a third way: integrating the feeling and reality of security.

The feeling and reality of security are different, but they're closely related. We make the best security trade-offs—and by that I mean trade-offs that give us genuine security for a reasonable cost—when our feeling of security matches the reality of security. It's when the two are out of alignment that we get security wrong.

In the past, I've criticized palliative security measures that only make people *feel* more secure as "security theater." But used correctly, they can be a way of raising our feeling of security to more closely match the reality of security. One example is the tamper-proof packaging that started to appear on over-the-counter drugs in the 1980s, after a few highly publicized random poisonings. As a countermeasure, it didn't make much sense. It's easy to poison many foods and over-the-counter medicines right through the seal—with a syringe, for example—or to open and reseal the package well enough that an unwary consumer won't detect it. But the tamper-resistant packaging brought people's perceptions of the risk more in line with the actual risk: minimal. And for that reason the change was worth it.

Of course, security theater has a cost, just like real security. It can cost money, time, capabilities, freedoms, and so on, and most of the time the costs far outweigh the benefits. And security theater is no substitute for real security. Furthermore, too much security theater will raise people's feeling of security to a level greater than the reality, which is also bad. But used in conjunction with real security, a bit of well-placed security theater might be exactly what we need to both be and feel more secure.

References

1. Schneier, B.: Beyond Fear: Thinking Sensibly About Security in an Uncertain World. Springer, Heidelberg (2003)
2. Ropeik, D., Gray, G.: Risk: A Practical Guide for Deciding What's Really Safe and What's Really Dangerous in the World Around You, Houghton Mifflin (2002)
3. Glassner, B.: The Culture of Fear: Why Americans are Afraid of the Wrong Things. Basic Books (1999)
4. Slovic, P.: The Perception of Risk. Earthscan Publications Ltd (2000)
5. Gilbert, D.: If only gay sex caused global warming. In: Los Angeles Times (July 2, 2006)
6. Kluger, J.: How Americans Are Living Dangerously. Time (November 26, 2006)
7. Johnson, S.: Mind Wide Open: Your Brain and the Neuroscience of Everyday Life, Scribner (2004)
8. Gilbert, D.: If only gay sex caused global warming. Los Angeles Times, (July 2, 2006)
9. Norman, D.A.: Being Analog. The Invisible Computer, ch. 7. MIT Press, Cambridge (1998), http://www.jnd.org/dn.mss/being_analog.html

10. Kahneman, D.: A Perspective on Judgment and Choice. American Psychologist 58(9), 697–720 (2003)
11. Gigerenzer, G., Todd, P.M., et al.: Simple Heuristics that Make us Smart. Oxford University Press, Oxford (1999)
12. Kahneman, D., Tversky, A.: Prospect Theory: An Analysis of Decision Under Risk. Econometrica 47, 263–291 (1979)
13. Tversky, A., Kahneman, D.: The Framing of Decisions and the Psychology of Choice. Science 211, 453–458 (1981)
14. Tversky, A., Kahneman, D.: Evidential Impact of Base Rates. In: Kahneman, D., Slovic, P., Tversky, A. (eds.) Judgment Under Uncertainty: Heuristics and Biases, pp. 153–160. Cambridge University Press, Cambridge (1982)
15. Kahneman, D.J., Knetsch, J.L., Thaler, R.H.: Experimental Tests of the Endowment Effect and the Coase Theorem. Journal of Political Economy 98, 1325–1348 (1990)
16. Knetsch, J.L.: Preferences and Nonreversibility of Indifference Curves. Journal of Economic Behavior and Organization 17, 131–139 (1992)
17. Tversky, A., Kahneman, D.: Advances in Prospect Theory: Cumulative Representation of Subjective Uncertainty. Journal of Risk and Uncertainty 5:xx, 297–323 (1992)
18. Adams, J.: Cars, Cholera, and Cows: The Management of Risk and Uncertainty. In: CATO Institute Policy Analysis 335 (1999)
19. Rosenhan, D.L., Messick, S.: Affect and Expectation. Journal of Personality and Social Psychology 3, 38–44 (1966)
20. Weinstein, N.D.: Unrealistic Optimism about Future Life Events. Journal of Personality and Social Psychology 39, 806–820 (1980)
21. Kahneman, D., Ritov, I., Schkade, D.: Economic preferences or attitude expressions? An analysis of dollar responses to public issues. Journal of Risk and Uncertainty 19, 220–242 (1999)
22. Winkielman, P., Zajonc, R.B., Schwarz, N.: Subliminal affective priming attributional interventions. Cognition and Emotion 11(4), 433–465 (1977)
23. Gilbert, D.: If only gay sex caused global warming. Los Angeles Times (July 2, 2006)
24. Wilson, R.S., Arvai, J.L.: When Less is More: How Affect Influences Preferences When Comparing Low-risk and High-risk Options. Journal of Risk Research 9(2), 165–178 (2006)
25. Cohen, J.: The Privileged Ape: Cultural Capital in the Making of Man. Parthenon Publishing Group (1989)
26. Slovic, P.: The Perception of Risk. Earthscan Publications Ltd (2000)
27. Paulos, J.A.: Innumeracy: Mathematical Illiteracy and Its Consequences, Farrar, Straus, and Giroux (1988)
28. Tversky, A., Kahneman, D.: Judgment under Uncertainty: Heuristics and Biases. Science 185, 1124–1130 (1974)
29. Schneier, B.: Beyond Fear: Thinking Sensibly About Security in an Uncertain World. Springer, Heidelberg (2003)
30. Glassner, B.: The Culture of Fear: Why Americans are Afraid of the Wrong Things. Basic Books (1999)
31. Tversky, A., Kahneman, D.: Availability: A Heuristic for Judging Frequency. Cognitive Psychology 5, 207–232 (1973)
32. Carroll, J.S.: The Effect of Imagining an Event on Expectations for the Event: An Interpretation in Terms of the Availability Heuristic. Journal of Experimental Social Psychology 14, 88–96 (1978)

33. Reyes, R.M., Thompson, W.C., Bower, G.H.: Judgmental Biases Resulting from Differing Availabilities of Arguments. Journal of Personality and Social Psychology 39, 2–12 (1980)
34. Sherman, S.J., Cialdini, R.B., Schwartzman, D.F., Reynolds, K.D.: Imagining Can Heighten or Lower the Perceived Likelihood of Contracting a Disease: The Mediating Effect of Ease of Imagery. Personality and Social Psychology Bulletin 11, 118–127 (1985)
35. Morewedge, C.K., Gilbert, D.T., Wilson, T.D.: The Least Likely of Times: How Memory for Past Events Biases the Prediction of Future Events. Psychological Science 16, 626–630 (2005)
36. Sunstein, C.R.: Terrorism and Probability Neglect. Journal of Risk and Uncertainty 26, 121–136 (2003)
37. Plous, S.: The Psychology of Judgment and Decision Making. McGraw-Hill, New York (1993)
38. Taylor, S.E., Fiske, S.T.: Point of View and Perceptions of Causality. Journal of Personality and Social Psychology 32, 439–445 (1975)
39. Slovic, P., Fischhoff, B., Lichtenstein, S.: Rating the Risks. Environment 2, 14–20, 36–39 (1979)
40. Tversky, A., Kahneman, D.: Extensional vs Intuitive Reasoning: The Conjunction Fallacy in Probability Judgment. Psychological Review 90, 293–315 (1983)
41. Tversky, A., Kahneman, D.: Judgments of and by Representativeness. In: Kahneman, D., Slovic, P., Tversky, A. (eds.) Judgment Under Uncertainty: Heuristics and Biases, Cambridge University Press, Cambridge (1982)
42. Kahneman, D., Tversky, A.: On the Psychology of Prediction. Psychological Review 80, 237–251 (1973)
43. Kahneman, D., Frederick, S.: Representativeness Revisited: Attribute Substitution in Intuitive Judgement. In: Gilovich, T., Griffin, D., Kahneman, D. (eds.) Heuristics and Biases, pp. 49–81. Cambridge University Press, Cambridge (2002)
44. Gilovich, T., Vallone, R., Tversky, A.: The Hot Hand in Basketball: On the Misperception of Random Sequences. Cognitive Psychology 17, 295–314 (1985)
45. Thaler, R.H.: Toward a Positive Theory of Consumer Choice. Journal of Economic Behavior and Organization 1, 39–60 (1980)
46. Tversky, A., Kahneman, D.: The Framing of Decisions and the Psychology of Choice. Science 211, 253–258 (1981)
47. Thayer, R.: Mental Accounting Matters. In: Camerer, C.F., Loewenstein, G., Rabin, M. (eds.) Advances in Behavioral Economics, Princeton University Press, Princeton (2004)
48. Thayer, R.: Mental Accounting and Consumer Choice. Marketing Science 4, 199–214 (1985)
49. Heath, C., Soll, J.B.: Mental Accounting and Consumer Decisions. Journal of Consumer Research 23, 40–52 (1996)
50. Ali, M.: Probability and Utility Estimates for Racetrack Bettors. Journal of Political Economy 85, 803–815 (1977)
51. Thayer, R.: Some Empirical Evidence on Dynamic Inconsistency. Economics Letters 8, 201–207 (1981)
52. Loewenstein, G., Prelec, D.: Anomalies in Intertemporal Choice: Evidence and Interpretation. Quarterly Journal of Economics, 573–597 (1992)
53. Loewenstein, G.: Anticipation and the Valuation of Delayed Consumption. Economy Journal 97, 666–684 (1987)

54. Benzion, U., Rapoport, A., Yagel, J.: Discount Rates Inferred from Decisions: An Experimental Study. Management Science 35, 270–284 (1989)
55. Simonson, I.: The Effect of Purchase Quantity and Timing on Variety-Seeking Behavior. Journal of Marketing Research 17, 150–162 (1990)
56. Tversky, A., Kahneman, D.: Judgment under Uncertainty: Heuristics and Biases. Science 185, 1124–1131 (1974)
57. Schurman, H., Presser, S.: Questions and Answers in Attitude Surveys: Experiments on Wording Form, Wording, and Context. Academic Press, London (1981)
58. Cialdini, R.B.: Influence: The Psychology of Persuasion. HarperCollins (1998)

An (Almost) Constant-Effort
Solution-Verification
Proof-of-Work Protocol Based on Merkle Trees

Fabien Coelho

CRI, École des mines de Paris,
35, rue Saint-Honoré, 77305 Fontainebleau CEDEX, France
fabien.coelho@ensmp.fr

Abstract. Proof-of-work schemes are economic measures to deter denial-of-service attacks: service requesters compute moderately hard functions the results of which are easy to check by the provider. We present such a new scheme for solution-verification protocols. Although most schemes to date are probabilistic unbounded iterative processes with high variance of the requester effort, our Merkle tree scheme is deterministic with an almost constant effort and null variance, and is computation-optimal.

1   Introduction

Economic measures to contain denial-of-service attacks such as spams were first suggested by Dwork and Naor [1]: a computation stamp is required to obtain a service. Proof-of-work schemes are dissymmetric: the computation must be moderately hard for the requester, but easy to check for the service provider. Applications include having uncheatable benchmarks [2], helping audit reported metering of web sites [3], adding delays [4,5], managing email addresses [6], or limiting abuses on peer-to-peer networks [7,8]. Proofs may be purchased in advance [9]. These schemes are formalized [10], and actual financial analysis is needed [11,12] to evaluate their real impact. There are two protocol flavors:

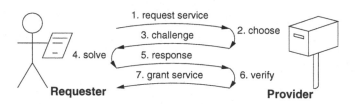

Fig. 1. Challenge-Response Protocol

Challenge-response protocols in Figure 1 assume an interaction between client and server so that the service provider chooses the problem, say an item with

S. Vaudenay (Ed.): AFRICACRYPT 2008, LNCS 5023, pp. 80–93, 2008.

some property from a finite set, and the requester must retrieve the item in the set. The solution is known to exist, the search time distribution is basically uniform, the solution is found on average when about half of the set has been processed, and standard deviation is about $\frac{1}{2\sqrt{3}} \approx 0.3$ of the mean.

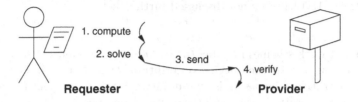

Fig. 2. Solution-Verification Protocol

Solution-verification protocols in Figure 2 do not assume such a link. The problem must be self-imposed, based somehow on the service description, say perhaps the intended recipient and date of a message. The target is usually a probabilistic property reached by iterations. The verification phase must check both the problem choice *and* the provided solution. Such iterative searches have a constant probability of success at each trial, resulting in a shifted geometrical distribution, the mean is the inverse of the success probability, and the standard deviation nearly equals the mean. The resulting distribution has a long tail as the number of iterations to success is not bounded: about every 50 searches an unlucky case requires more than 4 times the average number of iterations to complete (the probability of not succeeding in 4 times the average is about $e^{\frac{-4a}{a}} = e^{-4} \approx \frac{1}{50}$).

We present a new proof-of-work solution-verification scheme based on Merkle trees with an almost constant effort and null variance for the client. When considering a Merkle tree with N leaves, the solution costs about $2N$, $P \cdot \ln(N)$ data is sent, and the verification costs $P \cdot \ln(N)$ with $P = 8 \cdot \ln_2(N)$ a good choice. This contribution is theoretical with a constant requester effort, which is thus bounded or of possibly low variance, but also practical as our scheme is computation-optimal and has an interesting work-ratio.

Section 2 discusses proof-of-work schemes suggested to date and analyzes their optimality and the computation distribution of solution-verification variants. Section 3 describes our new scheme based on Merkle trees built on top of the service description. This scheme is shown computation-optimal, but is not communication-optimal. The solution involves the computation of most of the tree, although only part of it is sent thanks to a feedback mechanism which selects only a subset of the leaves. Section 4 computes a cost lower bound for a proof, then outlines two attacks beneficial to the service requester. The effort of our scheme is constant, thus bounded and with a null variance. However we show an iterative attack, which is not upper-bounded, and which results in a small gain. Together with the demonstrated lower-bound cost of a proof, it justifies our *almost* claim.

2   Related Work

We introduce two optimality criteria to analyze proof-of-work schemes, then discuss solution-verification protocols suggested to date with respect to these criteria and to the work distribution on the requester side. Challenge-response only functions [13,14,15] are not discussed further here.

Let the *effort* $E(w)$ be the amount of computation of the requester as a function of provider work w, and the *work-ratio* the effort divided by the provider work. Proof-of-work schemes may be: (a) *communication-optimal* if the amount of data sent on top of the service description D is minimal. For solution-verification iterative schemes it is about ln(work-ratio) to identify the found solution: the work-ratio is the number of iterations performed over a counter to find a solution, and it is enough to just return the value of this counter for the provider to check the requester proof. For challenge-response protocols, it would be ln(search space size). This criterion emphasizes minimum impact on communications. (b) *computation-optimal* if the challenge or verification work is simply linear in the amount of communicated data, which it must at least be if the data is processed. This criterion mitigates denial-of-service attacks on service providers, as fake proof-of-works could require significant resources to disprove. A scheme meeting both criteria is deemed optimal.

Three proof-of-work schemes are suggested by Dwork and Naor [1]. One is a formula (integer square root modulo a large prime $p \equiv 3 \bmod 4$), as computing a square root is more expensive than squaring the result to check it. Assuming a naïve implementation, it costs $\ln(p)^3$ to compute, $\ln(p)$ to communicate, and $\ln(p)^2$ to check. The search cost is deterministic, but the $w^{1.5}$ effort is not very interesting, and is not optimal. Better implementations reduce both solution and verification complexities. If $p \equiv 1 \bmod 4$, the square root computation with the Tonelli-Shanks algorithm involves a non deterministic step with a geometrical distribution. The next two schemes present shortcuts which allow some participants to generate cheaper stamps. They rely on forging a signature without actually breaking a private key. One uses the Fiat-Shamir signature with a weak hash function for which an inversion is sought by iteration, with a geometrical distribution of the effort. The computation costs $E \cdot \ln(N)^2$, the communication $\ln(N)$ and the verification $\ln(N)^2$, where $N \gg 2^{512}$ is needed for the scheme

Table 1. Comparison of Solution-Verification POW

ref	effort	var	comm.	work	constraints
[1]1	$\ln(p)^3$	0	$\ln(p)$	$\ln(p)^2$	p large prime
[1]2	$E\ln(N)^2$	> 0	$\ln(N)$	$\ln(N)^2$	$N \gg 2^{512}$
[16]	E	$=$	$\ln(E)$	$\ln(E)$	
[17]	$E\ell$	$=$	$\ln(E)$	ℓ	typical $\ell = 2^{13}$
[18]	$E\ell$	$=$	$\ln(E)$	ℓ	$E \ll 2^\ell, \ell > 2^{10}$
[19]	E	$=$	$\ln(E)$	$\ln(E)$	
here	$2N$	0	$P\ln(N)$	$P\ln(N)$	$P = 8 \cdot \ln_2(N)$

security and the arbitrary effort E is necessarily much smaller than N; thus the scheme is not optimal. The other is the Ong-Schnorr-Shamir signature broken by Pollard, with a similar non-optimality and a geometrical distribution because of an iterative step.

Some schemes [16,3,20] seek *partial* hash inversions. Hashcash [16] iterates a hash function on a string involving the service description and a counter, and is optimal. The following stamp computed in 400 seconds on a 2005 laptop:

`1:28:170319:hobbes@comics::7b7b973c8bdb0cb1:147b744d`

allows to send an email to *hobbes* on March 19, 2017. The last part is the hexadecimal counter, and the SHA1 hash of the whole string begins with 28 binary zeros. Franklin and Malkhi [3] build a hash sequence that statistically catches cheaters, but the verification may be expensive. Wang and Reiter [20] allow the requester to tune the effort to improve its priority.

Memory-bound schemes [17,18,19] seek to reduce the impact of the computer hardware performance on computation times. All solution-verification variants are based on an iterative search which target a partial hash inversion, and thus have a geometrical distribution of success and are communication-optimal. However only the last of these memory-bound solution-verification schemes is computation-optimal.

Table 1 compares the requester cost and variance, communication cost, and provider checking cost, of solution-verification proof-of-work schemes, with the notations used in the papers.

3   Scheme

This section describes our (almost) constant-effort and null variance solution-verification proof-of-work scheme. The client is expected to compute a Merkle tree which depends on a service description, but is required to give only part of the tree for verification by the service provider. A feedback mechanism uses the root hash so that the given part cannot be known in advance, thus induces the client to compute most of the tree for a solution. Finally choice of parameters and a memory-computation implementation trade-off are discussed. The notations used thoroughly in this paper are summarized in Table 2. The whole scheme is outlined in Figure 3.

3.1   Merkle Tree

Let h be a cryptographic hash function from anything to a domain of size 2^m. The complexity of such functions is usually stepwise linear in the input length. For our purpose the input is short, thus computations only involve one step. Let D be a service description, for instance a string such as `hobbes@comics:20170319:0001`. Let $s = h(D)$ be its hash. Let $h_s(x) = h(x\|s)$ be a service-dependent hash. The Merkle binary hash tree [21] of depth d ($N = 2^d$) is computed as follows: (1)

Table 2. Summary of notations

Symbol	Definition
w	provider checking work
$E(w)$	requester effort
D	service description, a string
h	cryptographic hash function
m	hash function bit width
s	service hash is $h(D)$
h_s	service-dependent hash function
d	depth of Merkle binary hash tree
N	number of leaves in tree is 2^d
P	number of proofs expected
n_i	a node hash in the binary tree
n_0	root hash of the tree
r	leaf selector seed is $h_s^P(n_0)$

leaf digests $n_{N-1+i} = h_s(i)$ for i in $0 \ldots N-1$; (2) inner nodes are propagated upwards $n_i = h_s(n_{2i+1} \| n_{2i+2})$ for i in $N-2 \ldots 0$. Root hash n_0 is computed with $2N$ calls to h, half for leaf computations, one for service s, and the remainder for the internal nodes of the tree. The whole tree *depends* on the service description as s is used at every stage: reusing such a tree would require a collision of service description hashes.

3.2   Feedback

Merkle trees help manage Lamport signatures [22]: a *partial* tree allows to check quickly that some leaves belong to the full tree by checking that they actually lead to the root hash. We use this property to generate our proof of work: the requester returns such a partial tree to show that selected leaves belong to the tree and thus were indeed computed. However, what particular leaves are needed must not be known in advance, otherwise it would be easy to generate a partial tree just with those leaves and to provide random values for the other branches. Thus we select returned leaves based on the root hash, so that they depend on the whole tree computation.

The feedback phase chooses P evenly-distributed independent leaves derived from the root hash as partial proofs of the whole computation. A cryptographic approximation of such an *independent-dependent* derivation is to seed a pseudo-random number generator from root hash n_0 and to extract P numbers corresponding to leaves in P consecutive chunks of size $\frac{N}{P}$. These leaf numbers and the additional nodes necessary to check for the full tree constitute the proof-of-work. Figure 4 illustrates the data sent for 4 leaf-proofs (black) and the intermediate hashes that must be provided (grey) or computed (white) on a 256-leaf tree.

solution work by requester

    define service description: $D = $ `hobbes@comics:20170319:0001`

    compute service description hash: $s = h(D) = $ `36639b2165bcd7c724...`

    compute leaf hashes: for i in $0 \dots N - 1$: $n_{N-1+i} = h_s(i)$

    compute internal node hashes: for i in $N - 2 \dots 0$: $n_i = h_s(n_{2i+1} \| n_{2i+2})$

    compute generator seed $r = h_s^P(n_0)$

    derive leaf numbers in each P chunk   for j in $0 \dots P - 1$: $\ell_j = \mathcal{G}(r, j)$

communication from requester to provider

    send service description D

    send P leaf numbers ℓ_j for $j \in (0 \dots P - 1)$

    for each paths of selected leaves send intermediate lower tree node hashes

    that's $P \ln_2(\frac{N}{P})$ hashes of width m

verification work by provider

    check service description D *do I want to provide this service?*

    compute service hash $s = h(D)$

    compute root hash n_0 from ℓ_j and provided node hashes

    compute generator seed $r = h_s^P(n_0)$

    derive leaf numbers in each P chunk   for j in $0 \dots P - 1$: $\ell'_j = \mathcal{G}(r, j)$

    check whether these leaf numbers were provided $\forall j \in (0 \dots P - 1), \ell_j = \ell'_j$

Fig. 3. Scheme Outline

They are evenly distributed as one leaf is selected in every quarter of the tree, so balanced branches only meet near the root.

3.3 Verification

The service provider receives the required service description D, P leaf numbers, and the intermediate hashes necessary to compute the root of the Merkle tree which amount to about $P \cdot \ln_2(\frac{N}{P}) \cdot (m + 1)$ bits: $P \cdot \ln_2(\frac{N}{P})$ for the leaf numbers inside the chunks, and $P \cdot \ln_2(\frac{N}{P}) \cdot m$ for the intermediate hashes.

The server checks the consistency of the partial tree by recomputing the hashes starting from service hash s and leaf numbers and up to the root hash using the provided intermediate node hashes, and then by checking the feedback choice, *i.e.* that the root hash does lead to the provided leaves. This requires about $P \cdot \ln_2(N)$ hash computations for the tree, and some computations of the pseudo-random number generator. This phase is computation-optimal as each data is processed a fixed number of times by the hash function for the tree and generator computations.

Note that the actual root hash is not really needed to validate the Merkle tree: it is computed anyway by the verification and, if enough leaves are required, its value is validated indirectly when checking that the leaves are indeed the one derived from the root hash seeded generator.

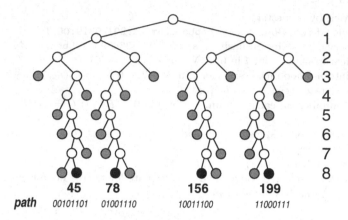

0
1
2
3
4
5
6
7
8

45   **78**   **156**   **199**
path *00101101* *01001110*   *10011100*   *11000111*

Fig. 4. Merkle tree proof ($P = 4$, $N = 2^8$)

3.4   Choice of Parameters

Let us discuss the random generator, the hash function h and its width m, the tree depth d ($N = 2^d$) and the number of proofs P.

The pseudo-random number generator supplies $P \cdot \ln_2(\frac{N}{P})$ bits ($14 = 22 - 8$ bits per proof for $N = 2^{22}$ and $P = 256 = 2^8$) to choose the evenly-distributed leaves. Standard generators can be seeded directly with the root hash. To add to the cost of an attack without impact on the verification complexity, the generator seed may rely further on h by using seed $r = h_s^P(n_0)$ (h_s composed P times over itself), so that about P hash computations are needed to test a partial tree, as discussed in Section 4.1. The generator itself may also use h, say with the j-th leaf in the j-th chunk chosen as $\ell_j = \mathcal{G}(r, j) = h_s(j\|r) \bmod \frac{N}{P}$ for j in $0 \dots P-1$.

The hash width may be different for the description, lower tree (close to the leaves), upper tree (close to the root), and generator. The description hash must avoid collisions which would lead to reusable trees; the generator hash should keep as much entropy as possible, especially as the seed is iterated P times; in the upper part of the tree, a convenient root hash should not be targetable, and the number of distinct root hashes should be large enough so that it is not worth precomputing them, as well as to provide a better initial entropy. A strong cryptographic hash is advisable in these cases. For the lower tree and leaves, the smaller m the better, as it drives the amount of temporary data and the proof size. Tabulating node hashes for reuse is not interesting because they all depend on s and if $2^{2m} \gg 2N$. Moreover it should not be easily invertible, so that a convenient hash cannot be targeted by a search process at any point. A sufficient condition is $2^m > 2N$: one hash inversion costs more than the *whole* computation. For our purpose, with $N = 2^{22}$, the lower tree hash may be folded to $m = 24$. The impact of choosing $m = \ln_2(N) + 2$ is not taken into account in our complexity analyses because h is assumed a constant cost for any practical tree depth: it would not change our optimality result to do so, but it would change the effort function to $e^{\sqrt[3]{w}}$.

The Merkle tree depth leads to the number of leaves N and the expected number of hash computations $2N$. The resource consumption required before the service is provided should depend on the cost of the service. For emails, a few seconds computation per recipient seems reasonable. With SHA1, depth $d = 22$ leads to 2^{23} hash calls and warrants this effort on my 2005 laptop. For other hash functions, the right depth depends on the performance of these functions on the target hardware. The number of leaves also induces the number of required proofs, hence the total proof size, as discussed hereafter.

The smaller the number of proofs, the better for the communication and verification involved, but if very few proofs are required a partial computation of the Merkle tree could be greatly beneficial to the requester. We choose $P = 8 \cdot \ln_2(N)$, maybe rounded up to a power of two to ease the even distribution. Section 4.2 shows that this value induces the service requester to compute most of the tree. With this number of proofs, the solution effort is $e^{\sqrt{w}}$ (verification work $w = \mathcal{O}(\ln(N)^2)$, and provider effort is $2N \approx e^{\sqrt{w}}$). It is not communication-optimal: proofs are a little bit large, for instance with SHA1 as a hash and with $N = 2^{22}$ it is about 11 KB (that is $256 \cdot (22 - 8) \cdot (24 + 1)$ bits), although around 22 bits are sufficient for a counter-based technique.

3.5   Memory-Computation Trade-off

The full Merkle tree needs about $2N \cdot m$ bits if it is kept in memory, to be able to extract the feedback hashes once the required leaves are known. A simple trade-off is to keep only the upper part of the tree, dividing the memory requirement by 2^t, at the price of $P \cdot 2^{t+1}$ hash computations to rebuild the subtrees that contain the proofs. The limit case recomputes the full tree once the needed leaves are known.

4   Attacks

In the above protocol, the requester uses $2N$ hash computations for the Merkle tree, but the provider needs only $P \cdot \ln_2(N) = 8 \cdot (lntwoN)^2$ to verify the extracted partial tree, and both side must run the generator. This section discusses attacks which reduce the requester work by computing only a fraction of the tree and being lucky with the feedback so that required leaves are available. We first compute a lower bound for the cost of finding a solution depending on the parameters, then we discuss two attacks.

4.1   Partial Tree

In order to cheat one must provide a matching partial tree, *i.e.*: (a) a valid partial tree starting from the service hashes *or the tree itself is rejected*; (b) with valid leaves choice based on the root hash *or the feedback fails*. As this tree is built from a cryptographic hash function, the successful attacker must have computed the provided partial Merkle tree root hash and its leaf derivations: otherwise the

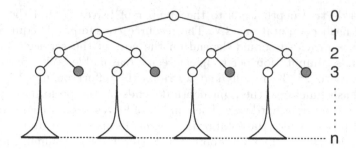

Fig. 5. Partial Merkle tree ($f = 0.5$, $P = 4$)

probability of returning a matching partial tree by chance is the same as finding a hash inversion.

Let us assume that the attacker builds a partial tree involving a fraction f of the leaves ($0 \leq f \leq 1$), where missing hash values are filled-in randomly, as outlined in Figure 5: evenly-distributed proofs result in 4 real hashes at depth 2, computed from 4 fake hashes (in grey) introduced at depth 3 to hide the non-computed subtrees, and 4 real hashes coming from the real subtrees. Half leaf hashes are really computed.

Once the root hash is available, the feedback leaves can be derived. If they are among available ones, a solution has been found and can be returned. The probability of this event is f^P. It is quickly reduced by smaller fractions and larger numbers of proofs. If the needed proof leaves are not all available, no solution was found. From this point, the attacker can either start all over again, reuse only part of the tree at another attempt, or alter the current tree. The later is the better choice. This tree alteration can either consist of changing a fake node (iteration at constant f), or of adding new leaves (extending f).

We are interested in the expected average cost of the search till a suitable root hash which points to available leaves is found. Many strategies are possible as iterations or extensions involving any subset of leaves can be performed in any order. However, each trial requires the actual root hash for a partial tree and running the generator. Doing so adds to the current total cost of the solution tree computation and to the cost of later trials.

4.2 Attack Cost Lower Bound

A conservative lower bound cost for a successful attack can be computed by assuming that for every added leaf the partial tree is tried without over-cost for the queue to reach the root nor for computing the seed more than once. We first evaluate an upper bound of the probability of success for these partial trees, which is then used to derive a lower bound for the total cost: Whatever the attack strategy, for our suggested number of proofs and a tree of depth 7 or more, a requester will have to compute at least 90% of the full Merkle tree on average to find an accepted proof of work.

Proof. If we neglect the even distribution of proof leaves, the probability of success at iteration i of constructing a tree (an i-th leaf is added in the tree) is $\rho_i = (\frac{i}{N})^P$, and the probability of getting there is $(1 - \sigma_{i-1})$ where σ_i is the cumulated probability of success up to i: $\sigma_0 = 0$, $\sigma_i = \sigma_{i-1} + (1 - \sigma_{i-1})\rho_i$, and $\sigma_N = 1$, as the last iteration solves the problem with $\rho_N = 1$. The $(1 - \sigma_{i-1})\rho_i$ term is the global probability of success at i: the computation got there (the problem was not solved before) and is solved at this very iteration. As it is lower than ρ_i:

$$\sigma_j \leq \sum_{i=0}^{j} \rho_i \leq \int_0^{\frac{j+1}{N}} N x^P \, dx = \frac{N}{P+1}\left(\frac{j+1}{N}\right)^{P+1} \tag{1}$$

If $c(i)$ is the increasing minimal cost of testing a tree with i leaves, the average cost $\mathcal{C}$ for the requester is:

$$\mathcal{C}(N, P) \geq \sum_{i=1}^{N} c(i)(1 - \sigma_{i-1})\rho_i = \sum_{i=1}^{N} c(i)(\sigma_i - \sigma_{i-1})$$

$$= \sum_{i=1}^{\ell-1} c(i)(\sigma_i - \sigma_{i-1}) + \sum_{i=\ell}^{N} c(i)(\sigma_i - \sigma_{i-1})$$

$$\geq 0 + c(\ell)(\sigma_N - \sigma_{\ell-1})$$

$$\geq c(\ell)(1 - \sigma_\ell)$$

The cost is bounded by cutting the summation at ℓ chosen as $\frac{\ell+1}{N} = (\frac{1}{N})^{\frac{1}{P+1}}$. The contributions below this limit are zeroed, and those over are minimized as $c(\ell) \geq 2\ell + P$ (the ℓ-leaf tree is built and the seed is computed once) and $(1 - \sigma_\ell)$ is bound with Equation (1) so that $(1 - \sigma_\ell) \geq (1 - \frac{1}{P+1}) = \frac{P}{P+1}$ hence, as $P \geq 2$:

$$\mathcal{C}(N, P) \geq \left(\frac{1}{N}\right)^{\frac{1}{P+1}} \frac{P}{P+1}(2N) \tag{2}$$

Figure 6 plots this estimation. The back-left corner is empty where the number of proofs is greater than the number of leaves. With $P = 8 \cdot \ln_2(N)$ and if $N \geq 2^7$, Equation (2) is simplified:

$$\mathcal{C}(N) \geq \left(\frac{1}{2}\right)^{\frac{1}{8}} \frac{8 \cdot \ln_2(N)}{8 \cdot \ln_2(N) + 1}(2N) \geq 0.9\,(2N)$$

Namely the average cost for the requester $\mathcal{C}(N)$ is larger than 90% of the $2N$ full tree cost.   QED.

4.3   Iterative Attack

Let us investigate a simple attack strategy that fills a fraction of the tree with fake hashes introduced to hide non computed leaves, and then iterates by modifying a fake hash till success, without increasing the number of leaves. The resulting

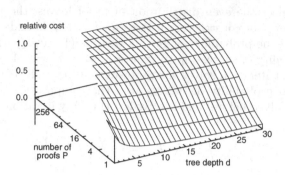

Fig. 6. Relative cost lower bound – Equation (2)

average cost is shown in Equation (3). The first term approximates the hash tree computation cost for the non-faked leaves and nodes, and is a minimum cost for the attack with a given fraction f: there are $N \cdot f$ leaves in the binary tree, and about the same number of internal nodes. The second term is the average iteration cost for a solution, by trying faked hash values from depth $\ln_2(P) + 1$ thanks to the even-distribution, and another P to derive the seed from the root hash; the resulting cost is multiplied by the average number of iterations which is the inverse of the probability of success at each trial.

$$\mathcal{C}_{\text{iter}}(f) \approx 2Nf + (P + \ln_2(P) + 1)\frac{1}{f^P} \tag{3}$$

If f is small, the second term dominates, and the cost is exponential. If f is close to 1, the first linear term is more important and the cost is close to the full tree computation. This effect is illustrated in Figure 7 for different number of proofs P: few proofs lead to very beneficial fractions: many proofs make the minimum of the functions close to the full tree computation.

$$\mathcal{F}(N, P) = \sqrt[P+1]{\frac{P(P + \ln_2(P) + 1)}{2N}} \tag{4}$$

Equation (4), the zero of the derivative of (3), gives the best fraction of this iterative strategy for a given size and number of proofs. $\mathcal{F}(2^{22}, 256) = 0.981$ and the cost is 0.989 of the full tree, to be compared to the 0.9 lower bound computed in Section 4.2. Whether a significantly better strategy can be devised is unclear. A conservative cost lower bound computed with a numerical simulation and for the same parameters gives a 0.961 multiplier. In order to reduce the effectiveness of this attack further, the hash-based generator may cost up to $P \cdot \ln_2(N)$ to derive seed r without impact on the overall verification complexity, but at the price of doubling the verification cost.

This successful attack justifies the *almost* constant-effort claim: either a full tree is computed and a solution is found with a null variance, or some partial-tree unbounded attack is carried out, maybe with a low variance, costing at least 90% of the full tree.

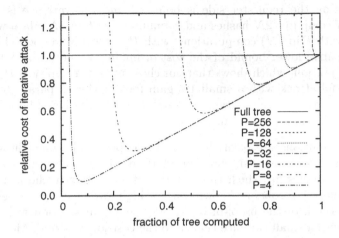

Fig. 7. Iterative cost for fraction f with $N = 2^{22}$

4.4 Skewed Feedback Attack

Let us study the impact of a non-independent proof selection by the pseudo-random number generator. This section simply illustrates the importance of the randomness of the generator. We assume an extreme case where the first bits of the root hash are used as a unique leaf index in the P chunks: the selected leaves would be $\{k, k + \frac{N}{P}, k + 2\frac{N}{P}, \ldots\}$. Then in the partial tree attack the requester could ensure that any leaf k computed in the first chunk have their corresponding shifted leaves in the other chunks available. Thus, when hitting one leaf in the first chunk, all other leaves follow, and the probability of a successful feedback is f instead of f^P. $N = 2^{22}$ and $P = 256$ lead to $0.002(2N)$, a 474 speedup of the attack efficiency.

5 Conclusion

Proof-of-work schemes help deter denial-of-service attacks on costly services such as email delivery by requiring moderately hard computations from the requester that are easy to verify by the provider. As solution-verification protocol variants do not assume any interaction between requesters and providers, the computations must be self-imposed, based somehow on the expected service. Most of these schemes are unbounded iterative probabilistic searches with a high variance of the requester effort. We have made the following contributions about proof-of-work schemes:

1. two definitions of optimality criteria: communication-optimal if the minimum amount of data is sent; computation-optimal if the verification is linear in the data sent;
2. a computation-optimal (but not communication-optimal) proof-of-work solution-verification scheme based on Merkle trees with a $e^{\sqrt{w}}$ effort, for which

the work on the requester side is bounded and the variance is null: the requester computes $2N$ hashes and communicates $P\ln_2(N)$ data which are verified with $P\ln_2(N)$ computations, with $P = 8\ln_2(N)$ a good choice;

3. a conservative lower bound of the cost of finding a solution at 90% of the full computation, which shows that our chosen number of proofs P is sound;
4. a successful attack with a small 1% gain for our chosen parameter values, which involves a large constant cost and a small iterative unbounded part, thus resulting in a low overall variance.

These contributions are both theoretical and practical. Our solution-verification scheme has a bounded, constant-effort solution. In contrast to iterative probabilistic searches for which the found solution is exactly checked, but the requester's effort is probably known with a high variance, we rather have a probabilistic check of the proof-of-work, but the actual solution work is quite well known with a small variance thanks to the cost lower bound. Moreover our scheme is practical, as it is computation-optimal thus not prone to denial-of-service attacks in itself as the verification work is propotional to the data sent by the requester. Also, although not optimal, the communication induces an interesting work-ratio. The only other bounded solution-verification scheme is a formula with a $w^{1.5}$ effort, which is neither communication nor computation-optimal. Whether a bounded fully optimal solution-verification scheme may be built is an open question.

Thanks

Françis Maisonneuve, Vincent Bachelot and Alexandre Aillos, helped to find a lower bound for the cost of computing a matching partial tree. Pierre Jouvelot, Ronan Keryell, François Irigoin, John B. Chapman and Zachary B. Ciccolo helped to improve the writing. Comments by anonymous reviewers were also very helpful.

References

1. Dwork, C., Naor, M.: Pricing via processing or combatting junk mail. In: Brickell, E.F. (ed.) CRYPTO 1992. LNCS, vol. 740, pp. 139–147. Springer, Heidelberg (1993)
2. Cai, J.-Y., Lipton, R.R., Sedgewick, R., Yao, A.C.C.: Towards uncheatable benchmarks. In: Eighth IEEE Annual Structure in Complexity Conference, San Diego, California, May 1993, pp. 2–11 (1993)
3. Franklin, M.K., Malkhi, D.: Auditable Metering with Lightweight Security. In: Luby, M., Rolim, J., Serna, M. (eds.) FC 1997. LNCS, vol. 1318, pp. 151–160. Springer, Heidelberg (1997)
4. Rivest, R., Shamir, A.: PayWord and MicroMint – Two Simple Micropayment Schemes. CryptoBytes 2(1) (1996)
5. Goldschlag, D.M., Stubblebine, S.G.: Publicly verifiable lotteries: Applications of delaying functions. In: Hirschfeld, R. (ed.) FC 1998. LNCS, vol. 1465, pp. 214–226. Springer, Heidelberg (1998)

6. Gabber, E., Jakobsson, M., Matias, Y., Mayer, A.J.: Curbing junk e-mail via secure classification. In: Hirschfeld, R. (ed.) FC 1998. LNCS, vol. 1465, pp. 198–213. Springer, Heidelberg (1998)
7. Rosenthal, D.S.H., Roussopoulos, M., Maniatis, P., Baker, M.: Economic Measures to Resist Attacks on a Peer-to-Peer Network. In: Workshop on Economics of Peer-to-Peer Systems, Berkeley, CA, USA (June 2003)
8. Garcia, F., Hoepman, J.-H.: Off-line Karma: A Decentralized Currency for Peer-to-peer and Grid Applications. In: Ioannidis, J., Keromytis, A.D., Yung, M. (eds.) ACNS 2005. LNCS, vol. 3531, pp. 364–377. Springer, Heidelberg (2005)
9. Abadi, M., Birrell, A., Mike, B., Dabek, F., Wobber, T.: Bankable postage for network services. In: Schmidt, D. (ed.) ESOP 2004. LNCS, vol. 2986, pp. 72–90. Springer, Heidelberg (2004)
10. Jakobsson, M., Juels, A.: Proofs of Work and Bread Pudding Protocols. Communications and Multimedia Security (September 1999)
11. Laurie, B., Clayton, R.: "Proof-of-Work" Proves Not to Work. In: WEAS 2004 (May 2004)
12. Liu, D., Camp, L.J.: Proof of Work can Work. In: Fifth Workshop on the Economics of Information Security (June 2006)
13. Rivers, R.L., Shamir, A., Wagner, D.: Time-lock puzzles and timed-release crypto. Technical Report MIT/LCS 684, MIT (1996)
14. Juels, A., Brainard, J.: Client Puzzles: A Cryptographic Defense Against Connection Depletion Attacks. In: Network and Distributed System Security (NDSS) (February 1999)
15. Waters, B., Juels, A., Halderman, J.A., Felten, E.W.: New client puzzle outsourcing techniques for DoS resistance. In: 11th ACM Conference on Computer and Communications Security (October 2004)
16. Back, A.: Hashcash package announce (March 1997),
    http://hashcash.org/papers/announce.txt
17. Abadi, M., Burrows, M., Manasse, M., Wobber, T.: Moderately Hard, Memory-bound Functions. In: 10th Annual Network and Distributed System Security Symposium (NDSS), San Diego, CA, USA (February 2003)
18. Dwork, C., Goldberg, A., Naor, M.: On memory-bound functions for fighting spam. In: Boneh, D. (ed.) CRYPTO 2003. LNCS, vol. 2729, pp. 426–444. Springer, Heidelberg (2003)
19. Coelho, F.: Exponential memory-bound functions for proof of work protocols. Research Report A-370, CRI, École des mines de Paris (September 2005); Also Cryptology ePrint Archive, Report (2005)/356
20. Wang, X., Reiter, M.: Defending against denial-of-service attacks with puzzle auctions. In: IEEE Symposium on Security and Privacy 2003 (May 2003)
21. Merkle, R.C.: Secrecy, Authentification, and Public Key Systems. PhD thesis, Stanford University, Dpt of Electrical Engineering (June 1979)
22. Lamport, L.: Constructing digital signatures from a one-way function. Technical Report SRI-CSL-98, SRI International Computer Science Laboratory (October 1979)

Robust Threshold Schemes Based on the Chinese Remainder Theorem

Kamer Kaya* and Ali Aydın Selçuk

Department of Computer Engineering
Bilkent University
Ankara, 06800, Turkey
{kamer,selcuk}@cs.bilkent.edu.tr

Abstract. Recently, Chinese Remainder Theorem (CRT) based function sharing schemes are proposed in the literature. In this paper, we investigate how a CRT-based threshold scheme can be enhanced with the robustness property. To the best of our knowledge, these are the first robust threshold cryptosystems based on a CRT-based secret sharing.

Keywords: Threshold cryptography, robustness, RSA, ElGamal, Paillier, Chinese Remainder Theorem.

1 Introduction

In threshold cryptography, *secret sharing* deals with the problem of sharing a highly sensitive secret among a group of n users so that only when a sufficient number t of them come together the secret can be reconstructed. *Function sharing* deals with evaluating the encryption/signature function of a cryptosystem without the involved parties disclosing their secret shares. A function sharing scheme (FSS) requires distributing the function's computation according to the underlying secret sharing scheme (SSS) such that each part of the computation can be carried out by a different user and then the partial results can be combined to yield the function's value without disclosing the individual secrets. Several SSSs [1,3,20] and FSSs [8,9,10,11,19,21] have been proposed in the literature.

Nearly all existing solutions for the function sharing problem have been based on the Shamir SSS [20]. Recently, Kaya and Selçuk [14] proposed several threshold function sharing schemes based on the Asmuth-Bloom SSS for the RSA [18], ElGamal [13] and Paillier [16] cryptosystems. These FSSs are the first examples of secure function sharing schemes based on Asmuth-Bloom secret sharing.

We say that a function sharing scheme is *robust* if it can withstand participation of corrupt users in the function evaluation phase. In a robust FSS, a detection mechanism is used to identify the corrupted partial results so that, the corrupted users can be eliminated. The FSSs proposed by Kaya and Selçuk [14]

---

* Supported by the Turkish Scientific and Technological Research Agency (TÜBİTAK) Ph.D. scholarship.

S. Vaudenay (Ed.): AFRICACRYPT 2008, LNCS 5023, pp. 94–108, 2008.

did not have the robustness property and, to the best of our knowledge, no CRT-based robust and secure function sharing scheme exists in the literature.

In this paper, we investigate how CRT-based threshold schemes can be enhanced with the robustness property. We first give a robust threshold function sharing scheme for the RSA cryptosystem. Then we apply the ideas to the ElGamal and Paillier decryption functions. For RSA and Paillier, we use the threshold schemes proposed by Kaya and Selçuk [14]. For ElGamal, we work with a modified version of the ElGamal decryption scheme by Wei et al. [22]. All of the proposed schemes are provably secure against a static adversary under the random oracle model [2].

In achieving robustness, we make use of a non-interactive protocol designed to prove equality of discrete logarithms [4,5,21]. The original interactive protocol was proposed by Chaum et al [5] and improved by Chaum and Pedersen [6]. Later, Shoup [21] and, Boudot and Traoré [4] developed a non-interactive version of the protocol.

The organization of the paper is as follows: In Section 2, we describe the Asmuth-Bloom SSS and the FSSs proposed by Kaya and Selçuk [14]. After describing a robust threshold RSA scheme and proving its security in Section 3, we apply the proposed idea to the Paillier and ElGamal cryptosystems in Section 4. Section 5 concludes the paper.

2 Function Sharing Based on the Asmuth-Bloom Secret Sharing

The Asmuth-Bloom SSS shares a secret among the parties using modular arithmetic and reconstructs it by the Chinese Remainder Theorem. Here we give the brief description of the scheme:

- *Dealer Phase:* To share a secret d among a group of n users with threshold t, the dealer does the following:
  - A set of pairwise relatively prime integers $m_0 < m_1 < m_2 < \ldots < m_n$ are chosen where $m_0 > d$ is prime,

$$\prod_{i=1}^{t} m_i > m_0 \prod_{i=1}^{t-1} m_{n-i+1}. \tag{1}$$

  - Let M denote $\prod_{i=1}^{t} m_i$. The dealer computes

$$y = d + Am_0$$

where A is a positive integer generated randomly subject to the condition that $0 \le y < M$.
  - The share of the ith user, $1 \le i \le n$, is

$$y_i = y \bmod m_i$$

i.e., the smallest nonnegative residue of y modulo m_i.

– *Combiner Phase:* Assume $\mathcal{S}$ is a coalition of t users to construct the secret. For any coalition $\mathcal{S}$, we define $M_{\mathcal{S}}$ as

$$M_{\mathcal{S}} = \prod_{i \in \mathcal{S}} m_i.$$

- Given the system

$$y \equiv y_i \pmod{m_i}$$

   for $i \in \mathcal{S}$, find y in $\mathbb{Z}_{M_{\mathcal{S}}}$ using the Chinese Remainder Theorem.
- Compute the secret as

$$d = y \bmod m_0.$$

According to the Chinese Remainder Theorem, y can be determined uniquely in $\mathbb{Z}_{M_{\mathcal{S}}}$. Since $y < M \leq M_{\mathcal{S}}$, the solution is also unique in $\mathbb{Z}_M$.

In the original Asmuth-Bloom scheme, m_0 is not needed until the last step of the combiner phase but still it is a public value. To avoid confusions, we emphasize that it will be secret for the robust FSSs proposed in this paper.

Kaya and Selçuk [14] modified the Asmuth-Bloom SSS by changing (1) as

$$\prod_{i=1}^{t} m_i > m_0^2 \prod_{i=1}^{t-1} m_{n-i+1}. \tag{2}$$

to make the Asmuth-Bloom SSS *perfect* in the sense that $t-1$ or fewer shares do not narrow down the key space and furthermore all candidates for the key are equally likely: Assume a coalition $\mathcal{S}'$ of size $t-1$ has gathered and let y' be the unique solution for y in $\mathbb{Z}_{M_{\mathcal{S}'}}$. According to (2), $M/M_{\mathcal{S}'} > m_0^2$, hence $y' + jM_{\mathcal{S}'}$ is smaller than M for $j < m_0^2$. Since $\gcd(m_0, M_{\mathcal{S}'}) = 1$, all $(y' + jM_{\mathcal{S}'}) \bmod m_0$ are distinct for $0 \leq j < m_0$ hence, d can be any integer from $\mathbb{Z}_{m_0}$. For each value of d, there are either $\lfloor M/(M_{\mathcal{S}'}m_0) \rfloor$ or $\lfloor M/(M_{\mathcal{S}'}m_0) \rfloor + 1$ possible values of y consistent with d, depending on the value of d. Hence, for two different integers in $\mathbb{Z}_{m_0}$, the probabilities of d being equal to these integers are almost equal. Note that $M/(M_{\mathcal{S}'}m_0) > m_0$ and given that $m_0 \gg 1$, all d values are approximately equally likely.

In the original Asmuth-Bloom SSS, the authors proposed an iterative process to solve the system $y \equiv y_i \pmod{m_i}$. Instead, a classical and non-iterative solution exists which is more suitable for function sharing in the sense that it does not require interaction between parties and has an additive structure convenient to share exponentiations [12].

1. Let $\mathcal{S}$ be a coalition of at least t users. Let $M_{\mathcal{S}\setminus\{i\}}$ denote $\prod_{j \in \mathcal{S}, j \neq i} m_j$ and $M'_{\mathcal{S},i}$ be the multiplicative inverse of $M_{\mathcal{S}\setminus\{i\}}$ in $\mathbb{Z}_{m_i}$, i.e.,

$$M_{\mathcal{S}\setminus\{i\}} M'_{\mathcal{S},i} \equiv 1 \pmod{m_i}.$$

First, the ith user computes

$$u_i = \left(y_i M'_{\mathcal{S},i} \bmod m_i\right) M_{\mathcal{S}\setminus\{i\}}. \tag{3}$$

2. y is computed as

$$y = \sum_{i \in S} u_i \bmod M_S.$$

3. The secret d is computed as

$$d = y \bmod m_0.$$

Even with these modifications, obtaining a threshold scheme by using Asmuth-Bloom SSS is not a straightforward task. Here we give the description of the proposed threshold RSA signature scheme [14].

- *Setup*: In the RSA setup phase, choose the RSA primes $p = 2p' + 1$ and $q = 2q' + 1$ where p' and q' are also large random primes. $N = pq$ is computed and the public key e and private key d are chosen from $\mathbb{Z}^*_{\phi(N)}$ where $ed \equiv 1 \pmod{\phi(N)}$. Use Asmuth-Bloom SSS for sharing d with a secret $m_0 = \phi(N) = 4p'q'$.
- *Signing*: Let w be the hashed message to be signed and suppose the range of the hash function is $\mathbb{Z}^*_N$. Assume a coalition S of size t wants to obtain the signature $s = w^d \bmod N$.

  - *Generating the partial results*: Each user $i \in S$ computes

  $$u_i = (y_i M'_{S,i} \bmod m_i) M_{S\setminus\{i\}}, \tag{4}$$
  $$s_i = w^{u_i} \bmod N.$$

  - *Combining the partial results*: The incomplete signature $\bar{s}$ is obtained by combining the s_i values

  $$\bar{s} = \prod_{i \in S} s_i \bmod N. \tag{5}$$

  - *Correction*: Let $\kappa = w^{-M_S} \bmod N$ be the *corrector*. The incomplete signature can be corrected by trying

  $$(\bar{s}\kappa^j)^e = \bar{s}^e(\kappa^e)^j \overset{?}{\equiv} w \pmod{N} \tag{6}$$

  for $0 \leq j < t$. Then the signature s is computed by

  $$s = \bar{s}\kappa^\delta \bmod N$$

  where δ denotes the value of j that satisfies (6).

- *Verification* is the same as the standard RSA verification where the verifier checks

$$s^e \overset{?}{\equiv} w \pmod{N}$$

The signature $\bar{s}$ generated in (5) is *incomplete* since we need to obtain $y = \sum_{i \in S} u_i \bmod M_S$ as the exponent of w. Once this is achieved, we have $w^y \equiv w^d \pmod{N}$ as $y = d + Am_0$ for some A where $m_0 = \phi(N)$.

Note that the equality in (6) must hold for some $j \leq t - 1$ since the u_i values were already reduced modulo M_S. So, combining t of them in (5) will give $d + am_0 + \delta M_S$ in the exponent for some $\delta \leq t - 1$. Thus in (5), we obtained

$$\bar{s} = w^{d+\delta M_S} \bmod N = sw^{\delta M_S} \bmod N = s\kappa^{-\delta} \bmod N$$

and for $j = \delta$, equation (6) will hold. Also note that the mappings $w^e \bmod N$ and $w^d \bmod N$ are bijections in $\mathbb{Z}_N$, hence there will be a unique value of $s = \bar{s}\kappa^j$ which satisfies (6).

Besides RSA, Kaya and Selçuk also applied this *combine-and-correct* approach to obtain threshold Paillier and ElGamal schemes [14] with Asmuth-Bloom secret sharing.

3   Robust Sharing of the RSA Function

To enhance the threshold cryptosystems with the robustness property, we use a non-interactive protocol proposed to prove equality of two discrete logarithms with respect to different moduli. The interactive protocol, which was originally proposed by Chaum et al [5] for the same moduli, was modified by Shoup and used to make a threshold RSA signature scheme robust [21]. He used Shamir's SSS as the underlying SSS to propose a practical and robust threshold RSA signature scheme. In Shamir's SSS, the secret is reconstructed by using Lagrange's polynomial evaluation formula and all participants use the same modulus which does not depend on the coalition. On the other hand, in the direct solution used in the abovementioned CRT-based threshold RSA scheme, the definition of u_is in (3) and (4) shows that we need different moduli for each user. For robustness, we need to check the correctness of u_i for each user i in the function evaluation phase. We modified the protocol in [21] for the case of different moduli as Boudot and Traoré [4] did to obtain efficient publicly verifiable secret sharing schemes.

To obtain robustness, we first modify the dealer phase of the Asmuth-Bloom SSS and add the constraint that

$$p_i = 2m_i + 1$$

be a prime for each $1 \leq i \leq n$. These values will be the moduli used to construct/verify the *proof of correctness* for each user. The robustness extension described below can be used to make the CRT-based threshold RSA signature scheme in Section 2 robust. We only give the additions for the robustness extension here since the other phases are the same.

- *Setup*: Use Asmuth-Bloom SSS for sharing d with $m_0 = \phi(N)$. Let g_i be an element of order m_i in $\mathbb{Z}_{p_i}^*$. Broadcast g_i and the public verification data

$$v_i = g_i{}^{y_i} \bmod p_i$$

for each user i, $1 \leq i \leq n$.

– *Generating the proof of correctness*: Let w be the hashed message to be signed and suppose the range of the hash function is $\mathbb{Z}_N^*$. Assume a coalition S of size t participated in the signing phase. Let $h : \{0,1\}^* \to \{0,\ldots,2^{L_1}-1\}$ be a hash function where L_1 is another security parameter. Let

$$w' = w^{M_{S\setminus\{i\}}} \bmod N,$$

$$v_i' = v_i^{M_{S,i}'} \bmod p_i,$$

$$z_i = y_i M_{S,i}' \bmod m_i.$$

Each user $i \in S$ first computes

$$W = w'^r \bmod N,$$

$$G = g_i^r \bmod p_i$$

where $r \in_R \{0,\ldots,2^{L(m_i)+2L_1}\}$. Then he computes the proof as

$$\sigma_i = h(w', g_i, s_i, v_i', W, G),$$

$$D_i = r + \sigma_i z_i \in \mathbb{Z}$$

and sends the proof (σ_i, D_i) along with the partial signature s_i.
– *Verifying the proof of correctness*: The proof (σ_i, D_i) for the ith user can be verified by checking

$$\sigma_i \overset{?}{=} h(w', g_i, s_i, v_i', w'^{D_i} s_i^{-\sigma_i} \bmod N, g_i^{D_i} v_i'^{-\sigma_i} \bmod p_i). \qquad (7)$$

Note that the above scheme can also be used to obtain a robust threshold RSA decryption scheme. Since RSA signature and decryption functions are mostly identical, we omit the details.

3.1   Security Analysis

Here we will prove that the proposed threshold RSA signature scheme is secure (i.e. existentially non-forgeable against an adaptive chosen message attack), provided that the RSA problem is intractable (i.e. RSA function is a one-way trapdoor function [7]). We assume a static adversary model where the adversary controls exactly $t - 1$ users and chooses them at the beginning of the attack. In this model, the adversary obtains all secret information of the corrupted users and the public parameters of the cryptosystem. She can control the actions of the corrupted users, ask for partial signatures of the messages of her choice, but she cannot corrupt another user in the course of an attack, i.e., the adversary is static in that sense.

First we will analyze the proof of correctness. For generating and verifying the proof of correctness, the following properties holds:

– *Completeness:* If the ith user is honest then the proof succeeds since

$$w'^{D_i} s_i^{-\sigma_i} = w'^r \bmod N,$$

$$g_i^{D_i} v_i'^{-\sigma_i} = g_i^r \bmod p_i.$$

– *Soundness:* To prove the soundness, we will use a lemma by Poupard and Stern [17] which states that if the prover knows $(a, b, \sigma, \sigma', D, D')$ such that $a^D b^\sigma \equiv a^{D'} b^{\sigma'} \pmod{K}$ for an integer K, then he knows the discrete logarithm of b in base a unless he knows the factorization of K. Let us define $\Psi : \mathbb{Z}^*_{Np_i} \to \mathbb{Z}^*_N \times \mathbb{Z}^*_{p_i}$ be the CRT isomorphism, i.e., $x \to (x \bmod N, x \bmod p_i)$ for $x \in \mathbb{Z}^*_{Np_i}$. Note that $\gcd(N, p_i) = 1$. Let $g = \Psi^{-1}(w', g_i)$, $v = \Psi^{-1}(s_i, v_i')$ and $\tau = \Psi^{-1}(W, G)$. Given W and G, if the ith user can compute valid proofs (σ, D) and (σ', D') then we have

$$\tau = g^D v^\sigma \bmod Np_i = g^{D'} v^{\sigma'} \bmod Np_i$$

and according to the lemma above, the ith user knows u_i unless he can completely factor Np_i. Since the factorization of N is secret we can say that if the proof is a valid proof then the discrete logarithms are equal in $\bmod m_i$ and the prover knows this discrete logarithm. Hence, an adversary cannot impersonate a user without knowing his share. Similar to Boudot and Treore [4], a range check on D_i might be necessary while verifying the proof of correctness to detect incorrect partial signatures from users with valid shares.

– *Zero-Knowledge Simulatability:* To prove the zero-knowledge simulatability, we will use the random oracle model for the hash function h and construct a simple simulator. When an uncorrupted user wants to create a proof (σ_i, D_i) for a message w and partial signature s_i, the simulator returns

$$\sigma_i \in_R \{0, \dots, 2^{L_1} - 1\}$$

and

$$D_i \in_R \{0, \dots, 2^{L(m_i) + 2L_1} - 1\}$$

and sets the value of the oracle at

$$(w', g_i, s_i, v_i', w'^{D_i} s_i^{-\sigma_i} \bmod N, g_i^{D_i} v_i'^{-\sigma_i} \bmod p_i)$$

as σ_i. Note that, the value of the random oracle is not defined at this point but with negligible probability. When a corrupted user queries the oracle, if the value of the oracle was already set the simulator returns that value otherwise it returns a random one. It is obvious that the distribution of the output of the simulator is statistically indistinguishable from the real output.

To reduce the security of the proposed threshold RSA signature scheme to the security of the standard RSA signature scheme, the following proof constructs another simulator.

Theorem 1. *Given that the standard RSA signature scheme is secure, the threshold RSA signature scheme is robust and secure under the static adversary model.*

Proof. To reduce the problem of breaking the standard RSA signature scheme to breaking the proposed threshold scheme, we will simulate the threshold protocol with no information on the secret where the output of the simulator is indistinguishable from the adversary's point of view. Afterwards, we will show that the secrecy of the private key d is not disrupted by the values obtained by the adversary. Thus, if the threshold RSA scheme is not secure, i.e., an adversary who controls $t - 1$ users can forge signatures in the threshold scheme, one can use this simulator to forge a signature in the standard RSA scheme.

Let $\mathcal{S}'$ denote the set of users controlled by the adversary. To simulate the adversary's view, the simulator first selects a random interval $I = [a, b)$ from $\mathbb{Z}_M$, $M = \prod_{i=1}^{t} m_i$. The start point a is randomly chosen from $\mathbb{Z}_M$ and the end point is computed as $b = a + m_0 M_{\mathcal{S}'}$. Then, the shares of the corrupted users are computed as $y_j = a \bmod m_j$ for $j \in \mathcal{S}'$. Note that, these $t-1$ shares are indistinguishable from random ones due to (1) and the improved perfectness condition. Although the simulator does not know the real value of d, it is guaranteed that for all possible d, there exists a $y \in I$ which is congruent to $y_j \pmod{m_j}$ and to $d \pmod{m_0}$.

Since we have a (t, n)-threshold scheme, given a valid RSA signature (s, w), the partial signature s_i for a user $i \notin \mathcal{S}'$ can be obtained by

$$s_i = s\kappa^{-\delta_{\mathcal{S}}} \prod_{j \in \mathcal{S}'} (w^{u_j})^{-1} \bmod N$$

where $\mathcal{S} = \mathcal{S}' \cup \{i\}$, $\kappa = w^{-M_{\mathcal{S}}} \bmod N$ and $\delta_{\mathcal{S}}$ is equal to either $\left\lfloor \frac{\sum_{j \in \mathcal{S}'} u_j}{M_{\mathcal{S}}} \right\rfloor + 1$ or $\left\lfloor \frac{\sum_{j \in \mathcal{S}'} u_j}{M_{\mathcal{S}}} \right\rfloor$. The value of $\delta_{\mathcal{S}}$ is important because it carries information on y. Let $U = \sum_{j \in \mathcal{S}'} u_j$ and $U_{\mathcal{S}} = U \bmod M_{\mathcal{S}}$. One can find whether y is greater than $U_{\mathcal{S}}$ or not by looking at $\delta_{\mathcal{S}}$:

$$y < U_{\mathcal{S}} \quad \text{if} \quad \delta_{\mathcal{S}} = \lfloor U/M_{\mathcal{S}} \rfloor + 1,$$
$$y \geq U_{\mathcal{S}} \quad \text{if} \quad \delta_{\mathcal{S}} = \lfloor U/M_{\mathcal{S}} \rfloor.$$

Since the simulator does not know the real value of y, to determine the value of $\delta_{\mathcal{S}}$, the simulator acts according to the interval randomly chosen at the beginning of the simulation.

$$\delta_{\mathcal{S}} = \begin{cases} \lfloor U/M_{\mathcal{S}} \rfloor + 1, & \text{if} \quad a < U_{\mathcal{S}} \\ \lfloor U/M_{\mathcal{S}} \rfloor, & \text{if} \quad a \geq U_{\mathcal{S}} \end{cases} \tag{8}$$

It is obvious that, the value of $\delta_{\mathcal{S}}$ is indistinguishable from the real case if $U_{\mathcal{S}} \notin I$. Now, we will prove that the $\delta_{\mathcal{S}}$ values computed by the simulator does not disrupt the indistinguishability from the adversary's point of view. First of all, there are $(n-t+1)$ possible $\delta_{\mathcal{S}}$ computed by using $U_{\mathcal{S}}$ since all the operations in the exponent depend on the coalition S alone. If none of the U_S values lies in I, the δ_S values observed by the adversary will be indistinguishable from a real execution of the protocol. Using this observation, we can prove that no information about the private key is obtained by the adversary.

Observing the $t-1$ randomly generated shares, there are $m_0 = \phi(N)$ candidates in I for y which satisfy $y_j = y \bmod m_j$ for all $j \in S'$. These m_0 candidates have all different remainders modulo m_0 since $\gcd(M_{S'}, m_0) = 1$. So, exactly one of the remainders is equal to the private key d. If $U_S \notin I$ for all S, given an s_i, the shared value y can be equal to any of these m_0 candidates hence any two different values of the secret key d will be indistinguishable from adversary's point of view. In our case, this happens with all but negligible probability. First, observe that $U_S \equiv 0 \bmod m_i$ and there are $m_0 M_{S'}/m_i$ multiples of m_i in I. Thus, the probability of $U_S \notin I$ for a coalition S is equal to $\left(1 - \frac{m_0 M_{S'}/m_i}{M_{S'}}\right) = \left(1 - \frac{m_0 M_{S'}}{M_S}\right)$. According to (1), $m_i > m_0{}^2$ for all i hence the probability of $U_S \notin I$ for all possible S is less than $\left(1 - \frac{1}{m_0}\right)^{n-t+1}$, which is almost surely 1 for $m_0 \gg n$.

The simulator computes the public verification data of the users in S' as $v_j = g^{y_j} \bmod p_j$ for $j \in S'$. For other users $i \notin S'$, the simulator chooses a random integer $y_i \in_R \mathbb{Z}_{m_i}$ and sets $v_i = g^{y_i} \bmod p_i$. Note that $\gcd(N, p_i) = 1$. So the public verification data generated by the simulator are computationally indistinguishable from the real ones.

Consequently, the output of the simulator is indistinguishable from a real instance from the adversary's point of view, and hence the simulator can be used to forge a signature in the standard RSA scheme if the threshold RSA scheme can be broken.   □

4   Robustness in Other CRT-Based Threshold Schemes

The robustness extension given in Section 3 can be applied to other CRT-based threshold schemes as well. Here we describe how to adapt the extension to the CRT-based threshold Paillier and ElGamal function sharing schemes.

4.1   Robust Sharing of the Paillier Decryption Function

Paillier's probabilistic cryptosystem [16] is a member of a different class of cryptosystems where the message is used in the exponent of the encryption operation. The description of the cryptosystem is as follows:

- *Setup*: Let $N = pq$ be the product of two large primes and $\lambda = \mathrm{lcm}(p-1, q-1)$. Choose a random $g \in \mathbb{Z}_{N^2}$ such that the order of g is a multiple of N. The public and private keys are (N, g) and λ, respectively.
- *Encryption*: Given a message $w \in \mathbb{Z}_N$, the ciphertext c is computed as

$$c = g^w r^N \bmod N^2$$

where r is a random number from $\mathbb{Z}_N$.

– *Decryption*: Given a ciphertext $c \in \mathbb{Z}_{N^2}$, the message w is computed as

$$w = \frac{L\left(c^{\lambda} \bmod N^2\right)}{L\left(g^{\lambda} \bmod N^2\right)} \bmod N$$

where $L(x) = \frac{x-1}{N}$, for $x \equiv 1 \pmod{N}$.

By using the combine-and-correct approach, Kaya and Selçuk proposed a threshold version of the Paillier's cryptosystem [14]. As in threshold RSA, the decryption coalition needs to compute an exponentiation, $s = c^{\lambda} \bmod N^2$, where the exponent λ is shared by Asmuth-Bloom SSS in the setup phase. Hence, similar to RSA, the partial result s_i of the ith user is equal to $s_i = c^{u_i} \bmod N^2$. The robustness extension can be applied to the Paillier cryptosystem as follows:

– *Setup*: Use Asmuth-Bloom SSS for sharing λ with $m_0 = \phi(N^2) = N\phi(N)$. Let $g_i \in \mathbb{Z}_{p_i}^*$ be an element with order m_i in $\mathbb{Z}_{p_i}^*$. Broadcast the public verification data g_i and

$$v_i = g_i^{y_i} \bmod p_i$$

for each user i, $1 \le i \le n$.
– *Generating the proof of correctness*: Let $h : \{0,1\}^* \to \{0, \dots, 2^{L_1} - 1\}$ be a hash function where L_1 is another security parameter. Let

$$c' = c^{M_{S \setminus \{i\}}} \bmod N^2,$$

$$v_i' = v_i^{M_{S,i}'} \bmod p_i,$$

$$z_i = y_i M_{S,i}' \bmod m_i.$$

Each user $i \in S$ first computes

$$W = c'^r \bmod N^2,$$
$$G = g_i^r \bmod p_i$$

where $r \in_R \{0, \dots, 2^{L(m_i)+2L_1}\}$. Then he computes the proof as

$$\sigma_i = h(c', g_i, s_i, v_i', W, G),$$

$$D_i = r + \sigma_i z_i \in \mathbb{Z}$$

and sends the proof (σ_i, D_i) along with the partial decryption s_i.
– *Verifying the proof of correctness*: The proof (σ_i, D_i) for the ith user can be verified by checking

$$\sigma_i \stackrel{?}{=} h(c', g_i, s_i, v_i', c'^{D_i} s_i^{-\sigma_i} \bmod N, g_i^{D_i} v_i'^{-\sigma_i} \bmod p_i). \tag{9}$$

If the ith user is honest then the proof succeeds since $c'^{D_i} s_i^{-\sigma_i} = c'^r \bmod N^2$ and $g_i^{D_i} v_i'^{-\sigma_i} = g_i^r \bmod p_i$. The soundness property can be proved with a proof similar to the proof of Theorem 1. Note that $\gcd(N^2, p_i) = 1$ for all users and $\phi(N^2) = N\phi(N)$ is secret. A similar proof can be given for the zero knowledge simulatability as the one in Section 3.1.

4.2   Robust Sharing of the ElGamal Decryption Function

The ElGamal cryptosystem [13] is another popular public key scheme with the following description:

- *Setup*: Let p be a large prime and g be a generator of $\mathbb{Z}_p^*$. Choose a random $\alpha \in \{1, \ldots, p-1\}$ and compute $\beta = g^\alpha \bmod p$. (β, g, p) and α are the public and private keys, respectively.
- *Encryption*: Given a message $w \in \mathbb{Z}_p$, the ciphertext $c = (c_1, c_2)$ is computed as

$$c_1 = g^k \bmod p,$$
$$c_2 = \beta^k w \bmod p$$

  where k is a random integer in $\{1, \ldots, p-1\}$.
- *Decryption*: Given a ciphertext c, the message w is computed as

$$w = (c_1{}^\alpha)^{-1} c_2 \bmod p.$$

Adapting our robustness extension to the threshold ElGamal scheme given in [14] is slightly more complicated than it is for the Paillier's cryptosystem, because $\phi(p) = p-1$ is public. A simple solution for this problem is to extend the modulus to $N = pq$ where $p = 2p'+1$ and $q = 2q'+1$ are safe primes. There exist versions of the ElGamal encryption scheme in the literature with a composite modulus instead of p. For example, Wei et al. [22] modified the standard ElGamal scheme to obtain a hidden-order ElGamal scheme. They proved that their scheme is as secure as each of the standard RSA and ElGamal cryptosystems. Here we give the description of a robust, CRT-based threshold scheme for Wei et al.'s version of the ElGamal encryption.

- *Setup*: In the ElGamal setup phase, choose $p = 2p' + 1$ and $q = 2q' + 1$ be large primes such that p' and q' are also prime numbers. Let $N = pq$ and let g_p and g_q be generators of $\mathbb{Z}_p^*$ and $\mathbb{Z}_q^*$, respectively. Choose $\alpha_p \in_R \mathbb{Z}_p^*$ and $\alpha_q \in_R \mathbb{Z}_q^*$ such that $\gcd(p-1, q-1) \mid (\alpha_p - \alpha_q)$. The secret key $\alpha \in \mathbb{Z}_{\lambda(N)}$ is the unique solution of the congruence system

$$\alpha \equiv \alpha_p \pmod{p-1},$$
$$\alpha \equiv \alpha_q \pmod{q-1}$$

where $\lambda(N) = 2p'q'$ is the Carmichael number of N. Similarly, the public key $\beta \in \mathbb{Z}_N$ is the unique solution of congruence system

$$\beta \equiv g_p{}^{\alpha_p} \pmod{p},$$
$$\beta \equiv g_q{}^{\alpha_q} \pmod{q}.$$

Let g be the unique solution of the congruence system

$$g \equiv g_p \pmod{p},$$
$$g \equiv g_q \pmod{q}$$

and α and (β, g, N) be the private and the public keys, respectively. Note that $\beta = g^\alpha \bmod N$. Use Asmuth-Bloom SSS for sharing the private key α with $m_0 = 2p'q'$. Let $g_i \in \mathbb{Z}_{p_i}^*$ be an element with order m_i in $\mathbb{Z}_{p_i}^*$. Broadcast the public verification data g_i and $v_i = g_i^{y_i} \bmod p_i$ for each user i, $1 \le i \le n$.

– *Encryption*: Given a message $w \in \mathbb{Z}_N$, the ciphertext $c = (c_1, c_2)$ is computed as

$$c_1 = g^k \bmod N,$$
$$c_2 = \beta^k w \bmod N$$

where k is a random integer from $\{1, \ldots, N-1\}$.

– *Decryption*: Let (c_1, c_2) be the ciphertext to be decrypted where $c_1 = g^k \bmod N$ for some $k \in \{1, \ldots, N-1\}$ and $c_2 = \beta^k w \bmod N$ where w is the message. The coalition S of t users wants to obtain the message $w = sc_2 \bmod N$ for the *decryptor* $s = (c_1^\alpha)^{-1} \bmod N$.

• *Generating the partial results*: Each user $i \in S$ computes

$$u_i = y_i M'_{S,i} M_{S \setminus \{i\}} \bmod M_S, \tag{10}$$
$$s_i = c_1^{-u_i} \bmod N,$$
$$\beta_i = g^{u_i} \bmod N. \tag{11}$$

• *Generating the proof of correctness*: Let $h : \{0,1\}^* \to \{0, \ldots, 2^{L_1} - 1\}$ be a hash function where L_1 is another security parameter. Let

$$c_1' = c_1^{M_{S \setminus \{i\}}} \bmod N,$$
$$v_i' = v_i^{M'_{S,i}} \bmod p_i,$$
$$z_i = y_i M'_{S,i} \bmod m_i.$$

Each user $i \in S$ first computes

$$W = c_1'^{\,r} \bmod N,$$
$$G = g_i^{\,r} \bmod p_i$$

where $r \in_R \{0, \ldots, 2^{L(m_i)+2L_1}\}$. Then he computes the proof as

$$\sigma_i = h(c_1', g_i, s_i, v_i', W, G),$$
$$D_i = r + \sigma_i z_i \in \mathbb{Z}$$

and sends the proof (σ_i, D_i) along with s_i.

• *Verifying the proof of correctness*: The proof (σ_i, D_i) for the ith user can be verified by checking

$$\sigma_i \stackrel{?}{=} h(c_1', g_i, s_i, v_i', c_1'^{\,D_i} s_i^{-\sigma_i} \bmod N, g_i^{\,D_i} v_i'^{\,-\sigma_i} \bmod p_i).$$

- *Combining the partial results*: The incomplete decryptor $\bar{s}$ is obtained by combining the s_i values

$$\bar{s} = \prod_{i \in S} s_i \bmod N.$$

- *Correction*: The β_i values will be used to find the exponent which will be used to correct the incomplete decryptor. Compute the incomplete public key $\bar{\beta}$ as

$$\bar{\beta} = \prod_{i \in S} \beta_i \bmod N. \tag{12}$$

Let $\kappa_s = c_1{}^{M_S} \bmod N$ and $\kappa_\beta = g^{-M_S} \bmod N$ be the *correctors* for s and β, respectively. The corrector exponent δ is obtained by trying

$$\bar{\beta}\kappa_\beta^j \stackrel{?}{\equiv} \beta \pmod{N} \tag{13}$$

for $0 \le j < t$.
- *Extracting the message*: Compute the message w as

$$s = \bar{s}\kappa_s{}^\delta \bmod N,$$

$$w = sc_2 \bmod N.$$

where δ denotes the value of j that satisfies (13).

As in the case of RSA, the decryptor $\bar{s}$ is *incomplete* since we need to obtain $y = \sum_{i \in S} u_i \bmod M_S$ as the exponent of c_1^{-1}. Once this is achieved, $(c_1^{-1})^y \equiv (c_1^{-1})^\alpha \pmod{N}$ since $y = \alpha + 2Ap'q'$ for some A.

When the equality in (13) holds we know that $\beta = g^\alpha \bmod N$ is the correct public key. This equality must hold for one j value, denoted by δ, in the given interval since the u_i values in (10) and (11) are first reduced modulo M_S. So, combining t of them will give $\alpha + am_0 + \delta M_S$ in the exponent in (12) for some $\delta \le t - 1$. Thus in (12), we obtained

$$\bar{\beta} = g^{\alpha + am_0 + \delta M_S} \bmod N \equiv g^{\alpha + \delta M_S} = \beta g^{\delta M_S} = \beta \kappa_\beta^{-\delta} \pmod{N}$$

and for $j = \delta$ equality must hold. Actually, in (12) and (13), our purpose is not to compute the public key since it is already known. We want to find the corrector exponent δ in order to obtain s, which is equal to the one used to obtain β. This equality can be seen as follows:

$$s \equiv c_1{}^{-\alpha} = \beta^{-r}$$

$$= \left(g^{-(\alpha + (\delta - \delta)M_S)} \right)^r$$

$$= c_1{}^{-(\alpha + am_0 + \delta M_S)} \left(c_1{}^{M_S} \right)^\delta = \bar{s}\kappa_s{}^\delta \pmod{N}$$

If the ith user is honest then the proof succeeds since $c_1'{}^{D_i} s_i{}^{-\sigma_i} = c_1'{}^r \bmod N$ and $g_i{}^{D_i} v_i'{}^{-\sigma_i} = g_i{}^r \bmod p_i$. The soundness property can be proved with a proof similar to the one in Section 3.1. Note that $\gcd(N, p_i) = 1$ for all users and $\lambda(N) = 2p'q'$ is secret. A similar proof can be given for the zero knowledge simulatability as the one in Section 3.1. We omit the security proof here since the structure of the simulator is very similar to the one in Theorem 1 of Section 3.1.

5 Conclusion

In this paper, we proposed robust threshold RSA, Paillier and ElGamal schemes based on the Asmuth-Bloom SSS. Previous solutions for robust function sharing schemes were based on the Shamir's SSSs [10,15,19,21]. To the best of our knowledge, the schemes described in this paper are the first robust and secure FSSs using a CRT-based secret sharing. The ideas presented in this paper can be used to obtain other robust FSSs based on the CRT.

References

1. Asmuth, C., Bloom, J.: A modular approach to key safeguarding. IEEE Trans. Information Theory 29(2), 208–210 (1983)
2. Bellare, M., Rogaway, P.: Random oracles are practical: a paradigm for designing efficient protocols. In: Proc. of First ACM Conference on Computer and Communications Security, pp. 62–73 (1993)
3. Blakley, G.: Safeguarding cryptographic keys. In: Proc. of AFIPS National Computer Conference (1979)
4. Boudot, F., Traoré, J.: Efficient publicly verifiable secret sharing schemes with fast or delayed recovery. In: Varadharajan, V., Mu, Y. (eds.) ICICS 1999. LNCS, vol. 1726, pp. 87–102. Springer, Heidelberg (1999)
5. Chaum, D., Evertse, J.H., Van De Graaf, J.: An improved protocol for demonstrating possesion of discrete logarithm and some generalizations. In: Price, W.L., Chaum, D. (eds.) EUROCRYPT 1987. LNCS, vol. 304, pp. 127–141. Springer, Heidelberg (1988)
6. Chaum, D., Pedersen, T.P.: Wallet databases with observers. In: Brickell, E.F. (ed.) CRYPTO 1992. LNCS, vol. 740, pp. 89–105. Springer, Heidelberg (1993)
7. Cramer, R., Shoup, V.: Signature schemes based on the strong RSA assumption. ACM Trans. Inf. Syst. Secur. 3(3), 161–185 (2000)
8. Desmedt, Y.: Some recent research aspects of threshold cryptography. In: Okamoto, E. (ed.) ISW 1997. LNCS, vol. 1396, pp. 158–173. Springer, Heidelberg (1998)
9. Desmedt, Y., Frankel, Y.: Threshold cryptosystems. In: Brassard, G. (ed.) CRYPTO 1989. LNCS, vol. 435, pp. 307–315. Springer, Heidelberg (1990)
10. Desmedt, Y., Frankel, Y.: Shared generation of authenticators and signatures. In: Feigenbaum, J. (ed.) CRYPTO 1991. LNCS, vol. 576, pp. 457–469. Springer, Heidelberg (1992)
11. Desmedt, Y., Frankel, Y.: Homomorphic zero-knowledge threshold schemes over any finite abelian group. SIAM Journal on Discrete Mathematics 7(4), 667–679 (1994)
12. Ding, C., Pei, D., Salomaa, A.: Chinese Remainder Theorem: Applications in Computing, Coding, Cryptography. World Scientific, Singapore (1996)
13. ElGamal, T.: A public key cryptosystem and a signature scheme based on discrete logarithms. IEEE Trans. Information Theory 31(4), 469–472 (1985)
14. Kaya, K., Selçuk, A.A.: Threshold cryptography based on Asmuth–Bloom secret sharing. Information Sciences 177(19), 4148–4160 (2007)
15. Lysyanskaya, A., Peikert, C.: Adaptive security in the threshold setting: From cryptosystems to signature schemes. In: Boyd, C. (ed.) ASIACRYPT 2001. LNCS, vol. 2248, pp. 331–350. Springer, Heidelberg (2001)

16. Paillier, P.: Public key cryptosystems based on composite degree residuosity classes. In: Stern, J. (ed.) EUROCRYPT 1999. LNCS, vol. 1592, pp. 223–238. Springer, Heidelberg (1999)
17. Poupard, G., Stern, J.: Security analysis of a practical on the fly authentication and signature generation. In: Nyberg, K. (ed.) EUROCRYPT 1998. LNCS, vol. 1403, pp. 422–436. Springer, Heidelberg (1998)
18. Rivest, R., Shamir, A., Adleman, L.: A method for obtaining digital signatures and public key cryptosystems. Comm. ACM 21(2), 120–126 (1978)
19. De Santis, A., Desmedt, Y., Frankel, Y., Yung, M.: How to share a function securely? In: Proc. of STOC 1994, pp. 522–533 (1994)
20. Shamir, A.: How to share a secret? Comm. ACM 22(11), 612–613 (1979)
21. Shoup, V.: Practical threshold signatures. In: Preneel, B. (ed.) EUROCRYPT 2000. LNCS, vol. 1807, pp. 207–220. Springer, Heidelberg (2000)
22. Wei, W., Trung, T., Magliveras, S., Hoffman, F.: Cryptographic primitives based on groups of hidden order. Tatra Mountains Mathematical Publications 29, 147–155 (2004)

An Authentication Protocol with Encrypted Biometric Data*

Julien Bringer and Hervé Chabanne

Sagem Sécurité

Abstract. At ACISP'07, Bringer *et al.* introduced a new protocol for achieving biometric authentication with a Private Information Retrieval (PIR) scheme. Their proposal is made to enforce the privacy of biometric data of users. We improve their work in several aspects. Firstly, we show how to replace the basic PIR scheme they used with Lipmaa's which has ones of the best known communication complexity. Secondly, we combine it with Secure Sketches to enable a strict separation between on one hand biometric data which remain the same all along a lifetime and stay encrypted during the protocol execution, and on the other hand temporary data generated for the need of the authentication to a service provider. Our proposition exploits homomorphic properties of Goldwasser-Micali and Paillier cryptosystems.

Keywords: Authentication, Biometrics, Privacy, Private Information Retrieval protocol, Secure Sketches.

1 Introduction

Biometric data are captured by sensors as physical or behavioral traits of individuals. They are used for identification or authentication. The underlying principle here is simple: during a preliminary phase called the enrollment, a template containing a biometric reference for an individual is acquired and stored and, then, is compared to new "fresh" acquisition of the same information during verification phase. Note that, at this point, no secrecy about biometric data is required for the verification to work. Moreover, as it is generally quite easy to have access to biometric traits – a face in the crowd, fingerprints on glass – it is wiser to treat biometric data as public.

To be clear, biometric acquisitions of the same trait do not give the same result each time. In fact, a big amount of changes has to be taken into account by specialized algorithms, called matching algorithms, during verification phase for recognizing different acquisitions of the same individuals.

Finally, in practice, biometric data obtained from the enrollment phase are usually stored in databases. For privacy reasons, this membership should not be revealed as this is a link between the service provider holding the database and the real person who want to authenticate.

---

* Work partially supported by french ANR RNRT project BACH.

S. Vaudenay (Ed.): AFRICACRYPT 2008, LNCS 5023, pp. 109–124, 2008.

1.1 Related Works

In order to integrate biometrics into cryptographic protocols, it is often proposed to replace traditional matching algorithms by error-correction procedures [3,4,13,16,17,25,26,28,35,36] and traditional templates by Secure Sketches. In these schemes, biometric features are treated to be secret and used for instance to extract keys for cryptographic purposes. This construction is of great interest to simplify the matching step but it does not answer to all security issues raised by biometrics so far, cf. Sect. 3.1.

There are also other works which deal with the secure comparison of data, e.g. [2,6,8,9,18,19,32]. Secure multi-party computation are used in [18,32] and homomorphic encryption schemes could help to compare directly encrypted data as it is the case in [6,8,9,32]. Specifically in [6], Bringer *et al.* propose to use these kinds of techniques by describing a PIR based on Goldwasser-Micali cryptosystem to increase privacy. However, the communication cost of their protocol is linear in the size of the database and they rely on biometric templates stored as cleartexts in a database.

1.2 Our Contributions

In this paper, we introduce a new protocol for biometric authentication. In our scheme, biometric data stay encrypted during all the computations. This is possible due to the integration of secure sketches into homomorphic cryptosystems. Moreover, confidentiality of requests made to the database is also obtained thanks to a Private Information Retrieval (PIR) protocol.We show how our results can be easily generalized to Lipmaa's PIR protocol [29]. Finally, it is worth noting that our proposition is proved secure in our security model.

1.3 Organization of This Work

The rest of the paper is organized as follows. In Section 2, we describe the security model we consider for (remote) biometric-based authentication. In Section 3, we introduce some basic notions on Secure Sketches and show how to deal with these primitives in an encrypted way. In Section 4, we give a review of the PIR protocol of Lipmaa and in Section 5, we introduce our new protocol for biometric authentication by combining techniques from the previous sections. In Section 6, we provide a security analysis of the protocol. Section 7 concludes the paper.

2   Biometric Authentication Model

Following the ideas of [6], we describe the system we consider and the associated security model.

2.1 Biometric Systems

For typical configurations, a biometric-based recognition scheme consists of an enrollment phase and a verification phase. To register a user U, a biometric template b is measured from U and stored in a token or a database. When a new biometric sample b' is captured from U, it is compared to the reference data via a matching function. According to a similarity measure m and some recognition threshold τ, b' will be accepted as a biometric capture of U if $m(b, b') \leq \tau$, else rejected. Should it be accepted, b and b' are called *matching* templates, and *non-matching* ones otherwise.

More formally, we will consider four types of components:

- Human user U_i, who uses his biometric to authenticate himself to a service provider.
- Sensor client $\mathcal{C}$, which extracts human user's biometric template using some biometric sensor and which communicates with the service provider and the database.
- Service provider $\mathcal{SP}$, which deals with human user's authentication. It may have access to a Hardware Security Model HSM [1] which stores the secret keys involved in the protocol.
- Database $\mathcal{DB}$, which stores biometric information for users.

This structure captures the idea of a centralized storage database which can be queried by many different applications and with several sensors. This helps the user to make remote authentication requests. For the simplicity of description, we assume that M users U_i ($1 \leq i \leq M$) register at the service provider $\mathcal{SP}$. Moreover, a user will use a pseudorandom username ID_i to manage potential multiple registrations on the same system.

We make the following classical *liveness assumption*:

Assumption 1. *We assume that, with a high probability, the biometric template captured by the sensor and used in the system is from a living human user. In other words, it is difficult to produce a fake biometric template that can be accepted by the sensor.*

As biometrics are public information, additional credentials are always required to establish security links in order to prevent some well-known attacks (e.g. replay attacks) and to relay liveness assumption from the sensor to its environment. Therefore we assume that the sensor client is always honest and trusted by all other components[1]. We also assume the following relationships.

Assumption 2. *With respect to the authentication service, service provider is trusted by human users to make the right decision, and database is trusted by human users and the service provider to store and provide the right biometric information. Only an outside adversary may try to impersonate an honest human user.*

---

[1] When the service provider or the database receives some fresh biometric information, it can confirm with a high probability that the fresh biometric information is extracted from the human user which has presented itself to the sensor client.

Assumption 3. *With respect to privacy concerns, both service provider and database are assumed to be malicious which means they may deviate from the protocol specification, but they will not collude. In reality, an outside adversary may also pose threats to the privacy concerns, however, it has no more advantage than a malicious system component.*

2.2 Security Model

We have two functionalities Enrollment and Verification, where Enrollment can be initiated only once to simulate the enrollment phase and Verification can be initiated for any user to start an authentication session for a polynomial number of times.

The security of a protocol will be evaluated via an experiment between an adversary and a challenger, where the challenger simulates the protocol executions and answers the adversary's oracle queries. Without specification, algorithms are always assumed to be polynomial-time.

Soundness. This first requirement is defined as follows.

Definition 1. *A biometric-based authentication scheme is defined to be sound if it satisfies the following requirement: The service provider will accept an authentication request if the sensor client sends (ID_i, b'_i) in an authentication request, where b_i and b'_i are matching data and b_i is the reference template registered for ID_i; and will reject it if they are non-matching data.*

Due to the nature of biometric measurements, the probability of success for non-matching data (respectively of reject for matching data) is not negligible. However, these issues – traditionally measured as False Acceptance Rate (FAR) and respectively False Reject Rate (FRR) – are related to biometric measurement, so are irrelevant to our privacy concerns, hence we make abstraction of this problem in the sequel. For instance we do not take into account these FAR and FRR when dealing with soundness analysis. It is in fact a constraint on the choice of the secure sketch construction (cf. Sect. 3.1).

Identity Privacy. Our main concern is the sensitive relationship between the pseudonyms of users – which can possibly have multiple registrations – and the reference biometric templates. In practice, a malicious service provider or a malicious database may try to probe these relationships.

Definition 2. *A biometric-based authentication scheme achieves identity privacy if $\mathcal{A} = (\mathcal{A}_1, \mathcal{A}_2)$ has only a negligible advantage in the following game, where the advantage is defined to be $|\Pr[e' = e] - \frac{1}{2}|$.*

$$\mathbf{Exp}_{\mathcal{A}}^{Identity\text{-}Privacy}$$

$$
\begin{vmatrix}
(i, ID_i, b_i^{(0)}, b_i^{(1)}, (ID_j, b_j)(j \neq i)) \leftarrow \mathcal{A}_1(1^\ell) \\
b_i = b_i^{(e)} \qquad\qquad\qquad \overset{R}{\leftarrow} \{b_i^{(0)}, b_i^{(1)}\} \\
\emptyset \qquad\qquad\qquad\qquad \leftarrow \mathsf{Enrollment}((ID_j, b_j)_j) \\
e' \qquad\qquad\qquad\qquad \leftarrow \mathcal{A}_2(1^\ell)
\end{vmatrix}
$$

Informally, the identity privacy means that, for any pseudorandom username, the adversary knows nothing about the corresponding biometric template. It also implies that the adversary cannot find any linkability between registrations.

The attack game can be formulated as: An adversary generates N pairs of username and relevant biometric template, but with two possible templates $(b_i^{(0)}, b_i^{(1)})$ for ID_i. Thereafter a challenger randomly chooses a template $b_i^{(e)}$ for the username ID_i, and simulates the enrollment phase to generate the parameters for the sensor client, the service provider, and the database. Then the adversary tries to guess which template has been selected for U_i by listening a polynomial amount of verifications. The particularity is that they must be Verification requests run by the sensor. This way, the adversary can neither learn nor control which biometric template is used on the sensor side.

Transaction Anonymity. We further want to guarantee that the database which is supposed to store biometric information, gets no information about which user is authenticating himself to the service provider or what is the authentication result.

Definition 3. *A biometric-based authentication protocol achieves transaction anonymity if a malicious database represented by an adversary $\mathcal{A} = (\mathcal{A}_1, \mathcal{A}_2, \mathcal{A}_3)$ has only a negligible advantage in the following game, where the advantage is defined to be $|\Pr[e' = e] - \frac{1}{2}|$.*

$$
\mathbf{Exp}_{\mathcal{A}}^{Transaction\text{-}Anonymity}
$$

$$
\begin{array}{rl}
(ID_j, b_j)(1 \le j \le N) & \leftarrow \mathcal{A}_1(1^\ell) \\
\emptyset & \leftarrow \mathsf{Enrollment}((ID_j, b_j)_j) \\
\{i_0, i_1\} & \leftarrow \mathcal{A}_2(Challenger, \mathsf{Verification}) \\
i_e & \xleftarrow{R} \{i_0, i_1\} \\
\emptyset & \leftarrow \mathsf{Verification}(i_e) \\
e' & \leftarrow \mathcal{A}_3(Challenger, \mathsf{Verification})
\end{array}
$$

This captures the requirement that the database can not distinguish an authentication request from a user U_{i_0} with one from U_{i_1}. In this experiment, we assume that the database is not able to learn the result of verifications.

3   Embedding Secure Sketches in Homomorphic Encryption

Before the description of our privacy-preserving biometric-authentication protocol, we first propose a way to manage secure sketches while encrypted.

3.1   Secure Sketches

Roughly speaking, a secure sketch scheme (SS, Rec) allows recovery of a hidden value from any element close to this hidden value. The goal is to manage noisy data, such as biometric acquisitions, in cryptographic protocols. This has been formalized by Dodis *et al.* [17] and the general idea is to absorb the differences occurring between two captures by viewing them as errors over a codeword. Many papers envisage applications of these techniques for cryptographic purposes in various contexts, e.g. remote biometric authentication [4] or authenticated key agreement [16].

Let $\mathcal{H}$ be a metric space with distance function d. A secure sketch allows to recover a string $w \in \mathcal{H}$ from any close string $w' \in \mathcal{H}$ thanks to a known data P which does not leak too much information about w.

Definition 4. *A $(\mathcal{H}, m, m', t)$–secure sketch is a pair of functions* (SS, Rec) *where the sketching function* SS *takes* $w \in \mathcal{H}$ *as input, and outputs a value in* $\{0, 1\}^*$, *called a sketch, such that for all random variables* W *over* $\mathcal{H}$ *with min-entropy* $\mathbf{H}_\infty(W) \geq m$, *we have the conditional min-entropy* $\overline{\mathbf{H}}_\infty(W \mid \mathsf{SS}(W)) \geq m'$.

The recovery function Rec *takes a sketch* P *and a vector* $w' \in \mathcal{H}$ *as inputs, and outputs a word* $w'' \in \mathcal{H}$, *such that for any* $P = \mathsf{SS}(w)$ *and* $\mathrm{d}(w, w') \leq t$, *it holds that* $w'' = w$.

When $\mathcal{F}$ is a finite field, then for some integer n, the set $\mathcal{F}^n$ equipped with the Hamming distance d_H is a Hamming space. Juels and Wattenberg [26] have proposed a very natural construction in this case by means of linear error-correcting code:

Definition 5 (Code-offset construction). *Given C an $[n, k, 2t + 1]$ binary linear code, the secure sketch scheme is a pair of functions* $(\mathsf{SS}_C, \mathsf{Rec}_C)$ *where*

- *the function* SS_C *takes w as input, and outputs the sketch $P = c \oplus w$, where c is taken at random from C.*
- *the function* Rec_C *takes w' and P as inputs, decodes $w' \oplus P$ into a codeword c', and then outputs $c' \oplus P$.*

Following [17], this yields a $(\mathcal{F}^n, m, m - (n - k) \log_2 q, t)$-secure sketch, which means that, given P, the entropy loss depends directly on the redundancy of the code. There is thus an obvious trade-off between the correction's capacity t of the code and the security of the secure sketch.

The authentication protocol which arises naturally from this construction follows.

- During the registration, we store $P = \mathsf{SS}_C(w) = c \oplus w$, where c is a random codeword in C, together with the hash value $H(c)$ of c (where H is a cryptographic hash function).
- To authenticate someone, we try to correct the corrupted codeword $w' \oplus P = c \oplus (w' \oplus w)$ and if we obtain a codeword c', we then check: $H(c') = H(c)$.

There are a lot of propositions of applications to real biometrics data: see for instance [7,34] for fingerprints, [27] for faces and [5,24] for irises. In all cases, the size of the code shall not be too small if we want to prevent an attacker from performing an exhaustive search on codewords and thus to recover biometric data.

Unfortunately, for biometric data, the security constraints of secure sketches are difficult to fulfill. And even with a large code's dimension, the security should be increase by additional means. First, we know that biometrics are not random data but their entropy is hard to measure, so that the consequences of entropy loss are not well understood in practice. Moreover, biometrics are widely considered as public data, thus when P and $H(c)$ are known, an attacker would easily check if it is associated to one of his own-made biometric database or try other kinds of cross-matching.

3.2   Review of the Goldwasser-Micali Scheme

The algorithms (Gen, Enc, Dec) of Goldwasser-Micali scheme [23] are defined as follows:

1. The key generation algorithm Gen takes a security parameter 1^ℓ as input, and generates two large prime numbers p and q, $n = pq$ and a non-residue x for which the Jacobi symbol is 1. The public key pk is (x, n), and the secret key sk is (p, q).
2. The encryption algorithm Enc takes a message $m \in \{0, 1\}$ and the public key (x, n) as input, and outputs the ciphertext c, where $c = y^2 x^m \mod n$ and y is randomly chosen from $\mathbb{Z}_n^*$.
3. The decryption algorithm Dec takes a ciphertext c and the private key (p, q) as input, and outputs the message m, where $m = 0$ if c is a quadratic residue, $m = 1$ otherwise.

It is well-known (cf. [23]) that, if the Quadratic Residuosity (QR) problem is intractable, then the Goldwasser-Micali scheme is semantically secure. In other words an adversary $\mathcal{A}$ has only a negligible advantage in the following game.

$$
\begin{aligned}
&\mathbf{Exp}_{\mathcal{E},\mathcal{A}}^{\text{IND-CPA}} \\
&\left|
\begin{aligned}
(sk, pk) &\leftarrow \mathsf{Gen}(1^\ell) \\
(m_0, m_1) &\leftarrow \mathcal{A}(pk) \\
c &\leftarrow \mathsf{Enc}(m_\beta, pk), \ \beta \xleftarrow{R} \{0, 1\} \\
\beta' &\leftarrow \mathcal{A}(m_0, m_1, c, pk)
\end{aligned}
\right.
\end{aligned}
$$

At the end of this game, the attacker's advantage $\mathbf{Adv}_{\mathcal{E},\mathcal{A}}^{\text{IND-CPA}}$ is defined to be

$$
\mathbf{Adv}_{\mathcal{E},\mathcal{A}}^{\text{IND-CPA}} = \left| Pr[\mathbf{Exp}_{\mathcal{E},\mathcal{A}}^{\text{IND-CPA}} = 1 | \beta = 1] - Pr[\mathbf{Exp}_{\mathcal{E},\mathcal{A}}^{\text{IND-CPA}} = 1 | \beta = 0] \right|.
$$

Moreover the encryption protocol possesses a nice homomorphic property, for any $m, m' \in \{0, 1\}$ the following equation holds.

$$
\mathsf{Dec}(\mathsf{Enc}(m, pk) \times \mathsf{Enc}(m', pk), sk) = m \oplus m'
$$

Note that the encryption algorithm encrypts one bit at a time, hence, in order to encrypt a binary string we will encrypt every bit individually. To simplify, we will denote the encryption of $m = (m_0, \ldots, m_{l-1})$ by $\mathsf{Enc}(m, pk) = (\mathsf{Enc}(m_0, pk), \ldots, \mathsf{Enc}(m_{l-1}, pk))$. The IND-CPA security under the quadratic residuosity assumption and the homomorphic property above remain naturally.

3.3 Encrypted Sketches

In a biometric authentication system, we generally want to store the enrolled data in the database $\mathcal{DB}$. In case of sketches, it is thus a problem as anybody can check the membership of a biometric data by having access to $\mathcal{DB}$. For privacy concerns, we prefer to store the sketches encrypted and make the comparison in a secure way. To solve this problem, we use a combination of the Goldwasser-Micali encryption scheme with the Code-Offset construction of Definition 5.

The main advantage is to use the correction functionality which allows to manage biometric data but without the need to fulfill the security constraints usually associated to secure sketches. For instance:

- The service provider $\mathcal{SP}$ generates a Goldwasser-Micali (pk, sk) key pair and publishes pk. In the following, we will denote a related encryption $\mathsf{Enc}(., pk)$ as $\sqsubset . \sqsupset$.
- At the enrollment, the user U_i registers to the service provider $\mathcal{SP}$ with b_i his reference biometric template. Then $P = \mathsf{SS}_\mathsf{C}(b_i) = c \oplus b_i$ is computed and $\sqsubset P \sqsupset$ is stored in $\mathcal{DB}$ for a random codeword c and $H(c)$ is stored by $\mathcal{SP}$ with H a cryptographic hash function.
- When U_i wants to authenticate to $\mathcal{SP}$, b' is captured, and $\sqsubset b' \sqsupset$ is sent to $\mathcal{DB}$. The database $\mathcal{DB}$ computes $\sqsubset P \sqsupset \times \sqsubset b' \sqsupset = \sqsubset c \oplus b_i \oplus b' \sqsupset = Z$ and sends it to $\mathcal{SP}$. Thereafter $\mathcal{SP}$ decrypts Z with its private key sk and decodes the output $c \oplus b_i \oplus b'$ to obtain a codeword c'. Finally, it checks if $H(c') = H(c)$.

Thanks to the homomorphic property of Goldwasser-Micali encryption, the service provider $\mathcal{SP}$ and the database $\mathcal{DB}$ never obtain information on the biometric data which stay encrypted. Moreover, the database learns nothing about c neither, as the computation is made in an encrypted way.

4   Description of the Lipmaa's PIR Protocol

We now make a brief description of the PIR protocol which will be used in our protocol.

4.1 Private Information Retrieval (PIR)

As introduced by Chor et al. [11,12], a PIR protocol allows a user to recover data from a database without leaking which data is currently request . Suppose a database $\mathcal{DB}$ is constituted with M bits $X = x_1, ..., x_M$. To be secure, the protocol should satisfy the following properties [22]:

- **Soundness:** When the user and the database follow the protocol, the result of the request is exactly the requested bit.
- **Request Privacy:** For all $X \in \{0, 1\}^M$, for $1 \leq i, j \leq M$, for any algorithm used by the database, it can not distinguish with a non-negligible probability the difference between the requests of index i and j.

Moreover, we have a Symmetric PIR (SPIR) when the user can not learn more information than the requested data itself. Among the known constructions of computational secure PIR, block-based PIR – i.e. working on block of bits – allow to reduce efficiently the cost. The best performances are from Gentry and Ramzan [21] and Lipmaa [29] with a communication complexity polynomial in the logarithm of M. Surveys of the subject are available in [20,30].

4.2 Review of the Paillier Cryptosystem

The Paillier cryptosystem [31] is defined as follows.

- The key generation algorithm Gen takes a security parameter 1^ℓ as input and generates an RSA integer $n = pq$. Let an integer g which order is a multiple of n modulo n^2. Then the public key is $pk = (n, g)$ and the private key is $sk = \lambda(n)$ where λ is the Carmichael function.
- The encryption algorithm Enc from a message $m \in \mathbb{Z}_n$ and the public key pk outputs $c = g^m r^n \mod n^2$, with r randomly chosen in $\mathbb{Z}_n^*$.
- The decryption algorithm Dec computes $m = \frac{L(c^{\lambda(n)} \mod n^2)}{L(g^{\lambda(n)} \mod n^2)} \mod n$ with L defined on $\{u < n^2 : u = 1 \mod n\}$ by $L(u) = \frac{u-1}{n}$.

The Paillier cryptosystem is known to be IND-CPA (following a similar experiment as in Sect. 3.2) if CR[n] (degree n decisional Composite Residue problem) is hard. This cryptosystem is homomorphic by construction: Dec(Enc(m, pk) × Enc(m', pk) $\mod n^2$, sk) = $m + m' \mod n$, and particularly

$$\mathsf{Dec}(\mathsf{Enc}(m, pk)^k \mod n^2, sk) = km \mod n.$$

In [14,15], Damgård and Jurik have shown, first that we can choose $g = 1 + n$ while keeping the same security. It simplifies the encryption and decryption as $g^m = 1 + mn \mod n^2$ and $c^{sk} = 1 + mn$ when sk is defined as $sk = 0 \mod \lambda(n)$ and $sk = 1 \mod n$. Moreover, they generalized the construction for a variable-length homomorphic cryptosystem by working modulo n^s with $s \geq 1$.

- The encryption of $m \in \mathbb{Z}_{n^s}^*$ becomes $[\![m]\!]_s = (1 + n)^m r^{n^s} \mod n^{s+1}$ with $r \in \mathbb{Z}_n^*$ (to simplify, the notation does not mention the public key, although it depends of its value).
- Decryption is deduced from successive applications of Paillier's decryption.

This generalization is used in the Lipmaa's PIR protocol [29], described below, in order to reduce the communication cost.

4.3 Lipmaa's Protocol

Here the database, denoted by $S = \mathcal{DB}$ is seen as a multidimensional array and the entries are associated to a vector of index. Let $L = \prod_{j=1}^{\lambda} l_j$ with integers l_j the size of S. For $i = (i_1, \ldots, i_\lambda)$ with $i_j \in \mathbb{Z}_{l_j}$, for $j = 1, \ldots, \lambda$, then

$$S[i] = S[i_1 \prod_{j=2}^{\lambda} l_j + i_2 \prod_{j=3}^{\lambda} l_j + \ldots + i_{\lambda-1} l_\lambda + i_\lambda + 1].$$

To answer to a request for the data of index $(q_1, \ldots, q_\lambda)$, the idea of the protocol is to decrease the dimension of S progressively by constructing a smaller database recursively till the last dimension. Let $S_0 = S$. The first iteration is to construct S_1 by defining $S_1(i_2, \ldots, i_\lambda)$ as the encryption of $S_0(q_1, i_2, \ldots, i_\lambda)$. This is repeated λ times and therefore $S_j(i_{j+1}, \ldots, i_\lambda)$ will be the encryption of $S_0(q_1, \ldots, q_j, i_{j+1}, \ldots, i_\lambda)$ for $j = 1, \ldots, \lambda$. At the end, the last element S_λ which is a λ times encryption of $S_0(q_1, \ldots, q_\lambda)$ is the answer to the user's request. To be a PIR, everything is made by concealing the index in several Damgård-Jurik encryptions.

We define the binary array δ with $\delta_{j,t} = 1$ if $q_j = t$, else 0 for $j = 1, \ldots, \lambda$ and $t \in \mathbb{Z}_{l_j}$. We suppose the user possesses a set of couple of keys for the Damgård-Jurik cryptosystem for various lengths and that the database knows the corresponding public keys. Then the user sends as its request, the encryptions $[\![\delta_{j,t}]\!]_{s+j-1}$, for all j, t. Then the database can exploit homomorphic properties like

$$([\![m_2]\!]_{s+\xi})^{[\![m_1]\!]_s} = [\![m_2 [\![m_1]\!]_s]\!]_{s+\xi}$$

and proceeds as follows.

- For $j = 1, \ldots, \lambda$, it computes $S_j(i_{j+1}, \ldots, i_\lambda) = \prod_{t \in \mathbb{Z}_{l_j}} [\![\delta_{j,t}]\!]_{s+j-1}^{S_{j-1}(t, i_{j+1}, \ldots, i_\lambda)}$
  for $i_{j+1} \in \mathbb{Z}_{l_{j+1}}, \ldots, i_\lambda \in \mathbb{Z}_{l_\lambda}$
- and outputs S_λ.

With successive decryptions, the user will recover the requested element. Indeed, starting from $j = \lambda$ to $j = 1$, as the sub-database S_j entries correspond to encryptions of entries $S_0(q_1, \ldots, q_j, *, \ldots, *)$, they are equal to

$$\left[\!\!\left[\sum_{t \in \mathbb{Z}_{l_j}} \delta_{j,t} S_{j-1}(t, i_{j+1}, \ldots, i_\lambda) \right]\!\!\right]_{s+j-1} = [\![S_{j-1}(q_j, i_{j+1}, \ldots, i_\lambda)]\!]_{s+j-1}.$$

So starting from $j = \lambda$ to $j = 1$, decryption of S_λ leads to $S_{\lambda-1}(q_\lambda)$, which decryption leads to $S_{\lambda-2}(q_{\lambda-1}, q_\lambda)$ and so on... It gives the results at the end.

The Request Privacy of this protocol is achieved thanks to the semantic security of the Damgård-Jurik cryptosystem used to encode the request's index, i.e. the $\delta_{j,t}$.

5   A Private Biometric Authentication Protocol with Secure Sketches

We describe here our biometric authentication protocol based on the use of the Lipmaa's protocol and the idea described in Sect. 3.3. It allows to achieve identity privacy with a small communication cost compared to previous construction and with biometric data protected in confidentiality during the entire process. Moreover, the use of secure sketches permits to recover a key binded to the enrolled biometric data which can then be reused in cryptographic applications provided by the service provider $\mathcal{SP}$.

We combine Goldwasser-Micali with Damgård-Jurik encryption in the Lipmaa's protocol in a way that it is still possible to use the homomorphic trick to compute the encryption of the output of the recovery Rec_C function of the Code-Offset construction. To simplify, we will describe our protocol with a database of dimension $\lambda = 1$, but it is easily applicable to any λ.

So here, we make use of a double encryption via Goldwasser-Micali and Paillier. We take benefit of their homomorphic properties in the principle below:

$$\llbracket \sqsubset s \sqsupset \rrbracket^{\sqsubset w \sqsupset} = \llbracket \sqsubset s \sqsupset \times \sqsubset w \sqsupset \rrbracket = \llbracket \sqsubset s \oplus w \sqsupset \rrbracket \tag{1}$$

5.1 Parameters

Let M be the number of enrolled user in the database $\mathcal{DB}$. The service provider $\mathcal{SP}$ is associated to two couples of keys (pk_{GM}, sk_{GM}) and (pk_P, sk_P) for the Goldwasser-Micali and the Paillier cryptosystems respectively. The corresponding encryption are denoted $\sqsubset . \sqsupset$ and $\llbracket . \rrbracket$ respectively. The public keys are published and the secret keys are stored inside the Hardware Security Module HSM.

The database $\mathcal{DB}$ contains the M encrypted sketches $\sqsubset SS_C(b_i) \sqsupset$, for $i = 1, \ldots, M$ with b_i the reference biometric template of the user U_i. The database $\mathcal{DB}$ also possesses the hash values $H(c_i)$ associated to $SS_C(b_i) = b_i \oplus c_i$ for all i.

Let a_i, $i = 1, \ldots, M$ be the vectors of $\mathcal{DB}$, then $a_{i,u} = \sqsubset \pi_u(SS_C(b_i)) \sqsupset$ for $u = 0, \ldots, l - 1$ and $a_{i,l} = H(c_i)$, where $\pi_u(x)$ denotes the u-th bit of a binary vector x.

5.2 Verification Phase

When the user U_i wants to authenticate itself to the service provider $\mathcal{SP}$, the steps are:

1. its new biometric template b' is encrypted by the sensor via Goldwasser-Micali: $\sqsubset b' \sqsupset$,
2. the sensor client $\mathcal{C}$ sends to $\mathcal{DB}$ a request constituted with the Paillier's ciphertexts $\llbracket \delta_k^u \rrbracket$, $k = 1, \ldots, M$, $u = 0, \ldots, l$ where $\delta_k^u = \sqsubset \pi_u(b') \sqsupset$ if $k = i$ and 0 else, for $u \le l - 1$ and $\delta_k^l = 1$ if $k = i$ and 0 else.
3. The database $\mathcal{DB}$ computes $\llbracket a_{i,u} \times \sqsubset \pi_u(b') \sqsupset \rrbracket = \prod_{k=1}^{M} \llbracket \delta_k^u \rrbracket^{a_{k,u}}$, for $u = 0, \ldots, l - 1$, i.e. via Eq. (1)

$$\llbracket \sqsubset \pi_u(SS_C(b_i) \oplus b') \sqsupset \rrbracket$$

and $\llbracket a_{i,l} \rrbracket = \prod_{k=1}^{M} \llbracket \delta_k^l \rrbracket^{a_{k,l}}$.
4. The database $\mathcal{DB}$ sends it to the service provider $\mathcal{SP}$ for $u = 0, \ldots, l$.
5. The HSM hence decrypts first via the Paillier decryption algorithm, then via the Goldwasser-Micali algorithm to recover $SS_C(b_i) \oplus b'$ and $H(c_i)$,
6. it decodes it to obtain a codeword c' and checks if $H(c') = H(c_i)$ to accept or reject the authentication,
7. the result is forwarded to the service provider $\mathcal{SP}$.

A natural extension is to use $H(c_i)$ as a key for cryptographic applications provided by $\mathcal{SP}$.

The construction can be generalized to the Lipmaa's protocol [29] with $\lambda > 1$ and several Damgård-Jurik systems. The idea is simply to use $l+1$-length vectors $\delta_{j,t}$ with coordinates $\delta^*_{j,t} = 1$ if $q_j = t$, else 0 for $j = 2, \ldots, \lambda$ and $t \in \mathbb{Z}_{l_j}$ but with the same modification as for item 2 above for $j = 1$, i.e. δ_{1,q_1} is the $l + 1$-length vector associated to the Goldwasser-Micali encryption of the fresh biometric data. In other words, the protocol above corresponds to the first iteration of the Lipmaa's protocol and the other iterations will then be as usual.

We see here that the request of the PIR protocol – which corresponds classically to encryption of 1 and 0 – is slightly modified to force the database to compute the multiplication of its own elements with the encryption of the fresh template. This doing, the operation is quite transparent. Note that this combination of Goldwasser-Micali scheme with a PIR protocol is possible only if the group law of the underlying homomorphic encryption scheme, here Paillier or Damgård-Jurik cryptosystem, is compatible. For instance, it is also applicable to the protocol of Chang [10].

One advantage of Lipmaa's protocol is to greatly decrease the communication complexity as opposed to the basic version with Paillier cryptosystem only – described above – which is linear on the size M of the database $\mathcal{DB}$. Lipmaa's protocol allows to achieve a communication cost in $O(\log^2 M)$. The parameters can be optimized further in some cases, see [29].

Remark 1. Here, as sketches are encrypted bit by bit, the storage cost of an encrypted sketch in the database $\mathcal{DB}$ is $l \times \log_2 n$ bits. For instance with $l = 512$ and n a 2048 bits RSA integer, it leads to about 128 kbytes per encrypted template (encrypted sketch and hashed codeword). Concerning the computation cost to answer a PIR request, $\mathcal{DB}$ performs about $(l + 1) \times M$ exponentiations modulo n^s (with some s; $s = 2$ with Paillier). It is thus a constraint on the size of the database.

6   Security Analysis

We show here that the protocol satisfies the security requirements of Sect. 2.2.

6.1   Soundness

When the system's components follow the protocol, the soundness is straightforward.

Lemma 1. *The protocol is sound under Definition 1 if the involved Secure Sketch* $(\mathsf{SS_C}, \mathsf{Rec_C})$ *is sound and if the PIR protocol is sound.*

In other words, we rely on the efficiency of secure sketches to fulfill this requirement while here we focus on increasing their security when related to a biometric system. For instance as in our protocol, biometric data and sketches are always encrypted via at least one semantically secure encryption scheme, it implies that the scheme provides a strong protection on templates as they can not be recovered by any adversary. It is also an interesting property, for better acceptability reason, although biometric data are assumed to be public.

6.2   Identity Privacy

The scheme is proved below to ensure identity privacy against non-colluding malicious service provider or malicious database, and any external adversary.

For this, we assume that the errors $b_i \oplus b_i'$ occurring between two matching biometric templates b_i, b_i' of any user U_i (registered or not) are indistinguishable among all the possible errors $b_j \oplus b_j'$. This is a quite reasonable assumption as errors can greatly vary depending on internal and external factors of measurement.

Lemma 2. *Our scheme achieves identity privacy against a malicious service provider or a malicious database under the semantic security of the Goldwasser-Micali scheme, i.e. under the QR assumption.*

Proof. It is clear that the database $\mathcal{DB}$ has no advantage in distinguishing the value of e in the experiment of Definition 2 as it has no access to any information about biometric templates thanks to Goldwasser-Micali encryption of the sketches. Any algorithm to obtain a valid guess of e with a non-negligible advantage would lead to an algorithm to break the semantic security of the Goldwasser-Micali encryption scheme.

Similarly for the service provider $\mathcal{SP}$, the only possible information would be obtained by using the secret keys. Under our assumption of non-collusion with the database $\mathcal{DB}$ and thanks to the honesty of the sensor client $\mathcal{C}$, we know that $\mathcal{SP}$ can only obtain data of the form $(c_j \oplus b_j \oplus b', H(c_j))$ for some j and some b' corresponding to the authentication request for user U_j with a fresh template b'. Hence, the only information on biometrics is at most – i.e. when decoding is successful – the difference $b_j \oplus b'$. From the assumption of indistinguishability of errors, the service provider $\mathcal{SP}$ does not learn information on b_j nor b'. It means that its best algorithm to guess e is to take it randomly. □

6.3   Transaction Anonymity

The anonymity of the verification's requests against the database $\mathcal{DB}$ is directly deduced from the Request Privacy property of the PIR protocol of Lipmaa, which is a consequence of the IND-CPA security of Damgård-Jurik cryptosystem.

Lemma 3. *Our scheme achieves transaction anonymity, following Definition 3, against a malicious database under the CR assumption.*

Of course, the scheme does not allow transaction anonymity against the service provider, because it can learn the value $H(c_i)$ for an authentication request of the user U_i, which can thus be used to track this user in future authentication requests. But it has a gain only on the final phase as the request is directly computed from the sensor's side. This means that if we renew regularly the enrolled data – the encrypted sketch and the corresponding hash value – the service provider will not be able to link the future authentication results with the previous ones. It leads to an interesting additional property as it can be seen as a way to forbid long-term tracking.

7   Conclusion

In this paper, we improve the biometric authentication scheme of Bringer *et al.* [6] but our goal is the same. We want to achieve biometric authentication while preserving the privacy of users. In particular, we modify the protocol of [6] to only have to deal with encrypted biometric data. To do so, we replace traditional matching algorithm by an error correction procedure thanks to the introduction of secure sketches. Moreover, we explain how our proposition can be integrated into one of the best Private Information Retrieval scheme due to Lipmaa.

There are still many performances issues to handle. In our proposal, encryption is performed bit by bit and one can look forward more efficient ways to encrypt biometric data. As another possible enhancement, following [33] computational aspects can take more importance than communication issues and introduction of different PIR schemes, for instance with multi-server, leading to smaller computational overheads should be considered.

Acknowledgments

The authors wish to thank the anonymous reviewers for their comments, David Pointcheval and Qiang Tang for their help in the definition of the security model.

References

1. Anderson, R., Bond, M., Clulow, J., Skorobogatov, S.: Cryptographic processors-a survey. Proceedings of the IEEE 94(2), 357–369 (2006)
2. Atallah, M.J., Frikken, K.B., Goodrich, M.l.T., Tamassia, R.: Secure biometric authentication for weak computational devices. In: S. Patrick, A., Yung, M. (eds.) FC 2005. LNCS, vol. 3570, pp. 357–371. Springer, Heidelberg (2005)
3. Boyen, X.: Reusable cryptographic fuzzy extractors. In: Atluri, V., Pfitzmann, B., McDaniel, P.D. (eds.) CCS 2004: Proceedings of the 11th ACM conference on Computer and communications security, pp. 82–91. ACM Press, New York (2004)
4. Boyen, X., Dodis, Y., Katz, J., Ostrovsky, R., Smith, A.: Secure remote authentication using biometric data. In: Cramer, R.J.F. (ed.) EUROCRYPT 2005. LNCS, vol. 3494, pp. 147–163. Springer, Heidelberg (2005)
5. Bringer, J., Chabanne, H., Cohen, G., Kindarji, B., Zémor, G.: Optimal iris fuzzy sketches. In: IEEE First International Conference on Biometrics: Theory, Applications and Systems, BTAS 2007 (2007)
6. Bringer, J., Chabanne, H., Izabachène, M., Pointcheval, D., Tang, Q., Zimmer, S.: An application of the Goldwasser-Micali cryptosystem to biometric authentication. In: Pieprzyk, J., Ghodosi, H., Dawson, E. (eds.) ACISP 2007. LNCS, vol. 4586, pp. 96–106. Springer, Heidelberg (2007)

7. Bringer, J., Chabanne, H., Kindarji, B.: The best of both worlds: Applying secure sketches to cancelable biometrics. Science of Computer Programming (to appear, Presented at WISSec 2007)
8. Bringer, J., Chabanne, H., Pointcheval, D., Tang, Q.: Extended private information retrieval and its application in biometrics authentications. In: Bao, F., Ling, S., Okamoto, T., Wang, H., Xing, C. (eds.) CANS 2007. LNCS, vol. 4856, pp. 175–193. Springer, Heidelberg (2007)
9. Bringer, J., Chabanne, H., Tang, Q.: An application of the Naccache-Stern knapsack cryptosystem to biometric authentication. In: AutoID, pp. 180–185. IEEE, Los Alamitos (2007)
10. Chang, Y.-C.: Single database private information retrieval with logarithmic communication. In: Wang, H., Pieprzyk, J., Varadharajan, V. (eds.) ACISP 2004. LNCS, vol. 3108, pp. 50–61. Springer, Heidelberg (2004)
11. Chor, B., Goldreich, O., Kushilevitz, E., Sudan, M.: Private information retrieval. In: FOCS, pp. 41–50 (1995)
12. Chor, B., Kushilevitz, E., Goldreich, O., Sudan, M.: Private information retrieval. J. ACM 45(6), 965–981 (1998)
13. Crescenzo, G.D., Graveman, R., Ge, R., Arce, G.: Approximate message authentication and biometric entity authentication. In: Patrick, A.S., Yung, M. (eds.) FC 2005. LNCS, vol. 3570, pp. 240–254. Springer, Heidelberg (2005)
14. Damgård, I., Jurik, M.: A generalisation, a simplification and some applications of Paillier's probabilistic public-key system. In: Kim, K.-c. (ed.) PKC 2001. LNCS, vol. 1992, pp. 119–136. Springer, Heidelberg (2001)
15. Damgård, I., Jurik, M.: A length-flexible threshold cryptosystem with applications. In: Safavi-Naini, R., Seberry, J. (eds.) ACISP 2003. LNCS, vol. 2727, pp. 350–364. Springer, Heidelberg (2003)
16. Dodis, Y., Katz, J., Reyzin, L., Smith, A.: Robust fuzzy extractors and authenticated key agreement from close secrets. In: Dwork, C. (ed.) CRYPTO 2006. LNCS, vol. 4117, pp. 232–250. Springer, Heidelberg (2006)
17. Dodis, Y., Reyzin, L., Smith, A.: Fuzzy extractors: How to generate strong keys from biometrics and other noisy data. In: Cachin, C., Camenisch, J.L. (eds.) EUROCRYPT 2004. LNCS, vol. 3027, pp. 523–540. Springer, Heidelberg (2004)
18. Du, W., Atallah, M.J.: Secure multi-party computation problems and their applications: a review and open problems. In: NSPW 2001: Proceedings of the 2001 workshop on New security paradigms, pp. 13–22. ACM Press, New York (2001)
19. Feigenbaum, J., Ishai, Y., Malkin, T., Nissim, K., Strauss, M.J., Wright, R.N.: Secure multiparty computation of approximations. ACM Transactions on Algorithms 2(3), 435–472 (2006)
20. Gasarch, W.: A survey on private information retrieval, http://www.cs.umd.edu/gasarch/pir/pir.html
21. Gentry, C., Ramzan, Z.: Single-database private information retrieval with constant communication rate. In: Caires, L., Italiano, G.F., Monteiro, L., Palamidessi, C., Yung, M. (eds.) ICALP 2005. LNCS, vol. 3580, pp. 803–815. Springer, Heidelberg (2005)
22. Gertner, Y., Ishai, Y., Kushilevitz, E., Malkin, T.: Protecting data privacy in private information retrieval schemes. In: STOC, pp. 151–160 (1998)
23. Goldwasser, S., Micali, S.: Probabilistic encryption and how to play mental poker keeping secret all partial information. In: Proceedings of the Fourteenth Annual ACM Symposium on Theory of Computing, San Francisco, California, USA, May 5-7, 1982, pp. 365–377. ACM, New York (1982)

24. Hao, F., Anderson, R., Daugman, J.: Combining crypto with biometrics effectively. IEEE Transactions on Computers 55(9), 1081–1088 (2006)
25. Juels, A., Sudan, M.: A fuzzy vault scheme. Des. Codes Cryptography 38(2), 237–257 (2006)
26. Juels, A., Wattenberg, M.: A fuzzy commitment scheme. In: ACM Conference on Computer and Communications Security, pp. 28–36 (1999)
27. Kevenaar, T.A.M., Schrijen, G.J., van der Veen, M., Akkermans, A.H.M., Zuo, F.: Face recognition with renewable and privacy preserving binary templates. In: AUTOID 2005: Proceedings of the Fourth IEEE Workshop on Automatic Identification Advanced Technologies, pp. 21–26. IEEE Computer Society, Washington (2005)
28. Linnartz, J.-P.M.G., Tuyls, P.: New shielding functions to enhance privacy and prevent misuse of biometric templates. In: Kittler, J., Nixon, M.S. (eds.) AVBPA 2003. LNCS, vol. 2688, pp. 393–402. Springer, Heidelberg (2003)
29. Lipmaa, H.: An oblivious transfer protocol with log-squared communication. In: Zhou, J., López, J., Deng, R.H., Bao, F. (eds.) ISC 2005. LNCS, vol. 3650, pp. 314–328. Springer, Heidelberg (2005)
30. Ostrovsky, R., Skeith III., W.E.: A survey of single database PIR: Techniques and applications. Cryptology ePrint Archive: Report 2007/059 (2007)
31. Paillier, P.: Public-key cryptosystems based on composite degree residuosity classes. In: Stern, J. (ed.) EUROCRYPT 1999. LNCS, vol. 1592, pp. 223–238. Springer, Heidelberg (1999)
32. Schoenmakers, B., Tuyls, P.: Efficient binary conversion for Paillier encrypted values. In: Vaudenay, S. (ed.) EUROCRYPT 2006. LNCS, vol. 4004, pp. 522–537. Springer, Heidelberg (2006)
33. Sion, R., Carbunar, B.: On the computational practicality of private information retrieval. In: Network and Distributed System Security Symposium NDSS (2007)
34. Tuyls, P., Akkermans, A.H.M., Kevenaar, T.A.M., Schrijen, G.J., Bazen, A.M., Veldhuis, R.N.J.: Practical biometric authentication with template protection. In: Kanade, T., Jain, A., Ratha, N.K. (eds.) AVBPA 2005. LNCS, vol. 3546, pp. 436–446. Springer, Heidelberg (2005)
35. Tuyls, P., Goseling, J.: Capacity and examples of template-protecting biometric authentication systems. In: Maltoni, D., Jain, A.K. (eds.) BioAW 2004. LNCS, vol. 3087, pp. 158–170. Springer, Heidelberg (2004)
36. Tuyls, P., Verbitskiy, E., Goseling, J., Denteneer, D.: Privacy protecting biometric authentication systems: an overview. In: EUSIPCO 2004 (2004)

Authenticated Encryption Mode for Beyond the Birthday Bound Security

Tetsu Iwata

Dept. of Computational Science and Engineering,
Nagoya University
Furo-cho, Chikusa-ku, Nagoya, 464-8603, Japan
iwata@cse.nagoya-u.ac.jp,
http://www.nuee.nagoya-u.ac.jp/labs/tiwata/

Abstract. In this paper, we propose an authenticated encryption mode for blockciphers. Our authenticated encryption mode, CIP, has provable security bounds which are better than the usual birthday bound security. Besides, the proven security bound for authenticity of CIP is better than any of the previously known schemes. The design is based on the encrypt-then-PRF approach, where the encryption part uses a key stream generation of CENC, and the PRF part combines a hash function based on the inner product and a blockcipher.

Keywords: Blockcipher, modes of operation, authenticated encryption, security proofs, birthday bound.

1   Introduction

Provable security is the standard security goal for blockcipher modes, i.e., encryption modes, message authentication codes, and authenticated encryption modes. For encryption modes, CTR mode and CBC mode are shown to have provable security [1]. The privacy notion we consider is called indistinguishability from random strings [24]. In this notion, the adversary is in the adaptive chosen plaintext attack scenario, and the goal is to distinguish the ciphertext from the random string of the same length. The nonce-based treatment of CTR mode was presented by Rogaway [22], and it was proved that, for any adversary against CTR mode, the success probability is at most $O(\sigma^2/2^n)$ under the assumption that the blockcipher is a secure pseudorandom permutation (PRP), where n is the block length and σ denotes the total ciphertext length in blocks that the adversary obtains. The security bound is known as the *birthday bound*.

Authenticity is achieved by message authentication codes, or MACs. Practical examples of MACs that have provable security include PMAC [7], EMAC [21], and OMAC [10]. We consider the pseudorandom function, or PRF [3], for authenticity which provably implies the adversary's inability to make a forgery. In this notion, the adversary is in the adaptive chosen plaintext attack scenario, and the goal is to distinguish the output of the MAC from that of the random function. It was proved that, for any adversary against PMAC, EMAC, and OMAC, the success probability is at most $O(\sigma^2/2^n)$.

S. Vaudenay (Ed.): AFRICACRYPT 2008, LNCS 5023, pp. 125–142, 2008.

An authenticated encryption mode is a scheme for both privacy and authenticity. It takes a plaintext M and provides both privacy and authenticity for M. There are a number of proposals: the first efficient construction was given by Jutla and the mode is called IAPM [13], OCB mode was proposed by Rogaway et. al. [24], CCM mode [27,12] is the standard of IEEE, EAX mode [6] is based on the generic composition, CWC mode [15] combines CTR mode and Wegman-Carter MAC, and GCM mode [19,20] is the standard of NIST. Other examples include CCFB mode [17], and XCBC [8]. All these modes have provable security with the standard birthday bound.

There are several proposals on MACs that have beyond the birthday bound security. For example, we have RMAC [11] and XOR MAC [2], and there are other proposals which are not based on blockciphers. On the other hand, few proposals are known for encryption modes and authenticated encryption modes. CENC [9] is an example of an encryption mode, and its generalization called NEMO was proposed in [16]. For authenticated encryption modes, CHM [9] is the only example we are aware of.

We view that the beyond the birthday bound security as the standard goal for future modes. AES is designed to be secure even if the adversary obtains nearly 2^{128} input-output pairs, and many other blockciphers have similar security goal. On the other hand, CTR mode, OMAC, or GCM have to re-key before 2^{64} blocks of plaintexts are processed, since otherwise the security is lost. This situation is unfortunate as the security of the blockcipher is significantly lost once it is plugged into the modes, and the current state-of-the-art, CTR mode, OMAC, or GCM, do not fully inherit the security of the blockcipher.

In this paper, we propose an authenticated encryption mode called CIP, CENC with Inner Product hash, to address the security issues in GCM, and CHM. GCM, designed by McGrew and Viega, was selected as the standard of NIST. It is based on CTR mode and Wegman-Carter MAC, and it is fully parallelizable. Likewise, CHM uses CENC for encryption part and Wegman-Carter MAC for PRF part,

While CHM has beyond the birthday bound security, its security bound for authenticity includes the term $M_{\max}/2^\tau$, where $M_{\max}$ is the maximum block length of messages, and τ is the tag length. It is a common practice to use small tag length to save communication cost or storage. For example, one may use $\tau = 32$ or 64 with 128-bit blockciphers. However the term, $M_{\max}/2^\tau$, is linear in $M_{\max}$, the bound soon becomes non-negligible if τ is small. For example, with $\tau = 32$, if we encrypt only one message of 2^{22} blocks (64MBytes), the security bound is $1/1024$, which is not acceptable in general. Therefore, beyond the birthday bound security has little impact when τ is small. GCM also has the same issue, and its security bounds for both privacy and authenticity have the term of the form $M_{\max}/2^\tau$.

Our design goal of CIP is to have beyond the birthday bound security, but we insist that it can be used even with small tag length. Besides, we want security proofs with the standard PRP assumption, and we maintain the full parallelizability. CIP follows the encrypt-then-PRF approach [4], which is shown to be a sound way to construct an authenticated encryption mode. We use CENC [9] for encryption part, since it achieves beyond the birthday bound security with very small cost compared to CTR mode. PRF part is a hash function that combines the inner product hash and the blockcipher, which may be seen as the generalization of PMAC [7] to reduce the number of blockcipher calls and still have full parallelizability.

CIP takes a parameter ϖ called frame width, which is supposed to be a small integer (e.g., $2 \leq \varpi \leq 8$). Our default recommendation is $\varpi = 4$, and with other default parameters, to encrypt a message of l blocks, CIP requires $257l/256$ blockcipher calls for encryption, and l multiplications and $l/\varpi = l/4$ blockcipher calls for PRF, while $\varpi = 4$ blocks of key stream has to be precomputed and stored. CIP requires about l/ϖ more blockcipher calls compared to GCM or CHM. For security, if we use the AES, CIP can encrypt at most 2^{64} plaintexts, and the maximum length of the plaintext is 2^{62} blocks (2^{36}GBytes), and the security bounds are, roughly, $\tilde{\sigma}^3/2^{245} + \tilde{\sigma}/2^{119}$ for privacy, and $\tilde{\sigma}^3/2^{245} + \tilde{\sigma}/2^{118} + 2/2^{\tau}$ for authenticity. This implies $\tilde{\sigma}$ should be sufficiently smaller than 2^{81} blocks (2^{55}GBytes). In particular, the only term that depends on tag length τ is $2/2^{\tau}$, and thus it does not depend on the message length. Therefore, CIP can be used even for short tag length. CIP has security bounds that are better than any of the known schemes we are aware of.

2   Preliminaries

Notation. If x is a string then $|x|$ denotes its length in bits, and $|x|_n$ is its length in n-bit blocks, i.e., $|x|_n = \lceil |x|/n \rceil$. If x and y are two equal-length strings, then $x \oplus y$ denotes the xor of x and y. If x and y are strings, then $x\|y$ or xy denote their concatenation. Let $x \leftarrow y$ denote the assignment of y to x. If X is a set, let $x \xleftarrow{R} X$ denote the process of uniformly selecting at random an element from X and assigning it to x. For a positive integer n, $\{0,1\}^n$ is the set of all strings of n bits. For positive integers n and ϖ, $(\{0,1\}^n)^{\varpi}$ is the set of all strings of $n\varpi$ bits, and $\{0,1\}^*$ is the set of all strings (including the empty string). For positive integers n and m such that $n \leq 2^m - 1$, $\langle n \rangle_m$ is the m-bit binary representation of n. For a bit string x and a positive integer n such that $|x| \geq n$, first(n,x) and last(n,x) denote the first n bits of x and the last n bits of x, respectively. For a positive integer n, 0^n and 1^n denote the n-times repetition of 0 and 1, respectively.

Let Perm(n) be the set of all permutations on $\{0,1\}^n$. We say P is a random permutation if $P \xleftarrow{R} $ Perm(n). The blockcipher is a function $E : \{0,1\}^k \times \{0,1\}^n \rightarrow \{0,1\}^n$, where, for any $K \in \{0,1\}^k$, $E(K, \cdot) = E_K(\cdot)$ is a permutation on $\{0,1\}^n$. The positive integer n is the block length, and k is the key length. Similarly, Func(m,n) denotes the set of all functions from $\{0,1\}^m$ to $\{0,1\}^n$, and R is a random function if $R \xleftarrow{R} $ Func(m,n).

The frame, nonce, and counter. CIP takes a positive integer ϖ as a parameter, and it is called a frame width. For fixed positive integer ϖ (say, $\varpi = 4$), a

ϖ-block string is called a frame. Throughout this paper, we assume $\varpi \geq 1$. A nonce N is a bit string, where for each pair of key and plaintext, it is used only once. The length of the nonce is denoted by ℓ_{nonce}, and it is at most the block length. We also use an n-bit counter, $\texttt{ctr}$. This value is initialized based on the value of the nonce, then it is incremented after each blockcipher invocation. The function for increment is denoted by $\text{inc}(\cdot)$. It takes an n-bit string x (a counter) and returns the incremented x. We assume $\text{inc}(x) = x + 1 \bmod 2^n$, but other implementations also work, e.g., with LFSRs if $x \neq 0^n$.

3   Specification of CIP

In this section, we present our authenticated encryption scheme, CIP. It takes five parameters: a blockcipher, a nonce length, a tag length, and two frame widths.

Fix the blockcipher $E : \{0,1\}^k \times \{0,1\}^n \to \{0,1\}^n$, the nonce length ℓ_{nonce}, the tag length τ, and the frame widths ϖ and w. We require that $\log_2(\lceil k/n \rceil + \varpi) < \ell_{\text{nonce}} < n$, and $1 \leq \tau \leq n$.

CIP consists of two algorithms, the encryption algorithm (CIP.Enc) and the decryption algorithm (CIP.Dec). These algorithms are defined in Fig. 1. The encryption algorithm, CIP.Enc, takes the key $K \in \{0,1\}^k$, the nonce $N \in \{0,1\}^{\ell_{\text{nonce}}}$, and the plaintext M to return the ciphertext C and the tag Tag $\in \{0,1\}^\tau$. We have $|M| = |C|$, and the length of M is at most $2^{n-\ell_{\text{nonce}}-2}$ blocks. We write $(C, \text{Tag}) \leftarrow \text{CIP.Enc}_K(N, M)$. The decryption algorithm, CIP.Dec, takes K, N, C and Tag to return M or a special symbol $\perp$. We write $M \leftarrow \text{CIP.Dec}_K(N, C, \text{Tag})$ or $\perp \leftarrow \text{CIP.Dec}_K(N, C, \text{Tag})$. Both algorithms internally use the key setup algorithm (CIP.Key), a hash function (CIP.Hash), and the keystream generation algorithm (CIP.KSGen).

We use the standard key derivation for key setup. The input of CIP.Key is the blockcipher key $K \in \{0,1\}^k$, and the output is $(K_H, T_H) \in \{0,1\}^k \times (\{0,1\}^n)^\varpi$, where $T_H = (T_0, \ldots, T_{\varpi-1})$, and

- K_H is the first k bits of

$$E_K(\langle 0 \rangle_{\ell_{\text{nonce}}} \| 1^{n-\ell_{\text{nonce}}}) \| \cdots \| E_K(\langle \lceil k/n \rceil - 1 \rangle_{\ell_{\text{nonce}}} \| 1^{n-\ell_{\text{nonce}}}),$$

and
- $T_i \leftarrow E_K(\langle \lceil k/n \rceil + i \rangle_{\ell_{\text{nonce}}} \| 1^{n-\ell_{\text{nonce}}})$ for $0 \leq i \leq \varpi - 1$.

These keys are used for CIP.Hash, which is defined in Fig. 2 (See also Fig. 8 for an illustration). It takes the key $(K_H, T_H) \in \{0,1\}^k \times (\{0,1\}^n)^\varpi$, and the input $x \in \{0,1\}^*$, and the output is a hash value Hash $\in \{0,1\}^n$. The inner product is done in the finite field $\text{GF}(2^n)$ using a canonical polynomial to represent field elements. The suggested canonical polynomial is the lexicographically first polynomial among the irreducible polynomials of degree n that have a minimum number of nonzero coefficients. For $n = 128$ the indicated polynomial is $x^{128} + x^7 + x^2 + x + 1$. CIP.KSGen, defined in Fig. 3, is equivalent to CENC in [9], and

Algorithm CIP.Enc$_K(N, M)$	**Algorithm** CIP.Dec$_K(N, C, \mathrm{Tag})$
100 $(K_H, T_H) \leftarrow$ CIP.Key(K)	200 $(K_H, T_H) \leftarrow$ CIP.Key(K)
101 $l \leftarrow \lceil |M|/n \rceil$	201 $l \leftarrow \lceil |C|/n \rceil$
102 $\mathbf{ctr} \leftarrow (N \| 0^{n-\ell_{\mathrm{nonce}}})$	202 $\mathbf{ctr} \leftarrow (N \| 0^{n-\ell_{\mathrm{nonce}}})$
103 $S \leftarrow$ CIP.KSGen$_K(\mathbf{ctr}, l+1)$	203 $S \leftarrow$ CIP.KSGen$_K(\mathbf{ctr}, l+1)$
104 $S_H \leftarrow \mathrm{first}(n, S)$	204 $S_H \leftarrow \mathrm{first}(n, S)$
105 $S_{\mathrm{mask}} \leftarrow \mathrm{last}(n \times l, S)$	205 $\mathrm{Hash}' \leftarrow$ CIP.Hash$_{K_H, T_H}(C)$
106 $C \leftarrow M \oplus \mathrm{first}(|M|, S_{\mathrm{mask}})$	206 $\mathrm{Tag}' \leftarrow \mathrm{first}(\tau, \mathrm{Hash}' \oplus S_H)$
107 $\mathrm{Hash} \leftarrow$ CIP.Hash$_{K_H, T_H}(C)$	207 **if** $\mathrm{Tag}' \neq \mathrm{Tag}$ **then return** $\bot$
108 $\mathrm{Tag} \leftarrow \mathrm{first}(\tau, \mathrm{Hash} \oplus S_H)$	208 $S_{\mathrm{mask}} \leftarrow \mathrm{last}(n \times l, S)$
109 **return** (C, Tag)	209 $M \leftarrow C \oplus \mathrm{first}(|C|, S_{\mathrm{mask}})$
	210 **return** M

Fig. 1. Definition of CIP.Enc (left), and CIP.Dec (right). CIP.KSGen is defined in Fig. 3, and CIP.Hash is defined in Fig. 2.

Algorithm CIP.Hash$_{K_H, T_H}(x)$
100    $x \leftarrow x \| 10^{n-1-(|x| \bmod n)}$
101    $l \leftarrow |x|/n;\ (x_0, \ldots, x_{l-1}) \leftarrow x;\ \ell \leftarrow \lceil l/\varpi \rceil$
102    $\mathrm{Hash} \leftarrow 0^n$
103   **for** $i \leftarrow 0$ **to** $\ell - 2$ **do**
104        $A_i \leftarrow (x_{i\varpi}, \ldots, x_{(i+1)\varpi-1}) \cdot (T_0, \ldots, T_{\varpi-1})$
105        $\mathrm{Hash} \leftarrow \mathrm{Hash} \oplus E_{K_H}(A_i \oplus \langle i \rangle_n)$
106    $A_{\ell-1} \leftarrow (x_{(\ell-1)\varpi}, \ldots, x_{l-1}) \cdot (T_0, \ldots, T_{l-(\ell-1)\varpi-1})$
107    $\mathrm{Hash} \leftarrow \mathrm{Hash} \oplus E_{K_H}(A_{\ell-1} \oplus \langle \ell-1 \rangle_n)$
108   **return** Hash

Fig. 2. Definition of CIP.Hash. The inner product in lines 104 and 106 is in GF(2^n).

is parameterized by E and w. It takes the blockcipher key K, counter value $\mathbf{ctr}$, and an integer l as inputs, and the output is a bit string S of l blocks. See Fig. 9 for an illustration.

Discussion and default parameters. CIP takes five parameters, the blockcipher $E : \{0,1\}^k \times \{0,1\}^n \to \{0,1\}^n$, the nonce length ℓ_{nonce}, the tag length τ, and the frame widths ϖ and w. With these parameters, CIP can encrypt at most $2^{\ell_{\mathrm{nonce}}}$ plaintexts, and the maximum length of the plaintext is $2^{n-\ell_{\mathrm{nonce}}-2}$ blocks.

Our default parameters are, E is any blockcipher such that $n \geq 128$, $\ell_{\mathrm{nonce}} = n/2$, and $\tau \geq 32$. The values of ϖ and w affect the efficiency and security. Specifically, to hash ϖ blocks of input, CIP.Hash requires ϖ blocks of keys for inner product, ϖ multiplications and one blockcipher call. Thus, large ϖ implies that per message block computation is reduced, while it increases the pre-computation time and register for keys. Therefore there is a trade-off between security, per block efficiency and the pre-computation time/resister size. ϖ is supposed to be a small integer (e.g., $2 \leq \varpi \leq 8$), and our default recommendation is $\varpi = 4$. w follows the recommendation of CENC, and its default value is 256.

```
Algorithm CIP.KSGen_K(ctr, l)
100    for j ← 0 to ⌈l/w⌉ − 1 do
101        L ← E_K(ctr)
102        ctr ← inc(ctr)
103        for i ← 0 to w − 1 do
104            S_{wj+i} ← E_K(ctr) ⊕ L
105            ctr ← inc(ctr)
106            if wj + i = l − 1 then
107                S ← (S_0‖S_1‖ ··· ‖S_{l−1})
108                return S
```

Fig. 3. Definition of CIP.KSGen, which is equivalent to CENC [9]

With these parameters, if we use the AES, CIP can encrypt at most 2^{64} plaintexts, and the maximum length of the plaintext is 2^{62} blocks (2^{36} GBytes), and the security bounds are $\tilde{\sigma}^3/2^{245} + \tilde{\sigma}/2^{119}$ for privacy, and $\tilde{\sigma}^3/2^{245} + \tilde{\sigma}/2^{118} + 2/2^\tau$ for authenticity, where $\tilde{\sigma}$ is (roughly) the total number of blocks processed by one key. This implies $\tilde{\sigma}$ should be sufficiently smaller than 2^{81} blocks (2^{55} GBytes).

Information theoretic version. We will derive our security results in the information theoretic setting and in the computational setting. In the former case, a random permutation is used instead of a blockcipher, where we consider that CIP.Key takes a random permutation P as its input, and uses P to derive K_H and T_H by "encrypting" constants. Therefore, P is used in CIP.KSGen (lines 101 and 104 in Fig. 3), and (K_H, T_H) derived from P is used in CIP.Hash. We still use a real blockcipher in lines 105 and 107 in Fig. 2 even in the information theoretic version.

4   Security of CIP

CIP is an authenticated encryption (AE) scheme. We first present its security definitions, and then present our security results.

Security of blockciphers. We follow the PRP notion for blockciphers that was introduced in [18]. An adversary is a probabilistic algorithm with access to one or more oracles. Let A be an adversary with access to an oracle, either the encryption oracle $E_K(·)$ or a random permutation oracle $P(·)$, and returns a bit. We say A is a *PRP-adversary* for E, and define

$$\mathbf{Adv}_E^{\text{prp}}(A) \stackrel{\text{def}}{=} \left| \Pr(K \stackrel{R}{\leftarrow} \{0,1\}^k : A^{E_K(·)} = 1) - \Pr(P \stackrel{R}{\leftarrow} \text{Perm}(n) : A^{P(·)} = 1) \right|.$$

For an adversary A, A's running time is denoted by *time(A)*. The running time is its actual running time (relative to some fixed RAM model of computation) and its description size (relative to some standard encoding of algorithms). The details of the big-O notation for the running time reference depend on the RAM model and the choice of encoding.

Privacy of CIP. We follow the security notion from [6]. Let A be an adversary with access to an oracle, either the encryption oracle $\text{CIP.Enc}_K(\cdot, \cdot)$ or $\mathcal{R}(\cdot, \cdot)$, and returns a bit. The $\mathcal{R}(\cdot, \cdot)$ oracle, on input (N, M), returns a random string of length $|\text{CIP.Enc}_K(N, M)|$. We say that A is a PRIV-adversary for CIP. We assume that any PRIV-adversary is nonce-respecting. That is, if $(N_0, M_0), \ldots, (N_{q-1}, M_{q-1})$ are A's oracle queries, then $N_0, \ldots, N_{q-1}$ are always distinct, regardless of oracle responses and regardless of A's internal coins. The advantage of PRIV-adversary A for $\text{CIP} = (\text{CIP.Enc}, \text{CIP.Dec})$ is

$$\mathbf{Adv}^{\text{priv}}_{\text{CIP}}(A) \stackrel{\text{def}}{=} \left| \Pr(K \stackrel{R}{\leftarrow} \{0,1\}^k : A^{\text{CIP.Enc}_K(\cdot, \cdot)} = 1) - \Pr(A^{\mathcal{R}(\cdot, \cdot)} = 1) \right|.$$

Privacy results on CIP. Let A be a nonce-respecting PRIV-adversary for CIP, and assume that A makes at most q oracle queries, and the total plaintext length of these queries is at most σ blocks, i.e., if A makes exactly q queries $(N_0, M_0), \ldots, (N_{q-1}, M_{q-1})$, then $\sigma = \lceil |M_0|/n \rceil + \cdots + \lceil |M_{q-1}|/n \rceil$, the total number of blocks of plaintexts. We have the following information theoretic result.

Theorem 1. *Let $Perm(n)$, ℓ_{nonce}, τ, ϖ, and w be the parameters for CIP. Let A be a nonce-respecting PRIV-adversary making at most q oracle queries, and the total plaintext length of these queries is at most σ blocks. Then*

$$\mathbf{Adv}^{\text{priv}}_{\text{CIP}}(A) \leq \frac{wr^2\tilde{\sigma}^2}{2^{2n-4}} + \frac{w\tilde{\sigma}^3}{2^{2n-3}} + \frac{r^2}{2^{n+1}} + \frac{w\tilde{\sigma}}{2^n}, \tag{1}$$

where $r = \lceil k/n \rceil + \varpi$ and $\tilde{\sigma} = \sigma + q(w+1)$.

The proof of Theorem 1 is given in the next section. From Theorem 1, we have the following complexity theoretic result.

Corollary 1. *Let E, ℓ_{nonce}, τ, ϖ, and w be the parameters for CIP. Let A be a nonce-respecting PRIV-adversary making at most q oracle queries, and the total plaintext length of these queries is at most σ blocks. Then there is a PRP-adversary B for E making at most $2\tilde{\sigma}$ oracle queries, $\text{time}(B) = \text{time}(A) + O(n\tilde{\sigma})$, and $\mathbf{Adv}^{\text{prp}}_E(B) \geq \mathbf{Adv}^{\text{priv}}_{\text{CIP}}(A) - wr^2\tilde{\sigma}^2/2^{2n-4} - w\tilde{\sigma}^3/2^{2n-3} - r^2/2^{n+1} - w\tilde{\sigma}/2^n$, where $r = \lceil k/n \rceil + \varpi$ and $\tilde{\sigma} = \sigma + q(w+1)$.*

The proof is standard (e.g., see [9]), and omitted.

Authenticity of CIP. A notion of authenticity of ciphertext for AE schemes was formalized in [24,23] following [14,5,4]. Let A be an adversary with access to an encryption oracle $\text{CIP.Enc}_K(\cdot, \cdot)$ and returns a tuple, (N^*, C^*, Tag^*), called a forgery attempt. We say that A is an AUTH-adversary for CIP. We assume that any AUTH-adversary is nonce-respecting, where the condition applies only to the adversary's encryption oracle. Thus a nonce used in an encryption-oracle query may be used in a forgery attempt. We say A forges if A returns (N^*, C^*, Tag^*) such that $\text{CIP.Dec}_K(N^*, C^*, \text{Tag}^*) \not\rightarrow \perp$ but A did not make a query (N^*, M^*)

to $\text{CIP.Enc}_K(\cdot, \cdot)$ that resulted in a response (C^*, Tag^*). That is, adversary A may never return a forgery attempt (N^*, C^*, Tag^*) such that the encryption oracle previously returned (C^*, Tag^*) in response to a query (N^*, M^*). Then the advantage of AUTH-adversary A for $\text{CIP} = (\text{CIP.Enc}, \text{CIP.Dec})$ is

$$\mathbf{Adv}_{\text{CIP}}^{\text{auth}}(A) \stackrel{\text{def}}{=} \Pr(K \stackrel{R}{\leftarrow} \{0,1\}^k : A^{\text{CIP.Enc}_K(\cdot,\cdot)} \text{ forges}).$$

Authenticity results on CIP. Let A be an AUTH-adversary for CIP, and assume that A makes at most q oracle queries (including the final forgery attempt), and the total plaintext length of these queries is at most σ blocks. That is, if A makes queries $(N_0, M_0), \dots, (N_{q-2}, M_{q-2})$, and returns the forgery attempt (N^*, C^*, Tag^*), then $\sigma = \lceil |M_0|/n \rceil + \cdots + \lceil |M_{q-2}|/n \rceil + \lceil |C^*|/n \rceil$. We have the following information theoretic result.

Theorem 2. *Let $\text{Perm}(n)$, ℓ_{nonce}, τ, ϖ, and w be the parameters for CIP. Let A be a nonce-respecting AUTH-adversary making at most q oracle queries, and the total plaintext length of these queries is at most σ blocks. Then, for some D,*

$$\mathbf{Adv}_{\text{CIP}}^{\text{auth}}(A) \leq \frac{wr^2\tilde{\sigma}^2}{2^{2n-4}} + \frac{w\tilde{\sigma}^3}{2^{2n-3}} + \frac{r^2}{2^{n+1}} + \frac{w\tilde{\sigma}}{2^n} + \frac{\sigma}{2^{n-1}} + \frac{2}{2^\tau} + \mathbf{Adv}_E^{\text{prp}}(D) \quad (2)$$

where $r = \lceil k/n \rceil + \varpi$, $\tilde{\sigma} = \sigma + q(w+1)$, D makes at most 2σ queries, and $\text{time}(D) = O(n\sigma)$.

Note that the left hand side of (2) has $\mathbf{Adv}_E^{\text{prp}}(D)$, since we use a blockcipher in CIP.Hash, while there is no restriction on the running time of A.

The proof of Theorem 2 is given in Section 6. From Theorem 2, we have the following complexity theoretic result.

Corollary 2. *Let E, ℓ_{nonce}, τ, ϖ, and w be the parameters for CIP. Let A be a nonce-respecting AUTH-adversary making at most q oracle queries, and the total plaintext length of these queries is at most σ blocks. Then there is a PRP-adversary B for E making at most $2\tilde{\sigma}$ oracle queries, $\text{time}(B) = \text{time}(A) + O(n\tilde{\sigma})$, and $\mathbf{Adv}_E^{\text{prp}}(B) \geq \mathbf{Adv}_{\text{CIP}}^{\text{auth}}(A) - wr^2\tilde{\sigma}^2/2^{2n-4} - w\tilde{\sigma}^3/2^{2n-3} - r^2/2^{n+1} - w\tilde{\sigma}/2^n - \sigma/2^{n-1} - 2/2^\tau - \mathbf{Adv}_E^{\text{prp}}(n\sigma, 2\sigma)$, where $r = \lceil k/n \rceil + \varpi$, $\tilde{\sigma} = \sigma + q(w+1)$, and $\mathbf{Adv}_E^{\text{prp}}(n\sigma, 2\sigma)$ is the maximum of $\mathbf{Adv}_E^{\text{prp}}(D)$ over all D such that it makes at most 2σ queries, and $\text{time}(D) = O(n\sigma)$.*

The proof is standard (e.g., see [9]), and omitted.

5   Security Proof for Privacy of CIP

We first recall the following tool from [9]. Consider the function family F^+, which corresponds to one frame of CIP.KSGen, and it is defined as follows: Let $P \stackrel{R}{\leftarrow} \text{Perm}(n)$ be a random permutation, and fix the frame width w. Then $F^+ : \text{Perm}(n) \times \{0,1\}^n \rightarrow (\{0,1\}^n)^w$ is $F_P^+(x) = (y[0], \dots, y[w-1])$, where $y[i] = L \oplus P(\text{inc}^{i+1}(x))$ for $i = 0, \dots, w-1$ and $L = P(x)$.

Now let A be an adversary. This A is the PRF-adversary for F^+, but we give A additional information, i.e., we allow A to access the blockcipher itself. That is, A is given either a pair of oracles $(P(\cdot), F_P^+(\cdot))$, or a pair of random function oracles $(R_0(\cdot), R_1(\cdot))$, where $R_0 \in \mathrm{Func}(n, n)$ and $R_1 \in \mathrm{Func}(n, nw)$, with the following rules.

- If $W_i \in \{0,1\}^n$ is the i-th query for the first oracle (either $P(\cdot)$ or $R_0(\cdot)$), then $(\ell_{\mathrm{nonce}} + 1)$-th bit of W_i must be 1.
- If $x_j \in \{0,1\}^n$ is the j-th query for the second oracle (either $F_P^+(\cdot)$ or $R_1(\cdot)$), then $(\ell_{\mathrm{nonce}} + 1)$-th bit of x_j must be 0. That is, input/output samples from the first oracle are not used in $F_P^+(\cdot)$ oracle.
- A does not repeat the same query to its first oracle.
- Let $x_j \in \{0,1\}^n$ denote A's j-th query to its second oracle, and let $X_j = \{x_j, \mathrm{inc}(x_j), \mathrm{inc}^2(x_j), \ldots, \mathrm{inc}^w(x_j)\}$, i.e., X_j is the set of input to P in the j-th query. Now if A makes at most q calls to the second oracle, $X_j \cap X_{j'} = \emptyset$ must hold for any $0 \le j < j' \le q - 1$, regardless of oracle responses and regardless of A's internal coins.

Define $\mathbf{Adv}^{\mathrm{prf}}_{\mathrm{Perm}(n),F^+}(A)$ as

$$\left| \Pr(P \xleftarrow{R} \mathrm{Perm}(n) : A^{P(\cdot),F_P^+(\cdot)} = 1) \right.$$
$$\left. - \Pr(R_0 \xleftarrow{R} \mathrm{Func}(n, n), R_1 \xleftarrow{R} \mathrm{Func}(n, nw) : A^{R_0(\cdot),R_1(\cdot)} = 1) \right|$$

and we say A is a PRF-adversary for $(\mathrm{Perm}(n), F^+)$.

We have the following information theoretic result, whose proof is almost the same as that of [9, Theorem 5].

Proposition 1. *Let $\mathrm{Perm}(n)$ and w be the parameters for F^+. Let A be the PRF-adversary for $(\mathrm{Perm}(n), F^+)$, with the above restrictions, making at most r oracle queries to its first oracle and at most q oracle queries to its second oracle. Then*

$$\mathbf{Adv}^{\mathrm{prf}}_{\mathrm{Perm}(n),F^+}(A) \le \frac{r^2 q^2 (w+1)^3}{2^{2n-1}} + \frac{q^3 (w+1)^4}{2^{2n+1}} + \frac{r(r-1)}{2^{n+1}} + \frac{qw(w+1)}{2^{n+1}}.$$

Now Theorem 1 follows by using Proposition 1. To see this, by using the PRIV-adversary A for CIP as a subroutine, it is possible to construct a PRF-adversary B for $(\mathrm{Perm}(n), F^+)$. B first makes $\lceil k/n \rceil + \varpi$ calls to its first oracle and constructs K_H and T_H, and simulates line 103 of Fig. 1 as in Fig. 4 by making $\tilde{\sigma}/w$ calls to the second oracle.

6   Security Proofs for Authenticity of CIP

6.1   Properties of the Inner Product Hash

We first recall that the inner product hash is ϵ-AXU for small ϵ [26].

```
Algorithm CIP.KSGen.Sim(ctr, l)
100    for j ← 0 to ⌈l/w⌉ − 1 do
101        S_j ← F_P^+(ctr)
102        ctr ← inc^{w+1}(ctr)
103    S ← (S_0, ..., S_{⌈l/w⌉−1})
104    S ← first(n × l, S)
105    return S
```

Fig. 4. The simulation CIP.KSGen.Sim of CIP.KSGen using F^+

Proposition 2. *Let* $(x_0, \ldots, x_{\varpi-1}), (x'_0, \ldots, x'_{\varpi-1}) \in (\{0,1\}^n)^\varpi$ *be two distinct bit strings. Then for any* $y \in \{0,1\}^n$,

$$\Pr(T_H \overset{R}{\leftarrow} (\{0,1\}^n)^\varpi : (x_0, \ldots, x_{\varpi-1}) \cdot (T_0, \ldots, T_{\varpi-1})$$
$$\oplus (x'_0, \ldots, x'_{\varpi-1}) \cdot (T_0, \ldots, T_{\varpi-1}) = y) = 1/2^n.$$

Proof. We have $x_i \oplus x'_i \neq 0^n$ for some i. Therefore, the coefficient of T_i in $(x_0 \oplus x'_0) \cdot T_0 \oplus \cdots \oplus (x_{\varpi-1} \oplus x'_{\varpi-1}) \cdot T_{\varpi-1} = y$ is non-zero, and for any fixed $T_0, \ldots, T_{i-1}, T_{i+1}, \ldots, T_{\varpi-1}$, exactly one value of T_i satisfies the equality. □

If $y \neq 0^n$, a similar result holds for two bit strings of different block sizes.

Proposition 3. *Let* $\varpi \geq \varpi'$, *and let* $x = (x_0, \ldots, x_{\varpi-1}) \in (\{0,1\}^n)^\varpi$ *and* $x' = (x'_0, \ldots, x'_{\varpi'-1}) \in (\{0,1\}^n)^{\varpi'}$ *be two distinct bit strings. Then for any non-zero* $y \in \{0,1\}^n$,

$$\Pr(T_H \overset{R}{\leftarrow} (\{0,1\}^n)^\varpi : (x_0, \ldots, x_{\varpi-1}) \cdot (T_0, \ldots, T_{\varpi-1})$$
$$\oplus (x'_0, \ldots, x'_{\varpi'-1}) \cdot (T_0, \ldots, T_{\varpi'-1}) = y) = 1/2^n.$$

Proof. The condition can be written as: $(x_0 \oplus x'_0) \cdot T_0 \oplus \cdots \oplus (x_{\varpi'-1} \oplus x'_{\varpi'-1}) \cdot T_{\varpi'-1} \oplus x_{\varpi'} \cdot T_{\varpi'} \oplus \cdots \oplus x_{\varpi-1} \cdot T_{\varpi-1} = y$. If all the coefficients of T_i are zero, then this equation can not be true since y is non-zero. Therefore, we can without loss of generality assume that at least one of coefficients of T_i is non-zero. □

6.2   Properties of the CIP.Hash

We next analyze the properties of CIP.Hash.

Let $x, x' \in \{0,1\}^*$ be two distinct bit strings, where $|x| \geq |x'|$. We show (in Proposition 8) that for any y, $\Pr(\text{CIP.Hash}_{K_H, T_H}(x) \oplus \text{CIP.Hash}_{K_H, T_H}(x') = y)$ is small, where the probability is taken over the choices of K_H and T_H.

We begin by introducing the notation. Let $X \leftarrow x \| 10^{n-1-(|x| \bmod n)}$ and $X' \leftarrow x' \| 10^{n-1-(|x'| \bmod n)}$. We parse them into blocks as $X = (X_0, \ldots, X_{l-1})$ and $X' = (X'_0, \ldots, X'_{l'-1})$, where $l = |X|/n$ and $l' = |X'|/n$. Let $\ell = \lceil l/\varpi \rceil$ and $\ell' = \lceil l'/\varpi \rceil$. We write the i-th frame of X and X' as χ_i and χ'_i, respectively. That is, $\chi_i = (X_{i\varpi}, \ldots, X_{(i+1)\varpi-1})$ for $0 \leq i \leq \ell - 2$, $\chi_{\ell-1} = (X_{(\ell-1)\varpi}, \ldots, X_{l-1})$, $\chi'_i = (X'_{i\varpi}, \ldots, X'_{(i+1)\varpi-1})$ for $0 \leq i \leq \ell' - 2$, and $\chi'_{\ell'-1} = (X'_{(\ell'-1)\varpi}, \ldots, X'_{l'-1})$.

Further, let $\chi_i \cdot T_H = A_i$ for $0 \leq i \leq \ell - 2$, $\chi_{\ell-1} \cdot (T_0, \ldots, T_{l-(\ell-1)\varpi-1}) = A_{\ell-1}$, $\chi'_i \cdot T_H = A'_i$ for $0 \leq i \leq \ell' - 1$, and $\chi'_{\ell'-1} \cdot (T_0, \ldots, T_{l'-(\ell'-1)\varpi-1}) = A_{\ell'-1}$. That is, A_i and A'_i are the results of the inner product in lines 104 and 106 of Fig. 2.

In the following three propositions, we first show that, for some i, $A_i \oplus \langle i \rangle_n$ is unique with high probability in the multi-set $\{A_0 \oplus \langle 0 \rangle_n, \ldots, A_{\ell-1} \oplus \langle \ell-1 \rangle_n, A'_0 \oplus \langle 0 \rangle_n, \ldots, A'_{\ell'-1} \oplus \langle \ell'-1 \rangle_n\}$.

Proposition 4. *Suppose that $l = l'$. Then there are at least $2^{n\varpi}(1 - (2\ell-1)/2^n)$ choices of $T_H \in (\{0,1\}^n)^\varpi$ such that the following is true: for some $0 \leq i \leq \ell-1$,*

$$A_i \oplus \langle i \rangle_n \neq A_j \oplus \langle j \rangle_n \text{ for all } j \in \{0, \ldots, i-1, i+1, \ldots, \ell-1\}, \text{ and} \quad (3)$$

$$A_i \oplus \langle i \rangle_n \neq A'_j \oplus \langle j \rangle_n \text{ for all } j \in \{0, \ldots, \ell-1\}. \quad (4)$$

Proof. Since $|X| = |X'|$ and $X \neq X'$, we have $\chi_i \neq \chi'_i$ for some i. We show the proof in three cases, (a) $|\chi_{\ell-1}|_n = \varpi$, (b) $|\chi_{\ell-1}|_n < \varpi$ and $0 \leq i < \ell-1$, and (c) $|\chi_{\ell-1}|_n < \varpi$ and $i = \ell-1$.

We first consider case (a). For any fixed $j \in \{0, \ldots, i-1, i+1, \ldots, \ell-1\}$, the number of T_H that satisfies $A_i \oplus \langle i \rangle_n = A_j \oplus \langle j \rangle_n$ is at most $2^{n\varpi}/2^n$ from Proposition 2. Note that, if $\chi_i = \chi_j$, then there is no T_H that satisfies this condition since $\langle i \rangle_n \oplus \langle j \rangle_n \neq 0^n$. Therefore, we have at most $(\ell-1)2^{n\varpi}/2^n$ values of T_H such that $A_i \oplus \langle i \rangle_n = A_j \oplus \langle j \rangle_n$ holds for some $j \in \{0, \ldots, i-1, i+1, \ldots, \ell-1\}$.

Similarly, the number of T_H which satisfies $A_i \oplus \langle i \rangle_n = A'_j \oplus \langle j \rangle_n$ for some $j \in \{0, \ldots, \ell-1\}$ is at most $\ell 2^{n\varpi}/2^n$. This follows by using Proposition 2 for $j \neq i$, and for $j = i$, we use Proposition 2 and the fact that $\chi_i \neq \chi'_i$.

Therefore, we have at least $2^{n\varpi} - (2\ell-1)2^{n\varpi}/2^n = 2^{n\varpi}(1 - (2\ell-1)/2^n)$ choices of $T_H \in (\{0,1\}^n)^\varpi$ which satisfies (3) and (4).

We next consider case (b). From Proposition 2, we have at most $(2\ell-3)2^{n\varpi}/2^n$ values of T_H such that $A_i \oplus \langle i \rangle_n = A_j \oplus \langle j \rangle_n$ for some $j \in \{0, \ldots, i-1, i+1, \ldots, \ell-2\}$, or $A_i \oplus \langle i \rangle_n = A'_j \oplus \langle j \rangle_n$ for some $j \in \{0, \ldots, \ell-2\}$.

From Proposition 3, we have at most $2 \times 2^{n\varpi}/2^n$ values of T_H such that $A_i \oplus \langle i \rangle_n = A_{\ell-1} \oplus \langle \ell-1 \rangle_n$, or $A_i \oplus \langle i \rangle_n = A'_{\ell-1} \oplus \langle \ell-1 \rangle_n$. Note that $\langle i \rangle_n \oplus \langle \ell-1 \rangle_n \neq 0^n$.

Finally, we consider case (c). From Proposition 3, we have at most $(2\ell-2)2^{n\varpi}/2^n$ values of T_H such that $A_{\ell-1} \oplus \langle \ell-1 \rangle_n = A_j \oplus \langle j \rangle_n$ for some $j \in \{0, \ldots, \ell-2\}$, or $A_{\ell-1} \oplus \langle \ell-1 \rangle_n = A'_j \oplus \langle j \rangle_n$ for some $j \in \{0, \ldots, \ell-2\}$.

From Proposition 3 and since $\chi_{\ell-1} \neq \chi'_{\ell-1}$, we have at most $2^{n\varpi}/2^n$ values of T_H such that $A_{\ell-1} \oplus \langle \ell-1 \rangle_n = A'_{\ell-1} \oplus \langle \ell-1 \rangle_n$. □

Proposition 5. *Suppose that $l > l'$ and $\ell > \ell'$. Then there are at least $2^{n\varpi}(1 - (\ell + \ell' - 1)/2^n)$ choices of $T_H \in (\{0,1\}^n)^\varpi$ such that the following is true:*

$$A_{\ell-1} \oplus \langle \ell-1 \rangle_n \neq A_j \oplus \langle j \rangle_n \text{ for all } j \in \{0, \ldots, \ell-2\}, \text{ and} \quad (5)$$

$$A_{\ell-1} \oplus \langle \ell-1 \rangle_n \neq A'_j \oplus \langle j \rangle_n \text{ for all } j \in \{0, \ldots, \ell'-1\}. \quad (6)$$

Proof. The number is at most $2^{n\varpi}(1 - (\ell + \ell' - 1)/2^n)$, since if $|\chi_{\ell-1}|_n = \varpi$, the bound follows by using Proposition 2 for each j, and if $|\chi_{\ell-1}|_n < \varpi$, it follows from Proposition 3 and the fact that $\langle \ell-1 \rangle_n \oplus \langle j \rangle_n \neq 0^n$. □

Proposition 6. *Suppose that $l > l'$ and $\ell = \ell'$. Then there are at least $2^{n\varpi}(1 - (\ell + \ell' - 1)/2^n)$ choices of $T_H \in (\{0,1\}^n)^\varpi$ which satisfies both (5) and (6).*

Proof. The bound follows by the same argument as in the proof of Proposition 5. The exception is the event $A_{\ell-1} \oplus \langle \ell-1 \rangle_n \neq A'_{\ell-1} \oplus \langle \ell-1 \rangle_n$, which is equivalent to $A_{\ell-1} \neq A'_{\ell-1}$. In this case, we are interested in the equation $(x_{(\ell-1)\varpi}, \ldots, x_{l-1}) \cdot (T_0, \ldots, T_{l-(\ell-1)\varpi-1}) = (x'_{(\ell-1)\varpi}, \ldots, x'_{l'-1}) \cdot (T_0, \ldots, T_{l'-(\ell-1)\varpi-1})$. We see that the coefficient of $T_{l-(\ell-1)\varpi-1}$ is non-zero (because of padding). Therefore, exactly one value of $T_{l-(\ell-1)\varpi-1}$ satisfies the equality. $\qquad\square$

We now consider CIP.Hash that uses a random permutation instead of a block-cipher. Thus, instead of $K_H \xleftarrow{R} \{0,1\}^k$, we let $P \xleftarrow{R} \mathrm{Perm}(n)$, and write $\mathrm{CIP.Hash}_{P,T_H}(\cdot)$ instead of $\mathrm{CIP.Hash}_{K_H,T_H}(\cdot)$. Besides, we consider CIP.Hash, where its output bits are truncated to τ bits. The next result proves that this truncated version of CIP.Hash is ϵ-AXU for small ϵ.

Proposition 7. *Let x and x' be two distinct bit strings, where $\ell + \ell' - 1 \leq 2^{n-1}$. For any $1 \leq \tau \leq n$ and any $y \in \{0,1\}^\tau$,*

$$\Pr(P \xleftarrow{R} \mathrm{Perm}(n), T_H \xleftarrow{R} (\{0,1\}^n)^\varpi :$$
$$\mathrm{first}(\tau, \mathrm{CIP.Hash}_{P,T_H}(x) \oplus \mathrm{CIP.Hash}_{P,T_H}(x')) = y) \leq \frac{\ell + \ell' - 1}{2^n} + \frac{2}{2^\tau}.$$

Proof. We first choose and fix any T_H. If there is no i such that $A_i \oplus \langle i \rangle_n$ is unique in the multi-set $\{A_0 \oplus \langle 0 \rangle_n, \ldots, A_{\ell-1} \oplus \langle \ell-1 \rangle_n, A'_0 \oplus \langle 0 \rangle_n, \ldots, A'_{\ell'-1} \oplus \langle \ell' - 1 \rangle_n\}$, then we give up the analysis and regard this as $\mathrm{CIP.Hash}_{P,T_H}(x) \oplus \mathrm{CIP.Hash}_{P,T_H}(x)) = y$ occurs. The probability is at most $(\ell + \ell' - 1)/2^n$ from Proposition 5, 6, and 7, and the first term follows.

Next, we assume for some i, $A_i \oplus \langle i \rangle_n$ is unique in the multi-set. Now since we have fixed T_H, all the inputs to P are now fixed. We next fix the outputs of P except for $A_i \oplus \langle i \rangle_n$. At most $(\ell + \ell' - l)$ input-output pairs are now fixed, and therefore, we have at least $2^n - (\ell + \ell' - l)$ choices for the output of $A_i \oplus \langle i \rangle_n$. Out of these $2^n - (\ell + \ell' - l)$ possible choices, at most $2^{n-\tau}$ values verify $\mathrm{first}(\tau, \mathrm{CIP.Hash}_{K_H,T_H}(x) \oplus \mathrm{CIP.Hash}_{K_H,T_H}(x)) = y$ since the unused $(n - \tau)$ bits may take any value. The probability of this event is at most $2^{n-\tau}/(2^n - (\ell + \ell' - l)) \leq 2/2^\tau$, and the second term follows. $\qquad\square$

We now derive the result with a blockcipher E.

Proposition 8. *Let x and x' be two distinct bit strings, where $\ell + \ell' - 1 \leq 2^{n-1}$. For any $1 \leq \tau \leq n$ and any $y \in \{0,1\}^\tau$, there exists a PRP-adversary A for E such that*

$$\Pr(K_H, T_H \xleftarrow{R} \{0,1\}^k \times (\{0,1\}^n)^\varpi : \mathrm{first}(\tau, \mathrm{CIP.Hash}_{K_H,T_H}(x)$$
$$\oplus \mathrm{CIP.Hash}_{K_H,T_H}(x')) = y) \leq \frac{\ell + \ell' - 1}{2^n} + \frac{2}{2^\tau} + \mathbf{Adv}_E^{\mathrm{prp}}(A),$$

where A makes at most $\ell + \ell'$ queries, and $\mathrm{time}(A) = O(n(\ell + \ell'))$.

Algorithm CIP.Sim1	**Algorithm** CIP.Sim1 (Cont.)
Setup:	**If** A **returns** (N^*, C^*, Tag^*):
100 $\quad (K_H, T_H) \xleftarrow{R} \text{CIP.Key}(R_0)$	300 $\quad l \leftarrow \lceil |C^*|/n \rceil$
If A **makes a query** (N_i, M_i):	301 $\quad \text{ctr} \leftarrow (N^* \| 0^{n-\ell_{\text{nonce}}})$
200 $\quad l \leftarrow \lceil |M_i|/n \rceil$	302 $\quad S \leftarrow \text{CIP.KSGen.Sim1}(\text{ctr}, l+1)$
201 $\quad \text{ctr} \leftarrow (N_i \| 0^{n-\ell_{\text{nonce}}})$	303 $\quad S_H \leftarrow \text{first}(n, S)$
202 $\quad S \leftarrow \text{CIP.KSGen.Sim1}(\text{ctr}, l+1)$	304 $\quad \text{Hash}' \leftarrow \text{CIP.Hash}_{K_H, T_H}(C^*)$
203 $\quad S_H \leftarrow \text{first}(n, S)$	305 $\quad \text{Tag}' \leftarrow \text{first}(\tau, \text{Hash}' \oplus S_H)$
204 $\quad S_{\text{mask}} \leftarrow \text{last}(n \times l, S)$	306 $\quad$ **if** $\text{Tag}' \neq \text{Tag}^*$ **then return** $\bot$
205 $\quad C_i \leftarrow M_i \oplus \text{first}(|M_i|, S_{\text{mask}})$	307 $\quad S_{\text{mask}} \leftarrow \text{last}(n \times l, S)$
206 $\quad \text{Hash}_i \leftarrow \text{CIP.Hash}_{K_H, T_H}(C_i)$	308 $\quad M^* \leftarrow C^* \oplus \text{first}(|C^*|, S_{\text{mask}})$
207 $\quad \text{Tag}_i \leftarrow \text{first}(\tau, \text{Hash}_i \oplus S_H)$	309 $\quad$ **return** M^*
208 $\quad$ **return** (C_i, Tag_i)	

Fig. 5. The simulation CIP.Sim1 of CIP. CIP.Hash is defined in Fig. 2.

Proof. Fix x, x' and y, and consider the following A: First, A randomly chooses $T_H \xleftarrow{R} (\{0,1\}^n)^\varpi$. Then A computes the hash values of x and x' following Fig. 2, except that, in lines 105 and 107, blockcipher invocations are replaced with oracle calls. The output of A is 1 iff the xor of their hash values is y. We see that A makes at most $\ell + \ell'$ queries, and

$$\Big| \Pr\Big(\text{first}(\tau, \text{CIP.Hash}_{P, T_H}(x) \oplus \text{CIP.Hash}_{P, T_H}(x')) = y \Big)$$
$$- \Pr\Big(\text{first}(\tau, \text{CIP.Hash}_{K_H, T_H}(x) \oplus \text{CIP.Hash}_{K_H, T_H}(x')) = y \Big) \Big|$$

is upper bounded by $\mathbf{Adv}_E^{\text{prp}}(A)$. $\qquad\square$

6.3 Proof of Theorem 2

We now present the proof of Theorem 2.

Proof (of Theorem 2). First, consider the simulation CIP.Sim1 in Fig. 5 of CIP, where K_H and T_H are generated by using CIP.Key(R_0), i.e., a random function $R_0 \in \text{Func}(n, n)$ is used to encrypt constants, and the keystream generation, CIP.KSGen.Sim1, works as follows: it is exactly the same as Fig. 4, except that it uses a random function $R_1 \in \text{Func}(n, nw)$ instead of F_P^+.

Let $\mathbf{Adv}_{\text{CIP.Sim1}}^{\text{auth}}(A)$ be the success probability of A's forgery, where the oracle is CIP.Sim1, i.e.,

$$\mathbf{Adv}_{\text{CIP.Sim1}}^{\text{auth}}(A) \stackrel{\text{def}}{=} \Pr(A^{\text{CIP.Sim1}} \text{ forges}),$$

where the probability is taken over the random coins in lines 100, 202, 302 and A's internal coins. We claim that

$$\Big| \mathbf{Adv}_{\text{CIP}}^{\text{auth}}(A) - \mathbf{Adv}_{\text{CIP.Sim1}}^{\text{auth}}(A) \Big| \tag{7}$$

$$\leq \frac{wr^2 \tilde{\sigma}^2}{2^{2n-4}} + \frac{w \tilde{\sigma}^3}{2^{2n-3}} + \frac{r^2}{2^{n+1}} + \frac{w \tilde{\sigma}}{2^n}. \tag{8}$$

To see this, suppose for a contradiction that (7) is larger than (8). Then, by using A as a subroutine, it is possible to construct a PRF-adversary B for $(\mathrm{Perm}(n), F^+)$ making at most r oracle queries to its first oracle and at most $\tilde{\sigma}/w$ oracle queries to its second oracle, where B simulates R_0 and R_1 in Fig. 5 by using its own oracles, and returns 1 if and only if A succeeds in forgery. This implies $\Pr(P \xleftarrow{R} \mathrm{Perm}(n) : B^{P(\cdot), F_P^+(\cdot)} = 1) = \mathbf{Adv}_{\mathrm{CIP}}^{\mathrm{auth}}(A)$ and $\Pr(R_0 \xleftarrow{R} \mathrm{Func}(n, n), R_1 \xleftarrow{R} \mathrm{Func}(n, nw) : B^{R_0(\cdot), R_1(\cdot)} = 1) = \mathbf{Adv}_{\mathrm{CIP.Sim1}}^{\mathrm{auth}}(A)$ and thus, $\mathbf{Adv}_{\mathrm{Perm}(n), F^+}^{\mathrm{prf}}(B)$ is larger than (8), which contradicts Proposition 1.

Now we modify CIP.Sim1 to CIP.Sim2 in Fig. 6.

1. Instead of using CIP.Key in line 100 in Fig. 5, we directly choose (K_H, T_H) randomly.
2. Instead of using CIP.KSGen.Sim1 in line 202 in Fig. 5, we choose an $(l+1)$-block random string. Therefore, we have $S \xleftarrow{R} \{0,1\}^{n(l+1)}$ in line 201 of Fig. 6. Also, we removed "$\texttt{ctr} \leftarrow (N_i \| 0^{n-\ell_{\mathrm{nonce}}})$" in line 201 of Fig. 5 because we do not need it.
3. We need a different treatment for a forgery attempt, since we allow the same nonce, i.e., $N^* \in \{N_0, \ldots, N_{q-2}\}$. We make two cases, case $N^* \notin \{N_0, \ldots, N_{q-2}\}$ and case $N^* = N_i$. In the former case, we simply choose a new random S_H in line 301 of Fig. 6. In the latter case, S_H for (N_i, M_i) has to be the same S_H for $(N^*, C^*, \mathrm{Tag}^*)$. Observe that $S_H = \mathrm{Hash}_i \oplus \mathrm{Tag}_i$, and thus, the simulation in line 306 of Fig. 6 is precise. Therefore, the simulation makes no difference in the advantage of A.
4. When A makes a query (N_i, M_i), we return the full n-bit tag, $\mathrm{Tag}_i \in \{0,1\}^n$, instead of a truncated one, while we allow τ-bit tag in the forgery attempt. This only increases the advantage of A.
5. If $\mathrm{Tag}' = \mathrm{Tag}^*$, we return $M^* = C^* \oplus \mathrm{first}(|C^*|, S_{\mathrm{mask}})$. Since the value of M^* has no effect on the advantage (as long as it is not the special symbol $\perp$), we let $M^* \leftarrow 0^{|C^*|}$. This makes no difference in the advantage of A.

Let $\mathbf{Adv}_{\mathrm{CIP.Sim2}}^{\mathrm{auth}}(A) \overset{\mathrm{def}}{=} \Pr(A^{\mathrm{CIP.Sim2}} \text{ forges})$, where the probability is taken over the random coins in lines 100, 201, 301 and A's internal coins. From the above discussion, we have

$$\mathbf{Adv}_{\mathrm{CIP.Sim1}}^{\mathrm{auth}}(A) \le \mathbf{Adv}_{\mathrm{CIP.Sim2}}^{\mathrm{auth}}(A). \tag{9}$$

Now we further modify CIP.Sim2 to CIP.Sim3 in Fig. 7.

1. We do not choose K_H and T_H until we need them (we need them after the forgery attempt).
2. Since C_i is the xor of M_i and a random string of length $|M_i|$, we let $C_i \xleftarrow{R} \{0,1\}^{|M_i|}$. The distribution of C_i is unchanged, and thus, this makes no difference in the advantage of A.
3. Similarly, since Tag_i includes S_H, which is a truly random string, we let $\mathrm{Tag}_i \xleftarrow{R} \{0,1\}^n$. The distribution of Tag_i is unchanged, and thus, this makes no difference in the advantage of A (Observe that we do not need K_H and T_H, and we can postpone the selection without changing the distribution of C_i and Tag_i).

Algorithm CIP.Sim2	**Algorithm** CIP.Sim2 (Cont.)
Setup:	**If** A **returns** (N^*, C^*, Tag^*):
100 $K_H \xleftarrow{R} \{0,1\}^k$; $T_H \xleftarrow{R} (\{0,1\}^n)^\varpi$	300 **if** $N^* \notin \{N_0, \dots, N_{q-2}\}$ **then**
If A **makes a query** (N_i, M_i):	301 $S_H \xleftarrow{R} \{0,1\}^n$
200 $l \leftarrow \lceil |M_i|/n \rceil$	302 $\text{Hash}' \leftarrow \text{CIP.Hash}_{K_H, T_H}(C^*)$
201 $S \xleftarrow{R} \{0,1\}^{n(l+1)}$	303 $\text{Tag}' \leftarrow \text{first}(\tau, \text{Hash}' \oplus S_H)$
202 $S_H \leftarrow \text{first}(n, S)$	304 **if** $N^* = N_i$ **then**
203 $S_{\text{mask}} \leftarrow \text{last}(n \times l, S)$	305 $\text{Hash}' \leftarrow \text{CIP.Hash}_{K_H, T_H}(C^*)$
204 $C_i \leftarrow M_i \oplus \text{first}(|M_i|, S_{\text{mask}})$	306 $\text{Tag}' \leftarrow \text{Hash}' \oplus \text{Hash}_i \oplus \text{Tag}_i$
205 $\text{Hash}_i \leftarrow \text{CIP.Hash}_{K_H, T_H}(C_i)$	307 $\text{Tag}' \leftarrow \text{first}(\tau, \text{Tag}')$
206 $\text{Tag}_i \leftarrow \text{Hash}_i \oplus S_H$	308 **if** $\text{Tag}' \neq \text{Tag}^*$ **then return** $\bot$
207 **return** (C_i, Tag_i)	309 $M^* \leftarrow 0^{|C^*|}$
	310 **return** M^*

Fig. 6. The simulation CIP.Sim2 of CIP

4. If $N^* \in \{N_0, \dots, N_{q-2}\}$, Tag' includes the random S_H, and we let $\text{Tag}' \xleftarrow{R} \{0,1\}^\tau$. The distribution of Tag' is unchanged.
5. If $N^* = N_i$, we need K_H and T_H. We choose them, and the rest is unchanged.

Since the distribution of (C_i, Tag_i) is unchanged, and there is no difference in the advantage of A, we have

$$\mathbf{Adv}^{\text{auth}}_{\text{CIP.Sim2}}(A) = \mathbf{Adv}^{\text{auth}}_{\text{CIP.Sim3}}(A), \tag{10}$$

where $\mathbf{Adv}^{\text{auth}}_{\text{CIP.Sim3}}(A) \overset{\text{def}}{=} \Pr(A^{\text{CIP.Sim3}} \text{ forges})$ and the probability is taken over the random coins in lines 100, 101, 201, 203 and A's internal coins.

We now fix A's internal coins and coins in lines 100 and 101. Then, the query-answer pairs $(N_0, M_0, C_0, \text{Tag}_0), \dots, (N_{q-2}, M_{q-2}, C_{q-2}, \text{Tag}_{q-2})$ and the forgery attempt (N^*, C^*, Tag^*) are all fixed, and we evaluate $\mathbf{Adv}^{\text{auth}}_{\text{CIP.Sim3}}(A)$ with the coins in lines 201, and 203 only. We evaluate it in the following two cases (Note that we are choosing K_H and T_H *after* fixing $N_i, C_i, \text{Tag}_i, N^*, C^*, \text{Tag}^*$).

- **Case** $N^* \notin \{N_0, \dots, N_{q-2}\}$: In this case, $\mathbf{Adv}^{\text{auth}}_{\text{CIP.Sim3}}(A) = 1/2^\tau$ since for any fixed Tag^*, $\Pr(\text{Tag}' \xleftarrow{R} \{0,1\}^\tau : \text{Tag}' = \text{Tag}^*) = 1/2^\tau$.
- **Case** $N^* = N_i$ and $C^* \neq C_i$: In this case, we have

$$\mathbf{Adv}^{\text{auth}}_{\text{CIP.Sim3}}(A) \leq \Pr(K_H \xleftarrow{R} \{0,1\}^k, T_H \xleftarrow{R} (\{0,1\}^n)^\varpi :$$
$$\text{first}(\tau, \text{CIP.Hash}_{K_H, T_H}(C^*) \oplus \text{CIP.Hash}_{K_H, T_H}(C_i)) = y),$$

where $y = \text{Tag}^* \oplus \text{Tag}_i$. This is at most $(\lceil |C^*|/n \rceil + \lceil |C_i|/n \rceil - 1)/2^n + 2/2^\tau + \mathbf{Adv}^{\text{prp}}_E(D)$ from Proposition 8, and this is upper bounded by $2\sigma/2^n + 2/2^\tau + \mathbf{Adv}^{\text{prp}}_E(D)$, where D makes at most 2σ queries, and $time(D) = O(n\sigma)$.

Therefore, we have

$$\mathbf{Adv}^{\text{auth}}_{\text{CIP.Sim3}}(A) \leq \frac{\sigma}{2^{n-1}} + \frac{2}{2^\tau} + \mathbf{Adv}^{\text{prp}}_E(D). \tag{11}$$

Finally, from (7), (9), (10), and (11), we have (2). $\square$

Algorithm CIP.Sim3

If A **makes a query** (N_i, M_i):

100    $C_i \stackrel{R}{\leftarrow} \{0,1\}^{|M_i|}$

101    $\text{Tag}_i \stackrel{R}{\leftarrow} \{0,1\}^n$

102   **return** (C_i, Tag_i)

Algorithm CIP.Sim3 (Cont.)

If A **returns** (N^*, C^*, Tag^*):

200   **if** $N^* \notin \{N_0, \ldots, N_{q-2}\}$ **then**

201      $\text{Tag}' \stackrel{R}{\leftarrow} \{0,1\}^\tau$

202   **if** $N^* = N_i$ **then**

203      $K_H \stackrel{R}{\leftarrow} \{0,1\}^k$; $T_H \stackrel{R}{\leftarrow} (\{0,1\}^n)^\varpi$

204      $\text{Hash}_i \leftarrow \text{CIP.Hash}_{K_H, T_H}(C_i)$

205      $S_H \leftarrow \text{Hash}_i \oplus \text{Tag}_i$

206      $\text{Hash}' \leftarrow \text{CIP.Hash}_{K_H, T_H}(C^*)$

207      $\text{Tag}' \leftarrow \text{Hash}' \oplus S_H$

208      $\text{Tag}' \leftarrow \text{first}(\tau, \text{Tag}')$

209   **if** $\text{Tag}' \neq \text{Tag}^*$ **then return** $\perp$

210    $M^* \leftarrow 0^{|C^*|}$

211   **return** M^*

Fig. 7. The simulation CIP.Sim3 of CIP

7 Conclusions

We presented an authenticated encryption mode CIP, CENC with Inner Product hash. It has provable security bounds which are better than the usual birthday bound security, and it can be used even when the tag length is short. Our proof is relatively complex, and it would be interesting to see the compact security proofs, possibly by following the "all-in-one" security definition in [25]. It would also be interesting to see schemes with improved security bound and/or efficiency.

Acknowledgement

The author would like to thank anonymous reviewers of Africacrypt 2008 for many insightful and useful comments.

References

1. Bellare, M., Desai, A., Jokipii, E., Rogaway, P.: A concrete security treatment of symmetric encryption. In: Proceedings of The 38th Annual Symposium on Foundations of Computer Science, FOCS 1997, pp. 394–405. IEEE, Los Alamitos (1997)
2. Bellare, M., Guerin, R., Rogaway, P.: XOR MACs: New methods for message authentication using finite pseudorandom functions. In: Coppersmith, D. (ed.) CRYPTO 1995. LNCS, vol. 963, pp. 15–28. Springer, Heidelberg (1995)
3. Bellare, M., Kilian, J., Rogaway, P.: The security of the cipher block chaining message authentication code. JCSS, 61(3), 362–399 (2000); Desmedt, Y.G. (ed.) CRYPTO 1994. LNCS, vol. 839, pp. 341–358. Springer, Heidelberg (1994)
4. Bellare, M., Namprempre, C.: Authenticated encryption: Relations among notions and analysis of the generic composition paradigm. In: Okamoto, T. (ed.) ASIACRYPT 2000. LNCS, vol. 1976, pp. 531–545. Springer, Heidelberg (2000)

5. Bellare, M., Rogaway, P.: Encode-then-encipher encryption: How to exploit nonces or redundancy in plaintexts for efficient cryptography. In: Okamoto, T. (ed.) ASIACRYPT 2000. LNCS, vol. 1976, pp. 317–330. Springer, Heidelberg (2000)

6. Bellare, M., Rogaway, P., Wagner, D.: The EAX mode of operation. In: Roy, B., Meier, W. (eds.) FSE 2004. LNCS, vol. 3017, pp. 389–407. Springer, Heidelberg (2004)

7. Black, J., Rogaway, P.: A block-cipher mode of operation for parallelizable message authentication. In: Knudsen, L.R. (ed.) EUROCRYPT 2002. LNCS, vol. 2332, pp. 384–397. Springer, Heidelberg (2002)

8. Gligor, V.G., Donescu, P.: Fast encryption and authentication: XCBC encryption and XECB authentication modes. In: Matsui, M. (ed.) FSE 2001. LNCS, vol. 2355, pp. 92–108. Springer, Heidelberg (2002)

9. Iwata, T.: New blockcipher modes of operation with beyond the birthday bound security. In: Robshaw, M.J.B. (ed.) FSE 2006. LNCS, vol. 4047, pp. 310–317. Springer, Heidelberg (2006), http://www.nuee.nagoya-u.ac.jp/labs/tiwata/

10. Iwata, T., Kurosawa, K.: OMAC: One-Key CBC MAC. In: Johansson, T. (ed.) FSE 2003. LNCS, vol. 2887, pp. 129–153. Springer, Heidelberg (2003)

11. Jaulmes, E., Joux, A., Valette, F.: On the security of randomized CBC-MAC beyond the birthday paradox limit: A new construction. In: Daemen, J., Rijmen, V. (eds.) FSE 2002. LNCS, vol. 2365, pp. 237–251. Springer, Heidelberg (2002)

12. Jonsson, J.: On the Security of CTR+CBC-MAC. In: Nyberg, K., Heys, H.M. (eds.) SAC 2002. LNCS, vol. 2595, pp. 76–93. Springer, Heidelberg (2003)

13. Jutla, C.S.: Encryption modes with almost free message integrity. In: Pfitzmann, B. (ed.) EUROCRYPT 2001. LNCS, vol. 2045, pp. 529–544. Springer, Heidelberg (2001)

14. Katz, J., Yung, M.: Unforgeable encryption and chosen ciphertext secure modes of operation. In: Schneier, B. (ed.) FSE 2000. LNCS, vol. 1978, pp. 284–299. Springer, Heidelberg (2001)

15. Kohno, T., Viega, J., Whiting, D.: CWC: A high-performance conventional authenticated encryption mode. In: Roy, B., Meier, W. (eds.) FSE 2004. LNCS, vol. 3017, pp. 408–426. Springer, Heidelberg (2004)

16. Lefranc, D., Painchault, P., Rouat, V., Mayer, E.: A generic method to design modes of operation beyond the birthday bound. In: Preproceedings of the 14th annual workshop on Selected Areas in Cryptography, SAC 2007 (2007)

17. Lucks, S.: The two-pass authenticated encryption faster than generic composition. In: Gilbert, H., Handschuh, H. (eds.) FSE 2005. LNCS, vol. 3557, pp. 284–298. Springer, Heidelberg (2005)

18. Luby, M., Rackoff, C.: How to construct pseudorandom permutations from pseudorandom functions. SIAM J. Comput. 17(2), 373–386 (1988)

19. McGrew, D., Viega, J.: The Galois/Counter mode of operation (GCM) (submission to NIST) (2004), http://csrc.nist.gov/CryptoToolkit/modes/

20. McGrew, D., Viega, J.: The security and performance of Galois/Counter mode of operation. In: Canteaut, A., Viswanathan, K. (eds.) INDOCRYPT 2004. LNCS, vol. 3348, pp. 343–355. Springer, Heidelberg (2004)

21. Petrank, E., Rackoff, C.: CBC MAC for real-time data sources. Journal of Cryptology 13(3), 315–338 (2000)

22. Rogaway, P.: Nonce-based symmetric encryption. In: Roy, B., Meier, W. (eds.) FSE 2004. LNCS, vol. 3017, pp. 348–358. Springer, Heidelberg (2004)

23. Rogaway, P.: Authenticated-encryption with associated-data. In: Proceedings of the ACM Conference on Computer and Communications Security, ACM CCS 2002, pp. 98–107. ACM, New York (2002)

24. Rogaway, P., Bellare, M., Black, J., Krovetz, T.: OCB: a block-cipher mode of operation for efficient authenticated encryption. ACM Trans. on Information System Security (TISSEC) 6(3), 365–403 (2003); Earlier version in Proceedings of the eighth ACM Conference on Computer and Communications Security, ACM CCS 2001, pp. 196–205, ACM, New York (2001)
25. Rogaway, P., Shrimpton, T.: Deterministic authenticated-encryption: A provable-security treatment of the keywrap problem. In: Vaudenay, S. (ed.) EUROCRYPT 2006. LNCS, vol. 4004, pp. 373–390. Springer, Heidelberg (2006)
26. Wegman, M.N., Carter, J.L.: New hash functions and their use in authentication and set equality. JCSS 22, 256–279 (1981)
27. Whiting, D., Housley, R., Ferguson, N.: Counter with CBC-MAC (CCM) (submission to NIST) (2002), http://csrc.nist.gov/CryptoToolkit/modes/

A   Figures

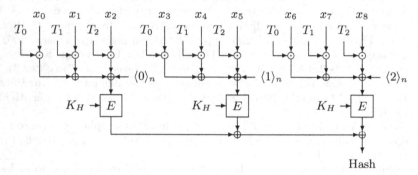

Fig. 8. Illustration of CIP.Hash. This example uses $\varpi = 3$, and $l = 9$.

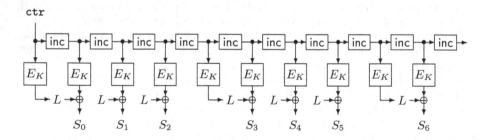

Fig. 9. Illustration of CIP.KSGen. This example uses $w = 3$ and outputs $l = 7$ blocks of keystream $S = (S_0, \ldots, S_6)$.

Cryptanalysis of the TRMS Signature Scheme of PKC'05

Luk Bettale, Jean-Charles Faugère, and Ludovic Perret

INRIA, Centre Paris-Rocquencourt, SALSA Project
UPMC, Univ Paris 06, LIP6
CNRS, UMR 7606, LIP6
104, avenue du Président Kennedy
75016 Paris, France
luk.bettale@free.fr, jean-charles.faugere@inria.fr,
ludovic.perret@lip6.fr

Abstract. In this paper, we investigate the security of the Tractable Rationale Maps Signature (TRMS) signature scheme [9] proposed at PKC'05. To do so, we present a hybrid approach for solving the algebraic systems naturally arising when mounting a signature-forgery attack. The basic idea is to compute Gröbner bases of several modified systems rather than a Gröbner basis of the initial system. We have been able to provide a precise bound on the (worst-case) complexity of this approach. For that, we have however assumed a technical condition on the systems arising in our attack; namely the systems are *semi-regular* [3,5]. This claim is supported by experimental evidences. Finally, it turns out that our approach is efficient. We have obtained a complexity bounded from above by 2^{57} to forge a signature on the parameters proposed by the designers of TRMS [9]. This bound can be improved; assuming an access to 2^{16} processors (which is very reasonable), one can actually forge a signature in approximately 51 hours.

1 Introduction

Multivariate Cryptography is the set of all the cryptographic primitives using multivariate polynomials. The use of algebraic systems in cryptography dates back to the mid eighties [15,29], and was initially motivated by the need for alternatives to number theoretic-based schemes. Indeed, although quite a few problems have been proposed to construct public-key primitives, those effectively used are essentially factorization (e.g. in RSA [33]) and discrete logarithm (e.g. in Diffie-Hellman key-exchange [16]). It has to be noted that multivariate systems enjoy low computational requirements; moreover, such schemes are not concerned with the quantum computer threat, whereas it is well known that number theoretic-based schemes like RSA, DH, or ECDH are [34].

Multivariate cryptography has become a dynamic research area, as reflected by the ever growing number of papers in the most famous cryptographic conferences. This is mainly due to the fact that an European project (NESSIE[1])

---

[1] https://www.cosic.esat.kuleuven.be/nessie/

S. Vaudenay (Ed.): AFRICACRYPT 2008, LNCS 5023, pp. 143–155, 2008.

has advised in 2003 to use such a scheme (namely, SFLASH [11]) in the smart-card context. Unfortunately, Dubois, Fouque, Shamir and Stern [14] discovered a sever flaw in the design of SFLASH, leading to an efficient cryptanalysis of this scheme. In this paper, we investigate the security of another multivariate signature scheme, the so-called Tractable Rationale Maps Signature (TRMS) [9].

1.1   Organization of the Paper. Main Results

After this introduction, the paper is organized as follows. In Section 2, we introduce the main concern of this paper, namely the Tractable Rationale Maps Signature (TRMS) scheme presented at PKC'05 [9]. Note that the situation of this scheme is a bit fuzzy. A cryptanalysis of a preprint/previous version [36] of such scheme has been presented at PKC'05 [25]. However, no attack against the version presented at PKC'05 [9] has been reported so far. In [25], the authors remarked that one can – more or less – split the public-key of [36] in two independent algebraic systems which can be solved efficiently. We tried to mount this attack on the TRMS version of PKC'05 [9] without success. Thus, it makes sense to study the security of [9]. By the way, the authors of [25] also proposed an "improved" version of the XL algorithm, the so-called *linear method*. We will not much detail this point in this paper, but this linear method is actually very similar to the F_5 [19] algorithm in its matrix form [20]. We briefly come back to this point in Section 3. We will explain why the linear method cannot be more efficient than F_5.

In Section 3, we will introduce the necessary mathematical tools (ideals, varieties and Gröbner bases), as well as the algorithmic tools (F_4/F_5), allowing to address the problem of solving algebraic systems. We will give the definition of *semi-regular* sequences which will be useful to provide a precise complexity bound on our attack. The reader already familiar with these notions can skip this part. However, we would like to emphasize that the material contained in this section is important for understanding the behavior of the attack presented in Section 4. By the way, the notion presented in this section will permit to compare F_5 [19] with the linear method of [25].

In Section 4, we present a hybrid approach for solving the algebraic systems arising when attacking TRMS. The basic idea is to compute Gröbner bases of several modified systems rather than one Gröbner basis of the (bigger) initial system. We have been able to provide a precise bound on the (worst-case) complexity of this approach. For that, we have assumed that the systems arising in our attack are semi-regular. This claim is supported by experimental evidences. This approach approach is efficient; we have obtained a complexity bounded from above by 2^{57} (fields operations) to forge a signature on the parameters proposed by the designers of TRMS [9]. This bound can be improved; assuming an access to 2^{16} processors (which is very reasonable), one can actually forge a signature in approximately 51 hours.

2   Tractable Rationale Maps Signature Schemes

To the best of our knowledge, multivariate public-key cryptosystems are mainly constructed from two different one-way functions. The first one, that we only mention for the sake of completeness is as follows. Let $\mathcal{I} = \langle f_1, \ldots, f_u \rangle$ be an ideal of the polynomial ring $\mathbb{K}[x_1, \ldots, x_n]$ ($\mathbb{K}$ is a finite field) then :

$$f_{\mathrm{PC}} : m \in \mathbb{K} \longmapsto e_{\mathcal{I}} + m \in \mathbb{K}[x_1, \ldots, x_n],$$

with $e_{\mathcal{I}}$ a random element of $\mathcal{I}$.

The one-way function f_{PC} gave rise to a family of public-key encryption schemes that are named *Polly Cracker cryptosystems* [23,27]. The public-key of such systems is an ideal $\mathcal{I} = \langle f_1, \ldots, f_u \rangle \subset \mathbb{K}[x_1, \ldots, x_n]$, and the secret-key (or trapdoor) is a zero $\mathbf{z} \in \mathbb{K}^n$ of $\mathcal{I}$. Although the security study of Polly Cracker-type systems led to interesting mathematical and algorithmic problems, several evidences have been presented showing that those schemes are not suited for the design of secure cryptosystems (for a survey, we refer the reader to [28]). Moreover, such systems suffer from efficiency problems, namely a poor encryption rate and a large public-key size.

From a practical point of view, the most interesting type of one-way function used in multivariate cryptography is based on the evaluation of a set of algebraic polynomials $\mathbf{p} = \big(p_1(x_1, \ldots, x_n), \ldots, p_u(x_1, \ldots, x_n) \big) \in \mathbb{K}[x_1, \ldots, x_n]^u$, namely :

$$f_{\mathrm{MI}} : \mathbf{m} = (m_1, \ldots, m_n) \in \mathbb{K}^n \longmapsto \mathbf{p}(\mathbf{m}) = \big(p_1(\mathbf{m}), \ldots, p_u(\mathbf{m}) \big) \in \mathbb{K}^u.$$

Here, the mathematical hard problem associated to this one-way function is :

Polynomial System Solving (PoSSo)

INSTANCE : polynomials $p_1(x_1, \ldots, x_n), \ldots, p_u(x_1, \ldots, x_n)$ of $\mathbb{K}[x_1, \ldots, x_n]$.

QUESTION : Does there exists $(z_1, \ldots, z_n) \in \mathbb{K}^n$ s. t. :

$$p_1(z_1, \ldots, z_n) = 0, \ldots, p_u(z_1, \ldots, z_n) = 0.$$

It is well known that this problem is NP-COMPLETE [24]. Note that PoSSo remains NP-COMPLETE even if we suppose that the input polynomials are quadratics. This restriction is sometimes called MQ [10].

To introduce a trapdoor, we start from a carefully chosen algebraic system :

$$\mathbf{f}(\mathbf{x}) = \big(f_1(x_1, \ldots, x_n), \ldots, f_u(x_1, \ldots, x_n) \big) \in \mathbb{K}[x_1, \ldots, x_n]^u,$$

which is *easy* to solve. That is, for all $\mathbf{c} = (c_1, \ldots, c_u) \in \mathbb{K}^u$, we have an efficient method for describing/computing the zeroes of :

$$f_1(x_1, \ldots, x_n) = c_1, \ldots, f_u(x_1, \ldots, x_n) = c_u.$$

In order to hide the specific structure of $\mathbf{f}$, we usually choose two linear transformations – given by invertible matrices – $(S, U) \in GL_n(\mathbb{K}) \times GL_u(\mathbb{K})$ and set

$$\big(p_1(\mathbf{x}), \ldots, p_u(\mathbf{x}) \big) = \big(f_1(\mathbf{x} \cdot S), \ldots, f_u(\mathbf{x} \cdot S) \big) \cdot U,$$

abbreviated by $\mathbf{p}(\mathbf{x}) = \mathbf{f}(\mathbf{x} \cdot S) \cdot U \in \mathbb{K}^u$ to shorten the notation.

The public-key of such systems will be the polynomials of $\mathbf{p}$ and the secret-key is the two matrices $(S, U) \in GL_n(\mathbb{K}) \times GL_u(\mathbb{K})$ and the polynomials of $\mathbf{f}$.

To generate a signature $\mathbf{s} \in \mathbb{K}^n$ of a digest $\mathbf{m} \in \mathbb{K}^u$, we compute $\mathbf{s}' \in \mathbb{K}^n$ such that $\mathbf{f}(\mathbf{s}') = \mathbf{m} \cdot U^{-1}$. This can be done efficiently due to the particular choice of $\mathbf{f}$. Finally, the signature is $\mathbf{s} = \mathbf{s}' \cdot S^{-1}$ since :

$$\mathbf{p}(\mathbf{s}) = \mathbf{f}(\mathbf{s}' \cdot S^{-1} \cdot S) \cdot U = \mathbf{m} \cdot U^{-1} \cdot U = \mathbf{m}.$$

To verify the signature $\mathbf{s} \in \mathbb{K}^n$ of the digest $\mathbf{m} \in \mathbb{K}^u$, we check whether the equality :

$$\text{``}\mathbf{p}(\mathbf{s}) = \mathbf{m}\text{''} \text{ holds.}$$

We would like to emphasize that most of the multivariate signature schemes proposed so far (e.g. [11,26,37]), including TRMS [9], follow this general principle.

The specificity of TRMS lies in the way of constructing the inner polynomials $\mathbf{f}(\mathbf{x}) = \big(f_1(x_1, \ldots, x_n), \ldots, f_u(x_1, \ldots, x_n)\big) \in \mathbb{K}[x_1, \ldots, x_n]^u$. The designers of TRMS propose to use so-called *tractable rational maps*, which are of the following form :

$$f_1 = r_1(x_1)$$
$$f_2 = r_2(x_2) \cdot \frac{g_2(x_1)}{q_2(x_1)} + \frac{h_2(x_1)}{s_2(x_1)}$$
$$\vdots$$
$$f_k = r_k(x_k) \cdot \frac{g_k(x_1, \ldots, x_{k-1})}{q_k(x_1, \ldots, x_{k-1})} + \frac{h_k(x_1, \ldots, x_{k-1})}{s_k(x_1, \ldots, x_{k-1})}$$
$$\vdots$$
$$f_n = r_k(x_n) \cdot \frac{g_k(x_1, \ldots, x_{n-1})}{q_k(x_1, \ldots, x_{n-1})} + \frac{h_k(x_1, \ldots, x_{n-1})}{s_k(x_1, \ldots, x_{n-1})}$$

where for all $i, 2 \le i \le n, g_i, q_i, h_i, s_i$ are polynomials of $\mathbb{K}[x_1, \ldots, x_n]$, and $i, 2 \le i \le n, r_i$ is a permutation polynomial on $\mathbb{K}$. Remember that r_i is a univariate polynomial. As explained in [9], tractable rational maps can be explicitly inverted on a well chosen domain. We will not detail this point, as well as how the polynomials g_i, q_i, h_i, s_i and r_i are constructed. This is not relevant for the attack that we will present. We refer the reader to the initial paper [9]. We just mention that we finally obtain quadratic polynomials for the f_is. We quote below the set of parameters recommended by the authors :

- $\mathbb{K} = \mathbb{F}_{2^8}$
- $n = 28$ and $u = 20$

We will show that this set of parameters does not guaranty a sufficient level of security.

3 Gröbner Basics

In order to mount a signature-forgery attack against TRMS, we have to address the problem of solving an algebraic system of equations. To date, Gröbner bases [6,7] provide the most efficient algorithmic solution for tackling this problem. We introduce here these bases and some of their useful properties (allowing in particular to find the zeroes of an algebraic system). We also describe efficient algorithms permitting to compute Gröbner bases. We will touch here only a restricted aspect of this theory. For a more thorough introduction, we refer the reader to [1,12].

3.1 Definition – Property

We start by defining two mathematical objects naturally associated to Gröbner bases : *ideals* and *varieties*. We shall call *ideal generated* by $p_1, \ldots, p_u \in \mathbb{K}[x_1, \ldots, x_n]$ the set :

$$\mathcal{I} = \langle p_1, \ldots, p_u \rangle = \left\{ \sum_{k=1}^{u} p_k \cdot h_k : h_1, \ldots, h_k \in \mathbb{K}[x_1, \ldots, x_n] \right\} \subseteq \mathbb{K}[x_1, \ldots, x_n].$$

We will denote by :

$$V_{\mathbb{K}}(\mathcal{I}) = \left\{ \mathbf{z} \in \mathbb{K}^n : p_i(\mathbf{z}) = 0, \text{ for all } i, 1 \leq i \leq u \right\},$$

the *variety associated* to $\mathcal{I}$, i.e. the common zeros – over $\mathbb{K}$ – of $p_1, \ldots, p_u$.

Gröbner bases offer an explicit method for describing varieties. Informally, a Gröbner basis of an ideal $\mathcal{I}$ is a generating set of $\mathcal{I}$ with "good" algorithmic properties. These bases are defined with respect to *monomial ordering*. For instance, the *Lexicographical* (Lex) and *Degree Reverse Lexicographical* (DRL) orderings – which are widely used in practice – are defined as follows :

Definition 1. *Let $\alpha = (\alpha_1, \ldots, \alpha_n)$ and $\beta = (\beta_1, \ldots, \beta_n) \in \mathbb{N}^n$. Then:*
- *$x_1^{\alpha_1} \cdots x_n^{\alpha_n} \succ_{\text{Lex}} x_1^{\beta_1} \cdots x_n^{\beta_n}$ if the left-most nonzero entry of $\alpha - \beta$ is positive.*
- *$x_1^{\alpha_1} \cdots x_n^{\alpha_n} \succ_{\text{DRL}} x_1^{\beta_1} \cdots x_n^{\beta_n}$ if $\sum_{i=1}^{n} \alpha_i > \sum_{i=1}^{n} \beta_i$, or $\sum_{i=1}^{n} \alpha_i = \sum_{i=1}^{n} \beta_i$ and the right-most nonzero entry of $\alpha - \beta$ is negative.*

Once a (total) monomial ordering is fixed, we define :

Definition 2. *We shall denote by $\mathrm{M}(n)$ the set of all monomials in n variables, and $\mathrm{M}_d(n)$ the set of all monomials in n variables of degree $d \geq 0$. We shall call **total degree** of a monomial $x_1^{\alpha_1} \cdots x_n^{\alpha_n}$ the sum $\sum_{i=1}^{n} \alpha_i$. The **leading monomial** of $p \in \mathbb{K}[x_1, \ldots, x_n]$ is the largest monomial (w.r.t. some monomial ordering $\prec$) among the monomials of p. This leading monomial will be denoted by $\mathrm{LM}(p, \prec)$. The **degree** of p, denoted $\deg(p)$, is the total degree of $\mathrm{LM}(p, \prec)$.*

We are now in a position to define more precisely Gröbner bases.

Definition 3. *A set of polynomials* $G \subset \mathbb{K}[x_1, \ldots, x_n]$ *is a* **Gröbner basis** – *w.r.t. a monomial ordering* $\prec$ – *of an ideal* $\mathcal{I} \subseteq \mathbb{K}[x_1, \ldots, x_n]$ *if, for all* $p \in \mathcal{I}$, *there exists* $g \in G$ *such that* $\mathrm{LM}(g, \prec)$ *divides* $\mathrm{LM}(p, \prec)$.

Gröbner bases computed for a lexicographical ordering (Lex-Gröbner bases) permit to easily describe varieties. A Lex-Gröbner basis of a *zero-dimensional system* (i.e. with a finite number of zeroes over the algebraic closure) is always as follows

$$\{f_1(x_1) = 0, f_2(x_1, x_2) = 0, \ldots, f_{k_2}(x_1, x_2) = 0, \ldots, f_{k_n}(x_1, \ldots, x_n)\}$$

To compute the variety, we simply have to successively eliminate variables by computing zeroes of univariate polynomials and back-substituting the results.

From a practical point of view, computing (directly) a Lex-Gröbner basis is much slower that computing a Gröbner basis w.r.t. another monomial ordering. On the other hand, it is well known that computing degree reverse lexicographical Gröbner bases (DRL-Gröbner bases) is much faster in practice. The FLGM algorithm [17] permits – in the zero-dimensional case – to efficiently solve this issue. This algorithm use the knowledge of a Gröbner basis computed for a given order to construct a Gröbner for another order. The complexity of this algorithm is polynomial in the number of solutions of the ideal considered. This leads to the following strategy for computing the solutions of a zero-dimensional system

$$p_1 = 0, \ldots, p_u = 0.$$

1. Compute a DRL-Gröbner basis G_{DRL} of $\langle p_1, \ldots, p_u \rangle$.
2. Compute a Lex-Gröbner basis of $\langle p_1, \ldots, p_u \rangle$ from G_{DRL} using FGLM.

This approach is sometimes called *zero-dim solving* and is widely used in practice. For instance, this is the default strategy used in the computer algebra system Magma[2] when calling the function **Variety**. In our context, the varieties will usually have only one solution. Thus, the cost of the zero-dim solving is dominated by the cost of computing a DRL-Gröbner basis. We now describe efficient algorithms for performing this task.

3.2   The F_4/F_5 Algorithms

The historical method for computing Gröbner bases is Buchberger's algorithm [6,7]. Recently, more efficient algorithms have been proposed, namely the F_4 and F_5 algorithms [18,19]. These algorithms are based on the intensive use of linear algebra techniques. Precisely, F_4 can be viewed as the "gentle" meeting of Buchberger's algorithm and Macaulay's ideas [30]. In short, the arbitrary choices – limiting the practical efficiency of Buchberger's algorithm – are replaced in F_4 by computational strategies related to classical linear algebra problems (mainly the computation of a row echelon form).

In [19], a new criterion (the so-called F_5 criterion) for detecting useless computations has been proposed. It is worth pointing out that Buchberger's algorithm spends 90% of its time to perform these useless computations. Under some

---

[2] http://magma.maths.usyd.edu.au/magma/

regularity conditions, it has been proved that all useless computations can be detected and avoided. A new algorithm, called F_5, has then been developed using this criterion and linear algebra methods. Briefly, F_5 (in its matrix form) constructs incrementally the following matrices in degree d :

$$
\begin{array}{c}
 & m_1 \succ m_2 \succ m_3 \ldots \\
A_d = \begin{array}{c} t_1 \cdot p_1 \\ t_2 \cdot p_2 \\ t_3 \cdot p_3 \\ \vdots \end{array} & \begin{bmatrix} \ldots & \ldots & \ldots & \ldots \\ \ldots & \ldots & \ldots & \ldots \\ \ldots & \ldots & \ldots & \ldots \\ \ldots & \ldots & \ldots & \ldots \end{bmatrix}
\end{array}
$$

where the indices of the columns are monomials sorted w.r.t. $\prec$ and the rows are products of some polynomials f_i by some monomials t_j such that $\deg(t_j f_i) \le d$. In a second step, row echelon forms of theses matrices are computed, i.e.

$$
\begin{array}{c}
 & m_1 \ m_2 \ m_3 \ldots \\
A'_d = \begin{array}{c} t_1 \cdot p_1 \\ t_2 \cdot p_2 \\ t_3 \cdot p_3 \\ \vdots \end{array} & \begin{bmatrix} 1 & 0 & 0 & \ldots \\ 0 & 1 & 0 & \ldots \\ 0 & 0 & 1 & \ldots \\ 0 & 0 & 0 & \ldots \end{bmatrix}
\end{array}
$$

For d sufficiently large, A'_d contains a Gröbner basis.

In [25], the authors proposed an "improved" version of the XL algorithm [10], the so-called *linear method*. This method is very similar to F_5 [19]. It can be proved [2] that the matrices constructed by F_5, with Lex, are sub-matrices of the matrices generated by the linear method. One can argue that the goal of F_5 and the linear method is not the same. Namely, F_5 computes Gröbner bases whereas the linear method computes varieties. Again, using the same arguments of [2], it can be proved that the linear method constructs intrinsically a Lex-Gröbner basis. As explained previously, we avoid to compute directly Lex Gröbner bases. We prefer to compute a DRL-Gröbner basis, and then use FGLM to obtain the Lex-Gröbner basis. Thus, the practical gain of F_5+FGLM versus the linear method will be even more important. This was already pointed out in [2].

Finally, the main idea of the linear method is to remove linear dependencies induced by trivial relation of the form $f \cdot g - g \cdot f$. This is actually the basic idea of F_5. Note that in F_5 the matrices are constructed "incrementally" to be sure of removing all the trivial linear dependencies. This is not the case for the linear method. To summarize one can say that the linear method is a degraded/devalued version of F_5 using the worst strategy for computing varieties.

We now come back to the complexity of F_5. An important parameter for evaluating this complexity is the *degree of regularity* which is defined as follows :

Definition 4. *We shall call* **degree of regularity** *of homogeneous polynomials $p_1, \ldots, p_u \in \mathbb{K}[x_1, \ldots, x_n]$, denoted $d_{\mathrm{reg}}(p_1, \ldots, p_m)$, the smallest integer $d \ge 0$ such that the polynomials of degree d in $\mathcal{I} = \langle p_1, \ldots, p_u \rangle$ generate – as a $\mathbb{K}$*

vectorial space – the set of all monomials of degree d in n variables (the number of such monomials $\#M_d(n)$ *is* [3] C^d_{n+d-1}*). In other words :*

$$\min \left\{ d \geq 0 : \dim_{\mathbb{K}}(\{f \in \mathcal{I} : \deg(f) = d\}) = \#M_d(n) \right\}.$$

For non-homogeneous polynomials $p_1, \ldots, p_u \in \mathbb{K}[x_1, \ldots, x_n]$*, the degree of regularity is defined by the degree of regularity of the homogeneous components of highest degree of the polynomials* $p_1, \ldots, p_u$*.*

This degree of regularity corresponds to the maximum degree reached during a Gröbner basis computation. The overall complexity of F_5 is dominated by the cost of computing the row echelon form of the last matrix $A_{d_{\mathrm{reg}}}$, leading to a complexity :

$$\mathcal{O}\left((m \cdot C^{d_{\mathrm{reg}}}_{n+d_{\mathrm{reg}}-1})^\omega \right),$$

with $\omega, 2 \leq \omega \leq 3$ being the linear algebra constant.

In general, it is a difficult problem to know *a priori* the degree of regularity. However, for *semi-regular sequences* [3,5,4] – that we are going to introduce – the behavior of the degree of regularity is well mastered.

Definition 5. *[3,5,4] Let* $p_1, \ldots, p_u \in \mathbb{K}[x_1, \ldots, x_n]$ *be homogeneous polynomials of degree* $d_1, \ldots, d_u$ *respectively. This sequence is* **semi-regular** *if :*

- $\langle p_1, \ldots, p_u \rangle \neq \mathbb{K}[x_1, \ldots, x_n]$,
- *for all* $i, 1 \leq i \leq u$ *and* $g \in \mathbb{K}[x_1, \ldots, x_n]$ *:*

$$\deg(g \cdot p_i) \leq d_{\mathrm{reg}} \ et \ g \cdot p_i \in \langle p_1, \ldots, p_{i-1} \rangle \Rightarrow g \in \langle p_1, \ldots, p_{i-1} \rangle.$$

We can extend the notion semi-regular sequence to non-homogeneous polynomials by considering the homogeneous components of highest degree of theses polynomials. We mention that the semi-regularity has been introduced by Bardet, Faugère, Salvy and Yang to generalize the notion of regularity [3,5,4].

It can be proved that no useless reduction to zero is performed by F_5 on semi-regular (resp. regular) sequences [3,5,4,19] , i.e. all the matrices A_d ($d < d_{\mathrm{reg}}$) generated in F_5 are of full rank. Moreover, the degree of regularity of a semi-regular sequence $(p_1, \ldots, p_u)$ of degree $d_1, \ldots, d_u$ respectively is given [3,5,4] by the index of the first non-positive coefficient of :

$$\sum_{k \geq 0} c_k \cdot z^k = \frac{\prod_{i=1}^m (1 - z^{d_i})}{(1 - z)^n}.$$

For instance, it has been proved that the degree [3,5] of regularity of a semi-regular system of $n - 1$ variables and n equations is asymptotically equivalent to $\left\lceil \frac{(n+1)}{2} \right\rceil$. The authors have recovered here a result obtained, with a different technique, by Szanto [35]. For a semi-regular system of n variables and n equations, we obtain a degree of regularity equal to $n + 1$, which is the well-known Macaulay bound. More details on these complexity analyses, and further complexity results can be found in [3,5,4].

---

[3] C^d_n if you consider the field equations.

4   Description of the Attack

In this part, we present our attack against TRMS [9]. Our goal is to forge a valid signature $\mathbf{s'} \in \mathbb{K}^n$ for a given digest $\mathbf{m} = (m_1, \ldots, m_u) \in \mathbb{K}^u$. In other words, we want to find an element of the variety :

$$V_{\mathbb{K}}(p_1 - m_1, \ldots, p_u - m_u) \subseteq \mathbb{K}^n,$$

with $p_1, \ldots, p_u \in \mathbb{K}[x_1, \ldots, x_n]$ the polynomials of a TRMS public-key. We recall that the parameters are $\mathbb{K} = \mathbb{F}_{2^8}, n = 28$ and $u = 20$.

Following the zero dim-solving strategy presented in Section 3, one can directly try to compute this variety. Unfortunately, there is at least two reasons for which such a direct approach cannot be efficient in this context. First, we have explicitly supposed that the field equations are included in the signature-forgery system. In our context, $\mathbb{K}$ is relatively large; leading to field equations of high degree. In particular, the degree of regularity of the system will be at least equal to $\#\mathbb{K}$. Thus, the computation of a Gröbner basis is impossible in practice.

Another limitation is due to the fact that the number of equations (u) is smaller that the number of variables (n). As a consequence, there is at least $(\#\mathbb{K})^{n-u}$ valid solutions to the signature-forgery system. Hence, even if you suppose that you have been able to compute a DRL-Gröbner basis, you will probably not be able to recover efficiently the Lex-Gröbner basis using FGLM.

A natural way to overcome these practical limitations is to randomly specialize (i.e. fix) $n - u$ variables, and remove the field equations. We will have to solve a system having the same number of variables and equations (u). For each specification of the $n - u$ variables, we can always find a solution of the new system yielding to a valid signature. We also mention that the specialized system will have very few solutions in practice. Thus, the cost of computing the variety will be now essentially the cost of computing a Gröbner basis.

The important observation here is that – after having specified $n-u$ variables – the new system will behave like a semi-regular system. We will present latter in this section experimental results supporting this claim. Note that such a behavior has been also observed, in a different context, in [38]. The degree of regularity of a semi-regular system of u variables and equations is equal to $u + 1$. In our context ($u = 20$), this remains out of the scope of the F_5 algorithm.

To decrease this degree of regularity, we can specialize $r \geq 0$ more variables (in addition of the $n - u$ variables already fixed). Thus, we will have to solve a systems of u equations with $u - r$ variables, which behave like semi-regular systems. This allows to decrease the degree of regularity, and thus the complexity of F_5. For instance, the degree of regularity of a semi-regular system of $u - 1$ variables and u equations is approximately equal to $\left\lceil \frac{(u+1)}{2} \right\rceil$. More generally, the degree of regularity is given by the index of the first non-positive coefficient of the series :

$$\frac{\prod_{i=1}^{u}(1 - z^2)}{(1 - z)^{u-r}}.$$

In the following table, we have quoted the degree of regularity observed in our experiments. Namely, the maximum degree reached during F_5 on systems obtained by fixing $n - u + r$ variables ($r \geq 0$) on signature-forgery systems. We have also quoted the theoretical degree of regularity of a semi-regular system of u equations in $u - r$ variables. These experiments strongly suggest that the systems obtained when mounting a specify+solve signature forgery attack against TRMS behave like semi-regular systems.

u	$u - r$	r	d_{reg} (theoretical)	d_{reg} (observed)
20	19	1	11	
20	18	2	9	9
20	17	3	8	8
20	16	4	7	7
20	15	5	6	6

By fixing variables, we obtain a significant gain on the complexity the F_5. On the other hand, as soon as $r > 0$, each specification of the r variables will not necessarily lead to an algebraic system whose set of solutions is not empty . But, we know that there exists a least one guess of the r variables (in practice exactly one) leading to a system whose zeroes allow to construct a valid signature. Thus, we have to perform an exhaustive search on the r new variables. In other words, instead of computing one Gröbner basis of a system of u equations and variables, we compute $(\#\mathbb{K})^r$ Gröbner bases of "easier" systems (u equations with $u - r$ variables). The complexity of this hybrid approach is bounded from above by:

$$\mathcal{O}\left((\#\mathbb{K})^r\left(m \cdot C_{u+d_{\text{reg}}-1}^{d_{\text{reg}}}\right)^\omega\right),$$

with $\omega, 2 \leq \omega \leq 3$ being the linear algebra constant. We have then to find an optimal tradeoff between the cost of F_5 and the number of Gröbner basis that we have to compute.

In the following table, practical results that we have obtained with F_5 when solving systems obtained by fixing $n - u + r$ variables ($r \geq 0$) on signature-forgery systems. We have quoted the experimental complexity of this approach (T) for different values of r (for that, we assumed that the r guesses are correct). We included the timings we obtained with F_5 (T_{F_5}) for computing one Gröbner basis, and the maximum number $((\#\mathbb{K})^r)$ of Gröbner bases that we have to compute. We also included the corresponding number of operations (field multiplications) Nop_{F_5} performed by F_5 for computing, and the total number N of operations of our attack (i.e. the cost of computing $2^{8 \cdot r}$ Gröbner bases). Finally, we have quoted the maximum memory, denoted Mem, used during the Gröbner basis computation. The experimental results have been obtained using a bi-pro Xeon 2.4 Ghz with 64 Gb. of Ram.

u	$u - r$	r	$(\#\mathbb{K})^r$	T_{F_5}	Mem	Nop_{F_5}	T
20	18	2	2^{16}	51h	41940 Mo	2^{41}	2^{57}
20	17	3	2^{24}	2h45min.	4402 Mo	2^{37}	2^{61}
20	16	4	2^{32}	626 sec.	912 Mo	2^{34}	2^{66}
20	15	5	2^{40}	46 sec.	368 Mo.	2^{30}	2^{70}

We observe that the optimal choice is for $r = 2$, for which you obtain a complexity bounded from above by 2^{57} to actually forge a signature on the parameters proposed by the designers of TRMS [9]. We also would like to emphasize that this approach is fully parallelizable (each computation of the $(\#\mathbb{K})^r$ Gröbner basis are totally independent). For instance, assuming an access to 2^{16} processors (which is very reasonable), the computation can be done in two days.

By extrapolating – from these experiments – the practical behavior of our approach for $r = 1$, we have estimated that one can forge a signature in approximately in 2^{53} (in terms of fields operations). As the consequence, the parameters of TRMS [9] should be increased to achieve a reasonable level of security. Further works need to be done for finding the optimal set of parameters.

References

1. Adams, W.W., Loustaunau, P.: An Introduction to Gröbner Bases. Graduate Studies in Mathematics, vol. 3, AMS (1994)
2. Ars, G., Faugère, J.-C., Imai, H., Kawazoe, M., Sugita, M.: Comparison Between XL and Gröbner Basis Algorithms. In: Lee, P.J. (ed.) ASIACRYPT 2004. LNCS, vol. 3329, pp. 338–353. Springer, Heidelberg (2004)
3. Bardet, M.: Etude des systèmes algébriques surdéterminés. Applications aux codes correcteurs et à la cryptographie. Thèse de doctorat, Université de Paris VI (2004)
4. Bardet, M., Faugère, J.-C., Salvy, B.: On the complexity of Grbner basis computation of semi-regular overdetermined algebraic equations. In: Proc. International Conference on Polynomial System Solving (ICPSS), pp. 71–75 (2004), http://www-calfor.lip6.fr/ICPSS/papers/43BF/43BF.htm
5. Bardet, M., Faugère, J.-C., Salvy, B., Yang, B.-Y.: Asymptotic Behaviour of the Degree of Regularity of Semi-Regular Polynomial Systems. In: Proc. of MEGA 2005, Eighth International Symposium on Effective Methods in Algebraic Geometry (2005)
6. Buchberger, B., Collins, G.-E., Loos, R.: Computer Algebra Symbolic and Algebraic Computation, 2nd edn. Springer, Heidelberg (1982)
7. Buchberger, B.: Gröbner Bases : an Algorithmic Method in Polynomial Ideal Theory. In: Recent trends in multidimensional systems theory, Reider ed. Bose (1985)
8. Berbain, C., Gilbert, H., Patarin, J.: QUAD: A Practical Stream Cipher with Provable Security. In: Vaudenay, S. (ed.) EUROCRYPT 2006. LNCS, vol. 4004, pp. 109–128. Springer, Heidelberg (2006)
9. Chou, C.-Y., Hu, Y.-H., Lai, F.-P., Wang, L.-C., Yang, B.-Y.: Tractable Rational Map Signature. In: Vaudenay, S. (ed.) PKC 2005. LNCS, vol. 3386, pp. 244–257. Springer, Heidelberg (2005)
10. Courtois, N., Klimov, A., Patarin, J., Shamir, A.: Efficient Algorithms for Solving Overdefined Systems of Multivariate Polynomial Equations. In: Preneel, B. (ed.) EUROCRYPT 2000. LNCS, vol. 1807, pp. 392–407. Springer, Heidelberg (2000)

11. Courtois, N., Goubin, L., Patarin, J.: SFLASH, a Fast Symmetric Signature Scheme for low-cost Smartcards – Primitive Specification and Supporting documentation, http://www.minrank.org/sflash-b-v2.pdf

12. Cox, D.A., Little, J.B., O'Shea, D.: Ideals, Varieties, and algorithms: an Introduction to Computational Algebraic Geometry and Commutative algebra. Undergraduate Texts in Mathematics. Springer, New York (1992)

13. Dubois, V., Fouque, P.-A., Stern, J.: Cryptanalysis of SFLASH with Slightly Modified Parameters. In: Naor, M. (ed.) EUROCRYPT 2007. LNCS, vol. 4515, Springer, Heidelberg (2007)

14. Dubois, V., Fouque, P.-A., Shamir, A., Stern, J.: Practical Cryptanalysis of SFLASH. In: Menezes, A. (ed.) CRYPTO 2007. LNCS, vol. 4622, Springer, Heidelberg (2007)

15. Diffie, W., Fell, H.J.: Analysis of a Public Key Approach Based on Polynomial Substitution. In: Williams, H.C. (ed.) CRYPTO 1985. LNCS, vol. 218, pp. 340–349. Springer, Heidelberg (1986)

16. Diffie, W., Hellman, M.E.: New Directions in Cryptography. IEEE Transactions on Information Theory IT 22, 644–654 (1976)

17. Faugère, J.C., Gianni, P., Lazard, D., Mora, T.: Efficient Computation of Zero-Dimensional Gröbner Bases by Change of Ordering. Journal of Symbolic Computation 16(4), 329–344 (1993)

18. Faugère, J.-C.: A New Efficient Algorithm for Computing Gröbner Basis: F_4. Journal of Pure and Applied Algebra 139, 61–68 (1999)

19. Faugère, J.-C.: A New Efficient Algorithm for Computing Gröbner Basis without Reduction to Zero: F_5. In: Proceedings of ISSAC, pp. 75–83. ACM Press, New York (2002)

20. Faugère, J.-C., Joux, A.: Algebraic Cryptanalysis of Hidden Field Equation (HFE) Cryptosystems using Gröbner bases. In: Boneh, D. (ed.) CRYPTO 2003. LNCS, vol. 2729, pp. 44–60. Springer, Heidelberg (2003)

21. Faugère, J.-C., Perret, L.: Polynomial Equivalence Problems: Algorithmic and Theoretical Aspects. In: Vaudenay, S. (ed.) EUROCRYPT 2006. LNCS, vol. 4004, pp. 30–47. Springer, Heidelberg (2006)

22. Faugère, J.-C., Perret, L.: Cryptanalysis of 2R⁻ schemes. In: Dwork, C. (ed.) CRYPTO 2006. LNCS, vol. 4117, pp. 357–372. Springer, Heidelberg (2006)

23. Fellows, M.R., Koblitz, N.: Combinatorial cryptosystems galore! Contemporary Math. 168, 51–61 (1994)

24. Garey, M.R., Johnson, D.B.: Computers and Intractability. A Guide to the Theory of NP-Completeness. W. H. Freeman, New York (1979)

25. Joux, A., Kunz-Jacques, S., Muller, F., Ricordel, P.-M.: Cryptanalysis of the Tractable Rational Map Cryptosystem. In: Vaudenay, S. (ed.) PKC 2005. LNCS, vol. 3386, pp. 258–274. Springer, Heidelberg (2005)

26. Kipnis, A., Patarin, J., Goubin, L.: Unbalanced Oil and Vinegar Signature Schemes. In: Stern, J. (ed.) EUROCRYPT 1999. LNCS, vol. 1592, pp. 206–222. Springer, Heidelberg (1999)

27. Koblitz, N.: Algebraic Aspects of Cryptography. In: Algorithms and Computation in Mathematics, vol. 3, Springer, Heidelberg (1998)

28. Levy–dit–Vehel, F., Mora, T., Perret, L., Traverso, C.: A Survey of Polly Cracker Systems (to appear)

29. Matsumoto, T., Imai, H.: Public Quadratic Polynomial-tuples for Efficient Signature-Verification and Message-Encryption. In: Günther, C.G. (ed.) EUROCRYPT 1988. LNCS, vol. 330, pp. 419–453. Springer, Heidelberg (1988)

30. Macaulay, F.S.: The Algebraic Theory of Modular Systems. Cambrige University Press, Cambrige (1916)
31. Patarin, J.: Hidden Fields Equations (HFE) and Isomorphisms of Polynomials (IP): two new families of asymmetric algorithms. In: Maurer, U.M. (ed.) EUROCRYPT 1996. LNCS, vol. 1070, pp. 33–48. Springer, Heidelberg (1996)
32. Patarin, J., Courtois, N., Goubin, L.: QUARTZ, 128-Bit Long Digital Signatures. In: Naccache, D. (ed.) CT-RSA 2001. LNCS, vol. 2020, pp. 282–297. Springer, Heidelberg (2001)
33. Rivest, R., Shamir, A., Adleman, L.: A Method for Obtaining Digital Signatures and Public-Key Cryptosystems. Communications of the ACM 21(2), 120–126 (1978)
34. Shor, P.W.: Polynomial-Time Algorithms for Prime Factorization and Discrete Logarithms on a Quantum Computer. SIAM J. Computing 26, 1484–1509 (1997)
35. Szanto, A.: Multivariate subresultants using jouanolous resultant matrices. Journal of Pure and Applied Algebra (to appear)
36. Wang, L., Chang, F.: Tractable Rational Map Cryptosystem.Cryptology ePrint archive, Report 2004/046, http://eprint.iacr.org
37. Wolf, C.: Multivariate Quadratic Polynomials in Public Key Cryptography. Ph.D. thesis, Katholieke Universiteit Leuven, B. Preneel (supervisor), 156+xxiv pages (November 2005)
38. Yang, B.-Y., Chen, J.-M., Courtois, N.T.: On Asymptotic Security Estimates in XL and Gröbner Bases-Related Algebraic Cryptanalysis. In: López, J., Qing, S., Okamoto, E. (eds.) ICICS 2004. LNCS, vol. 3269, pp. 401–413. Springer, Heidelberg (2004)

New Definition of Density on Knapsack Cryptosystems

Noboru Kunihiro

The University of Electro-Communications
1-5-1, Chofugaoka, Chofu-shi Tokyo, 182-8585, Japan*
kunihiro@k.u-tokyo.ac.jp

Abstract. Many knapsack cryptosystems have been proposed but almost all the schemes are vulnerable to lattice attack because of its low density. To prevent the lattice attack, Chor and Rivest proposed a low weight knapsack scheme, which made the density higher than critical density. In Asiacrypt2005, Nguyen and Stern introduced pseudo-density and proved that if the pseudo-density is low enough (even if the usual density is not low enough), the knapsack scheme can be broken by a single call of SVP/CVP oracle. However, the usual density and the pseudo-density are not sufficient to measure the resistance to the lattice attack individually. In this paper, we first introduce a new notion of density D, which naturally unifies the previous two densities. Next, we derive conditions for our density so that a knapsack scheme is vulnerable to lattice attack. We obtain a critical bound of density which depends only on the ratio of the message length and its Hamming weight. Furthermore, we show that if $D < 0.8677$, the knapsack scheme is solved by lattice attack. Next, we show that the critical bound goes to 1 if the Hamming weight decreases, which means that it is quite difficult to construct a low weight knapsack scheme which is supported by an argument of density.

Keywords: Low-Weight Knapsack Cryptosystems, Lattice Attack, (pseudo-)density, Shannon Entropy.

1 Introduction

1.1 Background

If quantum computers are realized, a factoring problem and a discrete logarithm problem can be solved in polynomial time [15] and some cryptosystems such as RSA or Elliptic Curve Cryptosystem will be totally broken. Many post-quantum schemes have been proposed. One of possible candidate is "knapsack cryptosystem", which is based on the difficulty of "subset sum problem". Since the subset sum problem is proved to be NP-hard, it is expected to not be solved by quantum computers. From the Merkle-Hellman's proposal of knapsack cryptosystem [9], many schemes have been proposed, but broken.

---

* Currently, with the University of Tokyo.

S. Vaudenay (Ed.): AFRICACRYPT 2008, LNCS 5023, pp. 156–173, 2008.
© Springer-Verlag Berlin Heidelberg 2008

Some attacks to knapsack cryptosystems have been done by obtaining the shortest vector of a corresponding lattice. First, the attacker constructs a lattice from public key and ciphertext. Then, he obtains the shortest vectors of the constructed lattice. Lagarias and Odlyzko introduces the concept of "density" and they proved that the knapsack scheme is broken with high probability if its density is low enough [7]. Coster et al. proposed the improved algorithm, which can solve the schemes with higher density [3].

Some designers of knapsack cryptosystems choose to reduce the Hamming weight of message in order to prevent low density. First, Chor-Rivest proposed the knapsack cryptosystem with relatively low Hamming weight of message [2]. They employ enumerative source encoding to decrease the Hamming weight of message. Then, they succeed to achieve relatively higher density. Unfortunately, it was broken by Schnorr and Hörner for some moderate parameters [13]. Furthermore, it was broken for all proposed parameters by Vaudenay [18] by its algebraic structures. In CRYPTO2000, Okamoto et al. proposed another low weight knapsack scheme [11]. We will call this scheme as OTU scheme in short. OTU scheme hides an algebraic structure unlike Chor-Rivest scheme and its security is improved.

Omura-Tanaka [12] and Izu et al. [6] analyzed the security of low weight knapsack cryptosystem. They pointed out that the low weight scheme can be broken by lattice attack even if the density is bigger than 1. In Asiacrypt2005, Nguyen and Stern [10] introduced another type of density: pseudo-density and proved that low weight knapsack cryptosystem will be broken with high probability by a single call of SVP/CVP-oracle if its pseudo-density is less than 1. They also showed that the pseudo-density of Chor-Rivest scheme and OTU scheme are less than 1 and the both schemes are vulnerable to lattice attack. Nguyen-Stern pointed out that the usual density alone is not sufficient to measure the resistance to lattice attack. Actually, after pointing out the above, they introduced pseudo-density.

We should notice that in this paper, "a scheme is broken by lattice attack or low density attack" means "the scheme is broken by a single call to SVP/CVP-oracle." Hence, this does not mean that "the scheme is totally broken."

1.2   Our Contributions

In this paper, we first reconsider the definition of *density*. We introduce a new notion of density D, which *unifies* the already proposed definitions of density: usual density d and pseudo-density κ. This means that (1) if the message is randomly generated, that is, the bit 1 and 0 in the message are randomly generated with probability $1/2$, our density is equivalent to usual density, (2) if the Hamming weight of the message is limited to be low, our density is equivalent to the pseudo-density. Interestingly, our density is identical to so-called *information rate*.

Next, we derive conditions for our density such that the knapsack scheme is solved by a single call of SVP/CVP-oracle. We reuse the framework of Nguyen-Stern's work [10] relating subset sum problem and lattice theory. Let n and

k be the message length and its Hamming weight. We derive a critical bound of density: $g_{CJ}(k/n)$, which means that the knapsack scheme is solved if $D <$ $g_{CJ}(k/n)$ by lattice attack. The function $g_{CJ}(p)$ is easily computable and is explicitly given. Notice that the critical bound depends only on the ratio of "1", that is, k/n. Our result shows that our density alone is sufficient to measure the resistance to lattice attack. Since $g_{CJ}(p) > 0.8677$ for any p, we can simply say that the knapsack scheme is solved by lattice attack if $D < 0.8677$.

Furthermore, we show that $\lim_{p\to 0} g_{CJ}(p) = 1$. This result leads to the conclusion that it is quite difficult to construct a secure "low-weight" knapsack scheme which are supported by an argument based on density. Notice that we do not use any concrete construction of knapsack cryptosystem in our discussion unlike Nguyen-Stern's analysis [10]. Instead of them, we will just use "unique decryptability".

Our analysis method is based on Nguyen-Stern's method [10]. We will point out the difference between our results and their results. One drawback of their results is that the definition of low weight is not clear. That is, it is not clear when we can decide whether the Hamming weight is low and the pseudo-density is applicable. In our analysis, we don't use a property that the Hamming weight is low. Hence, our analysis is valid for any parameter setting although Nguyen-Stern's result is valid only for low weight knapsack.

There are a big gap between SVP/CVP-oracle and lattice reduction algorithm, such as LLL. Hence, if we choose appropriate parameters, low weight cryptosystems may be supported by an argument based on difficulty of SVP/CVP.

Little attention has been given to the message expansion in low weight knapsack cryptosystem. Our results indicate that the message length before expansion is more important than expanded message length. Intuitively, we can say that since information does not increase by any deterministic processing, the difficulty of the problem will never increase.

1.3   Organization

The rest of paper is organized as follows. The next section contains the preliminaries. In Section 3, we redefine a density, which unifies already proposed density. Next, we derive the necessary conditions for secure knapsack scheme. In our analysis, we reuse the framework of Nguyen-Stern's work. We show that our density is sufficient to measure the resistance to lattice attack. In Section 4, we apply our results to Chor-Rivest and OTU schemes. In Section5, we show one another evidence to convince our results. We will discuss algorithms for directly solving the subset sum problem. We show that in the computational cost of some algorithms [2,16], the knapsack size is not important and message length before expansion is important. Section 6 concludes this paper. We also give a simple procedure based on our results for judgment whether a knapsack scheme is vulnerable or not. Our procedure is effective for any value of Hamming weight.

2   Preliminaries

2.1   Lattices

For a vector b, $\|b\|$ denotes the Euclidean norm of b. In this paper, we deals with integral lattice. A lattice is defined by a set of all integral linear combination of linearly independent vectors $b_1, b_2, \ldots, b_d$ in $\mathbb{Z}^n$:

$$L = \left\{ \sum_{i=1}^{d} n_i b_i \mid n_i \in \mathbb{Z} \right\}. \tag{1}$$

The lattice is closed under addition and subtraction, that is, if $x \in L$ and $y \in L$, then $x + y \in L$ and $x - y \in L$. The set of vectors: $b_1, b_2, \ldots, b_d$ is called a basis of lattice L. The dimension of L is d.

We introduce the following two well-known problems: Shortest Vector Problem (SVP) and Closest Vector Problem (CVP).

Definition 1 (Shortest Vector Problem (SVP)). *Given a basis of lattice L, find a shortest non-zero vector $v \in L$.*

Definition 2 (Closest Vector Problem (CVP)). *Given a basis of lattice L and $t \in \mathbb{Q}^n$, find a closest vector $w \in L$ to t or find a lattice vector minimizing $\|w - t\|$.*

It is known that CVP is an NP-hard problem [17] and SVP is NP-hard under randomized reductions [1]. However, it is known that some lattice reduction algorithm, such as LLL algorithm, solves SVP and CVP in practice if the dimension is moderate. Hence, it is important to judge whether the scheme is secure or not even if a single call to SVP/CVP-oracle is allowed.

2.2   Knapsack Cryptosystem and Two Definitions of Density

First, we define Subset Sum Problem.

Definition 3 (Subset Sum Problem). *Given a knapsack $\{a_1, a_2, \ldots, a_n\}$ of positive integers and a sum $s = \sum_{i=1}^{n} m_i a_i$, where $m_i \in \{0, 1\}$, recover $(m_1, \ldots, m_n)$.*

The knapsack cryptosystems are informally defined as follows.

Public Key: Knapsack $\{a_1, a_2, \ldots, a_n\}$, where each a_i is a positive integer. This knapsack must be difficult to solve without private key.

Private Key: A secret information which transforms the knapsack into easily solvable knapsack.

Encryption: A message $(m_1, \ldots, m_n) \in \{0, 1\}^n$ is encrypted into a ciphertext $C = \sum_{i=1}^{n} m_i a_i$.

Decryption: The message is recovered from the ciphertext C and the private key.

Solving Subset Sum Problem corresponds to passive attack to knapsack cryptosystems.

Let $k = \sum m_i$ be the Hamming weight of m_i's and $r = \sum m_i^2$. Since $m_i \in \{0,1\}$, $r = k$.

The usual density [7] is defined by

$$d = \frac{n}{\log A}, \tag{2}$$

where $A = \max a_i$. Lagarias and Odlyzko showed that if $d < 0.645\cdots$, then the subset sum problem can be solved with high probability by a single call to SVP-oracle. Furthermore, Coster et al. [3] improved the bound to $d < 0.9408\cdots$.

In Asiacrypt2005, Nguyen and Stern [10] introduced a new variant of density: "pseudo-density", which is effective for low weight knapsack cryptosystems. The pseudo-density is defined by

$$\kappa = \frac{r \log n}{\log A}. \tag{3}$$

They showed that if pseudo-density is less than 1, low weight knapsack scheme can be solved with a single call to SVP/CVP-oracle. Furthermore, they showed that the pseudo-densities of OTU scheme and Chor-Rivest scheme are less than 1.

It is important that the usual density and the pseudo-density *alone* are not sufficient to measure the resistance to the lattice attack individually. As the known results, we can say that Merkle-Hellman scheme [9] is vulnerable because of its low density and Chor-Rivest scheme [2] is vulnerable because of its low pseudo-density, respectively. On the other hand, if we use our density, we need not discuss individually. That is, we can simply conclude that both of Merkle-Hellman scheme and Chor-Rivest scheme are vulnerable since the our proposed density is low enough.

2.3 Success Probability of Reduction to SVP/CVP [10]

In this subsection, we review the success probability of reduction to SVP/CVP from the subset sum problem. The main part of this subsection comes from Nguyen-Stern's paper [10].

In evaluating the probability of reduction, the following value is important.

Definition 4. *We define $N(n,r)$ be the number of integer points in the n-dimensional sphere of radius $\sqrt{r}$ centered at the origin.*

It is known that $N(n,r)$ is known to be exponential in n if r is proportional to n. Especially, it is known that $N(n,n/2) \leq 2^{c_0 n}$, $N(n,n/4) \leq 2^{c_1 n}$ and $(c_0, c_1) = (1.54724, 1.0628)$. These analysis are done by Mazo-Odlyzko [8]. We should notice that $d < 1/c_0 = 0.6463$ is the critical density of Lagarias-Odlyzko's attack and $d < 1/c_1 = 0.9408$ is the critical density of Coster et al's attack.

Nguyen and Stern pointed out that Mazo-Odlyzko's analysis is not useful if $k \ll n$. Then, they gave a simple but effective upper bound of $N(n,r)$ as follows [10].

Lemma 1. *For any positive integers n and r, it holds that*

$$N(n, r) \leq 2^r \binom{n + r - 1}{r}. \tag{4}$$

By simple calculation, they obtained another (looking) bound of $N(n, r)$ [10].

Lemma 2. *For any positive integers n and r, it holds that*

$$N(n, r) \leq \frac{2^r e^{r(r-1)/(2n)} n^r}{r!}.$$

The *failure* probability of reduction to *CVP* that is, *failure* probability of solving knapsack scheme even if we can use CVP-oracle, is given as follows [10].

L is set of vectors $(z_1, \ldots, z_n) \in \mathbb{Z}^n$ such that $z_1 a_1 + z_2 a_2 + \cdots + z_n a_n = 0$. In this setting, the set is a lattice. The dimension of the lattice L is $n - 1$. Let $y_1, \ldots, y_n$ be integers such that $s = \sum_{i=1}^n y_i a_i$. These y_i's are computable in polynomial time.

Lemma 3. *Let $\boldsymbol{m} = (m_1, \ldots, m_n) \in \mathbb{Z}^n$. Let $a_1, \ldots, a_n$ be chosen uniformly and independently at random in $[0, A]$. Let $s = \sum_{i=1}^n m_i a_i$. Let $\boldsymbol{c}$ be the closest vector in L to $\boldsymbol{t} = (y_1, \ldots, y_n)$. Then, the probability $\Pr$ that $\boldsymbol{c} \neq (y_1 - m_1, \ldots, y_n - m_n)$ is given by*

$$\Pr < \frac{N(n, r)}{A}. \tag{5}$$

The *failure* probability of reduction to *SVP* is given as follows [10]. Let L and y_i's be the same as CVP case. Let $(\boldsymbol{b_1}, \ldots, \boldsymbol{b_{n-1}})$ be a basis of L. Let L' be the lattice spanned by $(1, y_1, \ldots, y_n) \in \mathbb{Z}^{n+1}$ and the $n - 1$ vectors $(0, \boldsymbol{b_i}) \in \mathbb{Z}^{n+1}$. Let $\boldsymbol{m'} = (1, m_1, \ldots, m_n) \in \mathbb{Z}^{n+1}$. Obviously, $\boldsymbol{m'} \in L'$ and its norm is relatively short. In this setting, we have the following lemma.

Lemma 4. *Let $\boldsymbol{m} = (m_1, \ldots, m_n) \in \mathbb{Z}^n$. Let $a_1, \ldots, a_n$ be chosen uniformly and independently at random in $[0, A]$. Let $s = \sum_{i=1}^n m_i a_i$. Let $\boldsymbol{s}$ be the shortest vector in L'. The probability $\Pr$ that $\boldsymbol{s} \neq \pm \boldsymbol{m'}$ is given by*

$$\Pr < (1 + 2\sqrt{1 + r^2}) \frac{N(n, r)}{A}. \tag{6}$$

Nguyen and Stern proved that the low weight knapsack schemes can be solved by a single call to an SVP/CVP-oracle by using Lemmas 2, 3 and 4.

2.4   Review of Information Theory

Both of Chor-Rivest scheme and OTU scheme expand a real m-bit random message into n-bit message with Hamming weight k as preprocessing of encryption. We should notice that message is usually expanded before encryption. Fig. 1 shows the process flow for actual low weight knapsack scheme. Concretely speaking, (1) m-bit message is expanded to n-bit message with Hamming weight k, (2) n-bit message is encrypted into a ciphertext, (3) the ciphertext is decrypted into n-bit message, (4) n-bit message is decoded to m-bit original message.

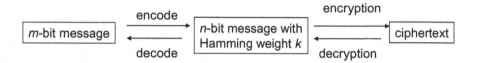

Fig. 1. Knapsack Cryptosystems with Preprocessing

We should notice that $n > m$ and hence the usual density become larger. Although both schemes recommend to use enumerative source encoding [5], we do not care about concrete encoding methods in this paper.

The space of m-bit messages is 2^m and the space of n-bit messages with Hamming weight k is given by $\binom{n}{k}$. There exists adequate encoding and decoding method which correspond each message on (almost) one-to-one. By increasing the message length from m to n, the Hamming weight of message decreases $m/2$ (in average) to k. We should notice that *information* will never increase by deterministic processing. Hence, you may think that expanded message has more information and knapsack scheme becomes more secure by decreasing Hamming weight. However, this is not correct since Shannon Entropy never increase by this kind of deterministic message expansion.

It is well known that the following inequalities hold between the number of combination and Shannon Entropy.

Lemma 5 ([4]). *For any positive integers n and k, it holds that*

$$\frac{1}{n+1}2^{nH(k/n)} \leq \binom{n}{k} \leq 2^{nH(k/n)}, \tag{7}$$

where $H(x)$ is an Entropy function: $H(x) = -x \log x - (1-x) \log(1-x)$. Note that the base of logarithm is 2.

Proof is given in pp. 284–285 of [4].

Remark 1. For large n, we have a rough approximation:

$$\binom{n}{k} \approx 2^{nH(k/n)}. \tag{8}$$

from Lemma 5. In an intuitive discussion, we will use Eq. (8) instead of Lemma 5 for easy understanding. However, we will use Lemma 5 in most of our analysis.

We will analyze the failure probability of reduction to SVP/CVP by using Lemmas 3 and 4 as in the analysis of Nguyen-Stern. The difference of our analysis from Nguyen-Stern's analysis is that we use Lemma 5 instead of Lemma 2. We obtain more precise bound than Nguyen-Stern's.

3   Necessary Conditions for Secure Knapsack Scheme

In this section, first, we introduce a new notion of density. Our density unifies previously proposed two density. Next, we reevaluate the security of low weight knapsack scheme using our density.

3.1   New Definition of Density

We define a new notion of density as follows.

Definition 5. *Let knapsack* $\{a_1, a_2, \ldots, a_n\}$ *and* $A = \max_{1 \le i \le n}\{a_i\}$. *Let* k *be the fixed Hamming weight of the solution of subset sum problem. Then, density* D *is defined by*

$$D = \frac{nH(p)}{\log A}, \text{where } p = k/n. \tag{9}$$

Our definition can be regarded as a kind of density of multiplying normalization factor $H(p)$ to usual density d. That is, $D = H(p)d$.

Next, we show that our density D unifies two already proposed density and includes them as a special case. That is to say, our density is almost equivalent to

(1) usual density when the message is randomly chosen, and
(2) pseudo-density when $k \ll n$.

We explain the details of each case.

(1) the case that the message is randomly chosen : By the law of large numbers, $k \approx n/2$ with overwhelming probability. Hence, for almost case, $p = 1/2$ and $H(p) = 1$. We have $D = n/\log A = d$.
(2) low weight case ($k \ll n$): Letting $k = n^\alpha$, where $\alpha < 1$. $nH(k/n)$ can be transformed as follows.

$$nH(k/n) = -n\left(\frac{k}{n}\log\frac{k}{n} + \frac{n-k}{n}\log\frac{n-k}{n}\right)$$

$$= k\log n - k\log k - (n-k)\log(1 - \frac{k}{n})$$

Here, Maclaurin's expansion of $\log_e(1-p)$ is given by

$$\log_e(1-x) = -\sum_{i=1}^{\infty}\frac{1}{i}x^i,$$

where e is the base of natural logarithm. The third term $-(n-k)\log(1-\frac{k}{n})$ can be evaluated by

$$
\begin{aligned}
-(n-k)\log\left(1-\frac{k}{n}\right) &= -k\left(\frac{1}{p}-1\right)\frac{\log_e(1-p)}{\log_e 2} \\
&= \frac{k}{\log_e 2}\left(\sum_{i=1}^{\infty}\frac{1}{i}x^{i-1} - \sum_{i=1}^{\infty}\frac{1}{i}x^i\right) \\
&= \frac{k}{\log_e 2}\left(1 + \sum_{i=1}^{\infty}\frac{1}{i+1}x^i - \sum_{i=1}^{\infty}\frac{1}{i}x^i\right) \\
&= \frac{k}{\log_e 2}\left(1 - \sum_{i=1}^{\infty}\frac{1}{i(i+1)}x^i\right)
\end{aligned}
$$

This term is obliviously upper bounded by $k/\log_e 2$. We have the lower bound of this term as

$$
\frac{k}{\log_e 2}(1 - \sum_{i=1}^{\infty}\frac{1}{i(i+1)}x^i) > \frac{k}{\log_e 2}(1 - \sum_{i=1}^{\infty}\frac{1}{i}x^i) = \frac{k}{\log_e 2}(1 + \log_e(1-p))
$$

Hence, this term is obviously negligible to the first term $k\log n$ for large n and small k. The first and second terms is evaluated by $(1-\alpha)k\log n$. Hence, if $\alpha \to 0$, it will go to $k\log n$ and if α is small enough, we can approximate $nH(k/n)$ as $k\log n$. Hence, we have

$$
D \approx \frac{k\log n}{\log A} = \kappa. \tag{10}
$$

3.2   Necessary Condition for Unique Decryptability

Unless the ciphertext space is larger than message space, the message cannot be recovered from ciphertext uniquely. The message space is given by $\binom{n}{k}$ and the ciphertext space is upper bounded by kA. Hence, a necessary condition for unique decryptability is given as follows.

$$
\binom{n}{k} \leq kA.
$$

From Lemma 5, we know $2^{nH(k/n)}/(n+1) \leq \binom{n}{k}$. By combining these two inequalities, we have the following lemma.

Lemma 6. *Let knapsack $\{a_1, a_2, \ldots, a_n\}$ and $A = \max_{1\leq i\leq n}\{a_i\}$. If the knapsack scheme is uniquely decryptable, it holds that*

$$
D < 1 + \frac{2}{n}. \tag{11}
$$

Proof. From Lemma 5, we have $nH(k/n) \leq \log A + \log k + \log(n+1)$. Then, we have

$$D = \frac{nH(k/n)}{\log A} \leq 1 + \frac{\log k(n+1)}{\log A}. \tag{12}$$

Let ϵ be $\frac{\log(k(n+1))}{\log A}$. Subsisting this into Eq.(12), we have

$$\epsilon \leq \frac{\log k(n+1)}{nH(p) - \log k(n+1)}.$$

ϵ is upper bounded by $2/n$ as follows.

$$\epsilon \leq \frac{2\log n}{nH(k/n)} = \frac{2\log n}{-k\log k + k\log n - (n-k)\log(n-k) + (n-k)\log n}$$
$$< \frac{2\log n}{n\log n} = \frac{2}{n}.$$

Hence, we have the lemma. □

Hence, if the scheme is uniquely decryptable, our density D must be $D \leq 1+2/n$. If n is large enough, the upper bound of D goes to 1.

3.3   Necessary Condition for Preventing Reduction to CVP

We have the following theorem in regard to the critical bound of lattice attack.

Theorem 1. *Assume that the a_i are chosen uniformly at random from $[0, A]$. If our density D satisfies*

$$D < \frac{H(p)}{p + (1+p)H(p/(1+p))}, \tag{13}$$

the knapsack scheme can be solved in a single call to CVP-oracle with high probability depending on n.

Remark 2. Eq. (13) is valid for higher weight knapsack in addition to low weight case. On the other hand, Nguyen-Stern's bound is valid for only low weight knapsack.

Proof. From Lemma 3, the failure probability Pr of reduction is given by $\Pr < N(n,k)/A$. For simplicity, we rewrite $N(n,k)$ by

$$N(n,k) = 2^{nf_n(p)}, \text{where } p = k/n. \tag{14}$$

From Lemma 3 and $D = nH(p)/\log A$, we have

$$\log \Pr < \log \frac{N(n,k)}{A} = nf_n(p) - \log A$$
$$= nf_n(p) - n\frac{H(p)}{D} = n\left(f_n(p) - \frac{H(p)}{D}\right). \tag{15}$$

If $f_n(p) - \frac{H(p)}{D}$ is negative, the reduction to CVP succeeds with high probability for large n. Then, the condition such that the reduction to CVP succeeds is given by

$$D < \frac{H(p)}{f_n(p)}. \tag{16}$$

Next, we evaluate $f_n(p)$ more precisely. If p is constant, the value of $f_n(p)$ is analyzed precisely by Mazo and Odlyzko [8]. Especially, $f_n(1/2) = 1.54724 \cdots$ and $f_n(1/4) = 1.0628 \cdots$. In these cases, $f_n(p)$ does not depend on n. Hereafter, we omit the subscript n and simply write $f(p)$ if not necessary. It is possible to obtain the exact value of $f(p)$ for each p by Mazo-Odlyzko's analysis. For $p = 1/2$, if D is less than $H(1/2)/f(1/2) = 0.6465$, the knapsack scheme can be solved in a single call to CVP-oracle. Hence, our result includes Lagarias-Odlyzko's result.

The drawback of using Mazo-Odlyzko's analysis is that it is difficult to calculate the exact value of $f_n(p)$. Next, we will obtain a simple bound of $f_n(p)$. From Lemmas 1 and 5, we have

$$N(n, k) \leq 2^k \binom{n+k}{k} \leq 2^k 2^{(n+k)H(\frac{k}{n+k})} = 2^{n(p+(1+p)H(p/(1+p)))}.$$

Hence, we have

$$f_n(p) \leq p + (1+p)H\left(\frac{p}{1+p}\right). \tag{17}$$

Therefore, the upper bound of $f_n(p)$ does not depend on n but only p. We should notice that this bound is not tighter than Mazo-Odlyzko's bound.

Summing up the above discussion, we obtain the theorem.   □

3.4   Necessary Condition for Preventing Reduction to SVP

From Lemma 4, the failure probability of reduction to SVP is given by

$$\Pr < (1 + 2\sqrt{1 + r^2})\frac{N(n, r)}{A}.$$

By the similar analysis of CVP case, we have the same bound. That is, if

$$D < \frac{H(p)}{p + (1+p)H(p/(1+p))},$$

we can solve knapsack scheme by a single call to SVP-oracle.

3.5   Necessary Conditions for Secure Scheme

From the above discussion, we have the necessary condition that the knapsack scheme is secure as follows:

$$\frac{H(p)}{p + (1+p)H(p/(1+p))} < D < 1 + \frac{2}{n}. \tag{18}$$

Now, we write

$$g_{LO}(p) \equiv \frac{H(p)}{p + (1+p)H(p/(1+p))}. \tag{19}$$

We can easily verify that

$$\lim_{p \to 0} g_{LO}(p) = 1. \tag{20}$$

This result leads to the following claim.

Claim. It is quite difficult to construct a low weight knapsack scheme which is supported by an argument of density.

The reason is as follows. If a knapsack scheme such that p asymptotically goes to zero is secure, its density D must be 1. This observation implies that it is quite difficult to construct such a scheme.

Substituting $p = 1/2$ into Eq. (13), we have

$$D < \frac{H(1/2)}{1/2 + 3/2H(1/3)} = 0.5326. \tag{21}$$

This condition is not so tight since the tight bound is given by $H(1/2)/f(1/2) = 0.6465$. But, we can easily calculate the bound Eq. (13).

The graph of $g_{LO}(p)$ is given by Fig. 2. As p increases, $g_{LO}(p)$ monotonically decreases. Hence, for any $p \leq 1/2$, $g_{LO}(p) \geq g_{LO}(1/2) = 0.5326$.

As already mentioned, our density D is normalization of usual density d. We can rewrite the above condition by using the usual density d:

$$\frac{1}{p + (1+p)H(p/(1+p))} < d < \frac{1 + 2/n}{H(p)}. \tag{22}$$

When $p \to 0$, the both of left and right terms goes to infinity. Hence, even if the density is infinitely large, it does not guarantee whether the scheme can prevent the lattice attack.

3.6 Improved Bound Based on CJLOSS

Coster et al. improved the bound of Lagarias-Odlyzko's attack. Then, their attack succeeds to break knapsack scheme with higher density knapsack [3]. They also pointed out that their idea can be applied to low weight knapsack scheme such as Chor-Rivest scheme. In this section, we apply this improvement to the case of our density. The analysis is based on Nguyen-Stern's analysis, which means that we use Theorem 3 in [10]. The failure probability of reduction to CVP is given as follows. Note that the lattice L is the same as Lemma 3.

Lemma 7. *Let* $\boldsymbol{m} = (m_1, \ldots, m_n) \in \mathbb{Z}^n$ *and* $s = \sum_{i=1}^{n} m_i a_i$. *We let* $\boldsymbol{c}$ *be the closest vector in* L *to* $\boldsymbol{t} = (y_1 - k/n, \ldots, y_n - k/n)$. *The probability* $\Pr$ *that* $\boldsymbol{c} \neq (y_1 - m_1, \ldots, y_n - m_n)$ *is given by*

$$\Pr < \frac{N(n, r - k^2/n)}{A}. \tag{23}$$

By the similar discussion in Section 3.3, we have

$$N(n, k - k^2/n) \leq 2^{n(p-p^2+(1+p-p^2)H(\frac{1}{1+p-p^2}))}. \tag{24}$$

Then we obtain the improved bound for low lattice attack as follows.

Theorem 2. *Assume that the a_i are chosen uniformly and at random from $[0, A]$. If our density D satisfies*

$$D < \frac{H(p)}{p - p^2 + (1 + p - p^2)H(1/(1 + p - p^2))} \equiv g_{CJ}(p) \tag{25}$$

the knapsack scheme can be solved in a single call to CVP-oracle.

Remark 3. If p is very small ($p \ll p^2$), the bound $g_{CJ}(p)$ is almost equal to $g_{LO(p)}$. Hence, in this case, the effect of this improvement is very small, which is also pointed out in Nguyen-Stern [10]. However, if p is not so small, this improvement is effective.

It also holds that $\lim_{p \to 0} g_{CJ}(p) = 1$. Substituting $p = 1/2$ into Eq. (25), we have

$$\frac{H(1/2)}{1/4 + 5/4H(1/5)} = 0.8677 < D. \tag{26}$$

This condition is not so tight since the tight bound is given by $H(1/2)/f(1/4) = 0.9408$. But, we can easily calculate the bound Eq. (25).

The graph of $g_{CJ}(p)$ is given by Fig. 2. As p increases, $g_{CJ}(p)$ also monotonically decreases. Hence, for any $p \leq 1/2$, $g_{CJ}(p) \geq g_{CJ}(1/2) = 0.8677$. This implies that the lattice attack is always applicable if $D < 0.8677$, which does not depend on the actual value p. We should notice that the above discussion is effective for any choice of public key and trapdoor.

Interestingly, $g_{CJ}(p)$ can be simply and roughly approximated by a constant: $g_{CJ}(p) = 0.87$ for $p > 0.08$. Furthermore, $g_{LO}(p)$ can be simply approximated by $g_{LO}(p) = 0.87 - \frac{2}{3}p$ for $p > 0.08$.

Remark 4. In SVP case, we have the same bound as the CVP case. So, we omit the details.

Remark 5. Since Eq. (25) does not give a tight bound, you may think that this bound is useless. However, it is not the case. If the density D of a knapsack scheme satisfies Eq. (25), we can immediately determine whether this scheme is vulnerable to the lattice attack. If not so, we calculate the exact value of $f(p)$ based on Mazo-Odlyzko's analysis and check whether $D < H(p)/f(p)$. Hence, we can use Eq. (25) as the condition of the first cutoff.

Remark 6. We expect that the critical bound can be improved to $D < 0.9408$ as like Coster et al.'s case. If we use exact analysis of $f(p)$, we will be able to prove it. It is enough to prove that $H(p)/f(p)$ monotonically decreases. We have not checked this so far.

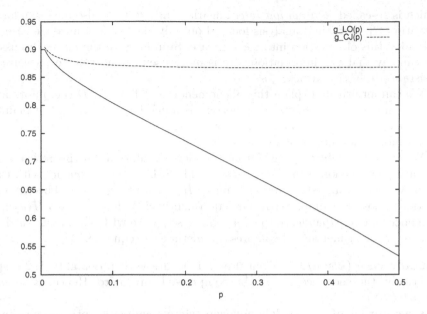

Fig. 2. Critical Bound for lattice attack: $g_{LO}(p)$ and $g_{CJ}(p)$

We would like to emphasize that our density alone is sufficient to measure the resistance to lattice attack for both random knapsack and low weigh knapsack although the usual density and pseudo-density alone are not sufficient.

3.7   Intuitive Meaning of Our New Density

In this subsection, we give an intuitive meaning of our introduced density. From the definition, our density D can be interpreted as

$$D = \frac{\text{the size of solution space}}{\text{the bit length of maximum value of knapsack}}. \tag{27}$$

Let m be a message length before expansion. Letting n be a message length after the expansion and k be the Hamming weight of expanded message, we have

$$\binom{n}{k} \approx 2^m. \tag{28}$$

From Eq. (8), we have $nH(p) \approx m$. Hence, our density D can be transformed as

$$D = \frac{m}{\log A}. \tag{29}$$

This can be interpreted as

$$D \approx \frac{\text{the bit length of message}}{\text{the bit length of ciphertext}}, \tag{30}$$

which is so-called *information rate*. Lagarias and Odlyzko also gave similar observation [7]. But, their analysis focused on only the random message case. We extended this observation into general case. Summing up the above discussion, we can say that the information rate is more useful indication to measure the resistance to lattice attack.

We can intuitively explain this phenomenon as follows. If D (i.e. information rate) is small, the bit length of ciphertext is much larger than that of plaintext and the ciphertext has more information about plaintext. Then, it tends to reveal more information about plaintext.

We try to explain the above from the view of information theoretical sense. From Eq. (8), we know the followings. (1) The n-bit binary sequence with Hamming weight k can be compressed into $nH(k/n)$-bit sequence. This transformation is lossless compression. Or, equivalently, (2) the random $nH(k/n)$-bit sequence can be expanded into n-bit binary sequence with Hamming weight k.

Next, we consider actual compression method of sequence.

random case ($k \approx n/2$)**.** When 0 and 1 occurs with probability $1/2$ respectively, the sequence cannot be compressed any more. Hence, $n = m = nH(1/2)$.

low weight case($k \ll n$)**.** The following simple coding is optimal encoding in extremely low weight case.

Step1 Record the place where the bit is "1".

Step2 Represent that place in the binary representation.

Step3 Concatenate all binary sequence obtained in Step2.

To represent the bit place, $\log n$-bit is needed. Since the number of bit whose value is "1" is k, the length of transformed sequence is $k \log n$. Hence, $m \approx k \log n$.

4   Application to Some Previously Proposed Schemes

In this section, we apply the results in Section 3 to some previously proposed low weight knapsack schemes: Chor-Rivest scheme [2] and OTU scheme [11]. Actually, these schemes are proved to be vulnerable to lattice attack by Nguyen-Stern [10]. However,we obtain more precise results than their results by using our analysis.

4.1   Application to Chor-Rivest Scheme

Chor and Rivest proposed a knapsack cryptosystem with low Hamming weight in 1988 [2]. Schnorr and Hörner broke this scheme for some parameters by using an improved lattice reduction technique. In 2001, Vaudenay [18] broke this scheme for all proposed parameters by using algebraic structures specific to the Chor-Rivest scheme. His attack is categorized as key-only attack and is not lattice attack. In this section, we apply our results to the Chor-Rivest scheme to clarify the difference between our analysis and Nguyen-Stern's analysis.

In Chor-Rivest scheme, A is set as $A = n^k$. Hence, letting $p = k/n$, we have

$$D = \frac{H(p)}{p \log n}.$$

If $D < g_{LO}(p)$, we can conclude that the Chor-Rivest scheme is vulnerable to the lattice attack. This condition can be transformed into

$$p \log n > p + (1 + p)H(p/(1 + p)).$$

We can easily verify that $p + (1+p)H(p/(1+p))$ is lower bounded by $H(p)$. Hence, we can simply write the condition as $\log n > H(p)/p$. On the other hand, from the unique decryptability, $\log n > H(p)/p$ is always satisfied. Hence, Chor-Rivest scheme is vulnerable to the lattice attack in any parameter setting.

Next, we show some numerical examples. Table 1 shows usual density d, pseudo-density κ, our density D, the critical bound $g_{LO}(p)$ and $g_{CJ}(p)$ for Chor-Rivest parameters in proposed in [2]. All parameters proposed in [2] is set as $d \geq 1$ to prevent lattice attack. However, since $D < g_{LO}(p)$ (and also $D < g_{CJ}(p)$) for any parameters, the reduction to SVP/CVP cannot be prevented.

Table 1. Application to the Chor-Rivest parameters

n	k	A	d	κ	D	$g_{LO}(p)$	$g_{CJ}(p)$
197	24	182 bit	1.08	1.005	0.58	0.79	0.87
211	24	185 bit	1.14	1.002	0.58	0.79	0.87
243	24	190 bit	1.28	1.001	0.59	0.80	0.87
256	25	200 bit	1.28	1.00	0.59	0.81	0.87

4.2 Application to OTU Scheme

Okamoto et al. showed the following two kinds of parameter setting [11].

1. For $c < 1$, $k = 2^{(\log n)^c}$.
2. $k = O(\frac{n}{\log n})$.

For two setting, if $n \to \infty$, $p = k/n \to 0$. They claimed that since density d is at least 1, (actually d is asymptotically ∞), the lattice attack can be prevented [11]. However, this claim is not correct.

We will show one typical numerical example. Consider the following parameter setting: $p = k/n = 1/10$, $A \approx 2^n$. In this setting, usual density is $d = 1$ and our density is $D = 0.47$. It holds that $D < g_{LO}(1/10) = 0.80$ (and also $D < g_{CJ}(1/10)$). So, in this setting, the reduction to SVP/CVP is not prevented. Hence, the security of OTU scheme is not supported by an argument based on density. It is strongly believed that if the lattice dimension is sufficiently high, SVP/CVP cannot be solved in polynomial time or practical computational time. Hence, the security of OTU should be based on the difficulty of SVP and CVP and the parameter for OTU should be carefully chosen such that SVP and CVP will not be solved.

5   Algorithms for Directly Solving Subset Sum Problem

Some algorithms for *directly* solving subset sum problem have been proposed. Schroppel and Shamir proposed the algorithm with computational time: $O(2^{n/2})$ and storage: $O(2^{n/4})$ [14]. Their algorithm does not use any inner structure.

Coppersmith proposed an algorithm which solves subset sum problem with promise that the Hamming weight of solution is known and limited to k [16]. This algorithm is based on meet-in-the middle technique. Its complexity is estimated by $O(\sqrt{\binom{n}{k}})$. From Eq. (8), it is approximated by $O(\sqrt{2^m})$. Chor and Rivest also proposed another Coppersmith-like algorithm [2]. Since the solution space is given by 2^m, the computational cost for exhaustive search is given by 2^m. Intuitively, their algorithm achieves the square root speed up. The above fact mentions that the superficial message length n (or the size of knapsack) is not important itself. The actual message length m (or the size of solution space) is rather important. The above discussion supports the validity of our proposed density.

6   Concluding Remarks

In this paper, we revisited the lattice attack to knapsack cryptosystems and introduced a new notion of density. Our density unifies naturally two density: usual density and pseudo-density. Then, we derived the necessary conditions for our density that lattice attack succeeds. We also showed that the lattice attack almost cannot be prevented by decreasing the Hamming weight of messages.

Our derived conditions are useful for judging whether a target knapsack scheme is vulnerable to lattice attack. The following is an explicit procedure.

Step 1. Calculate $D = nH(k/n)/\log A$ by n, k and A.
Step 2. If $D < 0.8677$, we decide the scheme is vulnerable to lattice attack and stop.
Step 3. If $D < g_{CJ}(k/n)$, we decide the scheme is vulnerable and stop.
Step 4. If $D < H(k/n)/f(p - p^2)$, we decide the scheme is vulnerable. Otherwise, we decide the scheme is secure against lattice attack.

The above procedure is valid for any values of Hamming weight. In Steps 1–3 of this procedure, we need not do complicated calculation. Hence, our result can be used for simple judgment.

Acknowledgment

We thank anonymous reviewers for many useful comments and suggestions. We also thank Phong Q. Nguyen, Kazuo Ohta and Keisuke Kitahara for helpful discussions.

References

1. Ajtai, M.: The Shortest Vector Problem in L^2 is NP-hard for Randomized Reductions. In: Proc. of 30th STOC, ACM, pp. 11–20 (1998)
2. Chor, B., Rivest, R.L.: A Knapsack-type Public Key Cryptosystem based on Arithmetic in Finite Fields. In: Blakely, G.R., Chaum, D. (eds.) CRYPTO 1984. LNCS, vol. 196, pp. 54–65. Springer, Heidelberg (1985)
3. Coster, M.J., Joux, A., LaMacchia, B.A., Odlyzko, A.M., Schnorr, C.P., Stern, J.: Improved Low-Density Subset Sum Algorithms. Computational Complexity 2, 111–128 (1992)
4. Cover, T.M., Thomas, J.A.: Elements of Information Theory. Wiley-Interscience, Chichester (1991)
5. Cover, C.M.: Enumerative Source Encoding. IEEE Trans. on Information Theory IT 19, 901–909 (1973)
6. Izu, T., Kogure, J., Koshiba, T., Shimoyama, T.: Low-Density Attack Revisited. Designs, Codes and Cryptography 43(1), 47–59 (2007) (Preliminary version appeared in 2004)
7. Lagarias, J.C., Odlyzko, A.M.: Solving Low-Density Subset Sum Problems. Journal of the Association for Computing Machinery 32(1), 229–246 (1985)
8. Mazo, J.E., Odlyzko, A.M.: Lattice Points in High-Dimensional Spheres. Monatsh Math. 110, 47–61 (1990)
9. Merkle, R.C., Hellman, M.E.: Hiding Information and Signatures in Trapdoor Knapsacks. IEEE Trans. on Information Theory 24, 525–530 (1978)
10. Nguyen, P., Stern, J.: Adapting Density Attacks to Low-Weight Knapsacks. In: Roy, B. (ed.) ASIACRYPT 2005. LNCS, vol. 3788, pp. 41–58. Springer, Heidelberg (2005)
11. Okamoto, T., Tanaka, K., Uchiyama, S.: Quantum Public-Key Cryptosystems. In: Bellare, M. (ed.) CRYPTO 2000. LNCS, vol. 1880, pp. 147–165. Springer, Heidelberg (2000)
12. Omura, K., Tanaka, K.: Density Attack to Knapsack Cryptosystems with Enumerative Source Encoding. IEICE Transactions on Fundamentals 87-A(6), 1564–1569 (2004)
13. Schnorr, C.P., Hörner, H.: Attacking the Chor-Rivest Cryptosystem by Improved Lattice Reduction. In: Guillou, L.C., Quisquater, J.-J. (eds.) EUROCRYPT 1995. LNCS, vol. 921, pp. 1–12. Springer, Heidelberg (1995)
14. Schroppel, R., Shamir, A.: A $T = O(2^{n/2})$, $S = O(2^{n/4})$ Algorithm for Certain NP-Complete Problems. SIAM J.Computing 10(3), 456–464 (1981)
15. Shor, P.W.: Algorithms for Quantum Computation: Discrete Log and Factoring. In: Proc. of 35th FOCS, pp. 124–134 (1994)
16. Stinson, D.R.: Some Baby-Step Giant-Step Algorithms for the Low Hamming Weight Discrete Logarithm Problem. Mathematics of Computation 71, 379–391 (2002)
17. van Emde Boas, P.: Another NP-complete Problem and the Complexity Computing Short Vectors in a Lattice. TR81-04, Mathematische Instituut, University of Amsterdam (1981)
18. Vaudenay, S.: Cryptanalysis of the Chor-Rivest Cryptosystem. Journal of Cryptology 14, 87–100 (2001)

Another Generalization of Wiener's Attack on RSA

Abderrahmane Nitaj

Laboratoire de Mathématiques Nicolas Oresme
Université de Caen, France
BP 5186, 14032 Caen Cedex, France
http://www.math.unicaen.fr/~nitaj,
nitaj@math.unicaen.fr

Abstract. A well-known attack on RSA with low secret-exponent d was given by Wiener in 1990. Wiener showed that using the equation $ed - (p - 1)(q - 1)k = 1$ and continued fractions, one can efficiently recover the secret-exponent d and factor $N = pq$ from the public key (N, e) as long as $d < \frac{1}{3}N^{\frac{1}{4}}$. In this paper, we present a generalization of Wiener's attack. We show that every public exponent e that satisfies $eX - (p - u)(q - v)Y = 1$ with

$$1 \le Y < X < 2^{-\frac{1}{4}}N^{\frac{1}{4}}, \ |u| < N^{\frac{1}{4}}, \ v = \left[-\frac{qu}{p - u} \right],$$

and all prime factors of $p - u$ or $q - v$ are less than 10^{50} yields the factorization of $N = pq$. We show that the number of these exponents is at least $N^{\frac{1}{2} - \varepsilon}$.

Keywords: RSA, Cryptanalysis, ECM, Coppersmith's method, Smooth numbers.

1 Introduction

The RSA cryptosystem invented by Rivest, Shamir and Adleman [20] in 1978 is today's most important public-key cryptosystem. The security of RSA depends on mainly two primes p, q of the same bit-size and two integers e, d satisfying $ed \equiv 1 \pmod{(p-1)(q-1)}$. Throughout this paper, we label the primes so that $q < p < 2q$. The RSA modulus is given by $N = pq$ and Euler's totient function is $\phi(N) = (p-1)(q-1)$. The integer e is called the public (or encrypting) exponent and d is called the private (or decrypting) exponent.

To reduce the decryption time or the signature-generation time, one may wish to use a short secret exponent d. This was cryptanalysed by Wiener [22] in 1990 who showed that RSA is insecure if $d < \frac{1}{3}N^{0.25}$. Wiener's method is based on continued fractions. These results were extended by Boneh and Durfee [3] in 1999 to $d < N^{0.292}$. The method of Boneh and Durfee is based on Coppersmith's results for finding small solutions of modular polynomial equations [6]. In 2004,

S. Vaudenay (Ed.): AFRICACRYPT 2008, LNCS 5023, pp. 174–190, 2008.

Blömer and May [2] presented a generalization of Wiener's attack by combining continued fractions and Coppersmith's method. They showed that RSA is insecure for every (N, e) satisfying $ex + y \equiv 0 \pmod{\phi(N)}$ with $x < \frac{1}{3}N^{1/4}$ and $|y| = O\left(N^{-3/4}ex\right)$.

In this paper, we present another generalization of Wiener's attack. Our method combines continued fractions, integer partial factorization, integer relation detection algorithms and Coppersmith's method. Let us introduce the polynomial

$$\psi(u, v) = (p - u)(q - v).$$

Observe that $\psi(1, 1) = (p - 1)(q - 1) = \phi(N)$, so ψ could be seen as a generalization of Euler's function. We describe an attack on RSA that works for all public exponents e satisfying

$$eX - \psi(u, v)Y = 1, \tag{1}$$

with integers X, Y, u, v such that

$$1 \leq Y < X < 2^{-\frac{1}{4}}N^{\frac{1}{4}}, \quad |u| < N^{\frac{1}{4}}, \quad v = \left[-\frac{qu}{p - u}\right],$$

with the extra condition that all prime factors of $p - u$ or $q - v$ are less than the Elliptic Curve Method of Factoring smoothness bound $B_{\text{ecm}} = 10^{50}$. Here and throughout this paper, we let $[x]$ and $\lfloor x \rfloor$ denote the nearest integer to the real number x and the fractional part of x.

Observe that when $u = 1$, we get $v = -1$ and rewriting (1) as

$$eX - (p - 1)(q + 1)Y = 1,$$

a variant of Wiener's attack enables us to compute p and q without assuming any additional condition on the prime divisors of $p - 1$ nor $q + 1$.

Our new method works as follows: We use the Continued Fraction Algorithm (see e.g. [11], p. 134) to find the unknowns X and Y among the convergents of $\frac{e}{N}$. Then we use Lenstra's Elliptic Curve Factorization Method (ECM) [14] to partially factor $\frac{eX-1}{Y}$. Afterwards, we use an integer relation detection algorithm (notably LLL [15] or PSLQ [7]) to find the divisors of the B_{ecm}-smooth part of $\frac{eX-1}{Y}$ in a short interval. Finally, we show that a method due to Coppersmith [6] can be applied. Moreover, we show that the number of keys (N, e) for which our method works is at least $N^{\frac{1}{2}-\varepsilon}$.

Organization of the paper. Section 2 presents well known results from number theory that we use. After presenting some useful lemmas in Section 3, and some properties of ψ in Section 4, we present our attack in Section 5 and in Section 6, we show that the number of keys (N, e) for which our method works is lower bounded by $N^{\frac{1}{2}-\varepsilon}$. We briefly conclude the paper in Section 7.

2   Preliminaries

2.1   Continued Fractions and Wiener's Attack

The continued fraction expansion of a real number ξ is an expression of the form

$$\xi = a_0 + \cfrac{1}{a_1 + \cfrac{1}{a_2 + \cfrac{1}{a_3 + \cdots}}}$$

where $a_0 \in \mathbb{Z}$ and $a_i \in \mathbb{N} - \{0\}$ for $i \geq 1$. The numbers $a_0, a_1, a_2, \cdots$ are called the partial quotients. As usual, we adopt the notation $\xi = [a_0, a_1, a_2, \cdots]$. For $i \geq 0$, the rationals $\frac{r_i}{s_i} = [a_0, a_1, a_2, \cdots, a_i]$ are called the convergents of the continued fraction expansion of ξ. If $\xi = \frac{a}{b}$ is rational with $\gcd(a, b) = 1$, then the continued fraction expansion is finite and the Continued Fraction Algorithm (see [11], p. 134) finds the convergents in time $O((\log b)^2)$. We recall a result on diophantine approximations (see Theorem 184 of [11]).

Theorem 1. *Suppose* $\gcd(a, b) = \gcd(x, y) = 1$ *and*

$$\left| \frac{a}{b} - \frac{x}{y} \right| < \frac{1}{2y^2}.$$

Then $\frac{x}{y}$ *is one of the convergents of the continued fraction expansion of* $\frac{a}{b}$.

Let us recall Wiener's famous attack on RSA with $N = pq$ and $q < p < 2q$. The idea behind Wiener's attack on RSA [22] with small secret exponent d is that for $d < \frac{1}{3}N^{1/4}$, the fraction e/N is an approximation to k/d and hence, using Theorem 1, k/d can be found from the convergents of the continued fraction expansion of e/N. Wiener's attack works as follows. Since $ed - k\phi(N) = 1$ with $\phi(N) = N - (p + q - 1)$ and $p + q - 1 < 3\sqrt{N}$ then $kN - ed = k(p + q - 1) - 1$. Therefore,

$$\left| \frac{k}{d} - \frac{e}{N} \right| = \frac{|k(p + q - 1) - 1|}{Nd} < \frac{3k\sqrt{N}}{Nd}.$$

Now, assume that $d < \frac{1}{3}N^{1/4}$. Since $k\phi(N) = ed - 1 < ed$ and $e < \phi(N)$, then $k < d < \frac{1}{3}N^{1/4}$. Hence

$$\left| \frac{k}{d} - \frac{e}{N} \right| < \frac{N^{3/4}}{Nd} = \frac{1}{dN^{1/4}} < \frac{1}{2d^2}.$$

From Theorem 1, we know that k/d is one of the convergents of the continued fraction expansion of e/N.

2.2   Coppersmith's Method

The problem of finding small modular roots of a univariate polynomial has been extensively studied by Coppersmith [6], Howgrave-Graham [13], May [17] and others. Let $f(x)$ be a monic univariate polynomial with integer coefficients of degree

δ. Let N be an integer of unknown factorization and $B = N^{1/\delta}$. The problem is to find all integers x_0 such that $|x_0| < B$ and $f(x_0) \equiv 0 \pmod{N}$. In 1997, Coppersmith presented a deterministic algorithm using $(2^\delta \log N)^{O(1)}$ bit operations to solve this problem. The algorithm uses lattice reduction techniques, and as an application, the following theorem was proved (see also [17], Theorem 11).

Theorem 2. *Let $N = pq$ be an RSA modulus with $q < p < 2q$. Given an approximation $\tilde{p}$ of p with $|p - \tilde{p}| < N^{\frac{1}{4}}$, N can be factored in time polynomial in $\log N$.*

2.3   Smooth Numbers

A few words about notation: let f and g be functions of x. The notation $f \asymp g$ denotes that $f(x)/g(x)$ is bounded above and below by positive numbers for large values of x. The notation $f = O(g)$ denotes that $\exists c$ such that $f(x) \leq cg(x)$. The notation $f \sim g$ denotes that $\lim_{x \to \infty} \frac{f(x)}{g(x)} = 1$.

Let y be a positive constant. A positive number n is y-smooth if all prime factors of n are less than y. As usual, we use the notation $\Psi(x, y)$ for the counting function of the y-smooth numbers in the interval $[1, x]$, that is,

$$\Psi(x, y) = \# \{n : 1 \leq n \leq x, \ n \text{ is } y\text{-smooth}\}.$$

The ratio $\Psi(x, y)/[x]$ may be interpreted as the probability that a randomly chosen number n in the interval $[1, x]$ has all its prime factors less than y. The function $\Psi(x, y)$ plays a central role in the running times of many integer factoring and discrete logarithm algorithms, including the Elliptic Curve Method (ECM) [14] and the number field sieve method (NFS) [16]. Let $\rho(u)$ be the Dickman-de Bruijn function (see [9]). In 1986, Hildebrand [12] showed that

$$\Psi(x, y) = x\rho(u) \left\{ 1 + O\left(\frac{\log(u+1)}{\log y} \right) \right\} \quad \text{where} \quad x = y^u \tag{2}$$

holds uniformly in the range $y > \exp\{(\log \log x)^{5/3+\varepsilon}\}$. Studying the distribution in short intervals of integers without large prime factors, Friedlander and Granville [8] showed that

$$\Psi(x + z, y) - \Psi(x, y) \geq c\frac{z}{x}\Psi(x, y), \tag{3}$$

in the range $x \geq z \geq x^{\frac{1}{2}+\delta}$, $x \geq y \geq x^{1/\gamma}$ and x is sufficiently large where δ and γ are positive constants and $c = c(\delta, \gamma) > 0$.

In order to study the distribution of divisors of a positive integer n, Hall and Tenenbaum [10] studied the counting function

$$U(n, \alpha) = \# \left\{ (d, d') : d|n, d'|n, \gcd(d, d') = 1, \left| \log \frac{d}{d'} \right| < (\log n)^\alpha \right\}, \tag{4}$$

where α is a real number. They proved that for any fixed $\alpha < 1$ and almost all n,

$$U(n, \alpha) \leq (\log n)^{\log 3 - 1 + \alpha + o(1)}, \tag{5}$$

where the $o(1)$ term tends to 0 as n tends to $+\infty$.

2.4   ECM

The Elliptic Curve Method (ECM) was originally proposed by H.W. Lenstra [14] in 1984 and then extended by Brent [4] and Montgomery [18]. It is suited to find small prime factors of large numbers. The original part of the algorithm proposed by Lenstra is referred to as Phase 1, and the extension by Brent and Montgomery is called Phase 2. ECM relies on Hasse's theorem: if p is a prime factor of a large number M, then an elliptic curve over $\mathbb{Z}/p\mathbb{Z}$ has group order $p + 1 - t$ with $|t| < 2\sqrt{p}$, where t depends on the curve. If $p + 1 - t$ is a smooth number, then ECM will probably succeed and reveal the unknown factor p. ECM is a sub-exponential factoring algorithm, with expected run time of

$$O\left(\exp\left\{\sqrt{(2 + o(1))\log p \log\log p}\right\} Mult(M)\right)$$

where the $o(1)$ term tends to 0 as p tends to $+\infty$ and $Mult(M)$ denotes the cost of multiplication mod M. The largest factor known to have been found by ECM is a 67-digit factor of the number $10^{381} + 1$, found by B. Dodson with P. Zimmerman's GMP-ECM program in August 2006 (see [23]). According to Brent's formula [5] $\sqrt{D} = (Y - 1932.3)/9.3$ where D is the number of decimal digits in the largest factor found by ECM up to a given date Y, a 70-digit factor could be found by ECM around 2010.

In Table 1, we give the running times obtained on a Intel(R) Pentium(R) 4 CPU 3.00 GHz to factor an RSA modulus $N = pq$ of size $2n$ bits with $q < p < 2q$ with ECM, using the algebra system Pari-GP[19].

Table 1. Running times for factoring $N = pq$ with $q < p < 2q$

n = Number of bits of q	60	70	80	90	100	110	120	130
T = Time in seconds	0.282	0.844	3.266	13.453	57.500	194.578	921.453	3375.719

Extrapolating Table 1, we find the formula

$$\log T = 2.609\sqrt{n} - 21.914 \quad \text{or equivalently} \quad T = \exp\left\{2.609\sqrt{n} - 21.914\right\},$$

where T denotes the running time to factor an RSA modulus $N = pq$ with $2n$ bits. Extrapolating, we can extract a prime factor of 50 digits (≈ 166 bits) in 1 day, 9 hours and 31 minutes. Throughout this paper, we then assume that ECM is efficient to extract prime factors up to the bound $B_{\text{ecm}} = 10^{50}$.

3   Useful Lemmas

In this section we prove three useful lemmas. We begin with a simple lemma fixing the sizes of the prime factors of the RSA modulus.

Lemma 1. *Let $N = pq$ be an RSA modulus with $q < p < 2q$. Then*

$$2^{-\frac{1}{2}}N^{\frac{1}{2}} < q < N^{\frac{1}{2}} < p < \sqrt{2}N^{\frac{1}{2}}.$$

Proof. Assume $q < p < 2q$. Multiplying by p, we get $N < p^2 < 2N$ or equivalently $N^{\frac{1}{2}} < p < \sqrt{2}N^{\frac{1}{2}}$. Since $q = \frac{N}{p}$, we obtain $2^{-\frac{1}{2}}N^{\frac{1}{2}} < q < N^{\frac{1}{2}}$ and the lemma follows.   □

Our second lemma is a consequence of Theorem 2 and Lemma 1.

Lemma 2. *Let $N = pq$ be an RSA modulus with $q < p < 2q$. Suppose $|u| < N^{\frac{1}{4}}$. If $p - u < N^{\frac{1}{2}}$ or $p - u > \sqrt{2}N^{\frac{1}{2}}$, then the factorization of N can be found in polynomial time.*

Proof. Assume $q < p < 2q$ and $|u| < N^{\frac{1}{4}}$. If $p - u < N^{\frac{1}{2}}$, then $p < N^{\frac{1}{2}} + u < N^{\frac{1}{2}} + N^{\frac{1}{4}}$. Combining this with Lemma 1, we obtain

$$N^{\frac{1}{2}} < p < N^{\frac{1}{2}} + N^{\frac{1}{4}}.$$

It follows that $\tilde{p} = N^{\frac{1}{2}}$ is an approximation of p with $0 < p - \tilde{p} < N^{\frac{1}{4}}$. By Theorem 2, we deduce that the factorization of N can be found in polynomial time.

Similarly, if $p - u > \sqrt{2}N^{\frac{1}{2}}$, then $p > \sqrt{2}N^{\frac{1}{2}} + u > \sqrt{2}N^{\frac{1}{2}} - N^{\frac{1}{4}}$ and using Lemma 1, we get

$$\sqrt{2}N^{\frac{1}{2}} > p > \sqrt{2}N^{\frac{1}{2}} - N^{\frac{1}{4}}.$$

It follows that $\tilde{p} = \sqrt{2}N^{\frac{1}{2}}$ satisfies $0 > p - \tilde{p} > -N^{\frac{1}{4}}$. Again, by Theorem 2, we conclude that the factorization of N can be found in polynomial time.   □

Our third lemma is a consequence of the Fermat Factoring Method (see e.g. [21]). We show here that it is an easy consequence of Theorem 2 and Lemma 1.

Lemma 3. *Let $N = pq$ be an RSA modulus with $q < p < 2q$. If $p - q < N^{\frac{1}{4}}$, then the factorization of N can be found in polynomial time.*

Proof. Assume $q < p < 2q$ and $p - q < N^{\frac{1}{4}}$. Combining with Lemma 1, we get

$$N^{\frac{1}{2}} < p < q + N^{\frac{1}{4}} < N^{\frac{1}{2}} + N^{\frac{1}{4}}.$$

It follows that $\tilde{p} = N^{\frac{1}{2}}$ is an approximation of p with $0 < p - \tilde{p} < N^{\frac{1}{4}}$. By Theorem 2, we conclude that the factorization of N can be found in polynomial time.   □

4   Properties of $\psi(u, v)$

Let $N = pq$ be an RSA modulus with $q < p < 2q$. The principal object of investigation of this section is the polynomial $\psi(u, v) = (p - u)(q - v)$ when p and q are fixed.

Lemma 4. *Let u be an integer with $|u| < N^{\frac{1}{4}}$. Put $v = \left[-\frac{qu}{p-u}\right]$. Then*

$$|\psi(u,v) - N| < 2^{-\frac{1}{2}}N^{\frac{1}{2}}.$$

Proof. Since v is the nearest integral value to $-\frac{qu}{p-u}$, then

$$-\frac{1}{2} \leq -\frac{qu}{p-u} - v < \frac{1}{2}.$$

Hence

$$q + \frac{qu}{p-u} - \frac{1}{2} \leq q - v < q + \frac{qu}{p-u} + \frac{1}{2}.$$

Multiplying by $p - u$, we get

$$N - \frac{1}{2}(p-u) \leq (p-u)(q-v) < N + \frac{1}{2}(p-u).$$

It follows that

$$|(p-u)(q-v) - N| \leq \frac{1}{2}(p-u).$$

Since $|u| < N^{\frac{1}{4}}$, then by Lemma 2, we can assume $p - u < \sqrt{2}N^{\frac{1}{2}}$ and we obtain

$$|(p-u)(q-v) - N| \leq 2^{-\frac{1}{2}}N^{\frac{1}{2}}.$$

This completes the proof.   □

Lemma 5. *Let u be an integer with $|u| < N^{\frac{1}{4}}$. Set $v = \left[-\frac{qu}{p-u}\right]$. Then $|v| \leq |u|$.*

Proof. Assume $q < p < 2q$ and $|u| < N^{\frac{1}{4}}$. By Lemma 3, we can assume that $p - q > N^{\frac{1}{4}}$. Then

$$u < N^{\frac{1}{4}} < p - q,$$

and $q < p - u$. Hence

$$|v| = \left[\frac{q|u|}{p-u}\right] \leq \frac{q|u|}{p-u} + \frac{1}{2} < |u| + \frac{1}{2}.$$

Since u and v are integers, then $|v| \leq |u|$ and the lemma follows.   □

Lemma 6. *Let u, u', be integers with $|u|, |u'| < N^{\frac{1}{4}}$. Define*

$$v = \left[-\frac{qu}{p-u}\right] \quad \text{and} \quad v' = \left[-\frac{qu'}{p-u'}\right].$$

If $v = v'$, then $|u' - u| \leq 1$.

Proof. Suppose $v' = v$. Then, from the definitions of v and v', we obtain

$$\left| \frac{qu'}{p - u'} - \frac{qu}{p - u} \right| < 1,$$

Transforming this, we get

$$|u' - u| < \frac{(p - u)(p - u')}{N}.$$

By Lemma 3 we can assume that $p - u < \sqrt{2} N^{\frac{1}{2}}$ and $p - u' < \sqrt{2} N^{\frac{1}{2}}$. Then

$$|u' - u| < \frac{\left(\sqrt{2} N^{\frac{1}{2}} \right)^2}{N} = 2.$$

Since u and u' are integers, the lemma follows. $\square$

Lemma 7. *Let u, u', be integers with $|u|, |u'| < N^{\frac{1}{4}}$. Define*

$$v = \left[-\frac{qu}{p - u} \right] \quad and \quad v' = \left[-\frac{qu'}{p - u'} \right].$$

If $\psi(u, v) = \psi(u', v')$, then $u = u'$.

Proof. Assume that $\psi(u, v) = \psi(u', v')$, that is $(p - u)(q - v) = (p - u')(q - v')$. If $v = v'$, then $p - u = p - u'$ and $u = u'$. Next, assume for contradiction that $v \neq v'$. Without loss of generality, assume that $u > u'$. Put $\psi = \psi(u, v) = \psi(u', v')$ and let $U(\psi, \alpha)$ as defined by (4), i.e.

$$U(\psi, \alpha) = \# \left\{ (d, d') : d | \psi, d' | \psi, \gcd(d, d') = 1, \left| \log \frac{d}{d'} \right| < (\log \psi)^{\alpha} \right\}.$$

Let $g = \gcd(p - u, p - u')$, $d = \frac{p - u}{g}$ and $d' = \frac{p - u'}{g}$. Hence $\gcd(d, d') = 1$. We have

$$\frac{d}{d'} = \frac{p - u}{p - u'} = 1 - \frac{u - u'}{p - u'}.$$

By Lemma 2, we can assume that $p - u > N^{\frac{1}{4}}$. For $N > 2^8$ we have

$$0 < \frac{u - u'}{p - u'} < \frac{2 N^{\frac{1}{4}}}{N^{\frac{1}{2}}} = 2 N^{-\frac{1}{4}} < \frac{1}{2}.$$

Using that $|\log(1 - x)| < 2x$ holds for $0 < x < \frac{1}{2}$ this yields

$$\left| \log \frac{d}{d'} \right| = \left| \log \left(1 - \frac{u - u'}{p - u'} \right) \right| < 2 \times \frac{u - u'}{p - u'} < 2 \sqrt{2} N^{-\frac{1}{4}} = (\log \psi)^{\alpha},$$

where

$$\alpha = \frac{\log \left(2 \sqrt{2} N^{-\frac{1}{4}} \right)}{\log(\log(\psi))}.$$

It follows that $U(\psi, \alpha) \geq 1$. On the other hand, we have

$$\alpha = \frac{\log\left(2\sqrt{2}N^{-\frac{1}{4}}\right)}{\log(\log(\psi))} \leq \frac{\log\left(2\sqrt{2}N^{-\frac{1}{4}}\right)}{\log\left(\log\left(N - 2^{-\frac{1}{2}}N^{\frac{1}{2}}\right)\right)} < 1 - \log 3,$$

where we used Lemma 4 in the medium step and $N > 2^7$ in the final step. Using the bound (5), we have actually

$$U(\psi, \alpha) \leq (\log \psi)^{\log 3 - 1 + \alpha + o(1)} \leq (\log N)^{\delta + o(1)},$$

where $\delta = \log 3 - 1 + \alpha < 0$ and we deduce $U(\psi, \alpha) = 0$, a contradiction. Hence $v = v'$, $u = u'$ and the lemma follows.   □

Lemma 8. *Let u, u' be integers with $|u|, |u'| < N^{\frac{1}{4}}$. Define*

$$v = \left[-\frac{qu}{p-u}\right] \quad and \quad v' = \left[-\frac{qu'}{p-u'}\right].$$

Assume that $\psi(u, v) < \psi(u', v')$. Let $[a_0, a_1, a_2, \cdots]$ be the continued fraction expansion of $\frac{\psi(u,v)}{\psi(u',v')}$. Then $a_0 = 0$, $a_1 = 1$ and $a_2 > 2^{-\frac{1}{2}}N^{\frac{1}{2}} - \frac{1}{2}$.

Proof. Let us apply the continued fraction algorithm (see e.g. of [11], p. 134). Assuming $\psi(u, u) < \psi(u', v')$, we get

$$a_0 = \left\lfloor \frac{\psi(u, v)}{\psi(u', v')} \right\rfloor = 0.$$

Next, we have

$$a_1 = \left\lfloor \frac{1}{\frac{\psi(u,v)}{\psi(u',v')} - a_0} \right\rfloor = \left\lfloor \frac{\psi(u', v')}{\psi(u, v)} \right\rfloor.$$

By Lemma 4, we have

$$0 < \psi(u', v') - \psi(u, v) \leq |\psi(u, v) - N| + |\psi(u', v') - N| < \sqrt{2}N^{\frac{1}{2}}. \quad (6)$$

Combining this with Lemma 4, we get

$$0 < \frac{\psi(u', v')}{\psi(u, v)} - 1 = \frac{\psi(u', v') - \psi(u, v)}{\psi(u, v)} < \frac{\sqrt{2}N^{\frac{1}{2}}}{\psi(u, v)} < \frac{\sqrt{2}N^{\frac{1}{2}}}{N - 2^{-\frac{1}{2}}N^{\frac{1}{2}}} < 1.$$

From this, we deduce $a_1 = 1$. Finally, combining (6) and Lemma 4, we get

$$a_2 = \left\lfloor \frac{1}{\frac{\psi(u',v')}{\psi(u,v)} - a_1} \right\rfloor = \left\lfloor \frac{\psi(u, v)}{\psi(u', v') - \psi(u, v)} \right\rfloor > \frac{N - 2^{-\frac{1}{2}}N^{\frac{1}{2}}}{\sqrt{2}N^{\frac{1}{2}}} = 2^{-\frac{1}{2}}N^{\frac{1}{2}} - \frac{1}{2}.$$

This completes the proof.   □

5   The New Attack.

In this section we state our new attack. Let $N = pq$ be an RSA modulus with $q < p < 2p$. Let e be a public exponent satisfying an equation $eX - \psi(u,v)Y = 1$ with integers X, Y, u, v such that

$$1 \leq Y < X < 2^{-\frac{1}{4}}N^{\frac{1}{4}}, \quad |u| < N^{\frac{1}{4}}, \quad v = \left[-\frac{qu}{p-u}\right],$$

and with the condition that all prime factors of $p-u$ or $q-v$ are $\leq B_{\text{ecm}} = 10^{50}$. Our goal is to solve this equation. As in Wiener's approach, we use the continued fraction algorithm to recover the unknown values X and Y.

Theorem 3. *Let $N = pq$ be an RSA modulus with $q < p < 2p$. Suppose that the public exponent e satisfies an equation $eX - \psi(u,v)Y = 1$ with*

$$|u| < N^{\frac{1}{4}}, \quad v = \left[-\frac{qu}{p-u}\right], \quad 1 \leq Y < X < 2^{-\frac{1}{4}}N^{\frac{1}{4}}.$$

Then $\frac{Y}{X}$ is one of the convergents of the continued fraction expansion of $\frac{e}{N}$.

Proof. Starting with the equation $eX - \psi(u,v)Y = 1$, we get

$$eX - NY = 1 - (N - \psi(u,v))Y.$$

Together with Lemma 4, this implies

$$\left|\frac{e}{N} - \frac{Y}{X}\right| = \frac{|1 - (N - \psi(u,v))Y|}{NX}$$

$$\leq \frac{1 + |(N - \psi(u,v))|\,Y}{NX}$$

$$\leq \frac{1 + 2^{-\frac{1}{2}}N^{\frac{1}{2}}Y}{NX}$$

$$\leq \frac{2 + \sqrt{2}N^{\frac{1}{2}}(X-1)}{2NX}.$$

Suppose we can upperbound the right-hand side term by $\frac{1}{2X^2}$, that is

$$\frac{2 + \sqrt{2}N^{\frac{1}{2}}(X-1)}{2NX} < \frac{1}{2X^2},$$

then, applying Theorem 1 the claim follows. Rearranging to isolate X, this leaves us with the condition

$$\sqrt{2}N^{\frac{1}{2}}X^2 - \left(\sqrt{2}N^{\frac{1}{2}} - 2\right)X - N < 0.$$

It is not hard to see that the condition is satisfied if $X < 2^{-\frac{1}{4}}N^{\frac{1}{4}}$. This gives us the theorem. $\square$

Afterwards, we combine ECM, integer relation detection algorithms and Coppersmith's method to factor $N = pq$.

Theorem 4. *Let $N = pq$ be an RSA modulus with $q < p < 2p$. Let B_{ecm} be the ECM-bound. Suppose that the public exponent $e < N$ satisfies an equation $eX - \psi(u, v)Y = 1$ with*

$$|u| < N^{\frac{1}{4}}, \quad v = \left[-\frac{qu}{p-u} \right], \quad 1 \leq Y < X < 2^{-\frac{1}{4}} N^{\frac{1}{4}}.$$

If $p - u$ or $q - v$ is B_{ecm}-smooth, then we can efficiently factor N.

Proof. By Theorem 3 we know that X and Y can be found among the convergents of the continued expansion of $\frac{e}{N}$. From X and Y, we get

$$\psi(u, v) = (p - u)(q - v) = \frac{eX - 1}{Y}.$$

Without loss of generality, suppose that $p - u$ is B_{ecm}-smooth. Using ECM, write $\frac{eX-1}{Y} = M_1 M_2$ where M_1 is B_{ecm}-smooth. Let $\omega(M_1)$ denote the number of distinct prime factors of M_1. Then the prime factorization of M_1 is of the form

$$M_1 = \prod_{i=1}^{\omega(M_1)} p_i^{a_i},$$

where the $a_i \geq 1$ are integers and the p_i are distinct primes $\leq B_{ecm}$. Since $p - u$ is B_{ecm}-smooth, then $p - u$ a divisor of M_1, so that

$$p - u = \prod_{i=1}^{\omega(M_1)} p_i^{x_i}, \tag{7}$$

where the x_i are integers satisfying $0 \leq x_i \leq a_1$. By Lemma 2, we can assume that $N^{\frac{1}{2}} < p - u < \sqrt{2} N^{\frac{1}{2}}$. Combining this with (7) and taking logarithms, we get

$$0 < \sum_{i=1}^{\omega(M_1)} x_i \log p_i - \frac{1}{2} \log N < \frac{1}{2} \log 2. \tag{8}$$

These inequalities are related to Baker's famous theory of linear forms in logarithms [1] and can be formulated as a nearly closest lattice problem in the 1-norm. They can be solved using the LLL [15] or the PSLQ algorithm [7]. The complexity of LLL and PSLQ depends on $\omega(M_1)$. Since Hardy and Ramanujan (see e.g. Theorem 431 of [11]), we know that, in average, $\omega(M_1) \sim \log \log M_1$ if M_1 is uniformly distributed. Since $X < 2^{-\frac{1}{4}} N^{\frac{1}{4}}$, we have for $e < N$

$$M_1 \leq \frac{eX - 1}{Y} < \frac{eX}{Y} \leq eX < N^{\frac{5}{4}},$$

This implies that the number of primes dividing M_1 satisfies

$$\omega(M_1) \sim \log\log M_1 \sim \log\log N.$$

Next, let us investigate the number of solutions of (8) which is related to the number of divisors of M_1. Let $\tau(M_1)$ denote the number of positive divisors of M_1. The prime decomposition of M_1 gives the exact value

$$\tau(M_1) = \prod_{i=1}^{\omega(M_1)} (1 + a_i).$$

By Dirichlet's Theorem, we know that if M_1 is uniformly distributed, then the average order of $\tau(M_1)$ is $\log M_1$ (see Theorem 319 of [11]). It follows that the average number of divisors of M_1 is

$$\tau(M_1) \sim \log(M_1) \sim \log(N).$$

This gives in average the number of solutions to the inequalities (8).

Next, let D be a divisor of M_1 satisfying (8). If D is a good candidate for $p - u$ with $|u| < N^{\frac{1}{4}}$, then applying Theorem 2, we get the desired factor p. This concludes the theorem. □

Notice that the running time is dominated by ECM since every step in our attack can be done in polynomial time and the number of convergents and divisors are bounded by $O(\log N)$.

6   The Number of Exponents for the New Method

In this section, we estimate the number of exponents for which our method works. Let $N = pq$ be an RSA modulus with $q < p < 2q$. The principal object of investigation of this section is the set

$$H(N) = \left\{ e : e < N, \ \exists u \in V(p), \ \exists X < 2^{-\frac{1}{4}} N^{\frac{1}{4}}, \ e \equiv X^{-1} \pmod{\psi(u,v)} \right\},$$

where

$$V(p) = \left\{ u : \ |u| < p^{\frac{1}{2}}, \ p - u \text{ is } B_{\text{ecm}}\text{-smooth} \right\}, \tag{9}$$

and $v = \left[-\frac{qu}{p-u} \right]$.

We will first show that every public exponent $e \in H(N)$ is uniquely defined by a tuple (u, X). We first deal with the situation when an exponent e is defined by different tuples (u, X) and (u, X').

Lemma 9. *Let $N = pq$ be an RSA modulus with $q < p < 2p$. Let u, v, X, X' be integers with $1 \leq X, X' < 2^{-\frac{1}{4}} N^{\frac{1}{4}}$ and $\gcd(XX', \psi(u,v)) = 1$ where $v = \left[-\frac{qu}{p-u} \right]$. Define*

$$e \equiv X^{-1} \pmod{\psi(u,v)} \quad \text{and} \quad e' \equiv X'^{-1} \pmod{\psi(u,v)}.$$

If $e = e'$, then $X = X'$.

Proof. Since $e \equiv X^{-1} \pmod{\psi(u,v)}$, there exists a positive integer Y such that $eX - \psi(u,v)Y = 1$ with $\gcd(X,Y) = 1$. Similarly, e' satisfies $e'X' - \psi(u,v)Y' = 1$ with $\gcd(X',Y') = 1$. Assume that that $e = e'$. Then

$$\frac{1 + \psi(u,v)Y}{X} = \frac{1 + \psi(u,v)Y'}{X'}.$$

Combining this with Lemma 4, we get

$$|XY' - X'Y| = \frac{|X' - X|}{\psi(u,v)} < \frac{2^{-\frac{1}{4}}N^{\frac{1}{4}}}{N - 2^{-\frac{1}{2}}N^{\frac{1}{2}}} < 1.$$

Hence $XY' = X'Y$ and since $\gcd(X,Y) = 1$, we get $X' = X$ and the lemma follows. □

Next, we deal with the situation when an exponent e is defined by different tuples (u,X) and (u',X') with $u \neq u'$ and $v = v'$.

Lemma 10. *Let $N = pq$ be an RSA modulus with $q < p < 2p$. Let u, u' be integers with $|u|, |u'| < N^{\frac{1}{4}}$. Let X, X' be integers with $1 \leq X, X' < 2^{-\frac{1}{4}}N^{\frac{1}{4}}$, $\gcd(X, \psi(u,v)) = 1$, $\gcd(X', \psi(u',v')) = 1$ where $v = \left[-\frac{qu}{p-u}\right]$ and $v' = \left[-\frac{qu'}{p-u'}\right]$. Define*

$$e \equiv X^{-1} \pmod{\psi(u,v)} \quad and \quad e' \equiv X'^{-1} \pmod{\psi(u',v')}.$$

If $v = v'$ and $e = e'$, then $X = X'$ and $u = u'$.

Proof. As in the proof of Lemma 9, rewrite e and e' as

$$e = \frac{1 + \psi(u,v)Y}{X} \quad and \quad e' = \frac{1 + \psi(u',v')Y'}{X'}.$$

Suppose $e = e'$. Then

$$|\psi(u',v')XY' - \psi(u,v)X'Y| = |X' - X|. \tag{10}$$

Assuming $v = v'$ and using $\psi(u,v) = (p-u)(q-v)$, $\psi(u',v') = (p-u')(q-v)$ in (10), we get

$$(q-v)\left|(p-u')XY' - (p-u)X'Y\right| = |X' - X|.$$

By Lemma 2, we have $q - v > 2^{-\frac{1}{2}}N^{\frac{1}{2}} - N^{\frac{1}{4}} > N^{\frac{1}{4}}$ and since $|X' - X| < 2^{-\frac{1}{4}}N^{\frac{1}{4}}$, we get

$$\begin{cases} X' - X & = 0, \\ (p-u')XY' - (p-u)X'Y = 0. \end{cases}$$

Hence $X = X'$ and $(p-u')Y' = (p-u)Y$. Suppose for contradiction that $u' \neq u$. Put $g = \gcd(p-u', p-u)$. Then g divides $(p-u) - (p-u') = u' - u$. Since $v = v'$, by Lemma 6 we have $|u' - u| \leq 1$, so $g = 1$. Hence $\gcd(p-u', p-u) = 1$ and $p - u$ divides Y'. Since $p - u > N^{\frac{1}{2}}$ and $Y' < X' < 2^{-\frac{1}{4}}N^{\frac{1}{4}}$, this leads to a contradiction, so we deduce that $u' = u$. This terminates the proof. □

Using the methods used to prove Lemma 9 and Lemma 10 plus some additional arguments, we shall prove the following stronger result.

Theorem 5. *Let $N = pq$ be an RSA modulus with $q < p < 2p$. Let u, u' be integers with $|u|, |u'| < N^{\frac{1}{4}}$. Let X, X' be integers with $1 \leq X, X' < 2^{-\frac{1}{4}}N^{\frac{1}{4}}$, $\gcd(X, \psi(u,v)) = 1$, $\gcd(X', \psi(u',v')) = 1$ where $v = \left[-\frac{qu}{p-u}\right]$ and $v' = \left[-\frac{qu'}{p-u'}\right]$. Define*

$$e \equiv X^{-1} \pmod{\psi(u,v)} \quad and \quad e' \equiv X'^{-1} \pmod{\psi(u',v')}.$$

If $e = e'$, then $u = u'$, $v = v'$ and $X = X'$.

Proof. Assume that $e = e'$. Then, as in the proof of Lemma 10, e and e' satisfy (10). We first take care of some easy cases.

If $u = u'$, then $v = v'$ and by Lemma 9, we get $X = X'$.

If $v = v'$, then by Lemma 10, we get $u = u'$ and $X = X'$.

Without loss of generality, suppose that $\psi(u,v) < \psi(u',v')$. Transforming (10), we get

$$\left|\frac{XY'}{X'Y} - \frac{\psi(u,v)}{\psi(u',v')}\right| = \frac{|X' - X|}{X'Y\psi(u',v')} \leq \frac{\max(X',X)}{X'Y\psi(u',v')} < \frac{1}{2(X'Y)^2},$$

where the final step is trivial since, for $N \geq 2^{10}$

$$2\max(X',X)X'Y < 2 \times \left(2^{-\frac{1}{4}}N^{\frac{1}{4}}\right)^3 < N - 2^{-\frac{1}{2}}N^{\frac{1}{2}} < \psi(u',v').$$

Combined with Theorem 1, this implies that $\frac{XY'}{X'Y}$ is one of the convergents of the continued fraction expansion of $\frac{\psi(u,v)}{\psi(u',v')}$. By Lemma 8, the first non trivial convergents are $\frac{1}{1}$ and $\frac{a_2}{a_2+1}$ where $a_2 > 2^{-\frac{1}{2}}N^{\frac{1}{2}} - \frac{1}{2}$. Observe that

$$a_2 + 1 > 2^{-\frac{1}{2}}N^{\frac{1}{2}} - \frac{1}{2} + 1 = 2^{-\frac{1}{2}}N^{\frac{1}{2}} + \frac{1}{2} > 2^{-\frac{1}{2}}N^{\frac{1}{2}} = \left(2^{-\frac{1}{4}}N^{\frac{1}{4}}\right)^2 > X'Y.$$

This implies that the only possibility for $\frac{XY'}{X'Y}$ to be a convergent of $\frac{\psi(u,v)}{\psi(u',v')}$ is $\frac{1}{1}$. This gives $XY' = X'Y$. Since $\gcd(X,Y) = \gcd(X',Y') = 1$ then $X = X'$ and $Y = Y'$. Replacing in (10), we get $\psi(u',v') = \psi(u,v)$ and by Lemma 7, we deduce $u = u'$. This completes the proof.   □

We now determine the order of the cardinality of the set $H(N)$. Recall that the elements of $H(N)$ are uniquely defined by the congruence

$$e \equiv X^{-1} \pmod{\psi(u,v)},$$

where $|u| < N^{\frac{1}{4}}$, $v = \left[-\frac{qu}{p-u}\right]$, $1 \leq X < 2^{-\frac{1}{4}}N^{\frac{1}{4}}$ and $\gcd(X, \psi(u,v)) = 1$. In addition, $p - u$ is B_{ecm}-smooth.

Theorem 6. *Let $N = pq$ be an RSA modulus with $q < p < 2p$. We have*

$$\#H(N) \geq N^{\frac{1}{2}-\varepsilon},$$

where ε is a small positive constant.

Proof. Assume $B_{ecm} < p - p^{\frac{1}{2}}$. Let us consider the set $V(p)$ as defined by (9). Put $x = p - p^{\frac{1}{2}}$, $z = 2p^{\frac{1}{2}}$ and $y = B_{ecm}$. Define $\delta > 0$ and $\gamma > 0$ such that

$$x^{\frac{1}{2}+\delta} \leq z, \qquad y = x^{1/\gamma}.$$

Then $x \geq z \geq x^{\frac{1}{2}+\delta}$, $x \geq y \geq x^{1/\gamma}$ and the conditions to apply (3) are fulfilled. On the other hand, we have $y > \exp\left\{(\log\log x)^{5/3+\varepsilon}\right\}$ for $x < \exp\left\{10^{7-\varepsilon}\right\}$ and the condition to apply (2) is fulfilled. Combining (3) and (2), we get

$$\#V(p) = \Psi(x+z, y) - \Psi(x, y) \geq c\frac{z}{x}\Psi(x, y) = cz\rho(\gamma)\left\{1 + O\left(\frac{\log(\gamma+1)}{\log(y)}\right)\right\},$$

where $c = c(\delta, \gamma) > 0$ and $\rho(\gamma)$ is the Dickman-de Bruijn ρ-function (see Table 2). Hence

$$\#V(p) \geq c\rho(\gamma) z = 2c\rho(\gamma) p^{\frac{1}{2}}.$$

Since trivially $\#V(p) < z = 2p^{\frac{1}{2}}$, we get $\#V(p) \asymp p^{\frac{1}{2}}$. Combining this with Table 2, we conclude that $\#V(p)$ is lower bounded as follows

$$\#V(p) \geq p^{\frac{1}{2}-\varepsilon'} = N^{\frac{1}{4}-\varepsilon_1},$$

with small constants $\varepsilon' > 0$ and $\varepsilon_1 > 0$.

Next, for every integer u with $|u| < N^{\frac{1}{4}}$ put

$$W(u) = \left\{X : 1 \leq X < 2^{-\frac{1}{4}}N^{\frac{1}{4}}, (X, \psi(u, v)) = 1\right\},$$

where $v = \left[-\frac{qu}{p-u}\right]$. Setting $m = \left\lfloor 2^{-\frac{1}{4}}N^{\frac{1}{4}}\right\rfloor$, we have

$$\#W(u) = \sum_{\substack{X=1 \\ (X,\psi(u,v))=1}}^{m} 1 = \sum_{d|\psi(u,v)} \mu(d)\left\lfloor\frac{m}{d}\right\rfloor \geq m\sum_{d|\psi(u,v)}\frac{\mu(d)}{d} = \frac{m\phi(\psi(u,v))}{\psi(u,v)}$$

where $\mu(.)$ is the Möbius function and $\phi(.)$ is the Euler totient function. We shall need the well known result (see Theorem 328 of [11]),

$$\frac{\phi(n)}{n} \geq \frac{C}{\log\log n},$$

where C is a positive constant. Applying this with $n = \psi(u, v)$ and using Lemma 4, we get

$$\#W(u) \geq \frac{Cm}{\log\log\psi(u,v)} \geq \frac{2^{-\frac{1}{4}}CN^{\frac{1}{4}}}{\log\log\left(N + 2^{-\frac{1}{2}}N^{\frac{1}{2}}\right)} = N^{\frac{1}{4}-\varepsilon_2},$$

with a small constant $\varepsilon_2 > 0$.

It remains to show that $\#H(n) \geq N^{\frac{1}{4}-\varepsilon}$ where ε is a positive constant. Indeed, for every $u \in V(p)$ there are at least $N^{\frac{1}{4}-\varepsilon_2}$ integers $X \in W(u)$. Hence

$$\#H(n) \geq \#V(p)\#W(u) \geq N^{\frac{1}{2}-\varepsilon_1-\varepsilon_2}.$$

Setting $\varepsilon = \varepsilon_1 + \varepsilon_2$, this completes the proof of the theorem.   □

Table 2. Table of values of $\rho(\gamma)$ with $(p - \sqrt{p})^{\frac{1}{\gamma}} = B\text{ecm} = 10^{50}$

Number of bits of p	256	512	1024	2048
$\gamma = \dfrac{\log(p - \sqrt{p})}{\log B\text{ecm}} \approx$	1.5	3	6.25	12.50
$\rho(\gamma) \approx (\text{see}[9])$	5.945×10^{-1}	4.861×10^{-2}	9.199×10^{-6}	1.993×10^{-15}

7   Conclusion

Wiener's famous attack on RSA with $d < \frac{1}{3}N^{0.25}$ shows that using the equation $ed - k(p-1)(q-1) = 1$ and a small d makes RSA insecure. In this paper, we performed a generalization of this attack. We showed that we can find any X and Y with $1 \leq Y < X < 2^{-0.25}N^{0.25}$ from the continued fraction expansion of e/N when they satisfy an equation

$$eX - Y(p - u)\left(q + \left[\frac{qu}{p-u}\right]\right) = 1,$$

and if $p - u$ or $q + [qu/(p-u)]$ is smooth enough to factor, then p and q can be found from X and Y. Our results illustrate that one should be very cautious when choosing some class of RSA exponent. Note that our attack, as well as all the attacks based on continued fractions do not apply to RSA with modulus N and small public exponents as the popular values $e = 3$ or $e = 2^{16}+1$ because the non-trivial convergents of $\frac{e}{N}$ are large enough to use diophantine approximation techniques, namely Theorem 1.

References

1. Baker, A.: Linear forms in the logarithms of algebraic numbers IV. Mathematika 15, 204–216 (1966)
2. Blömer, J., May, A.: A generalized Wiener attack on RSA. In: Bao, F., Deng, R., Zhou, J. (eds.) PKC 2004. LNCS, vol. 2947, pp. 1–13. Springer, Heidelberg (2004)
3. Boneh, D., Durfee, G.: Cryptanalysis of RSA with private key d less than $N^{0.292}$. In: Stern, J. (ed.) EUROCRYPT 1999. LNCS, vol. 1592, pp. 1–11. Springer, Heidelberg (1999)
4. Brent, R.P.: Some integer factorization algorithms using elliptic curves. Australian Computer Science Communications 8, 149–163 (1986)
5. Brent, R.P.: Recent progress and prospects for integer factorisation algorithms. In: Du, D.-Z., Eades, P., Sharma, A.K., Lin, X., Estivill-Castro, V. (eds.) COCOON 2000. LNCS, vol. 1858, pp. 3–22. Springer, Heidelberg (2000)

6. Coppersmith, D.: Small solutions to polynomial equations, and low exponent RSA vulnerabilities. Journal of Cryptology 10(4), 233–260 (1997)
7. Ferguson, H.R.P., Bailey, D.H.: A polynomial time, numerically stable integer relation algorithm. RNR Technical Report RNR-91-032, NASA Ames Research Center, Moffett Field, CA (December 1991)
8. Friedlander, J., Granville, A.: Smoothing "Smooth" Numbers. Philos. Trans. Roy. Soc. London Ser. 345, 339–347 (1993)
9. Granville, A.: Smooth numbers: computational number theory and beyond. In: Buhler, J., Stevenhagen, P. (eds.) Proc. MSRI Conf. Algorithmic Number Theory: Lattices, Number Fields, Curves, and Cryptography, Berkeley, Cambridge University Press, Cambridge (2000)
10. Hall, R.R., Tenenbaum, G.: Divisors. Cambridge Tracts in Mathematics, vol. 90. Cambridge University Press, Cambridge (1988)
11. Hardy, G.H., Wright, E.M.: An Introduction to the Theory of Numbers. Oxford University Press, London (1965)
12. Hildebrand, A.: On the number of positive integers $\leq x$ and free of prime factors $> y$. J. Number Theory 22, 289–307 (1986)
13. Howgrave-Graham, N.A.: Finding small roots of univariate modular equations revisited. In: Darnell, M.J. (ed.) Cryptography and Coding 1997. LNCS, vol. 1355, pp. 131–142. Springer, Heidelberg (1997)
14. Lenstra, H.W.: Factoring integers with elliptic curves. Annals of Mathematics 126, 649–673 (1987)
15. Lenstra, A.K., Lenstra, H.W., Lovasz, L.: Factoring polynomials with rational coefficients. Mathematische Annalen 261, 513–534 (1982)
16. Lenstra, A.K., Lenstra, H.W., Manasse, M.S., Pollard, J.M.: The number field sieve. In: Proc. 22nd Annual ACM Conference on Theory of Computing, Baltimore, Maryland, pp. 564–572 (1990)
17. May, A.: New RSA Vulnerabilities Using Lattice Reduction Methods, Ph.D. thesis, Paderborn (2003),
    http://www.informatik.tu-darmstadt.de/KP/publications/03/bp.ps
18. Montgomery, P.L.: Speeding the Pollard and elliptic curve methods of factorization. Mathematics of Computation 48, 243–264 (1987)
19. PARI/GP, version 2.1.7, Bordeaux (2007), http://pari.math.u-bordeaux.fr/
20. Rivest, R., Shamir, A., Adleman, L.: A Method for Obtaining Digital Signatures and Public-Key Cryptosystems. Communications of the ACM 21(2), 120–126 (1978)
21. de Weger, B.: Cryptanalysis of RSA with small prime difference, Applicable Algebra in Engineering. Communication and Computing 13(1), 17–28 (2002)
22. Wiener, M.: Cryptanalysis of short RSA secret exponents. IEEE Transactions on Information Theory 36, 553–558 (1990)
23. Zimmerman, P.: The ECMNET Project,
    http://www.loria.fr/~zimmerma/records/ecmnet.html

An Adaptation of the NICE Cryptosystem to Real Quadratic Orders

Michael J. Jacobson, Jr.[1,*], Renate Scheidler[2], and Daniel Weimer[3]

[1] Department of Computer Science, University of Calgary,
2500 University Drive NW, Calgary, Alberta T2N 1N4, Canada
jacobs@cpsc.ucalgary.ca
[2] Department of Mathematics & Statistics, University of Calgary,
2500 University Drive NW, Calgary, Alberta T2N 1N4, Canada
rscheidl@math.ucalgary.ca
[3] Charles River Development, 7 New England Executive Park,
Burlington MA 01803 USA
daniel.weimer@gmx.de

Abstract. In 2000, Paulus and Takagi introduced a public key cryptosystem called NICE that exploits the relationship between maximal and non-maximal orders in imaginary quadratic number fields. Relying on the intractability of integer factorization, NICE provides a similar level of security as RSA, but has faster decryption. This paper presents REAL-NICE, an adaptation of NICE to orders in real quadratic fields. REAL-NICE supports smaller public keys than NICE, and while preliminary computations suggest that it is somewhat slower than NICE, it still significantly outperforms RSA in decryption.

1 Introduction

The most well-known and widely used public-key cryptosystem whose security is related to the intractability of the integer factorization problem is the RSA scheme. A lesser known factoring-based system is the NICE (**N**ew **I**deal **C**oset **E**ncryption) scheme [13,18], a cryptosystem whose trapdoor decryption makes use of the relationship between ideals in the maximal and a non-maximal order of an imaginary quadratic number field. The security of NICE relies on the presumed intractability of factoring an integer of the form q^2p where p and q are prime, thereby providing a similar level of security as RSA, but with much faster decryption. NICE decryption has quadratic complexity, as opposed to RSA's cubic decryption complexity. This makes NICE particularly suited for devices with limited computing power or applications that require fast digital signature generation.

In this paper, we explain how to extend the NICE concept to real quadratic fields; this was first proposed in [19]. REAL-NICE exploits the same relationship between ideals in the maximal and a non-maximal quadratic order as NICE. Furthermore, just as in NICE, knowledge of the trapdoor information is provably

* Research by the first two authors supported by NSERC of Canada.

S. Vaudenay (Ed.): AFRICACRYPT 2008, LNCS 5023, pp. 191–208, 2008.
© Springer-Verlag Berlin Heidelberg 2008

equivalent to being able to factor the discriminant of the non-maximal order in random polynomial time. However, the security of REAL-NICE relies on the intractability of a somewhat different problem. In NICE, encryption hides the message ideal in its own exponentially large coset with respect to a certain subgroup of the ideal class group of the non-maximal order. In a real quadratic field, such a coset may be too small to prevent an exhaustive search attack. Instead, REAL-NICE encryption hides the message ideal in the generally exponentially large cycle of reduced ideals in its own ideal class in the non-maximal order.

While preliminary numerical data using prototype implementations suggest that REAL-NICE is somewhat slower than its imaginary counterpart NICE, REAL-NICE allows for the possibility of a smaller public key than NICE, at the expense of increased encryption effort. Moreover, both our NICE and REAL-NICE prototypes significantly outperformed a highly optimized public-domain implementation of RSA in decryption for all five NIST security levels [12]; for the two highest such levels, combined encryption and decryption was faster for both NICE and REAL-NICE compared to RSA.

The discrepancy in performance between NICE and REAL-NICE can be offset by using a more efficient encryption algorithm, called IMS encryption, for REAL-NICE. IMS encryption exploits the very fast baby step operation in the cycle of reduced ideals of a real quadratic order, an operation that has no imaginary analogue. Unfortunately, so far, the only known rigorous proof of security for IMS encryption needs to assume a very unfavourable parameter set-up. However, even under these adverse assumptions, IMS-REAL-NICE outperformed the original REAL-NICE system. It it is conceivable that a set-up could be established that makes IMS-REAL-NICE competitive to NICE without sacrificing security. IMS encryption and its security are the subject of future research.

2   Overview of Quadratic Orders

We begin with a brief overview of quadratic fields and their orders. Most of the material in this section can be found in [11] and Chapter 2, §7, of [4]; while the latter source considers mostly imaginary quadratic fields, much of the results are easily extendable to real quadratic fields as was done in [19].

Let $D \in \mathbb{Z}$, $D \neq 0, \pm 1$ be a squarefree integer. A *quadratic (number) field* is a field of the form $\mathcal{K} = \mathbb{Q}(\sqrt{D}) = \{a + b\sqrt{D} \mid a, b \in \mathbb{Q}\}$. $\mathcal{K}$ is an *imaginary*, respectively, *real* quadratic field if $D < 0$, respectively, $D > 0$. Set $\Delta_1 = 4D$ if $D \equiv 2$ or $3 \pmod 4$ and $\Delta_1 = D$ if $D \equiv 1 \pmod 4$, so $\Delta_1 \equiv 0$ or $1 \pmod 4$. Δ_1 is called a *fundamental discriminant*. For $f \in \mathbb{N}$, set $\Delta_f = f^2 \Delta_1$. The *(quadratic) order* of *conductor* f in $\mathcal{K}$ is the $\mathbb{Z}$-submodule $\mathcal{O}_{\Delta_f}$ of $\mathcal{K}$ of rank 2 generated by 1 and $f(\Delta_1 + \sqrt{\Delta_1})/2$; its *discriminant* is Δ_f. We speak of imaginary, respectively, real quadratic orders, depending on whether $\mathcal{K}$ is an imaginary, respectively, a real quadratic field. The *maximal order* of $\mathcal{K}$ is $\mathcal{O}_{\Delta_1}$; it contains all the orders of $\mathcal{K}$, and $f = [\mathcal{O}_{\Delta_1} : \mathcal{O}_{\Delta_f}]$ is the index of $\mathcal{O}_{\Delta_f}$ in $\mathcal{O}_{\Delta_1}$ as an additive subgroup.

Henceforth, let $f \in \mathbb{N}$ be any conductor. We denote by $\mathcal{O}_{\Delta_f}^*$ the group of units of the integral domain $\mathcal{O}_{\Delta_f}$, i.e. the group of divisors of 1 in $\mathcal{O}_{\Delta_f}$. The units of $\mathcal{O}_{\Delta_f}$, denoted by $\mathcal{O}_{\Delta_f}^*$, form an Abelian group under multiplication. If $\mathcal{K}$ is imaginary, then $\mathcal{O}_{\Delta_f}^*$ consists of the roots of unity in $\mathcal{K}$ and thus has 6, 4, or 2 elements, according to whether $\Delta_1 = -3$, $\Delta_1 = -4$, or $\Delta_1 < -4$. If $\mathcal{K}$ is real, then $\mathcal{O}_{\Delta_f}^*$ is an infinite cyclic group with finite torsion $\{1, -1\}$, whose unique generator ϵ_{Δ_f} exceeding 1 is the *fundamental unit* of $\mathcal{O}_{\Delta_f}$. In this case, the real number $R_{\Delta_f} = \log(\epsilon_{\Delta_f})$ is the *regulator* of $\mathcal{O}_{\Delta_f}$. Here, as usual, $\log(x)$ denotes the natural logarithm of $x > 0$.

An *(integral) $\mathcal{O}_{\Delta_f}$-ideal*[1] $\mathfrak{a}$ is a $\mathbb{Z}$-submodule of $\mathcal{O}_{\Delta_f}$ of rank 2 that is closed under multiplication by elements in $\mathcal{O}_{\Delta_f}$. A *fractional $\mathcal{O}_{\Delta_f}$-ideal* $\mathfrak{a}$ is a $\mathbb{Z}$-submodule of $\mathcal{K}$ of rank 2 such that $d\mathfrak{a}$ is an (integral) $\mathcal{O}_{\Delta_f}$-ideal for some $d \in \mathbb{N}$. A fractional $\mathcal{O}_{\Delta_f}$-ideal $\mathfrak{a}$ is *invertible* if there exists a fractional $\mathcal{O}_{\Delta_f}$-ideal $\mathfrak{b}$ such that $\mathfrak{a}\mathfrak{b} = \mathcal{O}_{\Delta_f}$, where the product of two fractional $\mathcal{O}_{\Delta_f}$-ideals $\mathfrak{a}, \mathfrak{b}$ is defined to consist of all finite sums of products of the form $\alpha\beta$ with $\alpha \in \mathfrak{a}$ and $\beta \in \mathfrak{b}$. The set of invertible fractional $\mathcal{O}_{\Delta_f}$-ideals, denoted by $\mathcal{I}(\mathcal{O}_{\Delta_f})$, is an infinite Abelian group under multiplication with identity $\mathcal{O}_{\Delta_f}$. A *principal* fractional $\mathcal{O}_{\Delta_f}$-ideal $\mathfrak{a}$ consists of $\mathcal{O}_{\Delta_f}$-multiples of some fixed element $\alpha \in \mathcal{K}^* = \mathcal{K} \setminus \{0\}$ that is said to *generate* (or be a *generator* of) $\mathfrak{a}$. We write $\mathfrak{a} = (\alpha) = \alpha\mathcal{O}_{\Delta_f}$. The principal fractional $\mathcal{O}_{\Delta_f}$-ideals form an infinite subgroup of $\mathcal{I}(\mathcal{O}_{\Delta_f})$ that is denoted by $\mathcal{P}(\mathcal{O}_{\Delta_f})$. The factor group $Cl(\mathcal{O}_{\Delta_f}) = \mathcal{I}(\mathcal{O}_{\Delta_f})/\mathcal{P}(\mathcal{O}_{\Delta_f})$ is a finite Abelian group under multiplication, called the *ideal class group* of $\mathcal{O}_{\Delta_f}$. Its order h_{Δ_f} is the *(ideal) class number* of $\mathcal{O}_{\Delta_f}$. For any $\mathcal{O}_{\Delta_f}$-ideal $\mathfrak{a}$, we denote the $\mathcal{O}_{\Delta_f}$-ideal class by $[\mathfrak{a}] \in Cl(\mathcal{O}_{\Delta_f})$.

For any element $\alpha = a + b\sqrt{D} \in \mathcal{K}$ $(a, b \in \mathbb{Q})$, the *conjugate* of α is $\overline{\alpha} = a - b\sqrt{D} \in \mathcal{K}$, and the *norm* of α is $N(\alpha) = \alpha\overline{\alpha} = a^2 - b^2 D \in \mathbb{Q}$. If $\alpha \in \mathcal{O}_{\Delta_1}$, then $N(\alpha) \in \mathbb{Z}$. The *norm* $N_{\Delta_f}(\mathfrak{a})$ of an (integral) $\mathcal{O}_{\Delta_f}$-ideal $\mathfrak{a}$ is the index of $\mathfrak{a}$ as an additive subgroup of $\mathcal{O}_{\Delta_f}$. When the context is clear, we will omit the subscript Δ_f from the ideal norm and simply write $N(\mathfrak{a})$. If we set $\overline{\mathfrak{a}} = \{\overline{\alpha} \mid \alpha \in \mathfrak{a}\}$, then $\mathfrak{a}\overline{\mathfrak{a}} = (N(\mathfrak{a}))$, the principal $\mathcal{O}_{\Delta_f}$-ideal generated by $N(\mathfrak{a})$. If $\mathfrak{a}$ is a principal $\mathcal{O}_{\Delta_f}$-ideal generated by $\alpha \in \mathcal{O}_{\Delta_f}$, then $N(\mathfrak{a}) = |N(\alpha)|$.

An integral $\mathcal{O}_{\Delta_f}$-ideal $\mathfrak{a}$ is *primitive* if the only positive integer d such that every element of $\mathfrak{a}$ is an $\mathcal{O}_{\Delta_f}$-multiple of d is $d = 1$. An $\mathcal{O}_{\Delta_f}$-ideal $\mathfrak{a}$ is *reduced* if it is primitive and there does not exist any non-zero $\alpha \in \mathfrak{a}$ with $|\alpha| < N(\mathfrak{a})$ and $|\overline{\alpha}| < N(\mathfrak{a})$. We summarize some important properties of reduced ideals; see for example [6,13,21] as well as Sections 2.1 and 2.2 of [19].

Theorem 1. *Let $\mathcal{O}_{\Delta_f}$ be an order in a quadratic number field $\mathcal{K} = \mathbb{Q}(\sqrt{D})$. Then the following hold:*

1. *Every ideal class of $Cl(\mathcal{O}_{\Delta_f})$ contains a reduced $\mathcal{O}_{\Delta_f}$-ideal.*
2. *If $\mathcal{K}$ is imaginary, then every ideal class of $Cl(\mathcal{O}_{\Delta_f})$ contains a unique reduced $\mathcal{O}_{\Delta_f}$-ideal. If $\mathcal{K}$ is real, then the number $r_{\mathbf{C}}$ of reduced ideals in any ideal class $\mathbf{C} \in Cl(\mathcal{O}_{\Delta_f})$ satisfies $R_{\Delta_f} / \log(f^2 D) \leq r_{\mathbf{C}} < 2R_{\Delta_f} / \log(2) + 1$.*
3. *If $\mathfrak{a}$ is a primitive $\mathcal{O}_{\Delta_f}$-ideal with $N(\mathfrak{a}) < \sqrt{|\Delta_f|}/2$, then $\mathfrak{a}$ is reduced.*
4. *If $\mathfrak{a}$ is a reduced $\mathcal{O}_{\Delta_f}$-ideal, then $N(\mathfrak{a}) < \sqrt{\Delta_f}$ if $\mathcal{K}$ is real and $N(\mathfrak{a}) < \sqrt{|\Delta_f|/3}$ is $\mathcal{K}$ is imaginary.*

---

[1] We always assume that integral and fractional ideals are non-zero.

For an $\mathcal{O}_{\Delta_f}$-ideal $\mathfrak{a}$, we denote by $\rho_{\Delta_f}(\mathfrak{a})$ any reduced $\mathcal{O}_{\Delta_f}$-ideal in the class of $\mathfrak{a}$. By the above theorem, if $\mathcal{K}$ is imaginary, then $\rho_{\Delta_f}(\mathfrak{a})$ is the unique reduced representative in the $\mathcal{O}_{\Delta_f}$-ideal class of $\mathfrak{a}$, whereas if $\mathcal{K}$ is real, then there are many choices for $\rho_{\Delta_f}(\mathfrak{a})$. Given any $\mathcal{O}_{\Delta_f}$-ideal $\mathfrak{a}$, a reduced ideal $\rho_{\Delta_f}(\mathfrak{a})$ in the equivalence class of $\mathfrak{a}$ can be found using at most $O\left(\log\left(N(\mathfrak{a})/\sqrt{\Delta_f}\right)\log(\Delta_f)\right)$ bit operations. Furthermore, in the real scenario, the entire cycle of reduced ideals in the $\mathcal{O}_{\Delta_f}$-ideal class of $\mathfrak{a}$ can then be traversed using a procedure called *baby steps*. Details on ideal reduction, baby steps, and other ideal arithmetic can be found in Section 7.

For any integer d, an integral $\mathcal{O}_{\Delta_f}$-ideal $\mathfrak{a}$ is said to be *prime to d* if $N(\mathfrak{a})$ is relatively prime to d. Of particular interest is the case $d = f$, as every $\mathcal{O}_{\Delta_f}$-ideal prime to f is invertible, and the norm map is multiplicative on the set of $\mathcal{O}_{\Delta_f}$-ideals prime to f. Denote by $\mathcal{I}(\mathcal{O}_{\Delta_f}, f)$ the subgroup of $\mathcal{I}(\mathcal{O}_{\Delta_f})$ generated by the $\mathcal{O}_{\Delta_f}$-ideals prime to f, by $\mathcal{P}(\mathcal{O}_{\Delta_f}, f)$ the subgroup of $\mathcal{I}(\mathcal{O}_{\Delta_f}, f)$ generated by the principal ideals (α) with $\alpha \in \mathcal{O}_{\Delta_f}$ and $N(\alpha)$ prime to f, and set $Cl(\mathcal{O}_{\Delta_f}, f) = \mathcal{I}(\mathcal{O}_{\Delta_f}, f)/\mathcal{P}(\mathcal{O}_{\Delta_f}, f)$. Then $Cl(\mathcal{O}_{\Delta_f}, f)$ is isomorphic to the class group $Cl(\mathcal{O}_{\Delta_f})$ of $\mathcal{O}_{\Delta_f}$; see Proposition 7.19, p. 143, of [4] and Theorem 2.16, p. 10, of [19].

Finally, we denote by $\mathcal{I}(\mathcal{O}_{\Delta_1}, f)$ the subgroup of $\mathcal{I}(\mathcal{O}_{\Delta_1})$ generated by the $\mathcal{O}_{\Delta_1}$-ideals prime to f, by $\mathcal{P}(\mathcal{O}_{\Delta_1}, f)$ the subgroup of $\mathcal{I}(\mathcal{O}_{\Delta_1}, f)$ generated by the principal $\mathcal{O}_{\Delta_1}$-ideals (α) with $\alpha \in \mathcal{O}_{\Delta_1}$ and $N(\alpha)$ prime to f, and define the factor group $Cl(\mathcal{O}_{\Delta_1}, f) = \mathcal{I}(\mathcal{O}_{\Delta_1}, f)/\mathcal{P}(\mathcal{O}_{\Delta_1}, f)$.

For the NICE cryptosystem in both real and imaginary quadratic orders, it will be important to move between $\mathcal{O}_{\Delta_f}$-ideals prime to f and $\mathcal{O}_{\Delta_1}$-ideals prime to f. More specifically, we have the following isomorphism (see Proposition 7.20, p. 144, of [4] and Theorem 3.2, p. 25, of [19]):

$$\phi : \mathcal{I}(\mathcal{O}_{\Delta_1}, f) \longrightarrow \mathcal{I}(\mathcal{O}_{\Delta_f}, f) \quad \text{via} \quad \phi(\mathfrak{A}) = \mathfrak{A} \cap \mathcal{O}_{\Delta_f}, \ \phi^{-1}(\mathfrak{a}) = \mathfrak{a}\mathcal{O}_{\Delta_1} \ . \quad (2.1)$$

The maps ϕ and ϕ^{-1} are efficiently computable if f and Δ_1 are known; for details, see Section 7. In fact, both the NICE and the REAL-NICE schemes use ϕ^{-1} as their underlying trapdoor one-way function, with public information Δ_f and trapdoor information f, where f is a prime. Note that ϕ and ϕ^{-1} preserve norms and primitivity. Furthermore, ϕ^{-1} preserves ideal principality, but ϕ does not. Thus, ϕ^{-1} induces a surjective homomorphism

$$\hat{\Phi} : Cl(\mathcal{O}_{\Delta_f}, f) \longrightarrow Cl(\mathcal{O}_{\Delta_1}, f) \quad \text{via} \quad \hat{\Phi}([\mathfrak{a}]) = [\phi^{-1}(\mathfrak{a})] = [\mathfrak{a}\mathcal{O}_{\Delta_1}] \ . \quad (2.2)$$

For proofs of these results, see pp. 144-146 of [4] and pp. 25-29 of [19].

The kernel of $\hat{\Phi}$, i.e. the subgroup of $Cl(\mathcal{O}_{\Delta_f}, f)$ of the form

$$\ker(\hat{\Phi}) = \{[\mathfrak{a}] \in Cl(\mathcal{O}_{\Delta_f}, f) \mid \phi^{-1}(\mathfrak{a}) \text{ is a principal } \mathcal{O}_{\Delta_1}\text{-ideal}\}$$

is of crucial importance to the NICE cryptosystem in imaginary quadratic orders, and also plays a role in its counterpart REAL-NICE in real quadratic orders. The size of this kernel is exactly the class number ratio $h_{\Delta_f}/h_{\Delta_1}$. For the cryptographically interesting case of prime conductor $f = q$, and disregarding the small cases $\Delta_1 = -3$ or -4 where $\mathcal{K}$ contains nontrivial roots of unity, the size of this kernel is given by

$$|\ker(\hat{\Phi})| = \frac{h_{\Delta_q}}{h_{\Delta_1}} = \begin{cases} q - (\Delta_1/q) & \text{if } \Delta_1 < -4, \\ q - (\Delta_1/q)R_{\Delta_1}/R_{\Delta_q} & \text{if } \Delta_1 > 0, \end{cases} \qquad (2.3)$$

where (Δ_1/q) denotes the Legendre symbol.

3   The Original NICE Cryptosystem

The original NICE cryptosystem [18,13] exploits the relationship between ideals in a maximal and a non-maximal imaginary quadratic order of prime conductor q as described in (2.1) and (2.2). The key observation is that images of $\mathcal{O}_{\Delta_q}$-ideals under the map ϕ^{-1} of (2.1) are efficiently computable if q is known, whereas without knowledge of the trapdoor information q (i.e. only knowledge of Δ_q), this task is infeasible and is in fact provably equivalent to being able to factor Δ_q in random polynomial time (see Theorem 2.1, pp. 13-14, of [18]).

The specifics of NICE are as follows:

Private Key: Two large primes p, q of approximately equal size with $p \equiv 3 \pmod 4$.

Public Key: $(\Delta_q, k, n, \mathfrak{p})$ where
- $\Delta_q = q^2 \Delta_1$ with $\Delta_1 = -p$;
- k and n are the bit lengths of $\lfloor \sqrt{|\Delta_1|}/4 \rfloor$ and $q - (\Delta_1/q)$, respectively;
- $\mathfrak{p}$ is a randomly chosen $\mathcal{O}_{\Delta_q}$-ideal with $[\mathfrak{p}] \in \ker(\hat{\Phi})$.

The key ideal $\mathfrak{p}$ can be found by generating a random element $\alpha \in \mathcal{O}_{\Delta_1}$ whose norm is not divisible by q, finding a $\mathbb{Z}$-basis of the principal $\mathcal{O}_{\Delta_1}$-ideal $\mathfrak{A} = (\alpha)$, and computing $\mathfrak{p} = \phi(\mathfrak{A})$. Note that the $\mathcal{O}_{\Delta_q}$-ideal $\mathfrak{p}$ itself is generally not principal, but its image $\phi^{-1}(\mathfrak{p})$ is a principal $\mathcal{O}_{\Delta_1}$-ideal.

Encryption: Messages are bit strings of bit length $k - t$, where t is a fixed parameter explained below. To encrypt a message m:
1. Embed m into a primitive $\mathcal{O}_{\Delta_q}$-ideal $\mathfrak{m}$ prime to q with $N_{\Delta_q}(\mathfrak{m}) \leq 2^k$ in such a way that $N_{\Delta_q}(\mathfrak{m})$ uniquely determines m.
2. Generate random $r \in_R \{1, 2, \ldots, 2^{n-1}\}$.
3. The ciphertext is the reduced $\mathcal{O}_{\Delta_q}$-ideal $\mathfrak{c} = \rho_{\Delta_q}(\mathfrak{m}\mathfrak{p}^r)$.

Note that since $2^{n-1} < q - (\Delta_1/q) < 2^n$, the range for r specified in step 2 ensures that $r < q - (\Delta_1/q) = |\ker(\hat{\Phi})|$. This is the optimal range, as $[\mathfrak{p}]^{q-(\Delta_1/q)}$ is the identity in $Cl(\mathcal{O}_{\Delta_q})$, i.e. the principal class. The cipher ideal $\mathfrak{c}$ is computed using standard ideal arithmetic; see Section 7 for details.

Decryption: To decrypt a ciphertext $\mathcal{O}_{\Delta_q}$-ideal $\mathfrak{c}$:
1. Compute $\mathfrak{M} = \rho_{\Delta_1}(\phi^{-1}(\mathfrak{c}))$.
2. Extract m from $N_{\Delta_1}(\mathfrak{M})$.

Note that $[\mathfrak{p}] \in \ker(\hat{\varPhi})$ and (2.2) together imply

$$[\mathfrak{M}] = [\rho_{\Delta_1}(\phi^{-1}(\mathfrak{c}))] = [\phi^{-1}(\mathfrak{c})] = \hat{\varPhi}([\mathfrak{c}]) = \hat{\varPhi}([\rho_{\Delta_q}(\mathfrak{m}\mathfrak{p}^r)])$$
$$= \hat{\varPhi}([\mathfrak{m}\mathfrak{p}^r]) = \hat{\varPhi}([\mathfrak{m}])\hat{\varPhi}([\mathfrak{p}])^r = \hat{\varPhi}([\mathfrak{m}]) = [\phi^{-1}(\mathfrak{m})] \ .$$

Since ϕ^{-1} is norm-preserving and $2^{k-1} \leq \lfloor\sqrt{|\Delta_1|}/4\rfloor < 2^k$, encryption step 1 yields

$$N_{\Delta_1}(\phi^{-1}(\mathfrak{m})) = N_{\Delta_q}(\mathfrak{m}) \leq 2^k \leq 2 \left\lfloor \frac{\sqrt{|\Delta_1|}}{4} \right\rfloor < \frac{\sqrt{|\Delta_1|}}{2} \ ,$$

where the last inequality follows since $\sqrt{|\Delta_1|}/4 \notin \mathbb{Z}$. By part 3 of Theorem 1, $\phi^{-1}(\mathfrak{m})$ is a reduced $\mathcal{O}_{\Delta_1}$-ideal. Thus, $\mathfrak{M}$ and $\phi^{-1}(\mathfrak{m})$ are reduced ideals in the same $\mathcal{O}_{\Delta_1}$-ideal class, so they must be equal by part 2 of Theorem 1. It follows that $N_{\Delta_1}(\mathfrak{M}) = N_{\Delta_1}(\phi^{-1}(\mathfrak{m})) = N_{\Delta_q}(\mathfrak{m})$, which by encryption step 1 uniquely determines m. Note also that $N_{\Delta_f}(\mathfrak{m}) = N_{\Delta_1}(\mathfrak{M}) < \sqrt{|\Delta_1|}/2 < q$, where the last inequality holds because p and q are of roughly the same size. It follows that both $\mathfrak{m}$ and $\mathfrak{M}$ are prime to q. Since $N(\mathfrak{c}) = N(\mathfrak{m})N(\mathfrak{p})^r$, $\mathfrak{c}$ is also prime to q.

Since the decrypter knows the conductor q of $\mathcal{O}_{\Delta_q}$, he can efficiently compute $\phi^{-1}(\mathfrak{c})$, and hence $\mathfrak{M}$ using standard reduction arithmetic. We explain how to compute images under ϕ^{-1} in Section 7.

To perform encryption step 1, one first selects a security parameter t; we explain below how large t should be chosen. The plaintext needs to be divided into message blocks of bit length $k - t$. To embed such a block m into a reduced $\mathcal{O}_{\Delta_q}$-ideal $\mathfrak{m}$ prime to q, the encrypter does the following:

1. Set $\overline{m} = m2^t$, obtaining an integer $\overline{m}$ of bit length k whose t low order bits are all 0.
2. Find the smallest prime l exceeding $\overline{m}$ such that $(\Delta_q/l) = 1$.
3. Set $\mathfrak{m}$ to be the $\mathcal{O}_{\Delta_q}$-ideal of norm l.

If $l \equiv 3 \pmod 4$, then a $\mathbb{Z}$-basis for the ideal $\mathfrak{m}$ can be found efficiently and deterministically. If $l \equiv 1 \pmod 4$, then there is a fast probabilistic method for performing step 3 above. For details, see again Section 7.

If $l \leq \overline{m}+2^t$, then $m \leq 2^{k-t}-1$ implies $N_{\Delta_f}(\mathfrak{m}) = l \leq \overline{m}+2^t = (m+1)2^t \leq 2^k$ as desired. Furthermore, the k high order bits of l agree with $\overline{m}$ and hence with m. Since $N_{\Delta_1}(\mathfrak{M}) = N_{\Delta_q}(\mathfrak{m}) = l$, m is easily obtained from $N_{\Delta_1}(\mathfrak{M})$ in decryption step 2 by truncating the first k bits from l. According to pp. 34–36 of [19], the probability that $l \leq \overline{m} + 2^t$ is bounded below by $P_t = 1 - 2^{-2^t/k}$. It follows that decryption step 2 is successful with high probability for t sufficiently large.

The security of NICE was analyzed in detail in [13], [7], and [19], and resides in the difficulty of factoring Δ_q. We only briefly review some facts here. Encryption under NICE can be viewed as masking the message ideal $\mathfrak{m}$ by multiplying it by a random ideal $\mathfrak{a} = \mathfrak{p}^r$ with $[\mathfrak{a}] \in \ker(\hat{\varPhi})$, thereby hiding it in its own coset $\mathfrak{m}\ker(\hat{\varPhi})$. The size of each such coset is equal to $|\ker(\hat{\varPhi})| = q - (\Delta_1/q)$. Obviously, q must be chosen large enough to make exhaustive search through any coset relative to $\ker(\hat{\varPhi})$ infeasible. Moreover, in order to guarantee a sufficiently large number of distinct elements of the form $\mathfrak{m}\mathfrak{p}^r$, or equivalently, a sufficiently large number of distinct powers $\mathfrak{p}^r$, we need to ensure that the subgroup in $\ker(\hat{\Phi})$ generated by the class $[\mathfrak{p}]$ is large. The order of this subgroup is a divisor of $q - (\Delta_1/q)$, so this quantity should be chosen prime or almost prime. Suppose that $q - (\Delta_1/q) = Ld$ where L is a large prime and $d \in \mathbb{N}$ is very small. Then the number of generators of the cyclic subgroup of $\ker(\hat{\Phi})$ of order L is $\phi(L) = L - 1$, where $\varphi(N)$ denotes the Euler totient function of $N \in \mathbb{N}$. So the probability that a random ideal $\mathfrak{p} \in \ker(\hat{\Phi})$ generates this subgroup is $(L - 1)/Ld \approx 1/d$ which is large if d is small. One expects d trials of an ideal $\mathfrak{p}$ to produce a desirable key ideal. For any such trial, checking that $\rho_{\Delta_q}(\mathfrak{p}^d) \neq \mathcal{O}_{\Delta_q}$ guarantees that $\mathfrak{p}$ generates a subgroup of $\ker(\hat{\Phi})$ of order L.

An algorithm for computing images of primitive $\mathcal{O}_{\Delta_q}$-ideals under ϕ^{-1} without knowledge of q would lead to the decryption of any message. However, according to Theorem 1 of [13], such an algorithm could be used as an oracle for factoring Δ_q in random polynomial time. Hence, the security of NICE is equivalent to factoring an integer of the form q^2p, so p and q need to be chosen sufficiently large to render the factorization of Δ_q via the elliptic curve method and the number field sieve infeasible. Using the estimate that factoring a 1024-bit RSA modulus is computationally equivalent to finding a 341-bit factor of a 3-prime modulus of the same size [10] yields the estimates in Table 1 for parameter sizes of Δ_q that are required to provide a level of security equivalent to block ciphers with keys of 80, 112, 128, 192, and 256 bits, respectively.

Table 1. NIST recommendations for parameter sizes of p and q

symmetric key size	80	112	128	192	256
Size of Δ_q	1024	2048	3072	8192	15360
Size of p and q	341	682	1024	2731	5120

To the best of our knowledge, revealing the public key and the form of Δ_q ($\Delta_q = -q^2p$ with primes p, q and $p \equiv 3 \pmod 4$) does not compromise the security of NICE. Finally, the chosen ciphertext attack of [7] is prevented by padding message blocks with t low order 0 bits for sufficiently large t. To ensure that this attack is approximately as costly as any other known attack method, t should be chosen according to the first row of Table 1 (the symmetric key size). NICE has also been extended, using standard techniques, to provide IND-CCA2 security in the random oracle model — see the NICE-X protocol presented in [1].

4   NICE in Real Quadratic Orders

Before we describe in detail REAL-NICE, our adaptation of NICE to orders in real quadratic fields, we highlight the main differences between REAL-NICE and NICE. The security of the original NICE scheme resides in the difficulty of identifying a specific representative in the coset of an $\mathcal{O}_{\Delta_q}$-ideal $\mathfrak{m}$ relative to $\ker(\hat{\Phi})$ without knowledge of the conductor q of the order $\mathcal{O}_{\Delta_q}$. In real quadratic orders, this problem is generally easy to solve via exhaustive search, since h_{Δ_q} can be a small multiple of h_{Δ_1}, resulting in a very small kernel of $\hat{\Phi}$ by (2.3). Instead, in the real case, an adversary needs to identify a specific reduced ideal in the cycle of reduced ideals in the $\mathcal{O}_{\Delta_q}$-ideal class $[\mathfrak{m}]$. It is therefore necessary to ensure that the number of reduced ideals in any $\mathcal{O}_{\Delta_q}$-ideal class is large.

At the same time, the decryption process of NICE no longer yields a unique reduced $\mathcal{O}_{\Delta_1}$-ideal. To extract m, we need to make sure that the $\mathcal{O}_{\Delta_1}$-ideal class of $\phi^{-1}(\mathfrak{c})$ contains very few reduced ideals, so they can all be quickly computed and the correct one identified by a predetermined unique bit pattern in its norm. During encryption, m is endowed with that same bit pattern. By part 2 of Theorem 1, the system parameters must therefore be chosen so that R_{Δ_1} is very small, while R_{Δ_q} is large.

Finally, the ideal $\mathfrak{p}$ need no longer be included in the public key; instead, a random ideal $\mathfrak{p}$ with $[\mathfrak{p}] \in \ker(\hat{\Phi})$ can be generated for each encryption.

The specifics of REAL-NICE are as follows:

Private Key: Two large primes p, q of approximately equal size with $p \equiv 1 \pmod{4}$.

Public Key: $(\Delta_q, k, n, \mathfrak{p})$ or (Δ_q, k, n), where
  - $\Delta_q = q^2 \Delta_1$ with $\Delta_1 = p$;
  - k and n are the bit lengths of $\lfloor \sqrt{\Delta_1}/4 \rfloor$ and $q - (\Delta_1/q)$, respectively;
  - $\mathfrak{p}$ is a randomly chosen $\mathcal{O}_{\Delta_q}$-ideal with $[\mathfrak{p}] \in \ker(\hat{\Phi})$; inclusion of $\mathfrak{p}$ in the public key is optional.

Here, p and q must be chosen so that R_{Δ_1} is small and R_{Δ_q} is large; details on how to select these primes will be provided in Section 5. If storage space for public keys is restricted, $\mathfrak{p}$ need not be included in the public key. Instead, a different ideal $\mathfrak{p}$ with $[\mathfrak{p}] \in \ker(\hat{\Phi})$ can be generated for each encryption, at the expense of increased encryption time. In the case where $\mathfrak{p}$ is included in the public key, it can be generated exactly as in the original NICE system. In Section 7, we describe an alternative method for finding $\mathfrak{p}$ that does not require knowledge of q and Δ_1 and can hence be used by the encrypter.

Encryption: Messages are bit strings of bit length $k - t - u$, where t and u are fixed parameters explained below. To encrypt a message m:
  1. Convert m to a string m' that uniquely determines m and contains a predetermined bit pattern of length u.
  2. Embed m' into a primitive $\mathcal{O}_{\Delta_q}$-ideal $\mathfrak{m}$ prime to q with $N_{\Delta_q}(\mathfrak{m}) \le 2^k$ in such a way that $N_{\Delta_q}(\mathfrak{m})$ uniquely determines m'.
  3. Generate random $r \in_R \{1, 2, \dots, 2^{n-1}\}$.
  4. If the public key does not include the ideal $\mathfrak{p}$, generate a random $\mathcal{O}_{\Delta_q}$-ideal $\mathfrak{p}$ with $[\mathfrak{p}] \in \ker(\hat{\Phi})$.
  5. The ciphertext is a reduced $\mathcal{O}_{\Delta_q}$-ideal $\mathfrak{c} = \rho_{\Delta_q}(\mathfrak{m}\mathfrak{p}^r)$.

Decryption: To decrypt a ciphertext $\mathcal{O}_{\Delta_q}$-ideal $\mathfrak{c}$:
1. Compute $\mathfrak{C} = \phi^{-1}(\mathfrak{c})$.
2. Find the reduced ideal $\mathfrak{M} \in [\mathfrak{C}]$ such that $N_{\Delta_1}(\mathfrak{M})$ contains the predetermined bit pattern of length u of encryption step 1.
3. Extract m' from $N_{\Delta_1}(\mathfrak{M})$ and m from m'.

Since the decrypter knows q, he can once again efficiently compute $\mathfrak{C}$; for details, see Section 7. As in the original NICE scheme, we see that $[\mathfrak{M}] = [\phi^{-1}(\mathfrak{m})]$. Unfortunately, we can no longer conclude from this that $\mathfrak{M} = \phi^{-1}(m)$, only that both are two among many reduced ideals prime to q in the same $\mathcal{O}_{\Delta_1}$-class. This is the reason why in contrast to encryption step 1 of NICE, the embedding of a message m into an $\mathcal{O}_{\Delta_q}$-ideal $\mathfrak{m}$ in REAL-NICE requires two steps. To ensure that $\mathfrak{M} = \phi^{-1}(\mathfrak{m})$ does in fact hold, m is endowed with a predetermined public bit pattern of length u to obtain m'. We argue below that this forces $\mathfrak{M} = \phi^{-1}(\mathfrak{m})$ with high probability, so $N_{\Delta_1}(\mathfrak{M}) = N_{\Delta_q}(\mathfrak{m})$ uniquely determines m' by encryption step 2, and hence m by encryption step 1.

More exactly, to perform encryption steps 1 and 2, one first selects the parameters t and u; we explain below how large t and u should be chosen. The plaintext needs to be divided into blocks of bit length $k - t - u$. To embed such a block m into a reduced $\mathcal{O}_{\Delta_q}$-ideal $\mathfrak{m}$ prime to q, one does the following:

1. Set $m' = m + 2^{k-t}$, obtaining an integer m' of bit length $k - t$ whose u high order bits are $100\cdots000$.
2. Set $\overline{m'} = m'2^t$, obtaining an integer $\overline{m'}$ of bit length k whose u high order bits are $100\cdots000$ and whose t low order bits are all 0.
3. Find the smallest prime l exceeding $\overline{m'}$ such that $(\Delta_q/l) = 1$.
4. Set $\mathfrak{m}$ to be the $\mathcal{O}_{\Delta_q}$-ideal of norm l.

The ideal $\mathfrak{m}$ is found exactly as in the NICE embedding procedure, and provided that $\mathfrak{M} = \phi^{-1}(\mathfrak{m})$, m' can again be extracted from $N_{\Delta_1}(\mathfrak{M}) = N_{\Delta_q}(\mathfrak{m}) = l$ with high probability by truncating the high order k bits from l. Then m is obtained from m' by simply discarding the u high order bits of m'.

Before we argue that, with high probability, the class of $[\mathfrak{C}]$ contains only one ideal whose norm contains our specified bit pattern (namely the ideal $\mathfrak{M} = \phi^{-1}(\mathfrak{m})$ of norm l), we explain how to find this ideal. In order to perform decryption step 2, the decrypter needs to traverse the set of reduced $\mathcal{O}_{\Delta_1}$-ideals in the class of $\mathfrak{C}$ to locate $\mathfrak{M}$. This is accomplished by applying repeated baby steps as described in Section 7, starting with the $\mathcal{O}_{\Delta_1}$-ideal $\mathfrak{C} \in [\mathfrak{M}]$. Since Δ_1 was chosen so that the class of $\mathfrak{C}$ contains very few reduced ideals, $\mathfrak{M}$ can be found efficiently. After each baby step, the decrypter performs a simple X-OR on the u high order bits of the ideal norm and the string $100\cdots000$, checking whether or not the resulting string consists of all 0's.

The decryption procedure will work with high probability under two conditions. Firstly, just as in NICE, the parameter t needs to be chosen as described in Section 3 to ensure that m' can be uniquely determined from $\mathfrak{M}$. We already saw that this succeeds with probability at least $P_t = 1 - 2^{-2^t/k}$. Secondly, u must be chosen large enough so that with high probability, the $\mathcal{O}_{\Delta_1}$-class of $\mathfrak{C}$ contains only one reduced $\mathcal{O}_{\Delta_1}$-ideal $\mathfrak{M}$ such that the u high order bits of $N_{\Delta_1}(\mathfrak{M})$ are $100\cdots000$. Using the analysis on p. 53, of [19], this probability is expected to be bounded below by $P_u = (1 - 2^{-u})^N$, where N is an upper bound on the number of reduced ideals in any class of $Cl(\mathcal{O}_{\Delta_1})$. We will see in Section 5 that Δ_1 can be chosen so that such an upper bound is of the form $\kappa_1 \log(\Delta_1)$ for some explicitly computable constant κ_1.

5   Choice of Parameters

The parameters for REAL-NICE clearly need to be selected with care to ensure both efficiency and security. As explained in Section 4, p and q must be chosen to satisfy the following conditions:

- R_{Δ_q} must be large enough to ensure a sufficiently large number of reduced ideals in any $\mathcal{O}_{\Delta_q}$-ideal class, thus rendering exhaustive search through any cycle of reduced $\mathcal{O}_{\Delta_q}$-ideals infeasible.
- R_{Δ_1} must be small enough to ensure a sufficiently small number of reduced ideals in any $\mathcal{O}_{\Delta_1}$-ideal class, thus rendering exhaustive search through any cycle of reduced $\mathcal{O}_{\Delta_1}$-ideals efficient.

We proceed in two steps. First, we explain how to ensure that the ratio $R_{\Delta_q}/R_{\Delta_1}$ is of order of magnitude q with high probability. Then we present a means of guaranteeing that R_{Δ_1} is small, i.e. bounded by a polynomial in $\log(\Delta_1)$.

The *unit index* of $\mathcal{O}_{\Delta_q}$ is the group index $[\mathcal{O}_{\Delta_1}^* : \mathcal{O}_{\Delta_q}^*]$, i.e. the smallest positive integer i such that $\epsilon_{\Delta_1}^i = \epsilon_{\Delta_f}$, or equivalently, $R_{\Delta_f} = iR_{\Delta_1}$. By (2.3), i divides $q - (\Delta_1/q)$, so forcing i to be large is another reason why $q - (\Delta_1/q)$ should be almost prime. Specifically, Theorem 5.8, p. 58, of [19] states that if $q - (\Delta_1/q) = Ld$ where L is a large prime and $d \leq \log(\Delta_1)^\kappa$ for some positive constant κ, then the probability that $i < L$ is bounded above by $\log(\Delta_1)^{2\kappa}/(\sqrt{\Delta_1} - 1)$. In other words, $i = R_{\Delta_q}/R_{\Delta_1} \geq L$ with overwhelming probability.

To verify that $i \geq L$ does in fact hold, it suffices to check that i does not divide d, i.e. that $\epsilon_{\Delta_1}^d \neq \epsilon_{\Delta_f}$. Suppose that R_{Δ_1} is sufficiently small so that $\epsilon_{\Delta_1} = U_1 + V_1\sqrt{\Delta_1}$ is computable; ϵ_{Δ_1} can be obtained from R_{Δ_1} using for example Algorithm 4.2 of [2]. Then any power $\epsilon_{\Delta_1}^j = U_j + V_j\sqrt{\Delta_1}$ can be efficiently evaluated using Lucas function arithmetic on U_j and V_j analogous to binary exponentiation; see Chapter 4, pp. 69-95, of [20].

Next, we illustrate how to choose Δ_1 so that R_{Δ_1} is small. In general, the regulator R_{Δ_f} of any real quadratic order $\mathcal{O}_{\Delta_f}$ is of magnitude $\sqrt{\Delta_f}$ which is far too large for our purposes. One possibility is to choose D to be a *Schinzel sleeper* [15], i.e. a positive squarefree integer of the form $D = D(x) = a^2x^2 + 2bx + c$ with $a, b, c, x \in \mathbb{Z}$, $a \neq 0$, and $b^2 - a^2c$ dividing $4\gcd(a^2, b)^2$. Schinzel sleepers were analyzed in detail in [3]; here, the regulator R_{Δ_1} is of order $\log(\Delta_1)$. More exactly, if a, b, c, x are chosen so that $\gcd(a^2, 2b, c)$ is squarefree and $D \equiv 0$ or $1 \pmod 4$ (so $\Delta_1 = D$), then by Theorem 5.4, p. 52, of [19], the number of reduced $\mathcal{O}_{\Delta_1}$-ideals in any class of $Cl(\mathcal{O}_{\Delta_1})$ is bounded above by $\kappa_1 \log(\Delta_1)$ for an explicitly computable constant κ_1 that depends only on a and b.

Finally, the fixed bit pattern in any message is again critical in defending REAL-NICE against the same chosen ciphertext attack [7] that was already mentioned in Section 3. This attack can be detected with the same probability with which the cipher ideal can be successfully decrypted. Therefore, it is suggested to choose $t + u$ according to first row of Table 1, keeping the probability of successful decryption via the chosen ciphertext attack consistent with the probability of success of any other known attack on REAL-NICE.

6   Security

Although the security of REAL-NICE is based on a different mathematical problem than NICE, namely locating an $\mathcal{O}_{\Delta_q}$-ideal within the cycle of reduced ideals in its own ideal class, as opposed to locating it in its own coset relative to $\ker(\hat{\Phi})$, the same security considerations apply. Assuming a passive adversary, both systems can be broken if and only if an adversary can efficiently compute images of $\mathcal{O}_{\Delta_q}$-ideals under the map ϕ^{-1} of (2.1) without knowledge of the trapdoor information q, a task that is provably equivalent to factoring in random polynomial time. More exactly, according to Theorem 2.1, pp. 13-14, of [18]:

Theorem 2. *Let $\Delta_1 \in \mathbb{N}$ be a fundamental discriminant and $\Delta_q = q^2 \Delta_1$ with q prime. Assume that there exists an algorithm $\boldsymbol{A}$ that computes for any primitive ideal $\mathfrak{a} \in \mathcal{I}(\mathcal{O}_{\Delta_q}, q)$ the primitive ideal $\mathfrak{A} = \phi^{-1}(\mathfrak{a}) \in \mathcal{I}(\mathcal{O}_{\Delta_1}, q)$ without knowledge of the conductor q of $\mathcal{O}_{\Delta_q}$. By using the algorithm $\boldsymbol{A}$ as an oracle, Δ_q can be factored in random polynomial time. The number of required queries to the oracle is polynomially bounded in $\log(\Delta_q)$.*

Hence, as with NICE, p and q must be chosen sufficiently large to render the factorization of Δ_q infeasible. Again, it is highly unlikely that knowledge of the public information would compromise the security of REAL-NICE. In addition, the specified bit pattern in the norm of the message ideal $\mathfrak{m}$ protects against the chosen ciphertext attack of [7]; once again, the length of this bit pattern should be chosen equal to the symmetric key size as specified in Table 1 to render this attack as expensive as any other known attack. As REAL-NICE is so similar to NICE, it should also be possible to adapt the methods of [1] to obtain IND-CCA2 security.

The fact that $\Delta_1 = a^2 x^2 + 2bx + c$ is chosen to be a Schinzel sleeper requires further analysis. It is recommended that the values a, b, c, x are kept secret and discarded after computing Δ_1. Care must also be taken how to select x in the Schinzel sleeper. Put $A = qa$, $B = q^2 b$, and suppose $B = SA + R$ with $0 \leq R < A$ (note that A, B, S, R are all unknown). Then by Theorem 4.1 of [3], the fraction A/R appears among the first $\kappa_2 \log(A)$ convergents of the continued fraction expansion of $\sqrt{\Delta_q}$ for some explicit positive constant κ_2, so there are only polynomially many possibilities for this fraction. If we find A/R and write it in lowest terms, i.e. $A/R = U/V$ with $\gcd(U, V) = 1$, then $q = \gcd(\Delta_q, U \lfloor \sqrt{\Delta_q} \rfloor + V)$ if x is sufficiently large, so Δ_q is factored. This factoring attack can be avoided if x is chosen sufficiently small, but at the same time large enough to guarantee sufficiently large Δ_1. More exactly, by Corollary 6.5, p. 72, of [19], it is sufficient to choose $x < 2^{-3w-1}q - 2^w$ where $a, b \leq 2^w$.

Suppose we wish to generate parameters Δ_1 and q of bit length s. If we choose a and b of some bit length $w \leq s/4 - 1$ and x of bit length $s/2 - w$, then ax has bit length $s/2$, $2bx$ has bit length $s/2 + 1$, and the condition $b^2 - a^2c$ divides $4\gcd(a^2, b)^2$ implies $|c| \leq 5b^2 < 2^{2w+3} \leq 2^{s/2+1}$, so $|c|$ has bit length at most $s/2 + 1$. Thus, $(ax)^2$ is the dominant term in $D(x)$, which then has bit length s. Since $q > 2^{s-1}$, it suffices to choose $x < 2^{s-3w-2} - 2^w$, so to obtain x of bit length at least $s/2 - w$, we require that $2^{s/2-w} < 2^{s-3w-2} - 2^w$. This is easily verified to always hold if $w \leq s/4 - 1$.

We also need to ensure that there are sufficiently many primes of desired size that occur as values of Schinzel sleepers. Let $\pi_F(n)$ denote the number of primes assumed by the polynomial $F(x) = ax^2 + bx + c$ for $0 \leq x \leq n$, with $a, b, c \in \mathbb{Z}$, $a > 0$, and $a + b, c$ not both even. The well-known Hardy-Littlewood conjecture [5] states in essence that $\pi_F(n) \sim \kappa_F n / \log(n)$, where κ_F is an explicitly computable constant than depends only on F. Under the assumption that prime values assumed by Schinzel polynomials behave similarly to those assumed by arbitrary quadratic polynomials, we conclude that the number of primes produced by Schinzel polynomials is large enough to render an exhaustive search for Δ_q infeasible. However, further study of this question is warranted.

Finally, we need to make sure that there are sufficiently many reduced $\mathcal{O}_{\Delta_q}$-ideals of the form $\mathfrak{c} = \rho_{\Delta_q}(\mathfrak{m}\mathfrak{p}^r)$ to ensure that cipher ideals cannot be found via exhaustive search. We already saw how to guarantee a large ratio $R_{\Delta_q}/R_{\Delta_1} = Ld$ where L is a large prime and $d \leq \log(\Delta_1)^\kappa$ for some positive constant κ. This ensures a large number of reduced ideals in each $\mathcal{O}_{\Delta_q}$-ideal class. For any $B \in \mathbb{N}$, any $\mathcal{O}_{\Delta_q}$-ideal $\mathfrak{p}$ with $[\mathfrak{p}] \in \ker(\hat{\Phi})$, and any reduced $\mathcal{O}_{\Delta_q}$-ideal $\mathfrak{m}$, consider the set of possible cipher ideals $\mathcal{C}_B = \{\rho_{\Delta_q}(\mathfrak{m}\mathfrak{p}^r) \mid 1 \leq r \leq B\}$. Then a sufficiently large choice of a generator $\alpha \in \mathcal{O}_{\Delta_q}$ of $\mathfrak{p}$ ensures that all the ideals in $\mathcal{C}_B$ are distinct. More exactly, according to Theorem 6.8, p. 81, of [19], if we choose α so that $\log(\alpha) \in I$ where

$$I = \,]\,(b+1)\log(4\Delta_q) + \log(2), \ \frac{L\log(\Delta_1)}{2^{b+2}} - b\log(2)\,] \tag{6.4}$$

and b is the bit length of B, then the set $\mathcal{C}_B$ has cardinality B.

Table 2 contains upper and lower bounds on $\log(\alpha)$ depending on the required level of security and the constant κ. The notation (v, w) in the column headers means that the set $\mathcal{C}_B$ contains at least 2^v different $\mathcal{O}_{\Delta_q}$-ideals, where Δ_q has bit length w. The columns "min" and "max" denote lower and upper bounds on the bit length of $\log(\alpha)$. The data show that it is feasible to choose α such that $\mathcal{C}_B$ is sufficiently large to satisfy the NIST security requirements.

7   Ideal Arithmetic and Algorithms

We review basic ideal arithmetic involving $\mathbb{Z}$-bases and provide the algorithms that are required in the REAL-NICE cryptosystem. See also [21,9] for details.

Table 2. Bounds the size of $\log \alpha$ depending on the level of security

κ	(80, 1024) min	max	(112, 2048) min	max	(128, 3092) min	max	(192, 8192) min	max	(256, 15360) min	max
1	17	259	18	568	20	1235	21	2536	22	4861
5	17	223	18	528	20	1191	21	2488	22	4809
10	17	178	18	478	20	1136	21	2428	22	4744
15	17	133	18	428	20	1081	21	2368	22	4679
20	17	88	18	378	20	1026	21	2308	22	4614

Let $\mathcal{K} = \mathbb{Q}(\sqrt{D})$ be a (real or imaginary) quadratic field and $\mathcal{O}_{\Delta_f}$ an order in $\mathcal{K}$ of conductor f. Set $\sigma = 1$ if $D \equiv 2, 3 \pmod{4}$ and $\sigma = 2$ if $D \equiv 1 \pmod{4}$, so $\Delta_1 = (2/\sigma)^2 D \equiv \sigma - 1 \pmod{4}$. Then every integral $\mathcal{O}_{\Delta_f}$-ideal $\mathfrak{a}$ is a $\mathbb{Z}$-module of the form

$$\mathfrak{a} = S\left(\frac{Q}{\sigma}\mathbb{Z} + \frac{P + f\sqrt{D}}{\sigma}\mathbb{Z}\right) ,$$

where $S, Q \in \mathbb{N}$, $P \in \mathbb{Z}$, σ divides Q, σQ divides $f^2 D - P^2$, and $\gcd(Q, 2P, (f^2 D - P^2)/Q) = \sigma$. Here, Q and S are unique and P is unique modulo Q, so we write $\mathfrak{a} = S(Q, P)$ for brevity. We have $N_{\Delta_f}(\mathfrak{a}) = S^2 Q/\sigma$. The ideal $\mathfrak{a}$ is primitive if and only if $S = 1$, in which case we simply write $\mathfrak{a} = (Q, P)$.

Suppose now that $D > 0$, so $\mathcal{K}$ is a real quadratic field. Recall that any $\mathcal{O}_{\Delta_f}$-ideal class contains a finite number of reduced ideals. A *baby step* moves from one such ideal to the next. More exactly, if $\mathfrak{a}_i = (Q_i, P_i)$ is a reduced $\mathcal{O}_{\Delta_f}$-ideal, then a reduced $\mathcal{O}_{\Delta_f}$-ideal $\mathfrak{a}_{i+1} = (Q_{i+1}, P_{i+1})$ in the $\mathcal{O}_{\Delta_f}$-ideal class of $\mathfrak{a}_i$ can be obtained using the formulas

$$q_i = \left\lfloor \frac{P_i + \sqrt{D}}{Q_i} \right\rfloor , \quad P_{i+1} = q_i Q_i - P_i, \quad Q_{i+1} = \frac{f^2 D - P_{i+1}^2}{Q_i} . \quad (7.5)$$

Note that $f^2 D = (\sigma/2)^2 \Delta_f$, so f need not be known here. Baby steps applied to any reduced $\mathcal{O}_{\Delta_f}$-ideal $\mathfrak{a}$ produce the entire cycle of reduced ideals in the $\mathcal{O}_{\Delta_f}$-ideal class of $\mathfrak{a}$. In practice, one uses a more efficient version of (7.5) that avoids the division in the expression for Q_{i+1}; see for example Algorithm 1 of [21].

We now give details on how to perform the different encryption and decryption steps, beginning with a method for finding a $\mathbb{Z}$-basis (Q, P) of the message ideal $\mathfrak{m}$ of prime norm l as required in encryption step 2 of REAL-NICE. Set $Q = 2l$, and let P' be a square root of Δ_q modulo l with $0 < P < l$. Such a square root exists since $(\Delta_q/l) = 1$ and can be found using standard probabilistic methods in at most an expected $O(\log(l)^3)$ bit operations. Now put $P = P'$ if P is odd and $P = l - P'$ if P is even. Then it is not hard to verify that $\mathfrak{m} = (Q, P)$ is a primitive $\mathcal{O}_{\Delta_q}$-ideal of norm $Q/2 = l$.

Given $\mathbb{Z}$-bases of two reduced $\mathcal{O}_{\Delta_f}$-ideals $\mathfrak{a}, \mathfrak{b}$, it is well-known how to compute a $\mathbb{Z}$-basis of a reduced $\mathcal{O}_{\Delta_f}$-ideal $\rho_{\Delta_f}(\mathfrak{a}\mathfrak{b})$ in the class of the (generally non-reduced and possibly not even primitive) product ideal $\mathfrak{a}\mathfrak{b}$ in $O(\log(D)^2)$

bit operations. This operation is called a *giant step*. Five different ways for effecting a giant step were described and compared in [9]; the most efficient one is Algorithm 6.8 on p. 111 (the NUCOMP algorithm). This method can be employed to compute the cipher ideal $\mathfrak{c}$ in REAL-NICE encryption step 5.

An ideal $\mathfrak{p} \in \ker(\hat{\Phi})$ as required in encryption step 4 can be determined during encryption or as part of the public key as follows. Generate a random element $x \in I$, with I as given in (6.4), and use Algorithm 5.2 of [14] to find the $\mathbb{Z}$-basis of a reduced principal $\mathcal{O}_{\Delta_q}$-ideal $\mathfrak{p}$ that has a generator $\alpha \in \mathcal{O}_{\Delta_q}$ with $\log_2(\alpha) \approx x/\log(2)$, so $\log(\alpha) \approx x$. This algorithm is essentially repeated squaring using giant steps and requires $O(\log(x)\log(\Delta_q)^2)$ bit operations. Since $\mathfrak{p}$ is principal and ϕ^{-1} preserves principality, $\phi^{-1}(\mathfrak{p})$ is a principal $\mathcal{O}_{\Delta_1}$-ideal, so $[\mathfrak{p}] \in \ker(\hat{\Phi})$.

In decryption step 1, the user needs to find the image $\mathfrak{C}$ of the cipher ideal $\mathfrak{c}$ under ϕ^{-1}. The functions ϕ^{-1} and ϕ can be efficiently computed if the conductor f is known. We briefly recall the procedures here; for details, see [6,13] as well as pp. 14 and 28 of [19]. Let $\mathfrak{a} = (Q, P_{\Delta_f})$ be any primitive $\mathcal{O}_{\Delta_f}$-ideal prime to f. Then $\mathfrak{A} = \phi^{-1}(\mathfrak{a}) = (Q, P_{\Delta_1})$ is a primitive $\mathcal{O}_{\Delta_1}$-ideal prime to f, where $P_{\Delta_1} \equiv xP_{\Delta_f} + ybQ/2 \pmod{Q}$. Here, $x, y \in \mathbb{Z}$ are given by $xf + yQ/\sigma = 1$, and b is the parity of Δ_f; note that if Q is odd, then $\sigma = 1$ and hence $b = 0$. Conversely, if $\mathfrak{A} = (Q, P_{\Delta_1})$ is a primitive $\mathcal{O}_{\Delta_1}$-ideal prime to f, then $\mathfrak{a} = \phi(\mathfrak{A}) = (Q, P_{\Delta_f})$ with $P_{\Delta_f} \equiv fP_{\Delta_1} \pmod{Q}$ is a primitive $\mathcal{O}_{\Delta_f}$-ideal prime to f.

8   Implementation and Run Times

We implemented prototypes of both NICE and REAL-NICE in C++ using GMP and the NTL library for large integer arithmetic [16]. Our numerical data were generated on an Athlon XP 2000+ with 512 MB RAM under the Linux Mandrake 9.1 operating system. In addition, we felt that a comparison to the RSA cryptosystem would be of interest, since the security of RSA also depends on integer factorization and, because it is so widely used in practice, highly-optimized implementations are readily available. We therefore determined run times for RSA using the open source implementation of OpenSSL. Note that our current implementations of NICE and REAL-NICE are first prototypes, whereas the RSA implementation in OpenSSL is highly optimized. Thus, our numerical results are somewhat skewed in favour of RSA.

We used the same parameter sizes for our RSA moduli and non-fundamental discriminants $\Delta_q = q^2\Delta_1$, with q and Δ_1 of approximately equal size, as they give the same level of security. We chose parameter sizes corresponding to the NIST recommended levels of security equivalent to block ciphers with keys of 80, 112, 128, 192, and 256 bits, as specified in Table 1. As public key cryptosystems are usually used for secure key exchange, the message lengths used in our set-up corresponded to these key sizes.

Both NICE and REAL-NICE require selecting a suitable fundamental discriminant Δ_1. For NICE, we simply chose $\Delta_1 = -p$ where p is a prime with $p \equiv 3 \pmod{4}$. In REAL-NICE, we chose $\Delta_1 = p$ where $p \equiv 1 \pmod{4}$ and

p is a Schinzel sleeper as described in Section 5. To find such a prime p of bit length s, we began by choosing random positive integers a and b of bit length $w = 12$; this length easily satisfies the requirement $w \leq s/4 - 1$ of Section 6 for all five NIST [12] security levels with s as given in the second row of Table 1. Then we attempted to determine an integer c such that $b^2 - a^2 c$ divides g^2 with $g = 2 \gcd(a^2, b)$. To find c, we searched the interval $S = [(b^2 - g^2)/a^2, (b^2 + g^2)/a^2]$ for an integer c satisfying the above divisibility condition. Note that if a and b are randomly generated, then $g = 1$ with high probability, leaving only a very small search interval S. Moreover, smaller values of a lead to a larger interval S. Hence, if for a given pair (a, b), no suitable c value was found, we decreased a by 1 — rather than generating a new random value a — and conducted a new search for c. We repeated this procedure until either $a = 1$ or a suitable value of c was found; in the former case, we discarded a and b and started over.

Once a suitable triple (a, b, c) was obtained, we generated successive random integers x of bit length at least $s/2 - 12$ until a value x was found such that $ax^2 + 2bx + c$ is a prime congruent to 1 (mod 4). After a certain number of unsuccessful trials at a value of x, we discarded the triple (a, b, c) and started over with a new choice of a and b. This method worked very well in practice.

To find a conductor q such that $q - (\Delta_1/q)$ is guaranteed to have a large prime factor, we first generated a random prime L close to $|\Delta_1|$ and checked exhaustively whether $q = jL + 1$ is prime and $(\Delta_1/q) = -1$ for $j = 2, 4, 6, \ldots$ If no prime was found for j up to some predetermined bound M, we discarded L and repeated the same procedure until a prime l with the desired properties was obtained.

For encryption under NICE and REAL-NICE, messages were embedded into an $\mathcal{O}_{\Delta_q}$-ideal of prime norm l such that the binary representation of l contained a fixed bit pattern of length $b_{\Delta_q} \in \{80, 112, 128, 192, 256\}$ corresponding to the level of security that was chosen for Δ_q; $b_{\Delta_q} = t$ in NICE and $b_{\Delta_q} = t + u$ in REAL-NICE. In addition, the 20 low order bits of l were set so that $(\Delta_q/l) = 1$. Consequently, messages were bit strings of length $k - b_{\Delta_q} - 20$.

Table 3 gives the average run times for NICE, REAL-NICE, and RSA for various parameter sizes. The run times were obtained by encrypting and decrypting 1000 randomly generated messages for each discriminant size. In addition to the timings for encryption, decryption and message embedding, Table 3 lists the minimal, the maximal and the average number of baby steps that were required to locate the $\mathcal{O}_{\Delta_1}$-ideal $\mathfrak{M}$ during decryption with REAL-NICE.

Our numerical results show that NICE out-performs REAL-NICE for both encryption and decryption. This is not surprising. Recall that in REAL-NICE, the ideal $\mathfrak{p}$ with $[\mathfrak{p}] \in \ker(\hat{\Phi})$ is not included in the public key, resulting in shorter keys. This is done at the expense of a considerable increase in encryption time due to the need for generating a new random ideal $\mathfrak{p}$ for each encryption. We also expect decryption times of REAL-NICE to be slower than those of NICE, due to the extra search through the cycle of reduced ideals in the class of the $\mathcal{O}_{\Delta_1}$-ideal $\mathfrak{C} = \phi^{-1}(\mathfrak{c})$, where $\mathfrak{c}$ is the cipher ideal. In fact, decryption showed the most significant difference in performance between NICE and REAL-NICE. When

Table 3. Average Run Times for NICE, REAL-NICE, and RSA

size(Δ_q)	1024	2048	3072	8192	15360
message length	80	112	128	192	256
block length	180	244	276	404	532
NICE					
encryption	0.02139s	0.06994s	0.12659s	0.68824s	2.35274s
decryption	0.00033s	0.00082s	0.00099s	0.00312s	0.00729s
embedding	0.00467s	0.01152s	0.01550s	0.04257s	0.08860s
REAL-NICE					
encryption	0.03532s	0.09944s	0.18096s	0.96830s	3.28507s
decryption	0.00210s	0.00468s	0.00757s	0.02811s	0.07735s
embedding	0.00531s	0.01152s	0.01547s	0.04289s	0.09770s
min. number of baby steps	1	1	1	2	2
max. number of baby steps	127	181	193	271	355
avg. number of baby steps	58.345	92.801	121.107	204.056	281.438
RSA					
encryption	0.0074s	0.0081s	0.0090s	0.0173s	0.0499s
decryption	0.0127s	0.0334s	0.0931s	1.1188s	7.8377s

considering the overall performance, NICE is up to 1.61 faster than REAL-NICE. However, we note that encryption in REAL-NICE can be replaced by a technique called *infrastructure multiple side-step* (IMS) encryption that could potentially make REAL-NICE competitive to NICE; a similar idea was used with considerable success in cryptographic protocols using real hyperelliptic curves [8], and is explained in the next section.

As expected, both NICE and REAL-NICE decryption significantly outperform RSA for all security levels. It is also noteworthy that when considering the overall performance, both NICE and REAL-NICE are faster than RSA for the two highest levels of security. This is surprising as our NICE and REAL-NICE implementations are first prototypes, whereas the implementation of RSA in the OpenSSL package is considered to be highly optimized.

9   Conclusion and Further Work

There exists a modified version of RSA due to Takagi [17] that would perhaps be more appropriate for comparison with NICE and REAL-NICE. Takagi's cryptosystem relies on the difficulty of factoring integers of the form $p^k q$ (similar to NICE and REAL-NICE) and has faster decryption than RSA. When using $k = 2$, Takagi reports decryption times that are three times faster than decryption using Chinese remaindering with a 768-bit modulus. Our main goal was to compare NICE and REAL-NICE with a highly-optimized implementation of the most widely-used factoring-based cryptosystem (namely RSA), but a comparison with Takagi's cryptosystem would clearly be of interest as well.

While the performance difference between NICE and REAL-NICE is at first glance disappointing, a method referred to as *infrastructure multiple side-step* (IMS) encryption can speed up REAL-NICE encryption time considerably, making the system potentially competitive with NICE. IMS is explained in detail in Section 7.2.1 of [19]. In both NICE and REAL-NICE, the $\mathcal{O}_{\Delta_q}$-ideal $\rho_{\Delta_q}(\mathfrak{p}^r)$ used to obtain the cipher ideal $\mathfrak{c}$ is evaluated using a standard binary exponentiation technique involving giant steps. That is, a square corresponds to a giant step of the form $\rho_{\Delta_q}(\mathfrak{a}^2)$, and a multiply to the giant step $\rho_{\Delta_q}(\mathfrak{a}\mathfrak{p})$, where $\mathfrak{a}$ is the intermediate ideal ($\mathfrak{a} = \rho_{\Delta_q}(\mathfrak{p}^r)$ at the end). In IMS encryption, no random exponent needs to be generated. Instead, a fixed number of square giant steps is chosen, and each square giant step is followed by a certain random number of baby steps (*multiple side steps*) in the cycle of reduced ideals (also referred to as the *infrastructure*) in the $\mathcal{O}_{\Delta_q}$-class of the message ideal $\mathfrak{m}$.

The complexity of a baby step is linear in $O(\log(\Delta_q))$ in terms of bit operations, whereas a giant step has quadratic complexity. Thus, if the number of square giant steps corresponds to the bit length of r, and the number of baby steps after each square & reduce operation is not too large, this results in a significant speed-up in encryption time. On the other hand, if the number of squarings or the number of side steps is too small, this may significantly decrease the number of possible values that the cipher ideal $\mathfrak{c}$ can take on, thereby rendering exhaustive search for $\mathfrak{c}$ potentially feasible. Preliminary numerical data in Section 7.3 of [19] showed that an IMS-prototype of REAL-NICE using even the most conservative security analysis outperformed the original REAL-NICE scheme. It is conceivable that the IMS parameters could be chosen to lead to significantly faster encryption times, while still ensuring the same level of security. Under these circumstances, IMS-REAL-NICE could be competitive to, or even outperform, NICE. This would make IMS-REAL-NICE potentially attractive in situations where fast decryption time is essential (e.g. for fast signature generation) and space is too restricted to hold the larger NICE keys. Clearly, the subject of IMS encryption requires further exploration.

The questions of whether there are sufficiently many prime Schinzel sleepers of a given bit length, and whether choosing Δ_1 to be a Schinzel sleeper presents a security risk, warrant further study. We also point out that it should be possible to adapt the IND-CCA2 secure version of NICE to REAL-NICE in order to provide the same level of security. These and other questions are the subject of future research.

References

1. Buchmann, J., Sakurai, K., Takagi, T.: An IND-CCA2 public-key cryptosystem with fast decryption. In: Kim, K.-c. (ed.) ICISC 2001. LNCS, vol. 2288, pp. 51–71. Springer, Heidelberg (2002)
2. Buchmann, J., Thiel, C., Williams, H.: Short representation of quadratic integers. In: Computational Algebra and Number Theory (Sydney, 1992). Math. Appl., vol. 325, pp. 159–185. Kluwer, Dordrecht (1995)

3. Cheng, K.H.F., Williams, H.C.: Some results concerning certain periodic continued fractions. Acta Arith. 117, 247–264 (2005)
4. Cox, D.A.: Primes of the Form $x^2 + ny^2$. John Wiley & Sons, Inc, New York (1989)
5. Hardy, G.H., Littlewood, J.E.: Partitio numerorum III: On the expression of a number as a sum of primes. Acta Math. 44, 1–70 (1923)
6. Hühnlein, D., Jacobson Jr., M.J., Paulus, S., Takagi, T.: A cryptosystem based on non-maximal imaginary quadratic orders with fast decryption. In: Nyberg, K. (ed.) EUROCRYPT 1998. LNCS, vol. 1403, pp. 294–307. Springer, Heidelberg (1998)
7. Jaulmes, E., Joux, A.: A NICE Cryptanalysis. In: Preneel, B. (ed.) EUROCRYPT 2000. LNCS, vol. 1807, pp. 382–391. Springer, Heidelberg (2000)
8. Jacobson Jr., M.J., Scheidler, R., Stein, A.: Cryptographic protocols on real hyperelliptic curves. Adv. Math. Commun. 1, 197–221 (2007)
9. Jacobson Jr., M.J., Sawilla, R.E., Williams, H.C.: Efficient Ideal Reduction in Quadratic Fields. Internat. J. Math. Comput. Sci. 1, 83–116 (2006)
10. Lenstra, A.K.: Unbelievable Security. In: Boyd, C. (ed.) ASIACRYPT 2001. LNCS, vol. 2248, pp. 67–86. Springer, Heidelberg (2001)
11. Mollin, R.A., Williams, H.C.: Computation of the class number of a real quadratic field. Util. Math. 41, 259–308 (1992)
12. National Institute of Standards and Technology (NIST), Recommendation for key management - part 1: General (revised). NIST Special Publication 800-57 (March 2007), http://csrc.nist.gov/groups/ST/toolkit/documents/SP800-57Part1_3-8-07.pdf
13. Paulus, S., Takagi, T.: A new public key cryptosystem over quadratic orders with quadratic decryption time. J. Cryptology 13, 263–272 (2000)
14. van der Poorten, A.J., te Riele, H.J.J., Williams, H.C.: Computer verification of the Ankeny-Artin-Chowla conjecture for all primes less than 100 000 000 000. Math. Comp. 70, 1311–1328 (2001)
15. Schinzel, A.: On some problems of the arithmetical theory of continued fractions. Acta Arith. 6, 393–413 (1961)
16. Shoup, V.: NTL: A Library for Doing Number Theory. Software (2001), Available at http://www.shoup.net/ntl
17. Takagi, T.: Fast RSA-type cryptosystem modulo $p^k q$. In: Krawczyk, H. (ed.) CRYPTO 1998. LNCS, vol. 1462, pp. 318–326. Springer, Heidelberg (1998)
18. Takagi, T.: A New Public-Key Cryptosystems with Fast Decryption. PhD Thesis, Technische Universität Darmstadt (Germany) (2001)
19. Weimer, D.: An Adaptation of the NICE Cryptosystem to Real Quadratic Orders. Master's Thesis, Technische Universität Darmstadt (Germany) (2004), http://www.cdc.informatik.tu-darmstadt.de/reports/reports/DanielWeimer.diplom.pdf
20. Williams, H.C.: Édouard Lucas and Primality Testing. John Wiley & Sons, New York (1998)
21. Williams, H.C., Wunderlich, M.C.: On the parallel generation of the residues for the continued fraction factoring algorithm. Math. Comp. 48, 405–423 (1987)

A Proof of Security in $O(2^n)$ for the Benes Scheme

Jacques Patarin

Université de Versailles
45 avenue des Etats-Unis, 78035 Versailles Cedex, France
jacques.patarin@prism.uvsq.fr

Abstract. In [1], W. Aiello and R. Venkatesan have shown how to construct pseudorandom functions of $2n$ bits $\rightarrow 2n$ bits from pseudorandom functions of n bits $\rightarrow n$ bits. They claimed that their construction, called "Benes" reaches the optimal bound ($m \ll 2^n$) of security against adversaries with unlimited computing power but limited by m queries in an Adaptive Chosen Plaintext Attack (CPA-2). This result may have many applications in Cryptography (cf [1,19,18] for example). However, as pointed out in [18] a complete proof of this result is not given in [1] since one of the assertions in [1] is wrong. It is not easy to fix the proof and in [18], only a weaker result was proved, i.e. that in the Benes Schemes we have security when $m \ll f(\epsilon) \cdot 2^{n-\epsilon}$, where f is a function such that $\lim_{\epsilon \to 0} f(\epsilon) = +\infty$ (f depends only of ϵ, not of n). Nevertheless, no attack better than in $O(2^n)$ was found. In this paper we will in fact present a complete proof of security when $m \ll O(2^n)$ for the Benes Scheme, with an explicit O function. Therefore it is possible to improve all the security bounds on the cryptographic constructions based on Benes (such as in [19]) by using our $O(2^n)$ instead of $f(\epsilon) \cdot 2^{n-\epsilon}$ of [18].

Keywords: Pseudorandom function, unconditional security, information-theoretic primitive, design of keyed hash functions, security above the birthday bound.

1 Introduction

In this paper we will study again the "Benes" Schemes of [1] and [18]. (The definition of the "Benes" Schemes will be given in Section 2). More precisely, the aim of this paper is to present a complete proof of security for the Benes schemes when $m \ll O(2^n)$ where m denotes the number of queries in an Adaptive Chosen Plaintext Attack (CPA-2) with an explicit O function. With this security result we will obtain a proof for the result claimed in [1] and this will also solve an open problem of [18], since in [18] only a weaker result was proved (security when $m \ll f(\epsilon) \cdot 2^{n-\epsilon}$ where f is a function such that $\lim_{\epsilon \to 0} f(\epsilon) = +\infty$). It is important to get precise security results for these schemes, since they may have many applications in Cryptography, for example in order to design keyed hash functions (cf [1]) or in order to design Information-theoretic schemes (cf [18]).

S. Vaudenay (Ed.): AFRICACRYPT 2008, LNCS 5023, pp. 209–220, 2008.

Here we will prove security "above the birthday bound", i.e. here we will prove security when $m \ll 2^n$ instead of the "birthday bound" $m \ll \sqrt{2^n}$ where m denotes the number of queries in an Adaptive Chosen Plaintext Attack (CPA-2). $\sqrt{2^n}$ is called the 'birthday bound' since when $m \ll \sqrt{2^n}$, if we have m random strings of n bits, the probability that two strings are equal is negligible. 2^n is sometimes called the 'Information bound' since security when $m \ll 2^n$ is the best possible security against an adversary that can have access to infinite computing power. In fact, in [18], it is shown that Benes schemes can be broken with $m = O(2^n)$ and with $O(2^n)$ computations. Therefore security when $m \ll O(2^n)$ is really the best security result that we can have with Benes schemes.

In [2], Bellare, Goldreich and Krawczyk present a similar construction that provides length-doubling for the input. However their construction is secure only against random queries and not against adaptively chosen queries. Benes schemes, in contrast, produce pseudorandom functions secure against adaptively chosen queries.

It is interesting to notice that there are many similarities between this problem and the security of Feistel schemes built with random round functions (also called Luby-Rackoff constructions), or the security of the Xor of two random permutations (in order to build a pseudorandom function from two pseudorandom permutations). The security of random Feistel schemes above the birthday bound has been studied for example in [13], [15], [17], and the security of the Xor of two random permutations above the birthday bound has been studied for example in [3], [8]. However the analysis of the security of the Benes schemes requires a specific analysis and the proof strategy used for Benes schemes is significantly different than for Feistel or the Xor of random permutations. In fact, our proof of security for Benes schemes in $m \ll O(2^n)$ is more simple than the proofs of security in $m \ll O(2^n)$ for Feistel schemes or the Xor of random permutations, since we will be able, as we will see, to use a special property of Benes schemes.

2   Notation

We will use the same notation as in [18].
- $I_n = \{0,1\}^n$ is the set of the 2^n binary strings of length n.
- F_n is the set of all functions $f : I_n \to I_n$. Thus $|F_n| = 2^{n \cdot 2^n}$.
- For $a, b \in I_n$, $a \oplus b$ stands for bit by bit exclusive or of a and b.
- For $a, b \in I_n$, $a||b$ stands for the concatenation of a and b.
- For $a, b \in I_n$, we also denote by $[a, b]$ the concatenation $a||b$ of a and b.
- Given four functions from n bits to n bits, $f_1, \ldots, f_4$, we use them to define the **Butterfly transformation** (see [1]) from $2n$ bits to $2n$ bits. On input $[L_i, R_i]$, the output is given by $[X_i, Y_i]$, with:

$$X_i = f_1(L_i) \oplus f_2(R_i) \text{ and } Y_i = f_3(L_i) \oplus f_4(R_i).$$

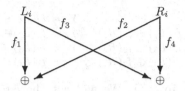

Fig. 1. Butterfly transformation

• Given eight functions from n bits to n bits, $f_1, \ldots, f_8$, we use them to define the **Benes transformation** (see [1]) (back-to-back Butterfly) as the composition of two Butterfly transformations. On input $[L_i, R_i]$, the output is given by $[S_i, T_i]$, with:

$$S_i = f_5(f_1(L_i) \oplus f_2(R_i)) \oplus f_6(f_3(L_i) \oplus f_4(R_i)) = f_5(X_i) \oplus f_6(Y_i)$$

$$T_i = f_7(f_1(L_i) \oplus f_2(R_i)) \oplus f_8(f_3(L_i) \oplus f_4(R_i)) = f_7(X_i) \oplus f_8(Y_i).$$

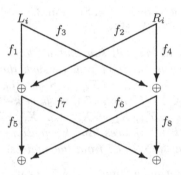

Fig. 2. Benes transformation (back-to-back Butterfly)

3   A Problem in the Proof of [1]

As showed in [18], there is a problem in the security proof of [1]. Let us recall what the problem is.

Definition 1. *We will say that we have "a circle in X, Y of length k" if we have k pairwise distinct indices such that $X_{i_1} = X_{i_2}$, $Y_{i_2} = Y_{i_3}$, $X_{i_3} = X_{i_4}, \ldots,$ $X_{i_{k-1}} = X_{i_k}$, $Y_{i_k} = Y_{i_1}$. We will say that we have "a circle in X, Y" if there is an even integer k, $k \geq 2$, such that we have a circle in X, Y of length k.*

Let $[L_1, R_1]$, $[L_2, R_2]$, $[L_3, R_3]$ and $[L_4, R_4]$ be four chosen inputs such that $L_1 = L_2$, $R_2 = R_3$, $L_3 = L_4$ and $R_4 = R_1$ (and $R_1 \neq R_2$ and $L_1 \neq L_3$). (Here we will say that we have "a circle in L, R" of length 4). Let p be the probability for these inputs to produce "a circle in X, Y" (or, in the language of [1], an "alternating cycle") after a Butterfly. In [1], page 318, it is claimed that "the probability that the top Butterfly produces an alternating cycle of length $2j$ is $\leq 2^{-2jn}$". So here this means $p \leq \frac{1}{2^{4n}}$. However we will see that $p \geq \frac{1}{2^{2n}}$. We have:

$$X_1 = f_1(L_1) \oplus f_2(R_1) \qquad\qquad Y_1 = f_3(L_1) \oplus f_4(R_1)$$
$$X_2 = f_1(L_2) \oplus f_2(R_2) = f_1(L_1) \oplus f_2(R_2) \qquad Y_2 = f_3(L_2) \oplus f_4(R_2) = f_3(L_1) \oplus f_4(R_2)$$
$$X_3 = f_1(L_3) \oplus f_2(R_3) = f_1(L_3) \oplus f_2(R_2) \qquad Y_3 = f_3(L_3) \oplus f_4(R_3) = f_3(L_3) \oplus f_4(R_2)$$
$$X_4 = f_1(L_4) \oplus f_2(R_4) = f_1(L_3) \oplus f_2(R_1) \qquad Y_4 = f_3(L_4) \oplus f_4(R_4) = f_3(L_3) \oplus f_4(R_1)$$

First possible circle in X, Y We will get the circle $X_1 = X_2$, $Y_2 = Y_3$, $X_3 = X_4$ and $Y_4 = Y_1$ if and only if $f_2(R_1) = f_2(R_2)$ and $f_3(L_1) = f_3(L_3)$ and the probability for this is exactly $\frac{1}{2^{2n}}$ (since $R_1 \neq R_2$ and $L_1 \neq L_3$).

Conclusion. The probability p to have a circle in X, Y of length 4 (i.e. the probability that the top Butterfly produces an alternating cycle of length 4 in the language of [1]) is $\geq \frac{1}{2^{2n}}$, so it is not $\leq \frac{1}{2^{4n}}$ as claimed in [1].

As we will see in this paper, this problem is not easily solved: a precise analysis will be needed in order to prove the security result $m \ll 2^n$.

4   "Lines" and "Circles" in X, Y

"Circles" in X, Y have been defined in Section 3. Similarly, (as in [18] p.104) we can define "Lines" in X, Y like this:

Definition 2. *If k is odd, we will say that we have "a line in X, Y of length k if we have $k + 1$ pairwise distinct indices such that $X_{i_1} = X_{i_2}$, $Y_{i_2} = Y_{i_3}$, $X_{i_3} = X_{i_4}$, ..., $Y_{i_{k-1}} = Y_{i_k}$, $X_{i_k} = X_{i_{k+1}}$. Similarly, if k is even, we will say that we have "a line in X, Y of length k" if we have $k+1$ pairwise distinct indices such that $X_{i_1} = X_{i_2}$, $Y_{i_2} = Y_{i_3}$, $X_{i_3} = X_{i_4}$, ..., $X_{i_{k-1}} = X_{i_k}$, $Y_{i_k} = Y_{i_{k+1}}$. So in a line in X, Y we have $k + 1$ indices, and k equations, in X or in Y, and these equations can be written "in a line" from the indices.*

Remark: with this definition, a "line in X, Y" always starts with a first equation in X. This will not be a limitation in our proofs. Of course we could also have defined lines in X, Y by accepting the first equation to be in X or in Y and then to alternate X and Y equations.

To get our security results, as for [1] and [18], we will start from this theorem:

Theorem 1. *The probability to distinguish Benes schemes, when $f_1, \ldots, f_8$ are randomly chosen in F_n, from random functions of $2n$ bits $\rightarrow 2n$ bits in CPA-2 is always less than or equal to p, where p is the probability to have a circle in X, Y.*

Proof of theorem 1

A proof of Theorem 1 can be found in [1] written in the language of "alternating cycles", or in [18] p.97, written with exactly these notations of "circles". In fact, this result can easily be proved like this:

With Benes, we have:

$$\forall i, 1 \leq i \leq m, \ Benes(f_1, \ldots, f_8)[L_i, R_i] = [S_i, T_i] \Leftrightarrow$$
$$\begin{cases} S_i = f_5(X_i) \oplus f_6(Y_i) \\ T_i = f_7(X_i) \oplus f_8(Y_i) \end{cases} \quad (1)$$

$$\text{with } \begin{cases} X_i = f_1(L_i) \oplus f_2(R_i) \\ Y_i = f_3(L_i) \oplus f_4(R_i) \end{cases}$$

When there are no circles in X, Y in each equation (1), we have a new variable $f_5(X_i)$ or $f_6(Y_i)$, and a new variable $f_7(X_i)$ or $f_8(Y_i)$, so if f_5, f_6, f_7, f_8 are random functions, the outputs S_i and T_i are perfectly random and independent from the previous S_j, T_j, $i < j$.

In this paper we will now evaluate p in a new way, in order to get stronger security result. For this we will introduce and study the properties of "first dependency lines".

5   First Dependencies

Definition 3. *A line in X, Y of length k will be called a "first dependency" line when all the equations in X, Y except the last one are independent and when the last one (i.e. the equation number k) is a consequence of the previous equations in X, Y.*

Example: If $L_1 = L_3$, $L_2 = L_4$, $R_1 = R_2$, $R_3 = R_4$, then $(X_1 = X_2)$, $(Y_2 = Y_3)$, $(X_3 = X_4)$ is a "first dependency line", since $(X_1 = X_2)$ and $(Y_2 = Y_3)$ are independent, but $(X_3 = X_4)$ is a consequence of $(X_1 = X_2)$.

Definition 4. *A circle in X, Y will be called a " circle with one dependency" when all the equations in the circle, except one are independent from the others, and when exactly one is a consequence of the others equations in X, Y.*

The key argument in our proof will be this (new) Theorem:

Theorem 2. *When f_1, f_2, f_3, f_4 are randomly chosen in F_n, the probability q_k to have a "first dependency line" in X, Y of length k satisfies $q_k \leq k^5 \dfrac{m^{k-1}}{2^{(k-1)n}}$*

Remark. Some possible improvements of this Theorem 2 (with a better coefficient than k^5) will be given in Section 7. However this version with a coefficient k^5 will be enough for us, in order to get a security for Benes in $O(2^n)$ as we will see in Section 6.

Proof of theorem 2

a) Rough Evaluation

Since we have $(k-1)$ independent equations in X or Y, when all the indices are fixed the probability to have all these equations is $\frac{1}{2^{(k-1)n}}$. Now, in order to choose the $k+1$ indices of the messages, we have less than m^{k+1} possibilities. Therefore, $q_k \leq \frac{m^{k+1}}{2^{(k-1)n}}$. Moreover, the last equation (in X or Y) is a consequence of the previous equations in X, Y. However, a dependency in these equations implies the existence of a circle in L, R on a subset of the indices involved in the dependency. [The proof is exactly the same as for Theorem 1 except that here we use L, R instead of X, Y and X, Y instead of S, T].

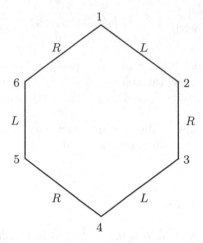

Fig. 3. Example of circle in L, R

Now if we have a circle in L, R of length α, α even, we know that $\frac{\alpha}{2}$ of the messages in the circle come from the other $\frac{\alpha}{2}$ messages.

For example, if $L_1 = L_2$, $R_2 = R_3$, $L_3 = L_4$, $R_4 = R_5$, $L_5 = L_6$, $R_6 = R_1$, we have a circle in L, R of length 6, and if we know the messages 1, 3, 5, then we know (L_1, R_1), (L_3, R_3), (L_5, R_5), and we can deduce (L_2, R_2), (L_4, R_4) and (L_6, R_6), since $(L_2, R_2) = (L_1, R_3)$, $(L_4, R_4) = (L_3, R_5)$ and $(L_6, R_6) = (L_5, R_1)$. In a circle in L, R of length α, we must have $\alpha \geq 4$, since $\alpha = 2$ gives $L_i = L_j$ and $R_i = R_j$, and therefore $i = j$. Therefore, if there is a circle in L, R we will be able to find $\frac{\alpha}{2}$ messages, $\frac{\alpha}{2} \geq 2$, from the other messages of the circle. So, in order to choose $k + 1$ indices of the messages in a first dependency line, we will have $O(m^{k-1})$ possibilities (instead of m^{k+1} possibilities since at least 2 messages will be fixed from the others), and therefore $q^k \leq \frac{O(m^{(k-1)})}{2^{(k-1)n}}$. We will now evaluate the term $O(m^{(k-1)})$ more precisely.

b) More precise evaluation

From a first dependency line in X, Y we have just seen that at least two messages of the line, let say messages $[L_a, R_a]$ and $[L_b, R_b]$ are such that $L_a = L_i$, $R_a = R_j$, $L_b = R_k$, $R_b = R_l$ with $i, j, k, l \notin \{a, b\}$. Moreover, we can choose b to be the last message of the line (since between the two last messages we have a dependency in X or in Y from the other equations in X and Y). Now for a we have less than k possibilities, and for i, j, k, l we have less than $(k - 1)^4$ possibilities. Therefore, for the choice of the $k + 1$ messages of the line we have less than $k(k - 1)^4 m^{k-1}$ possibilities, which is less than $k^5 m^{k-1}$. Therefore, $q_k \leq k^5 \frac{m^{k-1}}{2^{(k-1)n}}$ as claimed.

Remark. We can not always choose a and b to be the last two messages, because it is possible that we have an equality in L, or in R, between these two last messages. However, we can always choose b to be the last message, as we did here.

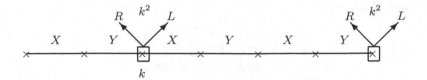

Fig. 4. An example of line in X, Y

Theorem 3. *When f_1, f_2, f_3, f_4 are randomly chosen in F_n, the probability q_k to have a "first dependency line" in X, Y of length k, or a "circle with one dependency" of length $k - 1$ (k odd) satisfies: $q_k \leq k^5 \dfrac{m^{k-1}}{2^{(k-1)n}}$.*

Proof of theorem 3
This is just a simple extension of Theorem 2. A circle of length $k - 1$ with one dependency can be seen as a special line of length k with the first index equal to the index number k, and the proof given for Theorem 2 extended to the classical lines in X, Y and to these special lines gives immediately Theorem 3.

6   Security of the Benes Schemes

Theorem 4. *When f_1, f_2, f_3, f_4 are randomly chosen in F_n, the probability p to have a circle in X, Y satisfies, if $m \leq \frac{2^n}{2}$*

$$p \leq \frac{m^2}{2^{2n}}\Big(\frac{1}{1 - \frac{m^2}{2^{2n}}}\Big) + \frac{m^2}{2^{2n}}\Big(\sum_{k=3}^{+\infty} \frac{k^5}{2^{(k-3)}}\Big)$$

and $\displaystyle\sum_{k=3}^{+\infty} \frac{k^5}{2^{(k-3)}} = 3^5 + \frac{4^5}{2} + \frac{5^5}{2^2} + \frac{6^5}{2^3} + \dots$ *converges to a finite value.*

Therefore, when $m \ll 2^n$, $p \simeq 0$, as wanted.

Proof of theorem 4
For each circle in X, Y of length k, k even, we have three possibilities:
   a) Either all the k equations in X, Y are independent. Then the probability to have a circle is less than or equal to $\frac{m^k}{2^{kn}}$.
   b) Or there exists a first dependency line of length strictly less than k in the equations in X, Y of the circle.
   c) Or the circle is a circle with exactly one dependency.
   Now from Theorems 2 and 3, we get immediately:

$$p \leq \Big(\frac{m^2}{2^{2n}} + \frac{m^4}{2^{4n}} + \frac{m^6}{2^{6n}} + \frac{m^8}{2^{8n}} + \dots\Big) + \sum_{k=3}^{+\infty} \frac{k^5 m^{k-1}}{2^{(k-1)n}}$$

Therefore, if $m \leq \frac{2^n}{2}$,

$$p \leq \frac{m^2}{2^{2n}} \left(\frac{1}{1 - \frac{m^2}{2^{2n}}} \right) + \frac{m^2}{2^{2n}} \left(\sum_{k=3}^{+\infty} \frac{k^5}{2^{(k-3)}} \right)$$

as claimed (since $\frac{m^{k-3}}{2^{(k-3)n}} \leq \frac{1}{2^{(k-3)}}$). Therefore, from Theorem 1, we see that we have proved the security of Benes when $m \ll O(2^n)$ against all CPA-2, with an explicit O function, as wanted.

7 Improving the k^5 Coefficient

By working a little more it is possible, as we will see now, to improve the k^5 coefficient in Theorem 2. First, we will see that it is possible to choose k^4 instead of k^5.

Theorem 5. *When f_1, f_2, f_3, f_4 are randomly chosen in F_n, the probability q_k to have a "first dependency line" in X, Y of length k satisfies $q_k \leq k^4 \dfrac{m^{k-1}}{2^{(k-1)n}}$.*

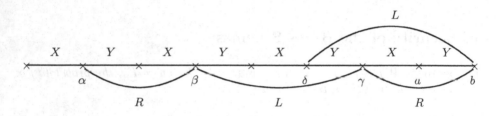

Fig. 5. Illustration of the proof in k^4 instead of k^5.

Proof of theorem 5

We will still denote by a and b the indices of the last equations (in X or in Y and dependent from the other equations). We can proceed like this:

a) We choose 4 indices α, β, γ, $\delta \notin \{b\}$ in the line X, Y. We have here less than k^4 possibilities to choose α, β, γ, δ.

b) We choose all the $k - 1$ messages of indices $\notin \{a, b\}$ in the line of length k. We have here less than m^{k-1} possibilities.

c) The messages of indices β and b will be fixed from the previous values from these equations: $R_\beta = R_\alpha$, $L_\beta = L_\gamma$, $R_b = R_\gamma$, $L_b = L_\delta$.

Therefore we have less than $k^4 m^{k-1}$ possibilities for the choice of the $k + 1$ messages in the first dependency line, so $q_k \leq k^4 \dfrac{m^{k-1}}{2^{(k-1)n}}$ as claimed.

As we will see now, we can get further improvements on the coefficient k^4 by looking at the type of circle in L, R that contains a and b.

Theorem 6. *With the same notation as in Theorem 5, we have: $q_k \leq \dfrac{1}{2^{(k-1)n}}$ $\left(3k m^{k-1} + k^6 m^{k-2} \right)$.*

Proof of theorem 6

We know that the last equation of the line ($X_a = X_b$ or $Y_a = Y_b$) is a consequence of the previous equations in X or Y. We also know that such a consequence is only possible if there is a circle in L, R that includes the two last points a and b. In a circle in L, R of length α, α even, we have seen that $\alpha \geq 4$ and that $\frac{\alpha}{2}$ points can be fixed from the others. We will consider two cases: $\alpha = 4$ and $\alpha \geq 6$.

Case 1: $\alpha = 4$. In this case, the circle in L, R is between a, b and two other points c, d such that the equation (in X or Y) in a, b is a consequence of the equation in c, d (in X or Y). Therefore, for $\{c, d\}$ we have at most $\frac{k}{2}$ possibilities (cf figure 6). Now when $\{a, b, c, d\}$ are fixed, for the circle in L, R we have at most 3×2 possibilities ($R_a = R_b$, R_c or R_d and when this equation in R is fixed, we have two possibilities for the equation in L). Therefore, we have at most $\frac{k}{2} \times 3 \times 2 \times m^{k-1}$ possibilities for a first dependency line in X, Y in this case 1.

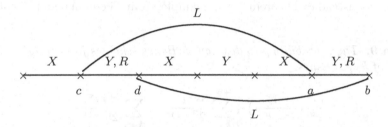

Fig. 6. Example of dependency generated by a circle of length 4 in L, R

Case 2: $\alpha \geq 6$. In this case, at least 2 indices can be fixed from the others, and by using exactly the same arguments as in the proof of Theorem 5 above, with two more points, we see immediately that we have at most $k^6 \cdot m^{k-2}$ possibilities for a first dependency line in X, Y in this case 2. By combining case 1 and case 2, we get immediately Theorem 6.

Theorem 7. *With the same notation as in Theorem 5, we have:*

$$q_k \leq \frac{1}{2^{(k-1)n}} \left(3km^{k-1} + 30k^2 m^{k-2} + k^8 m^{k-3} \right)$$

Proof of theorem 7

The proof is exactly the same as above: the term in $3km^{k-1}$ comes from circles in L, R of length 4, the term in $30k^2 m^{k-2}$ (i.e. $5! \frac{k}{2} \cdot \frac{k}{2} \cdot m^{k-2}$) comes from circles in L, R of length 6, and the term in $k^8 m^{k-3}$ from circles in L, R of length greater than or equal to 8.

Theorem 8. *With the same notation as in Theorem 5, we have: for all integer μ:*

$$q_k \leq \frac{1}{2^{(k-1)n}} \left(3km^{k-1} + 5!(\frac{k}{2})^2 m^{k-2} + 7!(\frac{k}{2})^3 m^{k-3} + 9!(\frac{k}{2})^4 m^{k-4} + \ldots \right.$$

$$\left. + (2\mu + 1)!(\frac{k}{2})^\mu m^{k-\mu} + k^{2\mu+4} m^{k-\mu-1} \right)$$

Alternatively, we also have:

$$q_k \le \frac{1}{2^{(k-1)n}} \Big(\sum_{\mu=1}^{+\infty} (2\mu+1)! (\frac{k}{2})^\mu m^{k-\mu} \Big)$$

Proof of theorem 8

The proof is exactly the same as above. The term $(2\mu+1)!(\frac{k}{2})^\mu m^{k-\mu}$ comes from the circles in L, R of length $2\mu+2$, and the term $k^{2\mu+4}m^{k-\mu-1}$ from the circles in L, R of length greater than or equal to $2\mu+4$, such that these circles in L, R generate the dependency $X_a = X_b$ (or $Y_a = Y_b$) from the previous equations in X, Y.

Application to the Benes schemes

We can immediately apply these results to the Benes schemes, by using these improved results instead of Theorem 2. For example, from Theorem 6 and Theorem 1 we get:

Theorem 9. *The probability p to distinguish Benes schemes from truly random functions of F_{2n} satisfies:*

$$p \le \frac{m^2}{2^{2n}} \Big(\frac{1}{1 - \frac{m^2}{2^{2n}}} \Big) + \sum_{k=3}^{+\infty} \frac{3km^{k-1}}{2^{(k-1)n}} + \sum_{k=5}^{+\infty} \frac{k^6 m^{k-2}}{2^{(k-1)n}}$$

and therefore if $m \le \frac{2^n}{2}$ we get:

$$p \le \frac{m^2}{2^{2n}} \Big(\frac{1}{1 - \frac{m^2}{2^{2n}}} \Big) + \frac{m^2}{2^{2n}} \Big(\sum_{k=3}^{+\infty} \frac{3k}{2^{(k-3)}} \Big) + \frac{m^3}{2^{4n}} \Big(\sum_{k=5}^{+\infty} \frac{k^6}{2^{(k-5)}} \Big) \quad (2)$$

In (2), we have again obtained a proof of security for the Benes schemes against all CPA-2 when $m \ll O(2^n)$. Moreover the O function obtained here is slightly better compared with the O function obtained with Theorem 4.

8   Modified Benes, i.e. Benes with $f_2 = f_3 = \mathrm{Id}$

If we take $f_2 = f_3 = \mathrm{Id}$ in the Benes schemes, we obtain a scheme called "Modified Benes" (see [1,18]). Then we have: $X_i = f_1(L_i) \oplus R_i$, $Y_i = L_i \oplus f_4(R_i)$ and the output $[S_i, T_i]$ is such that $S_i = f_5(X_i) \oplus f_6(Y_i)$ and $T_i = f_7(X_i) \oplus f_8(Y_i)$. It is conjectured that the security for Modified Benes is also in $O(2^n)$ but so far we just have a proof of security in $O(2^{n-\epsilon})$ for all $\epsilon > 0$ (see [18]). It is interesting to notice that the proof technique used in this paper for the regular Benes cannot be used for the Modified Benes, since, as we will see in the example below, for Modified Benes, unlike for regular Benes, the first 'dependent' equation can fix only one index instead of two. Example: If we have $L_1 = L_3$, $L_2 = L_4$, $R_1 \oplus R_2 \oplus R_3 \oplus R_4 = 0$, then we will get the 'line', $X_1 = X_2$, $Y_3 = Y_3$, $X_3 = X_4$ from only two independent equations in f, ($X_1 = X_2$ and $Y_2 = Y_3$), and the first 'dependent' equation, here $X_3 = X_4$, fixes only the index 4 from the previous indices (since $L_4 = L_2$ and $R_4 = R_1 \oplus R_2 \oplus R_3$). Therefore, a proof of security in $O(2^n)$ for the Modified Benes will be different, and probably more complex than our proof of security on $O(2^n)$ for the regular Benes.

9   Conclusion

W. Aiello and R. Venkatesan did a wonderful work by pointing out the great potentialities of the Benes schemes and by giving some very important parts of a possible proof. Unfortunately the complete proof of security when $m \ll 2^n$ for CPA-2 is more complex than what they published in [1] due to some possible attacks in L,R. However, in this paper we have been able to solve this open problem by improving the analysis and the results of [18]. The key point in our improved proof was to analyse more precisely what happens just after the first 'dependent' equations in X,Y (with the notation of Section 3), and to use the fact that in this case two 'indices' are fixed from the others. Therefore we have obtained the optimal security bound (in $O(2^n)$) with an explicit O function. This automatically improves the proved security of many schemes based on Benes, for example the schemes of [19].

References

1. Aiello, W., Venkatesan, R.: Foiling Birthday Attacks in Length-Doubling Transformations - Benes: a non-reversible alternative to Feistel. In: Maurer, U.M. (ed.) EUROCRYPT 1996. LNCS, vol. 1070, pp. 307–320. Springer, Heidelberg (1996)
2. Bellare, M., Goldreich, O., Krawczyk, H.: Stateless evaluation of pseudorandom functions: Security beyond the birthday barrier. In: Wiener, M.J. (ed.) CRYPTO 1999. LNCS, vol. 1666, Springer, Heidelberg (1999)
3. Bellare, M., Impagliazzio, R.: A Tool for Obtaining Tighter Security Analysis of Pseudorandom Based Constructions, with Applications to PRP to PRF Conversion,Cryptology ePrint archive: 19995/024: Listing for 1999
4. Damgård, I.: Design Principles of Hash Functions. In: Brassard, G. (ed.) CRYPTO 1989. LNCS, vol. 435, Springer, Heidelberg (1990)
5. Goldreich, O., Goldwasser, S., Micali, S.: How to Construct Random Functions. JACM 33, 792–807 (1986)
6. Luby, M.: Pseudorandomness and Its Cryptographic Applications. In: Princeton Computer Science Notes, Princeton University Press, Princeton
7. Luby, M., Rackoff, C.: How to construct pseudorandom permutations from pseudorandom functions. SIAM Journal on Computing 17(2), 373–386 (1988)
8. Lucks, S.: The Sum of PRP Is a Secure PRF. In: Preneel, B. (ed.) EUROCRYPT 2000. LNCS, vol. 1807, pp. 470–487. Springer, Heidelberg (2000)
9. Maurer, U.: A simplified and Generalized Treatment of Luby-Rackoff Pseudorandom Permutation Generators. In: Rueppel, R.A. (ed.) EUROCRYPT 1992. LNCS, vol. 658, pp. 239–255. Springer, Heidelberg (1993)
10. Maurer, U.: Information-Theoretic Cryptography. In: Wiener, M.J. (ed.) CRYPTO 1999. LNCS, vol. 1666, pp. 47–64. Springer, Heidelberg (1999)

11. Maurer, U.: Indistinguishability of Random Systems. In: Knudsen, L.R. (ed.) EUROCRYPT 2002. LNCS, vol. 2332, pp. 110–132. Springer, Heidelberg (2002)
12. Maurer, U., Pietrzak, K.: The security of Many-Round Luby-Rackoff Pseudo-Random Permutations. In: Biham, E. (ed.) EUROCRYPT 2003. LNCS, vol. 2656, pp. 544–561. Springer, Heidelberg (2003)
13. Naor, M., Reingold, O.: On the construction of pseudo-random permutations: Luby-Rackoff revisited. Journal of Cryptology 12, 29–66 (1997); In: Proc. 29th ACM Symp. on Theory of Computing, pp. 189–199 (1997) (extented abstract)
14. Patarin, J.: New results on pseudo-random permutation generators based on the DES scheme. In: Feigenbaum, J. (ed.) CRYPTO 1991. LNCS, vol. 576, pp. 301–312. Springer, Heidelberg (1992)
15. Patarin, J.: Improved security bounds for pseudorandom permutations. In: 4th ACM Conference on Computer and Communications Security, Zurich, April 1-4, 1997, pp. 142–150. ACM Press, New York (1997)
16. Patarin, J.: Luby-Rackoff: 7 rounds are Enough for $2^{n(1-\varepsilon)}$ Security. In: Boneh, D. (ed.) CRYPTO 2003. LNCS, vol. 2729, pp. 513–529. Springer, Heidelberg (2003)
17. Patarin, J.: Security of Random Feistel Schemes with 5 or more rounds. In: Franklin, M. (ed.) CRYPTO 2004. LNCS, vol. 3152, pp. 106–122. Springer, Heidelberg (2004)
18. Patarin, J., Montreuil, A.: Benes and Butterfly Schemes Revisited. In: Won, D.H., Kim, S. (eds.) ICISC 2005. LNCS, vol. 3935, pp. 92–116. Springer, Heidelberg (2006)
19. Patarin, J., Camion, P.: Design of near-optimal pseudorandom permutations in the information-theoretic model, Cryptology ePrint archive: 2005/153: Listing for 2005

Yet Another Attack on Vest

Pascal Delaunay[1,2] and Antoine Joux[2,3]

[1] THALES Land and Joint System
160 boulevard de Valmy
92704 Colombes, France
pascal.delaunay@fr.thalesgroup.com
[2] Université de Versailles Saint-Quentin-en-Yvelines
45, avenue des États-Unis
78035 Versailles Cedex, France
[3] DGA
antoine.joux@m4x.org

Abstract. We present a new side-channel attack against VEST, a set of four stream ciphers which reached the second phase of the eSTREAM project (the European stream cipher project). The proposed attacks target the counter part of the ciphers, composed of 16 short-length non-linear feedback shift registers (NLFSR) independently updated. Our aim is to retrieve the whole initial state of the counter (163 to 173 bits) which is a part of the keyed state. The first attack is directly adapted from previous works on differential side-channel attacks. The second attack is brand new. It involves a unique measurement thus it can be seen as a simple side-channel attack. However, it requires some signal processing so we call it *Refined Simple Power Analysis*. As we expect full recovery of the initial state with minimal complexity, one should carefully consider implementing any VEST cipher in an embedded device.

Keywords: Side Channel Attacks, VEST, Stream Ciphers, Fourier Transform.

1 Introduction

Since the introduction of Differential Power Analysis in 1998 [8], attacks and countermeasures of cryptographic algorithms performed on embedded devices have been deeply studied. While conventional attacks focus on the mathematical security of cryptographic algorithms, side-channel attacks target implementations on embedded devices to recover secret data. The limited investment and the low complexity are the major assets of these attacks thus the implementation of cryptographic algorithms in embedded or insecure devices is now carefully studied.

While side-channel attacks first targeted software implementations, recent results [15] adapted these attacks to hardware implementations in FPGA and ASIC. Although many attacks against block ciphers and public-key ciphers have been published so far (see [9,10] for example), few attacks target stream ciphers.

S. Vaudenay (Ed.): AFRICACRYPT 2008, LNCS 5023, pp. 221–235, 2008.

One explanation may be the lack of a standard stream cipher such as the standard AES for block ciphers. Anyway a lot of stream ciphers coexist and they are widely used for their high encryption and decryption speed.

From [16] we get a state of the art of side-channel attacks against stream ciphers. In [11] we discover two Differential Power Attacks (DPA) against two widely used algorithms: A5/1 (GSM communication encryption algorithm) and E0 (Bluetooth encryption algorithm).

According to the latter publication, side-channel attacks on stream ciphers are quite rare because *"The problem with DPA attacks against stream ciphers is that the key stream is computed independently from the plain text to be encrypted".* Actually we disagree with the authors on this assertion on the following points.

- Stream ciphers have an IV setup phase where a known IV is introduced into the cipher and is mixed with the secret key. Although no output is observable during this phase, the internal state depends on the known IV and the secret key.
- Why would one only be restrained to differential attacks ?
- Is the knowledge of some input data required to mount side channel attacks?

As an answer to the latter question, the authors of [6] proposed an attack targeting a single Galois Linear Feedback Shift Register with a simple power analysis and a fast correlation attack. Although this attack does not apply to any particular cipher, it presents an interesting opening by mixing two aspects of cryptanalysis : side-channel analysis and traditional attacks. In a recent paper([3]), the authors present two novel differential side-channel attacks against the eSTREAM ciphers GRAIN and TRIVIUM. They introduce well-chosen IV to decrease the noise level during the proposed attack.

In order to go further in this direction, we present a new side channel attack on a publicly submitted stream cipher called VEST [14]. VEST passed the phase I of the eSTREAM project but due to a practical attack described in [7], it was rejected during the second phase. The authors proposed a minor change in the design to resist to the attack in a second version of the cipher. However we study the behavior of a straightforward implementation of VEST and we point out two side-channel weaknesses in its architecture. Both allow an attacker to recover the initial state of the counter. Our attacks do not focus on the previously mentioned vulnerability thus they are intended to work on both versions of the cipher.

The present paper is organized as follows: in the next section we briefly describe the core components of VEST as submitted in [14]. Section 3 is devoted to Side-Channel Analysis. We shortly remind the different models of side-channel leakage in embedded devices and the main attacks discovered so far. In section 4 we present a classical differential side-channel analysis which focuses on the first half of the counter. Finally, in section 5, we highlight a vulnerability in the design of VEST which allows an attacker to mount a simple side-channel attack to recover the whole counter. We illustrate our approach with a time complexity analysis of our algorithms and theoretical results based on our experiments.

2   VEST Core Components

VEST is a family of four stream ciphers (VEST-4, VEST-8, VEST-16 and VEST-32) which was submitted to the ecrypt/eSTREAM project[1]. All VEST ciphers are constructed on the same basis but each one is intended to provide a different level of security (2^{80} for VEST-4, 2^{128} for VEST-8, 2^{160} for VEST-16 and 2^{256} for VEST-32).

The core of the cipher contains 4 different components:

- a counter, made of 16 non-linear feedback shift registers,
- a linear counter diffusor,
- an accumulator,
- a linear output filter.

The size of these inner components depends on the chosen cipher. In the following we describe the counter since our attacks specifically target this part. We shortly remind the other parts but we refer the reader to [14] to get full description.

2.1   The Counter

The counter is the autonomous part of VEST. It is made of 16 different registers of either 10 or 11-bit long. Each register is associated with a 5 to 1 non-linear function g_i. At each clock cycle, the update function introduces the XOR of the output of g_i with the shifted out bit o_i at the beginning of the register and shifts the other bits.

The registers have two modes of operation: the keying mode and the counter mode. In the keying mode, the NLFSR is disturbed by one bit k at each clock cycle while in the counter mode it is autonomously updated. Fig.1 summarizes the evolution of the NLFSR of VEST in both modes.

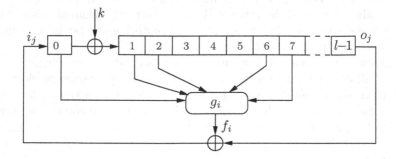

Fig. 1. NLFSR used in VEST

[1] harp//www.ecrypt.eu.org/stream

The authors of VEST proposed 32 non-linear functions g_i, 16 for the 10 bits long registers and 16 for the 11 bits long registers. For each cipher in the family, they specified a set of 16 NLFSR among these to define the counter part.

In VEST, the counter is a highly sensitive part. Indeed it serves as a random number generator for the whole cipher during the computation. It has to be unpredictable. Actually, this assertion is not satisfied: there is a flaw in the design. The authors of [7] highlighted a vulnerability in the design of the counter and they proposed an attack on the IV setup phase. They recovered 53 bits of the keyed state with 2^{22} IV setups. This attack made the cipher to be rejected in phase II of the eSTREAM project.

2.2 The Other Components

Each clock cycle, 16 bits of the counter are extracted. They pass through the linear counter diffusor. The latter maps the extracted bits to 10 bits and linearly combines them with its last state. The computed 10 bits long value enters the accumulator. Every clock cycle, the internal state of the accumulator goes into a substitution phase and a permutation phase. Then the first 10 bits of the result are XOR-ed with the result of the linear counter diffusor.

The last part of the cipher is the linear memoryless output combiner. It linearly combines the state of the accumulator to output M bits, $M \in \{4, 8, 16, 32\}$ depending on the chosen cipher.

3   Side Channel Attacks

In this section we describe some important facts about side-channel attacks. We firstly recall the different models of leakage and briefly remind the two major attacks. Finally, we survey the previous side-channel attacks on stream ciphers.

3.1 A Brief History

Side-channel attacks were first published by Kocher et al. in [8]. The authors linked the instantaneous power consumption of a smart card performing a cryptographic algorithm with the value of the data currently handled. They successfully recovered the secret key of a DES by monitoring the power consumption of the smart card performing the algorithm.

This area of research is being intensely studied. Many side-channel attacks have been discovered, essentially based on timing delays, power or electromagnetic leakage, even acoustic leakage for few of them. We refer the reader to [9,10] to find some examples of side-channel attacks and possible countermeasures on either block ciphers or asymmetric ciphers.

3.2 Models of Leakage

In order to assess the validity of side-channel attacks, we need to correlate the observed leakage with some information in the device. Many models [1,2] have been proposed since 1999 to answer this problem.

When dealing with registers storing values or flip-flops which update their content at each clock cycle, two models coexist: the Hamming weight model and the Hamming distance model. The former model linearly links the observed leakage W with the Hamming weight $\mathcal{H}$ of the data handled by the device (eq. (1)). The latter model links W with the Hamming distance $\mathcal{H}_d$ between the current and the previous handled data (eq. (2)).

$$W = a\mathcal{H} + b \tag{1}$$
$$W = a\mathcal{H}_d + b \tag{2}$$

When the device contains a majority of combinatorial logic, the most suitable model seems to be the transition count model. In this model, the leakage is linearly linked to the number of switchings that occur in CMOS cells during the computation. This switching activity is essentially due to timing delays occurring in a circuit. We refer the reader to [12] to get an overview of this model.

3.3   Types of Side-Channel Attacks

Although many different side-channel attacks have yet been published, we can group them into two major categories: *simple* and *differential* side-channel attacks.

Simple attacks recover secret data from a single curve of leakage. As an example, key-dependent operation attacks [13] and special value attacks [5] can be considered as simple attacks. Differential attacks exhibit biases in the leakage related to some secret data. A typical differential side-channel attack can be found in [8].

3.4   Side-Channel Attacks on Stream Ciphers

As we previously mentioned, side-channel attacks against stream ciphers are quite uncommon. As a matter of fact, previous publications on the subject are limited to [6,11,16,17]. In [11], the authors proposed two differential side-channel attacks against A5/1 and E0 which are respectively used in GSM communications and in the Bluetooth encryption process.

In [3], the authors exposed two differential attacks against the eSTREAM ciphers Grain and TRIVIUM. The attack against the first cipher occurs in three phases. The first two phases are based on a differential power analysis with carefully chosen IV to recover 34 and 16 bits of the secret key. These IV are chosen to minimize the power consumption of the rest of the cipher when computing the differential traces. Indeed some specific bits are fixed to obtain an identical power consumption in specific parts of the cipher. The third phase is a simple exhaustive search among the 30 remaining bits.

The attack against TRIVIUM is quite similar since it also uses specific IV to recover the secret key. These attacks seem really interesting and they are the first side-channel attacks against two phase III eSTREAM ciphers.

The attack described in [6] is quite different. The authors presented a simple power analysis on a n-bit length Galois LFSR that recovered a biased output sequence. They applied a modified fast correlation attack (fast correlation attack with bias weighting) on the sequence in order to recover the initial state of the LFSR.

Although this particular attack does not apply to any specific cipher, the authors showed that simple side-channel analysis can be applied to components of stream ciphers.

Based on these few publications, we decide to target a specific, yet publicly submitted stream cipher and we try to point out potential side-channel vulnerabilities in its design.

4   Differential Analysis of the Counter

From section 2 we know that VEST is made of 4 main components: the counter, the counter diffusor, the accumulator and the output combiner. Our attacks especially target the counter which is the only autonomous part of the cipher.

VEST is a hardware profile stream cipher thus when implemented in a FPGA or an ASIC, the registers of the NLFSR are synthesized as flip-flops which update their values at each clock cycle. However the non-linear functions are likely to be implemented with simple logic gates. In our attacks, we focus on the update of the registers. Hence the Hamming weight and the Hamming distance models described in section 3 are the most adequate for our analysis.

During the key setup phase, the key is introduced into the cipher. In order to use the same secret key for many encryptions, VEST offers a IV setup phase in which a known IV is introduced into the cipher. In this mode, NLFSR 0 to 7 are in keying mode while NLFSR 8 to 15 are in counter mode (see section 2 for more details). Hence the introduced IV only affects the first 8 NLFSR. This remark is the basis of our first attack.

Since the IV setup phase disturbs half of the NLFSR in the counter and the length of each disturbed NLFSR is 10 or 11 bits, we can mount a known plain text differential side-channel attack on each one with low complexity. Anyway, the results of this potential weakness highly depend on the differential characteristic of each NLFSR.

4.1   Differential Characteristic of a Short Length NLFSR

For two NLFSR chosen from [14], we perform the following test: for each possible initial state, we introduce 2 bytes of random data as described in the IV setup phase in [14]. Then we apply a theoretical differential side-channel attack at the end of this phase. We use the traditionnal selection function here, separate the curves where the theoretical leakage (Hamming weight or Hamming distance model) is among or above $n/2$ where n is the length of the targeted register.

In order to validate the differential attack, for each possible initial state we check whether the highest differential peak is obtained for the right initial state (validity of the differential attack). We also check whether the second highest peak is far lower than the first one (ghost peaks problem). Table 1 summarizes the results obtained for two distinct NLFSR of VEST, for 4000 IV of either 2 or 3 bytes long and 10000 IV of 2 bytes long. The feedback function is given as a 32 bits long word: the 5 input bits form a decimal number from 0 to 31 which is the index of the result in the 32 bits long word.

Table 1. Differential side-channel attacks on NLFSR of VEST

Feedback function	Length of the NLFSR	Length of IV	Number of IV	Validity of DPA	Closest ghost peak
0xDD1B4B41	11	2	4000	☑	11%
0xDD1B4B41	11	3	4000	☑	11%
0xDD1B4B41	11	2	10000	☑	8%
0x94E74373	10	2	4000	☑	10%
0x94E74373	10	3	4000	☑	9%
0x94E74373	10	2	10000	☑	6%

From this table we deduce some important facts. We can mount a differential side-channel attack with known IV on a short-length NLFSR. For both of the considered NLFSR, only 4000 IV are sufficient to recover the right initial state. Moreover, the *ghost peaks* problem is unlikely to happen: the closest highest peak is only $\frac{1}{10}^{th}$ of the highest for all possible initial states and for only 4000 IV. Moreover, that proportion decreases with more IV. Finally, increasing the length of the IV does not decrease the size of the ghost peaks. As a matter of fact, 2 bytes of IV looks like a good compromise in size and diffusion.

4.2 Application to the Counter Part of VEST

During the IV setup phase, the IV is introduced by the first 8 NLFSR. Thus we can only target these NLFSR with our differential attack. It is necessary to introduce at least 2 bytes of IV in each NLFSR to obtain a correct diffusion. Since each bit is inserted only once in one NLFSR, this attack requires at least 16 bytes of IV. Note that the authors of VEST do not specify any maximal length for the IV.

Each NLFSR is independent of the others: it involves a unique feedback polynomial and no cross-computation occurs. This ensures that when we target a single NLFSR, the contribution of the others can be seen as some random noise in the observed leakage. In other words, we require substantially more IV to keep the same level of signal. Table 2 summarizes the different experiments made for random initial states and 10000 to 35000 different IV.

As shown in the previous section, IV longer than 2 bytes do not necessarily increase the Signal to Noise Ration (SNR) while a larger number of IV helps decreasing the ghost peaks effect. If we need to increase the SNR, one solution would be to average the contribution of the untargeted NLFSR. This could be done by setting the bytes entering the targeted NLFSR while varying the other bytes. The complexity of the attack is $\mathcal{O}(N \times 2^{13})$ for N IV. It allows the recovery

Table 2. Differential side-channel attacks on VEST counters

Number of IV	Length of IV	Mean of closest ghost peaks	Highest ghost peak
10000	16	11.6%	14.2%
10000	24	12%	14.2%
20000	16	8.5%	10%
35000	16	6.2%	7.1%

of 83 bits of the initial state. Since we only recover half of the counter, we need to perform an additional attack to recover the key. Examples of such attacks can be found in [7].

As we are dealing with side-channel attacks, we try to find a new attack that can be applied to the whole counter part of VEST. Indeed we find some interesting points in [14] that allow an attacker to mount a simple side-channel attack on VEST with insignificant extra complexity.

5   Refined Simple Side-Channel Attack Based on the Fourier Transform

"*The period of each of the chosen NLFSR is guaranteed to be a predetermined prime number for any starting value. Such prime-period NLFSR when combined together result in a counter with a total period being a multiple of the individual periods of all 16 NLFSR*" ([4]). This assertion is actually true but, as we will see in the following, it also creates vulnerabilities that compromise the whole counter.

5.1   The Use of the Fourier Transform

First of all, the counter part of VEST is made of 16 NLFSR $\mathcal{N}_i$ which update their internal values at each clock cycle. The clock rate being known, we can simply separate a single long trace into multiple traces of one cycle each to sum them up. Moreover, for a given initial state, each NLFSR has a unique and predetermined prime period referenced in [14]. Hence a simple Fourier transform on a long trace will present remarkable peaks at some distinct frequencies f_i corresponding to the predetermined periods T_i.

From this simple operation, we recover the period T_i of each NLFSR $\mathcal{N}_i$ of the counter. In appendix F from [14] are listed all the possible periods for each NLFSR.

5.2   New NLFSR-Oriented Curves

Suppose we have a trace of leakage (power consumption, electro-magnetic emanations) $\mathcal{E}$ of N cycles of the VEST algorithm running into an embedded device:

$$\mathcal{E} = (\mathcal{E}_0 \ldots \mathcal{E}_{N-1})$$

Once T_i is identified with the Fourier transform, for each $i \in [\![0, 15]\!]$ we target the corresponding NLFSR $\mathcal{N}_i$ as follows. We construct the curve $\mathcal{C}_i$ of length T_i, defined as:

$$\mathcal{C}_i = (\mathcal{C}_{i,0} \ldots \mathcal{C}_{i,T_i-1}), \; \mathcal{C}_{i,j} = \sum_{k=0}^{\lfloor N/T_i \rfloor} \mathcal{E}_{j+k \times T_i}$$

The NLFSR $\mathcal{N}_i$ cycles every T_i steps. Hence we can separate its leakage induced in $\mathcal{C}_i$ from the leakage induced by the other NLFSR. We respectively define these parts A_i and B_i and we detail them hereafter.

5.3   The Biases

For the sake of simplicity, we consider the Hamming weight model as the leakage model in the following. We define $H(\mathcal{N}_i^j)$ as the Hamming weight of the targeted NLFSR $\mathcal{N}_i$ at time j. We firstly consider the A_i part. For each $j \in [\![0, T_i - 1]\!]$, we define

$$A_i = (A_{i,0} \ldots A_{i,T_i-1}), \; A_{i,j} = \sum_{k=0}^{N/T_i} H(\mathcal{N}_i^{j+k \times T_i})$$

Since $\mathcal{N}_i$ cycles every T_i steps, we have $H(\mathcal{N}_i^{j+k \times T_i}) = H(\mathcal{N}_i^j)$ and we obtain

$$\forall j \in [\![0, T_i - 1]\!], \; A_{i,j} = \frac{N}{T_i} H(\mathcal{N}_i^j)$$

Obviously, when we analyze the NLFSR $\mathcal{N}_i$ on its own curve $\mathcal{C}_i$, its Hamming weight is *linearly amplified* by $\frac{N}{T_i}$. Hence, from a single curve $\mathcal{C}_i$, we extract the evolution of the Hamming weight $H(\mathcal{N}_i)$ during T_i cycles. Anyway, 16 NLFSR run in parallel in the counter and this amplification solely targets $\mathcal{N}_i$. We need to analyse the behavior of the 15 remaining $\mathcal{N}_j$, $j \neq i$ with respect to T_i in the curve $\mathcal{C}_i$ (*i.e.* the B_i part).

As a matter of fact, no cross-computation occurs between the 15 remaining NLFSR. Moreover, each one has a unique and predetermined prime period for a fixed initial state. The update of each NLFSR is independent of the others and so it is for their Hamming weights. Thus the leakage of the 15 left NLFSR can be modeled as the sum of their own leakages. In other words, $B_i = \sum_{k=0, k \neq i}^{15} B_i^k$.

Since all the NLFSR are similar, in the following we model the theoretical leakage of one NLFSR $\mathcal{N}_k$ with respect to T_i, $i \neq k$. The other NLFSR will behave the very same way.

Even if $\mathcal{N}_k$ is made of simple flip-flops, we can not model its leakage with a normal distribution. Indeed, the Hamming weight $H(\mathcal{N}_k^{t+1})$ depends on $H(\mathcal{N}_k^t)$. Actually, $H(\mathcal{N}_k^{t+1}) = H(\mathcal{N}_k^t) \pm \{0, 1\}$ thus the variables are not independent. Anyway, since the length of each NLFSR is relatively small (10 to 11 bits long), we simply perform an exhaustive overview of its theoretical leakage B_i^k and deduce its contribution in the overall leakage B_i.

As an example, we choose $T_1 = 1009$ from [14] *i.e.* we target $\mathcal{N}_1$. We compute the theoretical behavior B_1^{17} of $\mathcal{N}_{17}$ for $N = 2^{20}$ cycles with respect to T_1 under the Hamming weight model. Note that these two NLFSR are recommended in VEST-4 and the potential periods of $\mathcal{N}_{17}$ are 503 and 521 which are close to $T_1/2$. For each possible initial state k of $\mathcal{N}_{17}$ we compute the theoretical leakage of $\mathcal{N}_{17}$ as explained above thus for each $j \in [\![0, T_1 - 1]\!]$ we compute the following values

$$B_{1,j}^{16}(k) = \sum_{m=0}^{N/T_1} H(\mathcal{N}_{17}^{j+m \times T_1})$$

These k curves correspond to the contribution of $\mathcal{N}_{17}$ in $\mathcal{C}_1$ for each possible initial state k. Unsurprisingly, when N increases, $\forall j \in [\![0, T_1 - 2]\!]$, $B_{1,j}^{17}(k) \approx B_{1,j+1}^{17}(k) \approx \frac{N}{T_1} \times \mu(\mathcal{N}_{17}^k)$ where $\mu(\mathcal{N}_{17}^k)$ is the mean of Hamming weights of $\mathcal{N}_{17}$ initialized with k during the period T_{17}.

Figure 2 represents the theoretical contributions of $\mathcal{N}_1$ (sharp curve) and $\mathcal{N}_{17}$ (flat curve) with respect to T_1 for random initial values. Note that we observe an equivalent behavior for the $2^{|\mathcal{N}_j|}$ possible initial states. Depending on the initial value, $\mu(\mathcal{N}_{17}^k)$ has 2 different values due to two possible periods $T_{17} = 503$ and $T_{17} = 521$. Anyway the curve remains almost flat which is the important point here. In other words for a sufficiently large number of output bits, the contribution of $\mathcal{N}_{17}$ in $\mathcal{C}_1$ (the curve B_1^{17}) is merely a constant whichever the initial state is.

We also simulate the experiments for the remaining NLFSR even if their unique and predetermined prime periods guarantee similar behaviors. The results validate our assumptions: for a sufficiently large number of output bits (2^{20} in our experiments), the contribution of $H(\mathcal{N}_j)$ in $\mathcal{C}_i$ is almost constant and the variations are insignificant compared to the variations of $H(\mathcal{N}_i)$. When we sum the leakages of the 16 NLFSR, the 15 untargeted ones add an almost constant value with a minimal standard deviation. Therefore the noise generated hardly interfere with the signal recovered from the targeted NLFSR. In other words, from a single curve of the leakage we construct the 16 curves $\mathcal{C}_i$. They correspond to the evolution of the Hamming weight of the 16 NLFSR which compose the counter.

5.4   Extracting Information from the Variations

The curve $\mathcal{C}_i$ emphasizes the evolution of $H(\mathcal{N}_i)$ during the period T_i and minimizes the effects of the other NLFSR. We consider the presence of a Gaussian noise induced by the whole chip. We model this noise by the normal distribution $\mathcal{N}(\mu_{nz}, \nu_{nz})$.

Now we focus on this evolution to extract information on the internal state of $\mathcal{N}_i$.

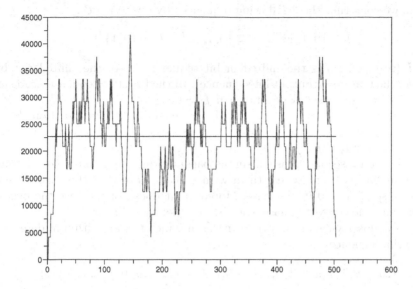

Fig. 2. Contribution of NLFSR $\mathcal{N}_0$ (blue) and $\mathcal{N}_{16}$ (black) with respect to T_0

For each $j \in [\![0, T_0 - 2]\!]$,

$$C_{i,j+1} - C_{i,j} \approx (A_{i,j+1} - A_{i,j}) + \mathcal{N}(\mu_{nz}, \nu_{nz}) \tag{3}$$

$$A_{i,j+1} - A_{i,j} = \lfloor N/T_i \rfloor \times \underbrace{(H(N_{i,j+1}) - H(N_{i,j}))}_{d_j} \tag{4}$$

Hence from $C_{i,j+1} - C_{i,j}$ we obtain information on d_j modulated with some noise. The difference between two successive points of C_i amplifies the difference of the Hamming weight of two successive states of N_i by $\lfloor N/T_i \rfloor$. As d_j can only take 3 values $\{0, 1, -1\}$ we can compute from eq. (3) a threshold t which recovers the value of d_j from $C_{i,j+1} - C_{i,j}$. This threshold depends on the level of the noise and the number of available samples. We will make no further analysis on the threshold since it is essentially based on the quality of the measurement and this study is theoretical.

This value d_j helps recovering the value i_j (resp. o_{j-1}) of the input bit i at time j (resp. the output bit o at time $j-1$), the only bits that can vary the Hamming weight of the register. Since $i_j = o_{j-1} \oplus f_{j-1}$, we also extract information on f_{j-1}, the value of the non-linear function f at time $j-1$. Information on these values is recovered as follows:

- if $C_{i,j+1} - C_{i,j} > t$ then $d_j = 1$, $i_j = 1$, $o_{j-1} = 0$ so $f_{j-1} = 1$ (the Hamming weight increases),
- if $C_{i,j+1} - C_{i,j} < -t$ then $d_j = -1$, $i_j = 0$, $o_{j-1} = 1$ so $f_{j-1} = 1$ (the Hamming weight decreases),
- if $C_{i,j+1} - C_{i,j} \in [-t, t]$, $d_j = 0$, the Hamming weight does not change, we can not guess the shifted in value but $i_j = o_{j-1}$ so $f_{j-1} = 0$.

Hence we construct the 3 following sequences by observing C_i:

$$I = \{0, 1, x\}^{T_i}, \; O = \{0, 1, x\}^{T_i}, \; F = \{0, 1\}^{T_i}$$

where I (resp. O, F) is the shifted in bit sequence (resp. the shifted out bit sequence, the non-linear output bit sequence). In the first two sequences x means *unable to decide* or *unset*, it occurs when $f_j = 0$.

When i_j is *unset* (*i.e.* $i_j = x$), we have an additional information. Since $i_j = o_{j-1} \oplus f_{j-1}$, $f_{j-1} = 0$ and $o_j = I_{j-|\mathcal{N}_i|+1}$ (the bits are simply shifted in the NLFSR) we have $i_j = i_{j-|\mathcal{N}_i|}$.

Therefore, on a second phase, a simple backtracking algorithm recovers some *unset* input bits i_j by matching them with $i_{j-|\mathcal{N}_i|}$. Note that this algorithm is valid because $f_{j-1} = 0$ in this case. Moreover, it does not increase the overall complexity of the recovery phase since its complexity is linear.

Fig.3 is a closer view of an attack on $\mathcal{N}_8$ in which the algorithm of detection predicts the sequence

$$\{x, 0, 1, x, 1, x, x, 1, 1, 0, 0, 1, 0, 0, 0, 1, 0, x, x, 1, x, 0, x, x, x, 0, \dots\}$$

The backtracking phase recovers the following sequence:

$$\{x, 0, 1, x, 1, x, x, 1, 1, 0, 0, 1, 0, 0, 0, 1, 0, 1, 1, 1, 0, 0, 0, 0, 0, 0, \dots\}$$

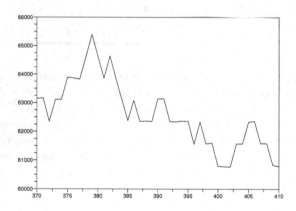

Fig. 3. Zoom on N_{24} at $t = 370$, $T_{24} = 677$

When no more input bits can be recovered within the first two phases, we extract the first subsequence of $|\mathcal{N}_i|$ bits from I with the least number of unknown bits, say t_i. Sometimes $t_i = 0$, thus we obtain a complete state of the NLFSR and the attack is over. Otherwise, we perform a correlation attack on $\mathcal{N}_i$ for the 2^{t_i} possible initial states and match the sequences with I, O and F. Only the correct initial state matches the recovered sequences with no error (see section 5.5 for more details).

The first subsequence of 10 bits (the length of N_{24}) with the least number of unpredicted bits of this sequence is $Z_{24} = \{1, 1, 0, 0, 1, 0, 0, 0, 1, 0\}$ which has no unpredicted bit. In this special case, our recovery algorithm compares the sequence of bits entering the NLFSR initialized with Z_{24} with the predicted non-linear output sequence F.

When the considered initial state is fully recovered, we simply rewind the NLFSR to obtain its initial value *i.e.* the state at the beginning of the curve. Proceeding similarly with the other NLFSR, we recover their respective initial states at the very same time $t = 0$. Thus we obtain the full state of the counter at this precise time.

Experiments on Fig.3 have been made using 2^{20} output bits. According to [4], it corresponds to 3 ms of encryption on the least efficient tested FPGA. It is worth noticing that this attack works on the whole counter part of VEST, unlike the previous attack.

The complexity of the Fourier transform is $\mathcal{O}(N \log N)$ for N output bits. For each NLFSR, the complexity of the attack is in $\mathcal{O}(N + T_i \, 2^{t_i})$ thus the whole complexity of the attack is $\mathcal{O}(N \log N + \sum_{i=0}^{15} (N + T_i \, 2^{T_i}))$. Since we deal with approximately 2^{20} or more output bits, this complexity is about $\mathcal{O}(N \log N)$, which is the complexity of the Fourier transform.

5.5   False Prediction

Two minor drawbacks arise from this attack:

- an incorrect prediction in I will automatically result in the rejection of the considered initial state,
- a sequence with an incorrect initial state can match the correct sequence on the predicted bits of F.

Depending on the number of collected samples and the noise level, our detection algorithm can predict incorrect values. In this case, the recovery algorithm does not work as intended.

However, we can tighten the threshold t by creating a new population of *discarded* input bits. In this case, we differentiate *unpredicted bits* i_j and o_{j-1} in which $f_{j-1} = 0$ and *discarded* bits in which i_j, o_{j-1} and f_{j-1} are unknown. The former population can be recovered by the second phase as explained above but the latter has to be put apart.

Obviously the population of predicted bits in I and O will decrease and the complexity of the attack will slightly increase. Anyway the confidence level of the predicted bits will also increase. The second phase of the algorithm ensures the recovery of some input bits with a high level of confidence since it is based on the predicted bits in the first phase. Thus the complexity of the attack does not significantly increase since t_i is still small.

This problem can also be avoided by observing a leakage trace with significantly more output bits. We previously saw that the leakage of the targeted NLFSR is linearly amplified by the number of samples N while the leakage of the other NLFSR acts as a constant. Thus the signal to noise ratio would increase in this scenario.

The second drawback is actually very unlikely to happen. Two different initial states can not generate the same output sequence: once the bits are in the NLFSR $\mathcal{N}_i$, they are just shifted and never modified until they are output. Thus if n output bits coincide (where n is the length of the NLFSR), the internal states are identical. In the case of *discarded input bits*, some bits are not predicted thus the correlation attack is only performed on the predicted output bits. The probability of an incorrect internal state to match all the t predicted bits exists but decreases if t increases. This problem can also be avoided by observing a longer output sequence.

6   Conclusion

We present two different side-channel attacks against VEST, a phase II candidate of the eSTREAM project. These attacks are based upon two weaknesses in the design of the cipher. The first vulnerability is the short length of the NLFSR. Since they are independent and only 10 or 11 bit long, we firstly apply a differential side-channel attack on each NLFSR used in the IV setup phase with random IV. The results confirm our assumptions: the initial state corresponds to the highest peak of leakage and the closest ghost peak is far lower. We applied this attack to the whole counter part and the results went unchanged. Although the 16 NLFSR in VEST run in parallel, they evolve independently. In this case, the leak induced by the untargeted NLFSR behaves like some random noise. We recovered 83 bits of the keyed state corresponding to the 8 NLFSR in which the IV is introduced.

The second pinpointed vulnerability is the small, unique and predetermined prime period of each NLFSR. We present a new simple side-channel attack exploiting this weakness to recover the whole keyed state of the counter. We highlight the evolution of the leakage of each NLFSR during its own period. Based on the Hamming weight model, we extract from this evolution some important bits of the targeted NLFSR such as the input bit and the value of the non-linear function. Then we apply a correlation attack on the extracted sequences to recover the initial state.

It is worth noticing that this new attack can be considered as a simple side-channel attack since it requires only a single trace. Moreover, it requires neither known plain text nor known cipher text and can only be performed with an Electro Magnetic leakage curve.

Contrary to the authors' assertion in [4], VEST contains vulnerabilities that can be exploited with side-channel analysis. In these conditions, implementing this cipher in an unprotected fashion should be avoided.

More generally, we do not encourage to introduce secret data in small length independent objects. Indeed, an attacker will directly target these specific parts with side-channel analysis to extract important information with low complexity. As an example, we provided two different attacks on VEST. Stream ciphers should be carefully implemented or synthesized as numerous side-channel attacks have not yet been discovered.

References

1. Agrawal, D., Archambeault, B., Rao, J., Rohatgi, P.: The EM Side-Channel(s). In: Kaliski Jr., B.S., Koç, Ç.K., Paar, C. (eds.) CHES 2002. LNCS, vol. 2523, pp. 29–45. Springer, Heidelberg (2003)
2. Brier, E., Clavier, C., Olivier, F.: Correlation Power Analysis with a Leakage Model. In: Joye, M., Quisquater, J.-J. (eds.) CHES 2004. LNCS, vol. 3156, pp. 16–29. Springer, Heidelberg (2004)
3. Fischer, W., Gammel, B.M., Kniffler, O., Velten, J.: Differential Power Analysis of Stream Ciphers. In: Abe, M. (ed.) CT-RSA 2007. LNCS, vol. 4377, pp. 257–270. Springer, Heidelberg (2006)
4. Gittins, B., Landman, H., O'Neil, S., Kelson, R.: VEST, a presentation on VEST Hardware Performance, Chip Area Measurements, Power Consumption Estimates (2005)
5. Goubin, L.: A refined power analysis attack on elliptic curve cryptosystems. In: Desmedt, Y.G. (ed.) PKC 2003. LNCS, vol. 2567, pp. 199–210. Springer, Heidelberg (2002)
6. Joux, A., Delaunay, P.: Galois LFSR, Embedded Devices and Side Channel Weaknesses. In: Barua, R., Lange, T. (eds.) INDOCRYPT 2006. LNCS, vol. 4329, pp. 436–451. Springer, Heidelberg (2006)
7. Joux, A., Reinhard, J.-R.: Overtaking Vest. In: Biryukov, A. (ed.) FSE 2007. LNCS, vol. 4593, pp. 60–75. Springer, Heidelberg (2007)
8. Kocher, P., Jaffe, J., Jun, B.: Differential Power Analysis. In: Wiener, M.J. (ed.) CRYPTO 1999. LNCS, vol. 1666, pp. 388–397. Springer, Heidelberg (1999)
9. Koç, Ç.K., Naccache, D., Paar, C. (eds.): CHES 2001. LNCS, vol. 2162. Springer, Heidelberg (2001)
10. Paar, C., Koç, Ç.K. (eds.): CHES 2000. LNCS, vol. 1965. Springer, Heidelberg (2000)
11. Lano, J., Mentens, N., Preneel, B., Verbauwhede, I.: Power Analysis of Synchronous Stream Ciphers with Resynchronization Mechanism. In: The State of the Art of Stream Ciphers (2004)
12. Mangard, S., Pramstaller, N., Oswald, E.: Successfully Attacking Masked AES Hardware Implementations. In: Rao, J.R., Sunar, B. (eds.) CHES 2005. LNCS, vol. 3659, pp. 157–171. Springer, Heidelberg (2005)
13. Messerges, T., Dabbish, E., Sloan, R.: Power analysis on modular exponentiation in smartcards. In: Koç, Ç.K., Paar, C. (eds.) CHES 1999. LNCS, vol. 1717, pp. 144–157. Springer, Heidelberg (1999)
14. O'Neil, S., Gittins, B., Landman, H.: VEST. Hardware-Dedicated Stream Ciphers (2005)
15. Ors, S.B., Oswald, E., Preneel, B.: Power-analysis attacks on an FPGA - first experimental results. In: D.Walter, C., Koç, Ç.K., Paar, C. (eds.) CHES 2003. LNCS, vol. 2779, pp. 35–50. Springer, Heidelberg (2003)
16. Rechberger, C., Oswald, E.: Stream ciphers and side channel analysis. In: SASC 2004, pp. 320–327 (2004)
17. Rechberger, C.: Side Channel Analysis of Stream Ciphers. Master's thesis, Institute for Applied Information Processing and Communications (IAIK), Graz University of Technology, Austria (2004)

Chosen IV Statistical Analysis for
Key Recovery Attacks on Stream Ciphers

Simon Fischer[1], Shahram Khazaei[2], and Willi Meier[1]

[1] FHNW, Windisch, Switzerland
[2] EPFL, Lausanne, Switzerland

Abstract. A recent framework for chosen IV statistical distinguishing analysis of stream ciphers is exploited and formalized to provide new methods for key recovery attacks. As an application, a key recovery attack on simplified versions of two eSTREAM Phase 3 candidates is given: For Grain-128 with IV initialization reduced to up to 180 of its 256 iterations, and for Trivium with IV initialization reduced to up to 672 of its 1152 iterations, it is experimentally demonstrated how to deduce a few key bits. Evidence is given that the present analysis is not applicable on Grain-128 or Trivium with full IV initialization.

Keywords: Stream ciphers, Chosen IV analysis, eSTREAM, Grain, Trivium

1 Introduction

Synchronous stream ciphers are symmetric cryptosystems which are suitable in software applications with high throughput requirements, or in hardware applications with restricted resources (such as limited storage, gate count, or power consumption). For synchronization purposes, in many protocols the message is divided into short frames where each frame is encrypted using a different publicly known initialization vector (IV) and the same secret key. Stream ciphers should be designed to resist attacks that exploit many known keystreams generated by the same key but different chosen IVs. In general, the key and the IV is mapped to the initial state of the stream cipher by an initialization function (and the automaton produces then the keystream bits, using an output and update function). The security of the initialization function relies on its mixing (or *diffusion*) properties: each key and IV bit should affect each initial state bit in a complex way. This can be achieved with a round-based approach, where each round consists of some nonlinear operations. On the other hand, using a large number of rounds or highly involved operations is inefficient for applications with frequent resynchronizations. Limited resources of hardware oriented stream ciphers may even preclude the latter, and good mixing should be achieved with simple Boolean functions and a well-chosen number of rounds. In [4, 8, 9, 6], a framework for chosen IV statistical analysis of stream ciphers is suggested to investigate the structure of the initialization function. If mixing is not perfect, then the initialization function has an algebraic normal form (ANF) which can

S. Vaudenay (Ed.): AFRICACRYPT 2008, LNCS 5023, pp. 236–245, 2008.

be distinguished from a uniformly random Boolean function. Particularly the co-efficients of high degree monomials in the IV (*i.e.* the product of many IV bits) are suspect to some biased distribution: it will take many operations before all these IV bits meet in the same memory cell. In [4], this question was raised: *"It is an open question how to utilize these weaknesses of state bits to attack the cipher."*. The aim of this paper is to contribute to this problem and present a framework to mount key recovery attacks. As in [4, 8] one selects a subset of IV bits as variables. Assuming all other IV values as well as the key fixed, one can write a keystream symbol as a Boolean function. By running through all possible values of these bits and generating a keystream output each time, one can compute the truth table of this Boolean function. Each coefficient in the algebraic normal form of this Boolean function is parametrized by the bits of the secret key. Based on the idea of *probabilistic neural bits* from [1], we now examine if every key bit in the parametrized expression of a coefficient does oc-cur, or more generally, how much influence each key bit does have on the value of the coefficient. If a coefficient depends on less than all key bits, this fact can be exploited to filter those keys which do not satisfy the imposed value for the coefficient. It is shown in [10] that for eSTREAM Phase 3 candidate Trivium with IV initialization reduced to 576 iterations, linear relations on the key bits can be derived for well chosen sets of variable IV bits. Our framework is more general, as it works with the concept of (probabilistic) neutral key bits, *i.e.* key bits which have no influence on the value of a coefficient with some (high) prob-ability. This way, we can get information on the key for many more iterations in the IV initialization of Trivium, and similarly for the eSTREAM Phase 3 can-didate Grain-128. On the other hand, extensive experimental evidence indicates clear limits to our approach: With our methods, it is unlikely to get information on the key faster than exhaustive key search for Trivium or Grain-128 with full IV initialization.

2   Problem Formalization

Suppose that we are given a fixed Boolean function $F(K, V) : \{0, 1\}^n \times \{0, 1\}^m \to \{0, 1\}$. An oracle chooses a random and unknown $K = (k_0, \ldots, k_{n-1})$ and returns us the value of $z = F(K, V)$ for every query $V = (v_0, \ldots, v_{m-1})$ of our choice (and fixed K). The function F could stand *e.g.* for the Boolean function which maps the key K and IV V of a stream cipher to the (let say) first output bit. Our goal as an adversary is to determine the unknown key K (or to distinguish F from a random function) in the chosen IV attack model only by dealing with the function F. If F mixes its inputs in a proper way, then one needs to try all 2^n possible keys in the worst case by sending $\mathcal{O}(n)$ queries to the oracle in order to find the correct key (since each query gives one bit information about the key for a balanced F). Here, we are going to investigate methods which can potentially lead to faster reconstruction of the key in the case where the function F does not properly mix its inputs. This could occur for example when the initialization phase of a stream cipher is performed through an iterated procedure for which the number of iterations has not been suitably chosen. On the other hand these methods may help to give the designers more insight to choose the required number of iterations. The existence of faster methods for finding the unknown key K highly depends on the structure of F. It may be even impossible to uniquely determine the key K. Let $F(K, V) = \sum_\kappa C_\kappa(V) K^\kappa$ where $K^\kappa = k_0^{\kappa_0} \cdots k_{n-1}^{\kappa_{n-1}}$ for the multi-index $\kappa = (\kappa_0, \ldots, \kappa_{n-1})$ (which can also be identified by its integer representation). Then the following lemma makes this statement more clear.

Lemma 1. *No adversary can distinguish between the two keys K_1 and K_2 for which $K_1^\kappa = K_2^\kappa$ for all $\kappa \in \{0, 1\}^n$ such that $C_\kappa(V) \neq 0$.*

Indeed, it is only possible to determine the values of $\{K^\kappa | \forall \kappa, C_\kappa(V) \neq 0\}$ which is not necessarily equivalent to determination of K. As a consequence of Lemma 1, the function F divides $\{0, 1\}^n$ into *equivalence classes*: $\mathcal{K}_1$, $\mathcal{K}_2, \ldots, \mathcal{K}_J$ (with $J \leq 2^n$). See Ex. 3 as an application on a reduced version of Trivium.

3   Scenarios of Attacks

The algebraic description of the function $F(K, V)$ is too complex in general to be amenable to direct analysis. Therefore, from the function $F(K, V)$ and with the partition $V = (U, W)$ we derive simpler Boolean functions $C(K, W)$ with the help of the oracle. In our main example, $C(K, W)$ is a coefficient of the algebraic normal form of the function deduced from F by varying over the bits in U only, see Sect. 4 for more details. If this function $C(K, W)$ does not have a well-distributed algebraic structure, it can be exploited in cryptanalytic attacks. Let us investigate different scenarios:

1. If $C(K, W)$ is imbalanced for (not necessarily uniformly) random W and many fixed K, then the function F (or equivalently the underlying stream cipher) with unknown K can be distinguished from a random one, see [4, 8, 9, 6].
2. If $C(K, W)$ is evaluated for some fixed W, then $C(K, W)$ is an expression in the key bits only. In [10], it was shown that in Trivium case for reduced iterations, linear relations on the key bits can be derived for a well chosen IV part.
3. If $C(K, W)$ has many key bits, which have (almost) no influence on the values of $C(K, W)$, a suitable approximation may be identified and exploited for key recovery attacks, see [1]. This is the target scenario of this paper and will be discussed in detail.

Scenario 1 has already been discussed in the introduction. In scenario 2, the underlying idea is to find a relation $C(K, W)$, evaluated for some fixed W, which depends only on a subset of t $(< n)$ key bits. The functional form of this relation can be determined with 2^t evaluations of $C(K, W)$. By trying all 2^t possibilities for the involved t key bits, one can filter those keys which do not satisfy the imposed relation. The complexity of this precomputation is 2^t times needed to compute $C(K,W)$, see Sect. 4. More precisely, if $p = \Pr\{C(K,W) = 0\}$ for the fixed W, the key space is filtered by a factor of $H(p) = p^2 + (1-p)^2$. For example, in the case of a linear function it is $p = H(p) = 1/2$. In addition, if several imposed relations on the key bits are available, it is easier to combine them to filter wrong keys if they have a simple structure, see $e.g.$ [10]. In scenario 3, the main idea is to find a function $A(L,W)$ which depends on a key part L of t bits, and which is correlated to $C(K,W)$ with correlation coefficient ε, that is $\Pr\{C(K,W) = A(L,W)\} = 1/2(1+\varepsilon)$. Then, by asking the oracle N queries we get some information (depending on the new equivalence classes produced by A) about t bits of the secret K in time $N2^t$ by carefully analyzing the underlying hypothesis testing problem. We will proceed by explaining how to derive such functions C from the coefficients of the ANF of F in Sect. 4, and how to find such functions A using the concept of probabilistic neutral bits in Sect. 5.

4   Derived Functions from Polynomial Description

The function F can be written in the form $F(K,V) = \sum_{\nu,\kappa} C_{\nu,\kappa} V^\nu K^\kappa$ with binary coefficients $C_{\nu,\kappa}$. We can make a partition of the IV according to $V = (U,W)$ and $\nu = (\alpha,\beta)$ with l-bit segments U and α, and $(m-l)$-bit segments W and β . This gives the expression $F(K,V) = \sum_{\alpha,\beta,\kappa} C_{(\alpha,\beta),\kappa} U^\alpha W^\beta K^\kappa = \sum_\alpha C_\alpha(K,W) U^\alpha$ where $C_\alpha(K,W) = \sum_{\beta,\kappa} C_{(\alpha,\beta),\kappa} W^\beta K^\kappa$. For every $\alpha \in \{0,1\}^l$, the function $C_\alpha(K,W)$ can serve as a function C derived from F. Here is a toy example to illustrate the notation:

Example 1. Let $n = m = 3$ and $F(K,V) = k_1 v_1 \oplus k_2 v_0 v_2 \oplus v_2$. Let $U := (v_0, v_2)$ of $l = 2$ bits and $W := (v_1)$ of $m-l = 1$ bit. Then $C_0(K,W) = k_1 v_1$, $C_1(K,W) = 0$, $C_2(K,W) = 1$, $C_3(K,W) = k_2$.   □

Note that an adversary with the help of the oracle can evaluate $C_\alpha(K,W)$ for the unknown key K at any input $W \in \{0,1\}^{m-l}$ for every $\alpha \in \{0,1\}^l$ by sending at most 2^l queries to the oracle. In other words, the partitioning of V has helped us to define a computable function $C_\alpha(K,W)$ for small values of l, even though the explicit form of $C_\alpha(K,W)$ remains unknown. To obtain the values $C_\alpha(K,W)$ for *all* $\alpha \in \{0,1\}^l$, an adversary asks for the output values of all 2^l inputs $V = (U,W)$ with the fixed part W. This gives the truth table of a Boolean function in l variables for which the coefficients of its ANF (*i.e.* the values of $C_\alpha(K,W)$) can be found in time $l2^l$ and memory 2^l using the Walsh-Hadamard transform. Alternatively, a *single* coefficient $C_\alpha(K,W)$ for a specific $\alpha \in \{0,1\}^l$ can be computed by XORing the output of F for all $2^{|\alpha|}$ inputs $V = (U,W)$ for which each bit of U is at most as large as the corresponding bit of α. This bypasses the need of 2^l memory.

One can expect that a subset of IV bits receives less mixing during the initialization process than other bits. These IV bits are called *weak*, and they would be an appropriate choice of U in order to amplify the non-randomness of C. However, it is an open question how to identify weak IV bits by systematic methods.

5 Functions Approximation

We are interested in the approximations of a given function $C(K, W) : \{0,1\}^n \times \{0,1\}^{m-l} \to \{0,1\}$ which depend only on a subset of key bits. To this end we make an appropriate partition of the key K according to $K = (L, M)$ with L containing t *significant* key bits and M containing the remaining $(n - t)$ *non-significant* key bits, and construct the function $A(L, W)$. We also use the term *subkey* to refer to the set of significant key bits. Such a partitioning can be identified by systematic methods, using the concept of probabilistic neutral bits from [1]:

Definition 1. *The neutrality measure of the key bit k_i with respect to the function $C(K, W)$ is defined as γ_i, where* $\mathrm{Pr} = \frac{1}{2}(1 + \gamma_i)$ *is the probability (over all K and W) that complementing the key bit k_i does not change the output of $C(K, W)$.*

In practice, we will set a threshold γ, such that all key bits with $|\gamma_i| < \gamma$ are included in the subkey L (*i.e.* the probabilistic neutral key bits are chosen according to the individual values of their neutrality measure). The approximation $A(L, W)$ could be defined by $C(K, W)$ with non-significant key bits M fixed to zero. Here is another toy example to illustrate the method:

Example 2. Let $n = m = 3$, $l = 2$ and $C(K, W) = k_0 k_1 k_2 v_0 v_1 \oplus k_0 v_1 \oplus k_1 v_0$. For uniformly random K and W, we find $\gamma_0 = 1/8$, $\gamma_1 = 1/8$, $\gamma_2 = 7/8$. Consequently, it is reasonable to use $L := (k_0, k_1)$ as the subkey. With fixed $k_2 = 0$, we obtain the approximation $A(L, W) = k_0 v_1 \oplus k_1 v_0$ which depends on $t = 2$ key bits only.   □

Note that, if M consists only of neutral key bits (with $\gamma_i = 1$), then the approximation A is exact, because $C(K, W)$ does not depend on these key bits. In [1] the notion of probabilistic neutral bits was used to derive an approximation function A in the case of $W = V$ and $C = F$ which lead to the first break of Salsa20/8.

6 Description of the Attack

In the precomputation phase of the attack, we need a suitable partitioning of the IV and the key (*i.e.* a function C and an approximation A). The weak IV bits are often found by a random search, while the weak key bits can be easily found with the neutrality measure for some threshold γ. Given C and A, we can find a small subset of candidates for the subkey L with a probabilistic guess-and-determine attack. In order to filter the set of all 2^t possible subkeys into a smaller set, we need to distinguish a correct guess of the subkey $\hat{L}$ from an incorrect one. Our ability in distinguishing subkeys is related to the correlation coefficient between $A(\hat{L}, W)$ and $C(K, W)$ with $K = (L, M)$ under the following two hypotheses. H_0 : the guessed part $\hat{L}$ is correct, and H_1 : the guessed part $\hat{L}$

is incorrect. More precisely, the values of ε_0 and ε_1 defined in the following play a crucial role:

$$\Pr_{W}\{A(\hat{L}, W) = C(K, W)|K = (\hat{L}, M)\} = \frac{1}{2}(1 + \varepsilon_0) \qquad (1)$$

$$\Pr_{\hat{L}, W}\{A(\hat{L}, W) = C(K, W)|K = (L, M)\} = \frac{1}{2}(1 + \varepsilon_1) . \qquad (2)$$

In general, both ε_0 and ε_1 are random variables, depending on the key. In the case that the distributions of ε_0 and ε_1 are separated, we can achieve a small non-detection probability p_{mis} and false alarm probability p_{fa} by using enough samples. In the special case where ε_0 and ε_1 are constants with $\varepsilon_0 > \varepsilon_1$, the optimum distinguisher is Neyman-Pearson [2]. Then, N values of $C(K, W)$ for different W (assuming that the samples $C(K, W)$ are independent) are sufficient to obtain $p_{\text{fa}} = 2^{-c}$ and $p_{\text{mis}} = 1.3 \times 10^{-3}$, where

$$N \approx \left(\frac{\sqrt{2c(1 - \varepsilon_0^2) \ln 2} + 3\sqrt{1 - \varepsilon_1^2}}{\varepsilon_1 - \varepsilon_0} \right)^2 . \qquad (3)$$

The attack will be successful with probability $1 - p_{\text{mis}}$ and the complexity is as follows: For each guess $\hat{L}$ of the subkey, the correlation ε of $A(\hat{L}, W) \oplus C(K, W)$ must be computed, which requires computation of the coefficients $A(\hat{L}, W)$ by the adversary, and computation of the coefficient $C(K, W)$ through the oracle, for the same N values of W, having a cost of $N2^l$ at most. This must be repeated for all 2^t possible guesses $\hat{L}$. The set of candidates for the subkey L has a size of about $p_{\text{fa}}2^t = 2^{t-c}$. The whole key can then be verified by an exhaustive search over the key part M with a cost of $2^{t-c}2^{n-t}$ evaluations of F. The total complexity becomes $N2^l2^t + 2^{t-c}2^{n-t} = N2^{l+t} + 2^{n-c}$. Using more than one function C or considering several chosen IV bits U may be useful to reduce complexity; however, we do not deal with this case here.

Remark 1. In practice, the values of ε_0 and ε_1 are key dependent. If the key is considered as a random variable, then ε_0 and ε_1 are also random variables. However, their distribution may not be fully separated, and hence a very small p_{mis} and p_{fa} may not be possible to achieve. We propose the following non-optimal distinguisher: first, we choose a threshold ε_0^* such that $p_\epsilon = \Pr\{\varepsilon_0 > \varepsilon_0^*\}$ has a significant value, e.g. $1/2$. We also identify a threshold ε_1^*, if possible, such that $\Pr\{\varepsilon_1 < \varepsilon_1^*\} = 1$. Then, we estimate the sample size using Eq. 3 by replacing ε_0 and ε_1 by ε_0^* and ε_1^*, respectively, to obtain $p_{\text{fa}} \leq 2^{-c}$ and effective non-detection probability $p_{\text{mis}} \cdot p_\epsilon \approx 1/2$. If ε_0^* and ε_1^* are close, then the estimated number of samples becomes very large. In this case, it is better to choose the number of samples intuitively, and then estimate the related p_{fa}.

Remark 2. It is reasonable to assume that a false subkey $\hat{L}$, which is close to the correct subkey, may lead to a larger value of ε. Here, the measure for being "close" could be the neutrality measure γ_i and the Hamming weight: if only a few key bits on positions with large γ_i are false, one would expect that ε is large.

However, we only observed an irregular (*i.e.* not continuous) deviation for very close subkeys. The effect on p_{fa} is negligible because subkeys with difference of low weight are rare.

7   Application to Trivium

The stream cipher Trivium [3] is one of the eSTREAM candidates with a 288-bit internal state consisting of three shift registers of different lengths. At each round, a bit is shifted into each of the three shift registers using a non-linear combination of taps from that and one other register; and then one bit of output is produced. To initialize the cipher, the $n = 80$ key bits and $m = 80$ IV bits are written into two of the shift registers, with the remaining bits being set to a fixed pattern. The cipher state is then updated $R = 18 \times 64 = 1152$ times without producing output in order to provide a good mixture of the key and IV bits in the initial state. We consider the Boolean function $F(K, V)$ which computes the first keystream bit after r rounds of initialization. In [4], Trivium was analyzed with chosen IV statistical tests and non-randomness was detected for $r = 10 \times 64, 10.5 \times 64, 11 \times 64, 11.5 \times 64$ rounds with $l = 13, 18, 24, 33$ IV bits, respectively. In [10], the key recovery attack on Trivium was investigated with respect to scenario 2 (see Sect. 3) for $r = 9 \times 64$. Here we provide more examples for key recovery attack with respect to scenario 3 for $r = 10 \times 64$ and $r = 10.5 \times 64$. In the following two examples, weak IV bits have been found by a random search. We first concentrate on equivalence classes of the key:

Example 3. For $r = 10 \times 64$ rounds, a variable IV part U with the $l = 10$ bit positions $\{34, 36, 39, 45, 63, 65, 69, 73, 76, 78\}$, and the coefficient with index $\alpha = 1023$, we could experimentally verify that the derived function $C_\alpha(K, W)$ only depends on $t = 10$ key bits L with bit positions $\{15, 16, 17, 18, 19, 22, 35, 64, 65, 66\}$. By assigning all 2^{10} different possible values to these 10 key bits and putting those L's which gives the same function $C_\alpha(K, W)$ (by trying enough samples of W), we could determine the equivalence classes for L with respect to C_α. Our experiment shows the existence of 65 equivalence classes: one with 512 members for which $k_{15}k_{16} + k_{17} + k_{19} = 0$ and 64 other classes with 8 members for which $k_{15}k_{16} + k_{17} + k_{19} = 1$ and the vector $(k_{18}, k_{22}, k_{35}, k_{64}, k_{65}, k_{66})$ has a fixed value. This shows that C_α provides $\frac{1}{2} \times 1 + \frac{1}{2} \times 7 = 4$ bits of information about the key in average.   □

Example 4. For $r = 10 \times 64$ rounds, a variable IV part U with the $l = 11$ bit positions $\{1, 5, 7, 9, 12, 14, 16, 22, 24, 27, 29\}$, and the coefficient with index $\alpha = 2047$, the derived function $C_\alpha(K, W)$ depends on all 80 key bits. A more careful look at the neutrality measure of the key bits reveals that $\max(\gamma_i) \approx 0.35$ and only 7 key bits have a neutrality measure larger than $\gamma = 0.18$, which is not enough to get a useful approximation $A(L, W)$ for an attack. However, we observed that $C_\alpha(K, W)$ is independent of the key for $W = 0$, and more generally the number of significant bits depends on $|W|$.   □

It is difficult to find a good choice of variable IV's for larger values of r, using a random search. The next example shows how we can go a bit further with some insight.

Example 5. Now we consider $r = 10.5 \times 64 = 10 \times 64 + 32 = 672$ rounds. The construction of the initialization function of Trivium suggests that shifting the bit positions of U in Ex. 4 may be a good choice. Hence we choose U with the $l = 11$ bit positions $\{33, 37, 39, 41, 44, 46, 48, 54, 56, 59, 61\}$, and $\alpha = 2047$. In this case, $C_\alpha(K, W)$ for $W = 0$ is independent of 32 key bits, and $p = \Pr\{C_\alpha(K, 0) = 1\} \approx 0.42$. This is already a reduced attack which is $1/H(p) \approx 1.95$ times faster than exhaustive search. □

The following example shows how we can connect a bridge between scenarios 2 and 3 and come up with an improved attack.

Example 6. Consider the same setup as in Ex. 5. If we restrict ourself to W with $|W| = 5$ and compute the value of γ_i conditioned over these W, then $\max_i(\gamma_i) \approx 0.68$. Assigning all key bits with $|\gamma_i| < \gamma = 0.25$ as significant, we obtain a key part L with the $t = 29$ bit positions $\{1, 3, 10, 14, 20, 22, 23, 24, 25, 26, 27, 28, 31, 32, 34, 37, 39, 41, 46, 49, 50, 51, 52, 57, 59, 61, 63, 68, 74\}$. Our analysis of the function $A(L, W)$ shows that for about 44% of the keys we have $\varepsilon_0 > \varepsilon_0^* = 0.2$ when the subkey is correctly guessed. If the subkey is not correctly guessed, we observe $\varepsilon_1 < \varepsilon_1^* = 0.15$. Then, according to Eq. 3 the correct subkey of 29 bits can be detected using at most $N \approx 2^{15}$ samples, with time complexity $N2^{l+t} \approx 2^{55}$. Note that the condition $N < \binom{69}{5}$ is satisfied here. □

8 Application to Grain

The stream cipher Grain-128 [7] consists of an LFSR, an NFSR and an output function $h(x)$. It has $n = 128$ key bits, $m = 96$ IV bits and the full initialization function has $R = 256$ rounds. We again consider the Boolean function $F(K, V)$ which computes the first keystream bit of Grain-128 after r rounds of initialization. In [4], Grain-128 was analyzed with chosen IV statistical tests. With $N = 2^5$ samples and $l = 22$ variable IV bits, they observed a non-randomness of the first keystream bit after $r = 192$ rounds. They also observed a non-randomness in the initial state bits after the full number of rounds. In [8], a non-randomness up to 313 rounds was reported (without justification). In this section we provide key recovery attack for up to $r = 180$ rounds with slightly reduced complexity compared with exhaustive search. In the following example, weak IV bits for scenario 2 have been found again by a random search.

Example 7. Consider $l = 7$ variable IV bits U with bit positions $\{2, 6, 8, 55, 58, 78, 90\}$. For the coefficient with index $\alpha = 127$ (corresponding to the monomial of maximum degree), a significant imbalance for up to $r = 180$ rounds can be detected: the monomial of degree 7 appears only with a probability of $p < 0.2$ for 80% of the keys. Note that in [4], the attack with $l = 7$ could only be applied to $r = 160$ rounds, while our improvement comes from the inclusion of weak IV bits. □

In the following examples, our goal is to show that there exists some reduced key recovery attack for up to $r = 180$ rounds on Grain-128.

Example 8. Consider again the $l = 7$ IV bits U with bit positions $\{2, 6, 8, 55, 58, 78, 90\}$. For $r = 150$ rounds we choose the coefficient with index $\alpha = 117$ and include key bits with neutrality measure less than $\gamma = 0.98$ in list of the significant key bits. This gives a subkey L of $t = 99$ bits. Our simulations show that $\varepsilon_0 > \varepsilon_0^\star = 0.95$ for about 95% of the keys, hence $p_{\text{mis}} = 0.05$. On the other hand, for 128 wrong guesses of the subkey with $N = 200$ samples, we never observed that $\varepsilon_1 > 0.95$, hence $p_{\text{fa}} < 2^{-7}$. This gives an attack with time complexity $N2^{t+l} + 2^n p_{\text{fa}} \approx 2^{121}$ which is an improvement of a factor of (at least) $1/p_{\text{fa}} = 2^7$ compared to exhaustive search.   □

Example 9. With the same choice for U as in Ex. 7 and 8, we take $\alpha = 127$ for $r = 180$ rounds. We identified $t = 110$ significant key bits for L. Our simulations show that $\varepsilon_0 > \varepsilon_0^\star = 0.8$ in about 30% of the runs when the subkey is correctly guessed. For 128 wrong guesses of the subkey with $N = 128$ samples, we never observed that $\varepsilon_1 > 0.8$. Here we have an attack with time complexity $N2^{t+l} + 2^n p_{\text{fa}} \approx 2^{124}$, *i.e.* an improvement of a factor of 2^4.   □

9   Conclusion

A recent framework for chosen IV statistical distinguishers for stream ciphers has been exploited to provide new methods for key recovery attacks. This is based on a polynomial description of output bits as a function of the key and the IV. A deviation of the algebraic normal form (ANF) from random indicates that not every bit of the key or the IV has full influence on the value of certain coefficients in the ANF. It has been demonstrated how this can be exploited to derive information on the key faster than exhaustive key search through approximation of the polynomial description and using the concept of probabilistic neutral key bits. Two applications of our methods through extensive experiments have been given: A reduced complexity key recovery for Trivium with IV initialization reduced to 672 of its 1152 iterations, and a reduced complexity key recovery for Grain-128 with IV initialization reduced to 180 of its 256 iterations. This answers positively the question whether statistical distinguishers based on polynomial descriptions of the IV initialization of a stream cipher can be successfully exploited for key recovery. On the other hand, our methods are not capable to provide reduced complexity key recovery of the eSTREAM Phase 3 candidates Trivium and Grain-128 with full initialization.

Acknowledgments

The first author is supported by the National Competence Center in Research on Mobile Information and Communication Systems (NCCR-MICS), a center of the Swiss National Science Foundation under grant number 5005-67322. The third author is supported by Hasler Foundation www.haslerfoundation.ch under project number 2005. We would like to thank the anonymous reviewers for their comments.

References

1. Aumasson, J.-P., Fischer, S., Khazaei, S., Meier, W., Rechberger, C.: New Features of Latin Dances: Analysis of Salsa, ChaCha, and Rumba. In: FSE 2008 (2008)
2. Cover, T., Thomas, J.A.: Elements of Information Theory. Wiley series in Telecommunication. Wiley, Chichester (1991)
3. De Cannière, C., Preneel, B.: TRIVIUM: A Stream Cipher Construction Inspired by Block Cipher Design Principles. In: Katsikas, S.K., López, J., Backes, M., Gritzalis, S., Preneel, B. (eds.) ISC 2006. LNCS, vol. 4176, pp. 171–186. Springer, Heidelberg (2006)
4. Englund, H., Johansson, T., Turan, M.S.: A Framework for Chosen IV Statistical Analysis of Stream Ciphers. In: Srinathan, K., Rangan, C.P., Yung, M. (eds.) INDOCRYPT 2007. LNCS, vol. 4859, Springer, Heidelberg (2007) See also Tools for Cryptoanalysis 2007
5. eSTREAM - The ECRYPT Stream Cipher Project - Phase 3. See, http://www.ecrypt.eu.org/stream
6. Filiol, E.: A New Statistical Testing for Symmetric Ciphers and Hash Functions. In: Deng, R.H., Qing, S., Bao, F., Zhou, J. (eds.) ICICS 2002. LNCS, vol. 2513, Springer, Heidelberg (2002)
7. Hell, M., Johansson, T., Maximov, A., Meier, W.: A Stream Cipher Proposal: Grain-128. In: ISIT 2006 (2006)
8. O'Neil, S.: Algebraic Structure Defectoscopy. In Cryptology ePrint Archive, Report 2007/378. See also, http://www.defectoscopy.com
9. Saarinen, M.-J.O.: Chosen-IV Statistical Attacks Against eSTREAM Ciphers. In: SECRYPT 2006 (2006)
10. Vielhaber, M.: Breaking ONE.FIVIUM by AIDA an Algebraic IV Differential Attack. In: Cryptology ePrint Archive, Report 2007/413 (2007)

Correlated Keystreams in MOUSTIQUE

Emilia Käsper[1], Vincent Rijmen[1,3], Tor E. Bjørstad[2], Christian Rechberger[3],
Matt Robshaw[4], and Gautham Sekar[1]

[1] K.U.Leuven, ESAT-COSIC
[2] The Selmer Center, University of Bergen
[3] Graz University of Technology
[4] France Télécom Research and Development

Abstract. MOUSTIQUE is one of the sixteen finalists in the eSTREAM stream cipher project. Unlike the other finalists it is a self-synchronising cipher and therefore offers very different functional properties, compared to the other candidates. We present simple related-key phenomena in MOUSTIQUE that lead to the generation of strongly correlated keystreams and to powerful key-recovery attacks. Our best key-recovery attack requires only 2^{38} steps in the related-key scenario. Since the relevance of related-key properties is sometimes called into question, we also show how the described effects can help speed up exhaustive search (without related keys), thereby reducing the effective key length of MOUSTIQUE from 96 bits to 90 bits.

Keywords: eSTREAM, MOUSTIQUE, related keys.

1 Introduction

eSTREAM [6] is a multi-year effort to identify promising new stream ciphers. Sponsored by the ECRYPT Network of Excellence, the project began in 2004 with proposals for new stream ciphers being invited from industry and academia. These proposals were intended to satisfy either a software-oriented or a hardware-oriented profile (or both if possible). The original call for proposals generated considerable interest with 34 proposals being submitted to the two different performance profiles. Among them was MOSQUITO [3], a self-synchronising stream cipher designed by Daemen and Kitsos.

As a self-synchronising stream cipher MOSQUITO was already a rather unusual submission. There was only one other self-synchronising stream cipher submitted, SSS [8]. Indeed it has long been recognised that the design of (secure) self-synchronising stream ciphers is a difficult task and attacks on SSS [5] and MOSQUITO [7] were proposed. As a result of the attack on MOSQUITO, a tweaked-variant of MOSQUITO, called MOUSTIQUE [4], was proposed for the second phase of analysis. This cipher is now one of the finalists in the eSTREAM project.

In this paper we describe a set of simple related-key pairs for MOUSTIQUE. Our observation illustrates unfortunate aspects of the tweaks in moving from MOSQUITO to MOUSTIQUE. They lead directly to a very strong distinguisher for the

S. Vaudenay (Ed.): AFRICACRYPT 2008, LNCS 5023, pp. 246–257, 2008.

keystream generated from two related keys; and further to a rather devastating key-recovery attack in the related-key setting [2]. In fairness it should be observed that related-key phenomena are not to everyone's taste [1]. Indeed, Daemen and Kitsos state that they make no claim for the resistance of the cipher to attackers that may manipulate the key. However, we take the view that related-key weaknesses might be viewed as certificational and that the very simple partition of the keyspace according to correlated keystreams is not particularly desirable. Aside from very efficient distinguishers and key-recovery attacks in the related-key setting, the related keys also lead to an improvement over exhaustive search in a non-related-key setting.

The paper is organised as follows. In the next section we describe MOUSTIQUE and make the observations that we need for attacks in Section 3. We then describe attacks on MOUSTIQUE in the related-key setting in Section 4 and use Section 5 to describe implications on key recovery in the standard (known keystream) setting. We summarise our experimental confirmation in Section 6 and close with our conclusions. Throughout we will use established notation.

2   Description of MOUSTIQUE

In this section we describe the parts of the MOUSTIQUE description that are relevant to our observations. More information can be found in [4]. MOUSTIQUE uses a key of 96 bits, denoted by k_j, with $0 \leq j \leq 95$. At each step MOUSTIQUE takes as input one bit of ciphertext and produces one bit of keystream.

MOUSTIQUE consists of two parts: a 128-bit conditional complementing shift register (CCSR) holding the state and a nonlinear output filter with 8 stages, see Figure 1.

2.1   The CCSR

The CCSR is divided into 96 cells, denoted by q^j with $1 \leq j \leq 96$. Each cell contains between 1 and 16 bits, denoted by q_i^j. The updating function of the CCSR is given by:

$$Q_0^j = g_x(q_0^{j-1}, k_{j-1}, 0, 0), \qquad j = 1, 2,$$
$$Q_i^j = g_x(q_i^{j-1}, k_{j-1}, q_i^v, q_i^w), \qquad 2 < j < 96, \forall i \text{ and } j = 96, i = 0 \qquad (1)$$
$$Q_i^{96} = g_2(q_i^{95}, q_0^{95-i}, q_i^{94}, q_1^{94-i}), \; i = 1, 2, \ldots 15.$$

The Q_i^j are the new values of the q_i^j after one iteration. The subscript indices are always taken modulo the number of bits in the particular cell. The values of x, v and w are defined in Table 1. A value 0 for v or w indicates that the ciphertext feedback bit is used as input. The g_x functions are defined as follows:

$$g_0(a, b, c, d) = a + b + c + d \qquad (2)$$
$$g_1(a, b, c, d) = a + b + c(d + 1) + 1 \qquad (3)$$
$$g_2(a, b, c, d) = a(b + 1) + c(d + 1) \qquad (4)$$

Addition and multiplication operations are over GF(2).

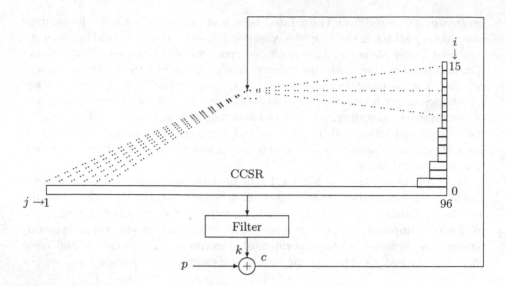

Fig. 1. State and filter of MOUSTIQUE. The only difference to MOSQUITO is that 1/3 of MOUSTIQUE state is now updated using a linear function g_0 to improve diffusion within the CCSR.

Table 1. The use of the functions g_0 and g_1 in the CCSR

Index	Function	v	w
$(j - i) \equiv 1 \bmod 3$	g_0	$2(j - i - 1)/3$	$j - 2$
$(j - i) \equiv 2 \bmod 3$	g_1	$j - 4$	$j - 2$
$(j - i) \equiv 3 \bmod 6$	g_1	0	$j - 2$
$(j - i) \equiv 0 \bmod 6$	g_1	$j - 5$	0

2.2   The Filter

The first stage of the filter compresses the 128 bits of the CCSR to 53 bits. First, the filter input $a^0 = (a_1^0, \ldots, a_{128}^0)$ is obtained by re-indexing the CCSR cells q_i^j in the following way:

$$
\begin{aligned}
a_i^0 &= q_0^i, & 1 \le i \le 96 \\
a_i^0 &= q_1^{i-8}, & 97 \le i \le 104 \\
a_i^0 &= q_2^{i-12}, & 105 \le i \le 108 \\
a_i^0 &= q_3^{i-16}, & 109 \le i \le 112 \\
a_{105+2i}^0 &= q_i^{95}, & 4 \le i \le 7 \\
a_{106+2i}^0 &= q_i^{96}, & 4 \le i \le 7 \\
a_{113+i}^0 &= q_i^{96}, & 8 \le i \le 15.
\end{aligned}
\tag{5}
$$

Then, the 53 bits of output are obtained by taking 53 applications of g_1:

$$a^1_{4i \bmod 53} = g_1(a^0_{128-i}, a^0_{i+18}, a^0_{113-i}, a^0_{i+1}), 0 \le i < 53. \tag{6}$$

The next four stages of the filter iteratively transform these 53 bits in a non-linear fashion. The sixth stage compresses the 53 bits to 12. Finally the last two stages exclusive-or these 12 bits together to produce a single bit of keystream. For simplicity, we omit the full description of the filter and refer the reader to the cipher specifications [4]. However, we note that the only non-linear filter component is the function g_1.

3   Observations on MOUSTIQUE

In this section we provide the basic observations that we will need in the paper. Some have already been observed in previous work [7].

3.1   Limited Impact of the IV

Observation 1. *The IV of* MOUSTIQUE *influences only the first 105 bits of the keystream.*

This is a consequence of the fact that the IV of MOUSTIQUE is used only to initialize the state, and as every self-synchronising stream cipher does, it gradually overwrites its state with ciphertext bits.

3.2   Differential Trails in the Filtering Function

As was done in the attack on MOSQUITO [7], we can make some simple observations on the filter function of MOUSTIQUE.

We note that the first stage of the filter is compressing and that no new information enters the filter after this stage. This leads to the first observation:

Observation 2. *Any two 128-bit CCSR states that produce the same 53-bit output after the first stage of filtering also produce an equal keystream bit.*

Recall that the first stage of the MOUSTIQUE output filter only uses the function $g_1(a, b, c, d) = a + b + c(d + 1) + 1$. So a consequence of this observation is that if we flip input c, the output of g_1 is unaffected with probability $p = \Pr[d = 1]$. Similarly, if we flip d, the output is unaffected with probability $p = \Pr[c = 0]$. To exploit this, we can observe the following:

Observation 3. *State bits* $q^1_0, \ldots, q^{17}_0$ *and* $q^{71}_0, \ldots, q^{75}_0$ *are used in the filter input only in one location, and only as the third or fourth input to the function* g_1.

Suppose we flip one of these 22 bits. The two outputs of g_1 are equal with probability p and, consequently, the two outputs of the filter will be equal with probability $0.5 + p/2$. If the inputs to g_1 are balanced, then we have $p = 0.5$ and the probability the output bit is unchanged is 0.75 (*i.e.* with bias $\varepsilon = 0.25$).

3.3 Impact of Key Bits on the CCSR

The chosen ciphertext attack on MOSQUITO [7] exploited slow diffusion within the CCSR and so the state-update function of MOSQUITO was tweaked. The state update of MOUSTIQUE uses a linear function g_0 for updating one third of the state bits. While this improves the worst-case diffusion, it exhibits weaknesses that we exploit to construct related-key pairs that result in highly correlated keystreams.

MOUSTIQUE only uses key bits in the state-update function of the CCSR. Each of the 96 key bits is added to one of the 96 bits q_0^j. The state-update function of the CCSR introduces diffusion in one direction only: a cell with index j does not depend on cells with indices $j' > j$. An immediate consequence is that key bit k_{95} affects state bit q_0^{96} only.

There are however more useful implications, which we introduce next. By expanding (1) and Table 1, we obtain the following equations:

$$Q_0^1 = c + k_0$$
$$Q_0^2 = q_0^1 + k_1 + 1$$
$$Q_0^3 = q_0^2 + k_2 + c(q_0^1 + 1) + 1$$
$$Q_0^4 = q_0^3 + k_3 + q_0^2 + q_0^2$$
$$Q_0^5 = q_0^4 + k_4 + q_0^1(q_0^3 + 1) + 1$$
$$Q_0^6 = q_0^5 + k_5 + q_0^1(c + 1) + 1$$
$$Q_0^7 = q_0^6 + k_6 + q_0^4 + q_0^5$$
$$Q_0^8 = q_0^7 + k_7 + q_0^4(q_0^6 + 1) + 1$$
$$Q_0^9 = q_0^8 + k_8 + c(q_0^7 + 1) + 1$$
$$Q_0^{10} = q_0^9 + k_9 + q_0^6 + q_0^8$$
$$Q_0^{11} = q_0^{10} + k_{10} + q_0^7(q_0^9 + 1) + 1$$
$$\vdots$$

Here c denotes the ciphertext feedback bit and we observe the following:

Observation 4. *In the computation of Q_0^4, the bit q_0^2 is cancelled. Only bit Q_0^3 depends on q_0^2.*

This leads to a related-key pair that for *any* ciphertext produces CCSR states with a one-bit difference. To see this, consider two instantiations of MOUSTIQUE running in decryption mode with the two keys denoted by k and k^*. Assume

$$k_i = k_i^* \text{ for } i \neq 1, 2 \text{ and}$$
$$k_i = k_i^* + 1 \text{ for } i = 1, 2.$$

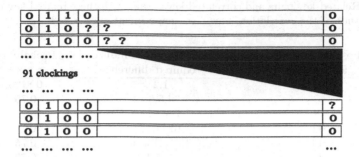

Fig. 2. CCSR differential propagation using related keys $k = (k_0, k_1, k_2, k_3, \ldots, k_{95})$ and $k^* = (k_0, k_1 + 1, k_2 + 1, k_3, \ldots, k_{95})$

We use both instantiations of MOUSTIQUE to decrypt the same ciphertext and observe the propagation of differences through the CCSR cells.

In the first iteration of the CCSR, the differences in k_1 and k_2 will cause differences in Q_0^2 and Q_0^3. After the second iteration, there will again be a difference in Q_0^2, but not in Q_0^3, because the incoming difference in q_0^2 cancels out the difference in k_2. What is left of course is the difference in q_0^3, which propagates through the CCSR and the filter stages. However, after 92 iterations, this unwanted difference has been propagated out of the CCSR. We obtain a steady state behavior: at every iteration, both CCSRs differ in bit q_0^2 only. Figure 2 illustrates the propagation of the related-key differential within the CCSR.

Since MOUSTIQUE has a *cipher function delay* of nine iterations, we can start deriving information from the keystream after nine more iterations. This will be demonstrated in the next section.

4   Related-Key Effects

4.1   Correlated Keystreams

There are several classes of related keys for MOUSTIQUE. We start with the simplest case which, coincidentally, appears to demonstrate the greatest bias.

First Related-Key Pairs. Consider two CCSR states with a difference only in bit q_0^2. According to Observation 4, this bit affects only one bit of the 53-bit output, namely bit a_8^1, which is computed as

$$a_8^1 = q_{14}^{96} + q_0^{19} + q_3^{96}(q_0^2 + 1) + 1.$$

Notice that if $q_3^{96} = 0$, the difference is extinguished and the two states produce equal output. If $q_3^{96} = 1$, the difference passes on and the two outputs will

Table 2. Related-key pairs and correlated keystreams. All these related-key pairs, and the magnitude of the correlation, have been experimentally verified.

Position j of the single bit difference	Key bits to flip to induce the required difference	Probability $z = z*$
2	1,2	0.8125
5	4,5,6	0.75
8	7,8,9,12	0.75
11	10,11,12	0.75
14	13,14,15,21	0.75
17	16,17,18	0.75
71	70,71,72	0.75
74	73,74,75	0.75

presumably collide with probability $\frac{1}{2}$. In fact q_3^{96} is computed using a non-balanced function g_2 and we have that $\Pr[q_3^{96} = 0] = \frac{5}{8}$.

So, after 105 cycles of IV setup, the two instances of MOUSTIQUE decrypting equal ciphertexts with related keys k and k^* will produce equal keystream bits z and z^* for which $\Pr[z = z^*] = \frac{5}{8} + \frac{3}{8} \times \frac{1}{2} = \frac{13}{16}$.

More Advanced Related-Key Pairs. We can extend the simple related keys already described. This allows us to obtain a range of related-key pairs that generate a 1-bit difference in the CCSR. Using Table 1, the following observation is easy to verify.

Observation 5. *If $j \leq 77$ and $j \equiv 2$ mod 3, then q_0^j occurs in the CCSR update only linearly.*

This implies that for each of $q_0^5, q_0^8, q_0^{11}, q_0^{14}, \ldots, q_0^{77}$, we can find a set of key bits such that by flipping these key bits simultaneously and iterating the scheme, a one-bit difference between the two CCSRs is retained in a single bit position q_0^j.

Among these 25 one-bit differences in the CCSR state, eight will also induce correlated keystream; these are bits $q_0^2, q_0^5, q_0^8, q_0^{11}, q_0^{14}, q_0^{17}, q_0^{71}$ and q_0^{74} (Observation 3). Table 2 lists the pairs of related keys that are generated along with the correlation in the associated keystream outputs. Since the correlation is extremely high, only a very small amount of keystream is required to reliably distinguish these related keystreams from a pair of random keystreams.

Furthermore, by simultaneously flipping relevant key bits for two or more indices j, we obtain a range of related keys with weaker correlation. The bias can be estimated by the Piling-Up Lemma; in the weakest case where all 8 keybit tuples are flipped, it is approximately $\varepsilon = 2^{-8.6}$. We have verified this estimate experimentally, and we now make the following conclusion.

Observation 6. *Each key of MOUSTIQUE produces correlated keystream with (at least) $2^8 - 1 = 255$ related keys, with the bias ranging from $\varepsilon = 2^{-1.7}$ to $\varepsilon = 2^{-8.6}$.*

4.2 Key-Recovery Attacks

A distinguisher can often be exploited to provide a key-recovery attack, and this is also the case here. Using (6) with $i = 42$, (5), and the definition of g_1 we have that

$$a_9^1 = q_0^{86} + q_0^{60} + q_0^{71}(q_0^{43} + 1) + 1.$$

As described in Section 4.1, if we take two instantiations of MOUSTIQUE and flip the key bits k_{70}, k_{71}, and k_{72} in one instantiation, then only q_0^{71} will change. This change can only propagate to the output if the bit q_0^{43} equals zero. Thus, a difference in the output of two copies of Moustique running with these related keys gives us one bit of information about the CCSR state (the value $q^{43} = 0$). Furthermore, the state bit q_0^{43} only depends on the first 43 bits of the key, which leads to an efficient divide-and-conquer attack as follows.

We first observe the output of two related instances of MOUSTIQUE, using some (arbitrary) ciphertext c and record the time values where the output bits differ. We then guess 43 key bits $k_0, \ldots, k_{42}$, compute the state bit q_0^{43} under the same ciphertext c, and check whether indeed $q_0^{43} = 0$ for all the recorded values. If there is a contradiction then we know that our guess for the 43-bit subkey was wrong. On average, only 8 bits of keystream are required to eliminate wrong candidates; and n bits of keystream eliminate a false key with probability $1 - 2^{-n/4}$.

The final attack requires a slight adjustment, as the existence of related keys introduces some false positives. Namely, certain related keys produce extinguishing differential trails that never reach q_0^{43}. For example, if the guessed key only differs from the correct key in the bits k_1 and k_2 then this difference affects q_0^2 only, and not q_0^{43}. Thus, the key with bits k_1 and k_2 flipped passes our test. The same holds for all combinations of the 14 values of j smaller than 43 and with $j \equiv 2 \bmod 3$; as well as bit k_{39} and pair k_{41}, k_{42}. Altogether, we have found that out of the 2^{43} key candidates, 2^{16} survive and after running our attack we still need to determine $96 - (43 - 16) = 69$ key bits. This can be done by exhaustive key search, and the 2^{69} complexity of this stage dominates the attack.

Notice that in the first stage, we know in advance which related keys give false positives. Thus, we only need to test one key in each set of 2^{16} related keys, and the complexity of the first stage is $2^{43-16} = 2^{27}$. The complexity of the second stage can be reduced if we were to allow the attacker access to multiple related keys.

In such a case, a second stage to the attack would use (6) with $i = 16$:

$$a_{11}^1 = q_3^{96} + q_0^{34} + q_1^{89}(q_0^{17} + 1) + 1.$$

The state bit q_0^{17} can be changed by flipping k_{16}, k_{17} and k_{18}. The state bit q_1^{89} depends on 89 key bits, of which we know already $43 - 16 = 27$ bits. In addition, we found 2^{31} related-key differentials that extinguish without ever reaching q_1^{89}. Hence, we need to test $2^{89-27-31} = 2^{31}$ keys to determine 31 more bits. In total we have then determined $27 + 31 = 58$ bits of the key and the remaining 38 bits

Table 3. The codewords of the $(7,4)$ Hamming code

c_0	0000000	c_4	0100110	c_8	1000101	c_{12}	1100011
c_1	0001011	c_5	0101101	c_9	1001110	c_{13}	1101000
c_2	0010111	c_6	0110001	c_{10}	1010010	c_{14}	1110100
c_3	0011100	c_7	0111010	c_{11}	1011001	c_{15}	1111111

can be determined by exhaustive search. The complexity of the attack can be estimated by $2^{27} + 2^{31} + 2^{38}$ which is dominated by the third brute-force phase.

We have verified the first two stages of the attack experimentally, and are indeed able to recover 58 bits of the key, given only 256 bits of keystream from two related-key pairs. Recovering the first 27 bits requires only a few minutes and 256 bits of output from a single related-key pair.

5   Accelerated Exhaustive Key Search

Next, we show how the existence of related keys in MOUSTIQUE can be used in cryptanalysis even if we cannot observe the output of the cipher in a related-key setting.

In Section 4, we observed that each key has eight related keys that produce strongly correlated output. In particular, the correlation can be detected from very short keystream. Thus, we can imagine the following attack scenario: given, say, 128 bits of cipher output from a key-IV pair (k, IV), compare this to the output of the cipher, using a candidate key k', the same IV and equal ciphertext. If the outputs are not correlated, eliminate key k' as well as its 8 related keys.

In order to compete with brute force, we need to be able to eliminate related keys efficiently. We now discuss two strategies representing different trade-offs between required keystream and computational complexity.

5.1   The Strong Correlation Attack

In the first approach we use the $(7, 4)$ Hamming code. As Hamming codes are perfect, we know that for each 7-bit string s, there exists a codeword c_i such that the Hamming distance between s and c_i is at most one. The codewords of the $(7, 4)$ Hamming code are listed in Table 3.

Now, for each codeword c_i, we fix candidate key bits $k_1, k_4, k_7, k_{10}, k_{13}, k_{16}, k_{70}$ to this codeword, and exhaustively search over the remaining 89 key bits. This strategy guarantees that we test either the correct key or one of the closely related keys given in Table 2. A related key can then be easily detected from the strong correlation of the two keystreams. For example, assume that the correct subkey is $(k_1, k_4, k_7, k_{10}, k_{13}, k_{16}, k_{70})$ and the closest codeword is $(k_1, k_4 + 1, k_7, k_{10}, k_{13}, k_{16}, k_{70})$. Then, according to Table 2, $k^* = (k_1, k_2, k_3, k_4 + 1, k_5 + 1, k_6 + 1, k_7, \ldots, k_{95})$ is a related key that has been selected for testing.

Our experiments suggest that 128 keystream bits are sufficient to detect correlation between the correct key k and a related candidate key k^* (see Sect. 6

for experimental results). Given that IV setup takes 105 cipher clocks, the total worst-case complexity of our attack is $(105 + 128) \cdot 2^4 \cdot 2^{89} \approx 2^{100.9}$ cipher clocks. In comparison, naive brute force requires on average 2 keystream bits to eliminate false candidates, so the complexity is $(105 + 2) \cdot 2^{96} = 2^{102.7}$ cipher clocks.

5.2   The Piling-Up Attack

Following Observation 6, we partition the keys into 2^{88} sets of 2^8 related keys and test only one key in each set. After 105 clocks of IV setup, the states corresponding to two related keys differ in at most 8 bits (given in Table 2). If

$$a^0_{40} = a^0_{43} = 1 \text{ and } a^0_{97} = a^0_{100} = a^0_{103} = a^0_{106} = a^0_{109} = a^0_{112} = 0, \qquad (7)$$

then none of these 8 bits influences a^1, the output of the first filter stage, and hence the keystream bits generated by two related keys are equal. Consequently, if, while testing a key k' we observe that the bit of keystream generated by k' differs from the bit of the observed keystream at a time when the candidate state satisfies (7), then we are sure that the key k we are looking for is not a related key of k' and we can discard k' as well as its $2^8 - 1$ related keys.

To estimate the amount of keystream needed to eliminate wrong keys, we note that two unrelated keystreams overlap with probability $\frac{1}{2}$, so we can use half of the available keystream to test for condition (7). As $\Pr[a^0_{112} = 0] = \frac{5}{8}$, while the remaining bits in (7) are balanced, condition (7) is true with probability $p = \frac{5}{8} \cdot \frac{1}{2^7}$. Thus, we need to generate on average $\frac{2}{p} = 409.6$ bits of keystream from one candidate key in order to rule out an entire class of 2^8 related keys. In total, the complexity of our attack can be estimated at $(105 + 409.6) \cdot 2^{88} = 2^{97.0}$ cipher clocks. Our experiments confirm this estimate and suggest that 5000-6000 bits of known keystream are sufficient to eliminate all false candidates with high confidence.

Both our strategies for accelerated exhaustive key search are rather simple and just as easily parallelisable as exhaustive search, so they are likely to provide an advantage over simple brute force in practice. The piling-up attack is an estimated 50 times faster than exhaustive key search, indicating that the effective key length of MOUSTIQUE is reduced to 90 bits instead of the claimed 96-bit security.

6   Experimental Verification

The results in this paper were verified using the source code for MOUSTIQUE that was submitted to eSTREAM [6]. All sets of key bits identified in Table 2 were tested with one thousand random keys and their related partners. The minimum, maximum, and average number of agreements between the two generated keystreams, over the first 128 bits, was recorded. Note that for un-correlated keystreams we would expect 64 matches.

Key bits to induce the required difference	Minimum # of matches	Maximum # of matches	Average # of matches
1,2	91	118	104.02
4,5,6	82	111	96.10
7,8,9,12	79	109	96.03
10,11,12	74	108	95.81
13,14,15,21	79	110	96.11
16,17,18	80	114	95.72
70,71,72	77	109	96.23
73,74,75	81	112	95.94

We then constructed a distinguisher by setting the agreement threshold to $t \geq 74$. We chose randomly 10 000 related-key pairs, all of which passed the test, indicating that the false negative rate is below 0.01%. In comparison, out of 10 000 128-bit keystreams obtained from random key pairs, 440 passed the test, so the false positive rate was below 5%. Thus, we can use our accelerated key search to eliminate 95% of the keys, and then brute-force the remaining candidates. The total complexity of the attack is still below that of naive exhaustive search, and the success rate is at least 99.99%.

7   Conclusions

In moving from MOSQUITO, it seems that the design of the self-synchronizing stream cipher MOUSTIQUE was established in a rather *ad hoc* way. While the tweaked design resists the chosen-ciphertext attack on MOSQUITO, we showed that it still exhibits weaknesses that lead to strong distinguishers in the related-key setting. Further, we presented two different strategies for exploiting those distinguishers in a key-recovery attack. The first strategy allows the attacker to recover the 96-bit secret key in 2^{69} steps, assuming that the attacker is able to observe the output of two instances of the cipher using the secret key and a related key. The complexity of this attack can be reduced to 2^{38} steps if the attacker is able to observe the output of three instances of the cipher using the secret key and two related keys. Both require a negligible amount of ciphertext, *e.g.* less than 256 bits.

We have also exploited the observations we made in a non-related-key attack. Our first attack breaks the cipher in around 2^{101} steps, using only 128 bits of known plaintext. If furthermore a few thousand keystream bits are known, the complexity is reduced to 2^{97} steps. In comparison, exhaustive search would take 2^{103} equivalent steps, indicating that MOUSTIQUE falls about 6 bits short of the claimed 96-bit security. While, admittedly, a 2^{97} attack is still far from being practical, it illustrates the relevance of related-key weaknesses in the standard (non-related-key) setting.

Acknowledgments

Emilia Käsper thanks the Computer Laboratory of the University of Cambridge for hosting her while this work was done.

This work was supported in part by the European Commission through the IST Programme under Contract IST-2002-507932 ECRYPT, the IAPP–Belgian State–Belgian Science Policy BCRYPT and the IBBT (Interdisciplinary institute for BroadBand Technology) of the Flemish Government. Emilia Käsper is also partially supported by the FWO-Flanders project nr. G.0317.06 Linear Codes and Cryptography.

References

1. Bernstein, D.J.: Related-key attacks: who cares? eSTREAM discussion forum (June 22, 2005), http://www.ecrypt.eu.org/stream/phorum/
2. Biham, E.: New Types of Cryptoanalytic Attacks Using related Keys (Extended Abstract). In: Helleseth, T. (ed.) EUROCRYPT 1993. LNCS, vol. 765, pp. 398–409. Springer, Heidelberg (1994) (extended Abstract)
3. Daemen, J., Kitsos, P.: The Self-Synchronising Stream Cipher MOSQUITO. eStream Report 2005/018, http://www.ecrypt.eu.org/stream/papers.html
4. Daemen, J., Kitsos, P.: The Self-Synchronising Stream Cipher MOUSTIQUE, http://www.ecrypt.eu.org/stream/mosquitop3.html
5. Daemen, J., Lano, J., Preneel, B.: Chosen Ciphertext Attack on SSS. eStream Report 2005/044), http://www.ecrypt.eu.org/stream/papers.html
6. ECRYPT. The eSTREAM project, http://www.ecrypt.eu.org/stream/
7. Joux, A., Muller, F.: Chosen-ciphertext attacks against Mosquito. In: Robshaw, M.J.B. (ed.) FSE 2006. LNCS, vol. 4047, pp. 390–404. Springer, Heidelberg (2006)
8. Rose, G., Hawkes, P., Paddon, M., Wiggers de Vries, M.: Primitive Specification for SSS. eStream Report 2005/028, http://www.ecrypt.eu.org/stream/papers.html

Stream Ciphers Using a Random Update Function: Study of the Entropy of the Inner State

Andrea Röck

Team SECRET, INRIA Paris-Rocquencourt, France
andrea.roeck@inria.fr,
http://www-rocq.inria.fr/secret/Andrea.Roeck

Abstract. Replacing random permutations by random functions for the update of a stream cipher introduces the problem of entropy loss. To assess the security of such a design, we need to evaluate the entropy of the inner state. We propose a new approximation of the entropy for a limited number of iterations. Subsequently, we discuss two collision attacks which are based on the entropy loss. We provide a detailed analysis of the complexity of those two attacks as well as of a variant using distinguished points.

Keywords: Entropy, Random Functions, Stream Cipher, Collisions.

1 Introduction

Recently, several stream ciphers have been proposed with a non-bijective update function. Moreover, in some cases the update function seems to behave like a random function as for the MICKEY stream cipher [BD05]. Using a random function instead of a random permutation induces an entropy loss in the state. An attacker might exploit this fact to mount an attack. Particularly, we will study some attacks which apply the approach of Time–Memory tradeoff [Hel80] and its variants [HS05]. At first we introduce the model with which we are going to work.

Stream cipher model. The classical model of an additive synchronous stream cipher (Fig. 1) is composed of an internal state updated by applying a function Φ. Then a filter function is used to extract the keystream bits from the internal state. To obtain the ciphertext we combine the keystream with the plaintext.

The particularity of our model is that Φ is a random mapping which allows us to make some statistical statements about the properties of the stream cipher.

Definition 1. *Let $\mathcal{F}_n = \{\varphi \mid \varphi : \Omega_n \to \Omega_n\}$ be the set of all functions which map a set $\Omega_n = \{\omega_1, \omega_2, \ldots, \omega_n\}$ of n elements onto itself. We say that Φ is a* random function *or a* random mapping *if it takes each value $\varphi \in \mathcal{F}_n$ with the same probability $Pr[\Phi = \varphi] = 1/n^n$.*

S. Vaudenay (Ed.): AFRICACRYPT 2008, LNCS 5023, pp. 258–275, 2008.

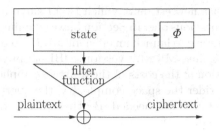

Fig. 1. Model of a simple stream cipher

For an extended definition of a random function we refer to the book of Kolchin [Kol86].

Let S_k be the random variable denoting the value of the state after k iterations of Φ, for $k \geq 0$. From the model in Fig. 1 we see that $S_k = \Phi(S_{k-1})$ where the value of Φ is the same for all iterations $k > 1$. The probability distribution of the initial state S_0 is $\{p_i\}_{i=1}^n$ such that

$$p_i = Pr[S_0 = \omega_i] \, .$$

If we do not state otherwise, we assume a uniform distribution thus $p_i = 1/n$ for all $1 \leq i \leq n$. By

$$p_i^\Phi(k) = Pr[\Phi^k(S_0) = \omega_i]$$

we describe the probability of the state being ω_i after applying k times Φ on the initial state S_0. If we write only $\mathrm{p}_i^\varphi(k)$ we mean the same probability but for a specific function $\varphi \in \mathcal{F}_n$. The notation above allows us to define the entropy of the state after k iterations of Φ

$$H_k^\Phi = \sum_{i=1}^n p_i^\Phi(k) \log_2 \left(\frac{1}{p_i^\Phi(k)} \right) \, .$$

If $p_i^\Phi(k) = 0$ we use the classical convention in the computation of the entropy that $0 \log_2(\frac{1}{0}) = 0$. This can be done, since a zero probability has no influence on the computation of the entropy. In this article we are interested in expectations where the average is taken over all functions $\varphi \in \mathcal{F}_n$. To differentiate between a value corresponding to a random mapping Φ, to a specific function φ, and the expectation of a value, taken over all functions $\varphi \in \mathcal{F}_n$, we will write in the following the first one normal (*e.g.* H_k^Φ), the second one upright (*e.g.* H_k^φ), and the last one bold (*e.g.* $\mathbf{H_k}$). For instance, the formula:

$$\mathbf{H_k} = \mathbf{E}(H_k^\Phi)$$

denotes the expected state entropy after k iterations.

The subsequent article is divided in two main sections. In Section 2, we discuss ways of estimating the state entropy of our model. We give a short overview of previous results from [FO90a] and [HK05] in Section 2.1. Subsequently in

Section 2.2, we present a new estimate which is, for small numbers of iterations, more precise than the previous one. In Section 3, we examine if it is possible to use the entropy loss in the state to launch an efficient attack on our model. We discuss two collision attacks against MICKEY version 1 [BD05] presented in [HK05]. We give a detailed evaluation of the costs of these attacks applied on our model. For this evaluation we consider the space complexity, the query complexity and the number of different initial states needed. By the space complexity we mean the size of the memory needed, by the query complexity we mean the number of times we have to apply the update function during the attack. For the first attack, we show that we only gain a factor on the space complexity by increasing the query complexity by the same factor. For the second attack, we demonstrate that, contrary to what is expected from the results in [HK05], the complexities are equivalent to a direct collision search in the initial values. In the end, we present a new variant of these attacks which allows to reduce the space complexity; however the query complexity remains the same.

2   Estimation of Entropy

The entropy is a measure of the unpredictability. An entropy loss in the state facilitates the guessing of the state for an adversary. In this section, we therefore discuss different approaches to estimate the expected entropy of the inner state.

2.1   Previous Work

Flajolet and Odlyzko provide, in [FO90a], a wide range of parameters of random functions by analyzing their functional graph. A functional graph of a specific

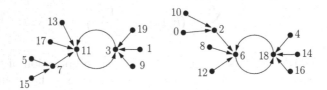

Fig. 2. Example of a functional graph for $\varphi : x \mapsto x^2 + 2 \pmod{20}$

function φ is a graph which has a directed edge from vertex x to vertex y if and only if $\varphi(x) = y$. An example for $\varphi(x) = x^2 + 2 \pmod{20}$ can be seen in Fig. 2. For functions on a finite set of elements, such a graph consists of one or more separated components, where each component is build by a cycle of trees, i.e. the nodes in the cycle are the root of a tree.

To find the expected value of a given parameter of a random function, Flajolet and Odlyzko construct the generating function of the functional graph associated with this parameter. Subsequently, they obtain an asymptotic value of the expectation by means of a singularity analysis of the generating function. All asymptotic values are for n going to $+\infty$. In the following, we present some examples of the examined parameters. The maximal tail length is, for each graph, the maximal number of steps before reaching a cycle. An r–node is a node in the graph with exactly r incoming nodes which is equivalent to a preimage of size r. By the image points we mean all points in the graph that are reachable after k iterations of the function. The asymptotic values of these parameters are:

- the expected number of cycle point $\mathbf{cp}(n) \sim \sqrt{\pi n / 2}$,
- the expected maximal tail length $\mathbf{mt}(n) \sim \sqrt{\pi n / 8}$,
- the expected number of r–nodes $\mathbf{rn}(n, r) \sim \frac{n}{r!e}$ and
- the expected number of image points after k iterations $\mathbf{ip}(n, k) \sim n(1 - \tau_k)$ where $\tau_0 = 0$ and $\tau_{k+1} = e^{-1+\tau_k}$.

For all these values, the expectation is taken over all functions in $\mathcal{F}_n$.

In [HK05], Hong and Kim use the expected number of image points to give an upper bound for the state entropy after k iterations of a random function. They utilize the fact that the entropy is always less or equal than the logarithm of the number of points with probability larger than zero. After a finite number of steps, each point in the functional graph will reach a cycle, and thus the number of image points can never drop below the number of cycle points. Therefore, the upper bound for the estimated entropy of the internal state

$$\mathbf{H_k} \leq \log_2(n) + \log_2(1 - \tau_k) \tag{1}$$

is valid only as long as $\mathbf{ip}(n, k) > \mathbf{cp}(n)$. We see that for this bound the loss of entropy only depends on k and not on n.

In Fig. 3 we compare, for $n = 2^{16}$, the values of this bound with the empirically derived average of the state entropy.

To compute this value we chose 10^4 functions, using the HAVEGE random number generator [SS03], and computed the average entropy under the assumption of a uniform distribution of the initial state. Even if n is not very big, it is sufficient to understand the relation between the different factors. We can see in the graph that if k stays smaller than $\mathbf{mt}(n)$ this bound stays valid and does not drop under $\log_2(\mathbf{cp}(n))$.

2.2 New Entropy Estimation

The expected number of image points provides only an upper bound (1) for the expected entropy. We found a more precise estimation by employing the methods stated in [FO90a].

For a given function $\varphi \in \mathcal{F}_n$, let ω_i be a node with r incoming nodes (an r–node). The idea is that this is equivalent to the fact that ω_i is produced by exactly r different starting values after one iteration. Thus, if the initial distribution of the state is uniform, this node has the probability $p_i^{\varphi}(1) = r/n$. The same idea works also for more than one iteration.

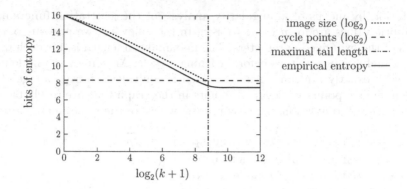

Fig. 3. Upper bound and empirical average of the entropy for $n = 2^{16}$

Definition 2. *For a fixed n let us choose a function $\varphi \in \mathcal{F}_n$. Let $\varphi^{-k}(i) = \{j | \varphi^k(\omega_j) = \omega_i\}$ define the* preimage *of i after k iterations of φ. By $\mathrm{rn}_k^{\varphi}(r) = \#\{i | |\varphi^{-k}(i)| = r\}$ we denote the number of points in the functional graph of φ which are reached by exactly r nodes after k iterations.*

For a random function Φ on a set of n elements, we define by $\mathbf{rn_k}(n, r)$ the expected value of $\mathrm{rn}_k^{\varphi}(r)$, thus

$$\mathbf{rn_k}(n, r) = \frac{1}{n^n} \sum_{\varphi \in \mathcal{F}_n} \mathrm{rn}_k^{\varphi}(r) .$$

A small example might help to better understand these definitions. For $n = 13$

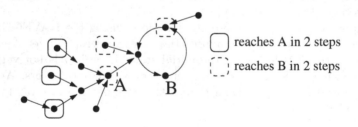

reaches A in 2 steps

reaches B in 2 steps

Fig. 4. Example of a functional graph to illustrate $\mathrm{rn}_k^{\varphi}(r)$

we consider a function φ with a functional graph as displayed in Fig. 4. The only points that are reached by $r = 3$ points after $k = 2$ iterations are A and B. Thus, in this case we have $\mathrm{rn}_2^{\varphi}(13, 3) = 2$. The value $\mathbf{rn_2}(13, 3)$ is then the average taken over all functions $\varphi \in \mathcal{F}_{13}$.

Using Def. 2 we can state the following theorem.

Theorem 1. *In the case of a uniform initial distribution the expected entropy of the inner state after k iterations is*

$$\mathbf{H_k} = \log_2(n) - \sum_{r=1}^{n} \mathbf{rn_k}(n,r) \, \frac{r}{n} \, \log_2(r) \,. \tag{2}$$

Proof. Let us fix a function φ. We use the idea that after k iterations of φ we have $\mathrm{rn}_k^\varphi(r)$ states with probability $\frac{r}{n}$. Thus, the entropy after k iterations for this specific function is

$$\mathrm{H}_k^\varphi = \sum_{r=1}^{n} \mathrm{rn}_k^\varphi(r) \frac{r}{n} \log_2\left(\frac{n}{r}\right)$$

$$= \log_2(n) \frac{1}{n} \sum_{r=1}^{n} r \, \mathrm{rn}_k^\varphi(r) - \sum_{r=1}^{n} \mathrm{rn}_k^\varphi(r) \frac{r}{n} \log_2(r) \,.$$

We ignore the case $r = 0$ since it corresponds to a probability zero, which is not important for the computation of the entropy. Each $1 \le j \le n$ appears exactly in one preimage of φ after k iterations. We can thus see directly from the definition of $\mathrm{rn}_k^\varphi(r)$ that $\sum_{r=1}^{n} r \, \mathrm{rn}_k^\varphi(r) = n$. Therefore, we can write

$$\mathrm{H}_k^\varphi = \log_2(n) - \sum_{r=1}^{n} \mathrm{rn}_k^\varphi(r) \frac{r}{n} \log_2(r) \,.$$

By using this equation, we can give the expected entropy after k iterations as

$$\mathbf{H_k} = \frac{1}{n^n} \sum_{\varphi \in \mathcal{F}_n} \mathrm{H}_k^\varphi$$

$$= \frac{1}{n^n} \sum_{\varphi \in \mathcal{F}_n} \left[\log_2(n) - \sum_{r=1}^{n} \mathrm{rn}_k^\varphi(r) \frac{r}{n} \log_2(r) \right]$$

$$= \log_2(n) - \frac{1}{n^n} \sum_{\varphi \in \mathcal{F}_n} \left[\sum_{r=1}^{n} \mathrm{rn}_k^\varphi(r) \frac{r}{n} \log_2(r) \right] \,.$$

Since we only have finite sums we can change the order:

$$\mathbf{H_k} = \log_2(n) - \sum_{r=1}^{n} \left[\frac{1}{n^n} \sum_{\varphi \in \mathcal{F}_n} \mathrm{rn}_k^\varphi(r) \right] \frac{r}{n} \log_2(r) \,.$$

We conclude our proof by applying Def. 2.

In the same way we can compute the entropy for any arbitrary initial distribution.

Theorem 2. *For a given n, let $P = \{p_1, p_2, \ldots, p_n\}$ define the distribution of the initial state. Then, the expected entropy of the state after k iterations is given by*

$$\mathbf{H}_{\mathbf{k}}^{\mathbf{P}} = \sum_{r=1}^{n} \mathbf{rn_k}(n, r) \frac{1}{\binom{n}{r}} \sum_{1 \leq j_1 < \cdots < j_r \leq n} (p_{j_1} + \cdots + p_{j_r}) \log_2 \frac{1}{p_{j_1} + \cdots + p_{j_r}}. \qquad (3)$$

Proof. Let us choose a specific φ and an index i. After k iterations of φ, the state ω_i has the probability $\sum_{j \in \varphi^{-k}(i)} p_j$. Therefore, the expected entropy after k iterations is given by

$$\mathbf{H}_{\mathbf{k}}^{\mathbf{P}} = \frac{1}{n^n} \sum_{\varphi \in \mathcal{F}_n} \sum_{i=1}^{n} \left(\sum_{j \in \varphi^{-k}(i)} p_j \right) \log_2 \frac{1}{\sum_{j \in \varphi^{-k}(i)} p_j}. \qquad (4)$$

For a given r we fix a set of indices $\{j_1, \ldots, j_r\}$. Without loss of generality we assume that they are ordered, e.i. $1 \leq j_1 < \cdots < j_r \leq n$. We now want to know how many times we have to count $(p_{j_1} + \cdots + p_{j_r}) \log_2 \frac{1}{p_{j_1} + \cdots + p_{j_r}}$ in (4). This is equivalent to the number of pairs (i, φ) where $\varphi^{-k}(i) = \{j_1, \ldots, j_r\}$.

From Def. 2 we know that $n^n \mathbf{rn_k}(n, r)$ is the number of pairs (i, φ) such that $|\varphi^{-k}(k)| = r$. Due to symmetry, each set of indices of size r is counted the same number of times in (4). There are $\binom{n}{r}$ such sets. Thus, $(p_{j_1} + \cdots + p_{j_r}) \log_2 \frac{1}{p_{j_1} + \cdots + p_{j_r}}$ is counted exactly $\frac{n^n \mathbf{rn_k}(n,r)}{\binom{n}{r}}$ times and we can write

$$\mathbf{H}_{\mathbf{k}}^{\mathbf{P}} = \frac{1}{n^n} \sum_{r=1}^{n} \frac{n^n \mathbf{rn_k}(n, r)}{\binom{n}{r}} \sum_{1 \leq j_1 < \cdots < j_r \leq n} (p_{j_1} + \cdots + p_{j_r}) \log_2 \frac{1}{p_{j_1} + \cdots + p_{j_r}},$$

which is equivalent to (3).

Theorem 1 can also be shown by using Theorem 2; however the first proof is easier to follow. Finally, we want to consider a further special case.

Corollary 1. *For a given n let the distribution of the initial state be $P_m = \{p_1, p_2, \ldots, p_n\}$. From the n possible initial values only m occur with probability exactly $\frac{1}{m}$. Without loss of generality we define*

$$p_i = \begin{cases} \frac{1}{m} & 1 \leq i \leq m \\ 0 & m < i \leq n. \end{cases}$$

In this case we get

$$\mathbf{H}_{\mathbf{k}}^{\mathbf{P}_m} = \sum_{r=1}^{n} \mathbf{rn_k}(n, r) \frac{1}{\binom{n}{r}} \sum_{\ell=0}^{r} \binom{m}{\ell} \binom{n-m}{r-\ell} \frac{\ell}{m} \log_2 \frac{m}{\ell}. \qquad (5)$$

Proof. For a given r, let us consider the sum $(p_{j_1} + \cdots + p_{j_r})$ for all possible index tuples $1 \leq j_1 < \cdots < j_r \leq n$. In $\binom{m}{\ell} \binom{n-m}{r-\ell}$ cases we will have $(p_{j_1} + \cdots + p_{j_r}) = \frac{\ell}{m}$ for $0 \leq \ell \leq r$. Thus, (5) follows directly from Theorem 2.

To approximate $\mathbf{rn_k}(n, r)$ for one iteration, we use directly the results for the expected number of r–nodes, given in [FO90a], since $\mathbf{rn_1}(n, r) = \mathbf{rn}(n, k) \sim \frac{n}{r!e}$. We can see that already for $k = 1$, a uniform initial distribution, and n large enough, there is a non negligible difference between our estimate (2)

$$\mathbf{H_1} \sim \log_2(n) - e^{-1} \sum_{r=1}^{n} \frac{1}{(r-1)!} \log_2(r) \approx \log_2(n) - 0.8272$$

and the upper bound (1)

$$\mathbf{H_1} \leq \log_2(n) + \log_2(1 - e^{-1}) \approx \log_2(n) - 0.6617 \,.$$

For more than one iteration we need to define a new parameter.

Theorem 3. *For $n \to \infty$ we can give the following asymptotic value*

$$\mathbf{rn_k}(n, r) \sim n \, c_k(r) \tag{6}$$

of the expected number of points in the functional graph which are reached by r points after k iterations, where

$$c_k(r) = \begin{cases} \frac{1}{r!e} & \text{for } k = 1 \\ D(k, r, 1) f_1(k)^{\frac{1}{e}} & \text{for } k > 1 \end{cases}$$

$$D(k, r, m) = \begin{cases} 1 & \text{for } r = 0 \\ 0 & \text{for } 0 < r < m \\ \sum_{u=0}^{\lfloor r/m \rfloor} \frac{c_{k-1}(m)^u}{u!} D(k, r - mu, m+1) & \text{otherwise} \end{cases}$$

and

$$f_1(k) = \begin{cases} 1 & \text{for } k = 1 \\ e^{e^{-1} f_1(k-1)} & \text{for } k > 1 \,. \end{cases}$$

Proof. The concept of this proof is that we see the functional graph as a combinatorial structure. We are going to build the generating function corresponding to this structure where we mark a desired parameter. By means of the singularity analysis of the generating function we obtain the asymptotic value of this parameter. The difficulty is to mark the right property in the generation function. The rest of the proof is just following the method described in [FO90a].

For an arbitrary structure, let a_n define the number of elements of this structure with size n for $n \geq 1$. Then, the exponential generating function of the infinite sequence $\{a_n\}_{n \geq 1}$ is defined as

$$A(z) = \sum_{n \geq 1} a_n \frac{z^n}{n!} \,.$$

By $[z_n] A(z)$ we mean the n'th coefficient a_n of $A(z)$. The nice property of a generating function is that many combinatorial constructions on the structure correspond to simple manipulation of the generating function. We refer the reader to [FS96] for a deeper introduction to the area of generating functions.

The functional graph of a function which maps a finite set onto itself can be described in the following recursive way:

$$FuncGraph = SET(Component),$$
$$Component = CYCLE(Tree),$$
$$Tree = Node \times SET(Tree).$$

Each of this constructions: $SET, CYCLE$ and $\times$ (concatenation) can be applied directly on a generating function.

We are interested in the average value of a specific parameter, where the average is taken over all functions of size n. For this purpose we need a bivariate generating function. Let $\mathcal{F} = \bigcup_{n \geq 1} \mathcal{F}_n$ be the set of all functions which map a finite set onto itself. For a specific $\varphi \in \mathcal{F}$, we denote by $|\varphi|$ the size n of the finite set. With $\xi(\varphi)$ we define a specific property of the function. In our case we are interested in $\xi_{r,k}(\varphi) = \mathrm{rn}_k^\varphi(|\varphi|, r)$. The bivariate generating function for this parameter, marked by the variable u, is then defined by

$$\xi_{r,k}(u, z) = \sum_{\varphi \in \mathcal{F}} u^{\xi_{r,k}(\varphi)} \frac{z^{|\varphi|}}{|\varphi|!}.$$

By $\xi_{r,k,n} = \sum_{\varphi \in \mathcal{F}_n} \xi_{r,k}(\varphi)$ we mean the sum of $\xi_{r,k}(\varphi)$ taken over all $\varphi \in \mathcal{F}_n$. Let

$$\Xi(z) = \sum_{n \geq 1} \xi_{r,k,n} \frac{z^n}{n!}$$

be the generating function for $\xi_{r,k,n}$. Thus, it is clear that the average value of $\xi^{k,r}$ is given by

$$\mathbf{E}(\xi^{k,r}|\mathcal{F}_n) = \frac{\xi_n^{k,r}}{n^n} = \frac{n!}{n^n} [z^n] \Xi(z).$$

To obtain the function $\Xi(z)$ we use that

$$\Xi(z) = \frac{\partial}{\partial u} \xi_{r,k}(u, z) \Big|_{u=1}.$$

Since in our case the evaluation of $[z^n]\Xi(z)$ is not directly possible, we can use a singularity analysis to get an asymptotic value for $n \to \infty$. More information about singularity analysis can be found in [FO90a] and [FO90b].

We now have to define the function $\xi_{r,k}(u, z)$. For this, we start by a tree. A node in a tree which is reached by r nodes after k iterations can be described by the concatenation of three elements:

1. A *node*.
2. A *SET* of *trees* where each tree has a depth smaller than $k - 1$.
3. A *concatenation* of j *trees* where the order of the concatenation is not important and where $1 \leq j \leq r$. Each of these trees has a depth larger or equal to $k - 1$ and their roots are reached by respectively $i_1, \ldots, i_j$ nodes after $k - 1$ iterations such that $i_1 + \cdots + i_j = r$.

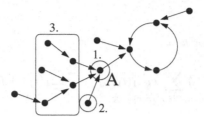

Fig. 5. Example of the structure explained in 1.-3. for the node A

In Fig. 5 these three elements are marked for the node A.

To write the corresponding generating function we need some notations:

The generating function of a set of trees of depth smaller than $k - 1$, as described in 2., is given by $f_1(k, z)$ where

$$f_1(k, z) = \begin{cases} 1 & \text{for } k = 1 \\ e^{z \, f_1(k-1, z)} & \text{for } k > 1 \, . \end{cases}$$

By $Par(r)$ we mean the integer partition of r, *i.e.* the set of all possible sequences $[i_1, \ldots, i_j]$ for $1 \leq j \leq r$ such that $1 \leq i_1 \leq \cdots \leq i_j \leq r$ and $i_1 + \cdots + i_j = r$. For example, for $r = 4$ we have $Par(4) = \{[1, 1, 1, 1], [1, 1, 2], [2, 2], [1, 3], [4]\}$.

Since the order of the concatenation in 3. is not important, we need a correction term $f_2([i_1, \ldots, i_j])$. If there are some $i_{x_1}, \ldots, i_{x_\ell}$ with $1 \leq x_1 < \cdots < x_\ell \leq j$ and $i_{x_1} = \cdots = i_{x_\ell}$ we have to multiply by a factor $1/\ell!$ to compensate this repeated appearance, e.g. $f_2([1, 1, 1, 1, 2, 2, 3]) = \frac{1}{4!2!1!}$.

Let $t_{r,k}(u, z)$ be the generation function of a tree. By $c_k(r, z)$ we define a variable such that

$$c_k(r, z) t_{r,k}(u, z)^r$$

is the generating function of a tree where the root is reached by r nodes after k iterations. For $k = 1$, such a tree has r children, where each child is again a tree. In terms of generating functions, this structure can be represented by $z \frac{t_{r,k}(u,z)^r}{r!}$. Thus, we get

$$c_1(r, z) = \frac{z}{r!} \, .$$

For $k > 1$ we can use the structure given in 1.-3. and our notations to write:

$$c_k(r, z)$$

$$= \frac{1}{t_{r,k}(u,z)^r} \; \overbrace{z}^{1.} \; \overbrace{f_1(k, z)}^{2.}$$

$$\overbrace{\sum_{[i_1,\ldots,i_j] \in Par(r)} \left[c_{k-1}(i_1, z) t_{r,k}(u, z)^{i_1} \right] \cdots \left[c_{k-1}(i_j, z) t_{r,k}(u, z)^{i_j} \right] f_2([i_1, \ldots, i_j])}^{3.}$$

$$= z f_1(k, z) \sum_{[i_1,\ldots,i_j] \in Par(r)} c_{k-1}(i_1, z) \cdots c_{k-1}(i_j, z) f_2([i_1, \ldots, i_j]) \, .$$

In total we get

$$
c_k(r, z) = \begin{cases} z/r!, \text{ for } k = 1 \\ z \, f_1(k, z) \sum_{[i_1,\ldots,i_j] \in Par(r)} c_{k-1}(i_1, z) \cdots c_{k-1}(i_j, z) \, f_2([i_1,\ldots,i_j]) \\ \text{for } k > 1 \,. \end{cases} \tag{7}
$$

We can now write the generation function of a tree where we mark with the variable u the nodes which are reached by r other nodes after k iterations.

$$
t_{r,k}(u, z) = z e^{t_{r,k}(u,z)} + (u - 1) t_{r,k}(u, z)^r c_k(r, z) \,,
$$

The first part describes a tree as a node concatenated with a set of trees. The second part correspond to our desired parameter. By applying the properties that a graph of a random function is a *set of components* where each component is a *cycle of trees* we get the generating function for a general functional graph

$$
\xi_{r,k}(u, z) = \frac{1}{1 - t_{r,k}(u, z)} \,.
$$

Now, we can follow the steps as described at the beginning of this proof. We will use the fact that the general generating function of a tree $t_{r,k}(1, z) = t(z) = z e^z$ has a singularity expansion

$$
t(z) = 1 - \sqrt{2}\sqrt{1 - ez} - \frac{1}{3}(1 - ez) + O((1 - ez)^{3/2})
$$

for z tending to e^{-1}. Finally, by applying the singularity analysis for $z \to e^{-1}$ we get

$$
\mathbf{E}(\xi_{r,k}|\mathcal{F}_n) \sim n \, c_k(r, e^{-1}) \,.
$$

Remark 1. In the construction of our generating function we only count the nodes in the *tree* which are reached by r points after k iterations (*e.g.* node A in Fig. 4). We ignore the nodes on the cycle (*e.g.* node B in Fig. 4). However, the average proportion of the number of cycle points in comparison to the image size after k iterations is

$$
\frac{\mathbf{cp}(n)}{\mathbf{ip}(n, k)} \sim \frac{\sqrt{\pi n / 2}}{n(1 - \tau_k)} \,.
$$

For a fixed k and $n \to \infty$ it is clear that this proportion gets negligible.

Thus, we can write

$$
\mathbf{rn_k}(n, r) \sim n \, c_k(r, e^{-1}) \,.
$$

The computation of $c_k(r, e^{-1})$ as defined in (7) is not very practical. In this paragraph, we will show that we can do it more efficiently using dynamic programming. For simplicity we write in the following $c_k(r, e^{-1}) = c_k(r)$ and $f_1(k, e^{-1}) = f_1(k)$. We define the new value $D(k, r, m)$ by

$$D(k, r, m) = \sum_{[i_1, \ldots, i_j] \in Par_{\geq m}(r)} c_{k-1}(i_1) \cdots c_{k-1}(i_j) f_2([i_1, \ldots, i_j])$$

where $Par_{\geq m}(r)$ is the set of all partitions of the integer r such that for each $[i_1, \ldots, i_j] \in Par_{\geq m}(r)$ must hold that $i_\ell \geq m$ for all $1 \leq \ell \leq j$. Using this, we can give the recursive definition of $D(k, r, m)$ and $c_k(r)$ as described in this theorem.

Proposition 1. *For fixed values R and K we can compute $c_k(r)$, as described in Theorem 3, for all $r \leq R$ and $k \leq K$ in a time complexity of $O\left(KR^2 \ln(R)\right)$.*

Proof. We use dynamic programming to compute $c_k(r)$.

The computation of $f_1(k)$ can be done once for all $k \leq K$ and then be stored. Thus, it has a time and space complexity of $O(K)$. For $k = 1$, if we start with $r = 1$ we can compute $c_1(r)$ for all $r \leq R$ in R steps. The same is true for $1 < k \leq K$ if we already know $D(k, r, 1)$ and $f_1(k)$.

The most time consuming factor is the computation of $D(k, r, m)$. For a given k', let us assume that we have already computed all $c_{k'-1}(r)$ for $1 \leq r \leq R$. In the computation of $D(k, r, m)$ we will go for r from 1 to R, and for m from r to 1. This means that

- For a given r' we already know all $D(k', r, m)$ with $r < r'$.
- For a fixed r' and m' we already know all $D(k', r', m)$ with $m > m'$.
- To compute

$$\sum_{u=0}^{\lfloor r/m \rfloor} \frac{c_{k-1}(m)^u}{u!} D(k, r - mu, m + 1)$$

we need $\lfloor r/m \rfloor$ steps.

Thus in total, for each $1 < k \leq K$ we need

$$\sum_{r=1}^{R} \sum_{m=1}^{r} \left\lfloor \frac{r}{m} \right\rfloor$$

steps to compute all $D(k, r, m)$. By using that

$$\sum_{m=1}^{r} \frac{1}{m} = ln(r) + C + O\left(\frac{1}{r}\right)$$

where $C = 0.5772 \ldots$ is the Euler constant, and

$$\sum_{r=1}^{R} r \ln(r) \leq \ln(R) \sum_{r=1}^{R} r$$

$$= \ln(R) \frac{R(R+1)}{2}$$

we obtain the final time complexity of $O(KR^2 \ln(R))$.

Let us go back to the expected entropy in (2). By using (6) we can write for $n \to \infty$

$$\mathbf{H_k} \sim \log_2(n) - \underbrace{\sum_{r=1}^{R} c_k(r)\ r\ \log_2(r)}_{(a)} - \underbrace{\sum_{r=R+1}^{n} c_k(r)\ r\ \log_2(r)}_{(b)}\ , \qquad (8)$$

where (a) represents an estimation of the entropy loss which does not depend on n and (b) is an error term. In Fig. 6, we see that the value $c_k(r)\ r\ \log_2(r)$

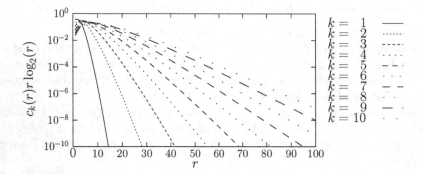

Fig. 6. The course of $c_k(r)\ r\ \log_2(r)$ for different values of k and r

decreases fast with growing r. However, for larger k this decrease becomes slower. If we want (b) to be negligible for larger k we also need a larger value for R. In Table 1, we compare our entropy estimator

$$H_k(R) = \log_2(n) - \sum_{r=1}^{R} c_k(r)\ r\ \log_2(r) \qquad (9)$$

with the estimated lower bound of the loss given by the expected number of image points (1) and the empirical results from the experiment presented in Fig. 3. From (6) and (8) we know that

$$\mathbf{H_k} \sim H_k(R)$$

for $n \to \infty$ and $R \to n$. We can see that for small k, in the order of a few hundred, we reach a much better approximation than the upper bound (1). For example, for most of the modern stream ciphers, the number of iterations for a key/IV–setup is in this order of magnitude. However, for increasing values of k we also need bigger values of R and, thus, this method gets computationally expensive. For $k = 100$ and $R = 1000$ the result of our estimate is about 0.02 larger than the empirical data. The fact that our estimate is larger shows that it is not due to the choice of R (it does not change a lot if we take $R = 2000$)

Table 1. Comparison of different methods to estimate the entropy loss

k		1	2	3	⋯	10	⋯	50	⋯	100
empirical data, $n = 2^{16}$		0.8273	1.3458	1.7254	⋯	3.1130	⋯	5.2937	⋯	6.2529
image points (1)		0.6617	1.0938	1.4186	⋯	2.6599	⋯	4.7312	⋯	5.6913
	$R = 50$	0.8272	1.3457	1.7254	⋯	3.1084	⋯	2.6894	⋯	1.2524
$H_k(R)$, (9)	$R = 200$	0.8272	1.3457	1.7254	⋯	3.1129	⋯	5.2661	⋯	5.5172
	$R = 1000$	0.8272	1.3457	1.7254	⋯	3.1129	⋯	5.2918	⋯	6.2729

but to the fact that our $n = 2^{16}$ is relatively small and, thus, the proportion of cycle points which is about

$$\frac{\sqrt{\pi n/2}}{n(1 - \tau_k)} \approx 0.253$$

is not negligible.

In this section we presented a new entropy estimator. We could show that if the number of iterations is not too big, it is much more precise than the upper bound given by the image size. In addition, the same method can be used for any arbitrary initial distribution.

3   Collision Attacks

In the previous section, we studied the loss of entropy in the inner state of our stream cipher model. In this section, we examine if it is possible to exploit this loss for a generic attack on our model. Hong and Kim present in [HK05] two attacks on the MICKEY stream cipher [BD05], based on the entropy loss in the state. This stream cipher has a fixed update function; however Hong and Kim state, due to empirical results, that the update function behaves almost like a random function with regard to the expected entropy loss and the expected number of image points. Thus, these attacks are directly applicable on our model. We will give a detailed complexity analysis of these attacks and will show that in the case of a real random function they are less efficient than what one might assume from the argumentation of Hong and Kim.

Let us take two different initial states S_0 and S_0' and apply the same function iteratively onto both of them. We speak about a collision if there exists k and k' such that $S_k = S_{k'}$, for $k \neq k'$, or $S_k = S_{k'}'$ for any arbitrary pair k, k'. The idea of Hong and Kim was that a reduced entropy leads to an increased probability of a collision. Once we have found a collision, we know that the subsequent output streams are identical. Due to the birthday paradox, we assume that with an entropy of m-bits we reach a collision, with high probability, by choosing $2^{\frac{m}{2}}$ different states.

The principle of the attacks is that we start from m different, randomly chosen, initial states and that we apply iteratively the same update function k times on each of them. In the end, we search for a collision and hope that our costs are less than for a search directly in the initial states. We will study the two attacks proposed in [HK05] as well as a variant using distinguished points. For each of these attacks we provide a detailed complexity analysis where we examine the query and the space complexity as well as the number of necessary initial states to achieve a successful attack with high probability. By the query complexity we mean the number of all states produced by the cipher during the attack which is equivalent to the number of times the updated function is applied. By the space complexity we mean the number of states we have to store such that we can search for a collision within them. Each time we compare the results to the attempt of finding a collision directly within the initial states which has a space and query complexity of $\sim \sqrt{n}$.

All these attacks consider only the probability of finding a collision in a set of states. This is not equivalent to an attack where we have $m - 1$ initial states prepared and we want the probability that if we take a new initial state, it will be one of the already stored. In such a scenario, the birthday paradox does not apply. We also never consider how many output bits we would really need to store and to recognize a collision, since this value depends on the specific filter function used. In the example of MICKEY, Hong and Kim states that they need about 2^8 bits.

3.1 States After k Iterations

The first attack of Hong and Kim takes randomly m different initial states, applies k times the same instance of Φ on each of them, and searches a collision in the m resulting states. Using (1) we know that the average entropy after k iterations is less than $\log_2(n) + \log_2(1 - \tau_k)$. Hong and Kim conjecture, based on experimental results, that this is about the same as $\log_2(n) - \log_2(k) + 1$. Thus, with high probability we find a collision if $m > 2^{(\log_2(n) - \log_2(k) + 1)/2} = \sqrt{2n/k}$.

This attack stores only the last value of the iterations and searches for a collision within this set. This leads to a space complexity of $m \sim \sqrt{2n/k}$ for large enough k. However, Hong and Kim did not mention that we have to apply k times Φ on each of the chosen initial states, which results in a query complexity of $mk \sim \sqrt{2kn}$. This means that we increase the query complexity by the same factor as we decrease the space complexity and the number of initial states.

The question remains, if there exists any circumstances under which we can use this approach without increasing the query complexity. The answer is yes, if the stream cipher uses a set of update functions which loose more than $2\log_2(k)$ bits of entropy after k iterations. Such a characteristic would imply that we do not use random functions to update our state, since they have different properties as we have seen before. However, the principle of the attack stays the same.

3.2 Including Intermediate States

The second attack in [HK05] is equivalent to applying $2k - 1$ times the same instance of Φ on m different initial states and searching for a collision in all intermediate states from the k-th up to the $(2k - 1)$-th iteration. Since after

$k - 1$ iterations we have about $\log_2(n) - \log_2(k) + 1$ bits of entropy, Hong and Kim assume that we need a m such that $mk \sim \sqrt{n/k}$. They state that this result would be a bit too optimistic since collisions within a row normally do not appear in a practical stream cipher. However, they claim that this approach still represents a realistic threat for the MICKEY stream cipher. We will show that for a random function, contrary to their conjecture, this attack has about the same complexities as a direct collision search in the initial states.

Let us take all the $2km$ intermediate states for the $2k - 1$ iterations. Let $Pr[A]$ define the probability that there is no collision in all the $2km$ intermediate states. If there is no collision in this set then there is also no collision in the km states considered by the attack. Thus, the probability of a successful attack is even smaller than $1 - Pr[A]$. By means of only counting arguments we can show the following proposition.

Proposition 2. *The probability of no collision in the $2km$ intermediate states is*

$$Pr[A] = \frac{n(n-1)\cdots(n-2k+1)}{n^{2km}},\tag{10}$$

where the probability is taken over all functions $\varphi \in \mathcal{F}_n$ and all possible choices of m initial states.

Proof. Let $Pr[I]$ be the probability of no collision in the m initial states. We can see directly that

$$Pr[A] = Pr[A \cap I]$$
$$= Pr[A|I]\, Pr[I]\,.$$

Let us assume that we have chosen m different initial states. This happens with a probability of

$$Pr[I] = \frac{\binom{n}{m}m!}{n^m}\,.\tag{11}$$

In this case we have

- n^n different instances $\varphi \in \mathcal{F}_n$ of our random functions, where each of them creates
- $\binom{n}{m}m!$ different tables. Each table can be produced more than once. There exists
- $n\,(n-1)\ldots(n-2km+1)$ different tables that contain no collisions. Each of them can be generated by
- $n^{n-(2k-1)m}$ different functions, since a table determines already $(2k-1)m$ positions of φ.

Thus, we get the probability

$$Pr[A|I] = \frac{n\,(n-1)\ldots(n-2km+1)\,n^{n-(2k-1)m}}{n^n\binom{n}{m}m!}\tag{12}$$

for $m > 0$ and $2km \leq n$. By combining (11) and (12) we can conclude our proof.

The probability of $Pr[A]$ given in (10) is exactly the probability of no collision in $2km$ random points, which means that we need at least an m such that $2km \sim \sqrt{n}$. This leads to a query complexity of $\sim \sqrt{n}$ and a space complexity of $\sim \sqrt{n}/2$.

3.3   Improvement with Distinguished Points

By applying the known technique of distinguished points [DQ88] we can reduce the space complexity in the second attack; however the query complexity stays the same.

By *distinguished points* (DPs) we mean a subset of Ω_n which is distinguished by a certain property, e.g. by a specific number of 0's in the most significant bits. In our new variant of the second attack we iterate Φ in each row up to the moment where we reach a DP. In this case we stop and store the DP. If we do not reach a DP after k_{MAX} iterations we stop as well but we store nothing. If there was a collision in any of the states in the rows where we reached a DP, the subsequent states would be the same and we would stop with the same DP. Thus it is sufficient to search for a collision in the final DPs.

Let d be the number of distinguished points in Ω_n. We assume that the ratio $c = \frac{d}{n}$ is large enough that with a very high probability we reach a DP before the end of the cycle in the functional graph. This means that the average number of iterations before arriving at a DP is much smaller than the expected length of the tail and the cycle together (which would be about $\sqrt{\frac{\pi n}{2}}$ due to [FO90a]). We assume that in this case the average length of each row would be in the range of $1/c$ like in the case of random points. We also suppose that that we need about $m/c \sim \sqrt{n}$ query points to find a collision, like in the previous case. This leads to a query complexity of $\sim \sqrt{n}$ and a space complexity of only $\sim c\sqrt{n}$. Empirical results for example for $n = 2^{20}$, $0.7 \leq \frac{\log_2(d)}{\log_2(n)} \leq 1$ and $k_{MAX} = \sqrt{n}$ confirm our assumptions.

A summary of the complexities of all attacks can be found in Table 2, where we marked by *(new)* the results that where not yet mentioned by Hong and Kim. In the case where we consider only the states after k iterations, we have to substantially increase the query complexity to gain in the space complexity and the number of initial states. We were able to show that even when we consider all intermediate states, the query complexity has a magnitude of $\sqrt{n}$. The variant using the distinguished points allows to reduce the space complexity by leaving the other complexities constant.

Table 2. Complexities of attacks

attack	# initial states	space complexity	query complexity
after k iterations, 3.1	$\sim \sqrt{2n/k}$	$\sim \sqrt{2n/k}$	$\sim \sqrt{2kn}$ *(new)*
with interm. states, 3.2	$\sim \sqrt{n/2k}$ *(new)*	$\sim \sqrt{n/2}$ *(new)*	$\sim \sqrt{n}$ *(new)*
with DPs, 3.3	$\sim c\sqrt{n}$ *(new)*	$\sim c\sqrt{n}$ *(new)*	$\sim \sqrt{n}$ *(new)*

4   Conclusion

In this article, we studied a stream cipher model which uses a random update function. We have introduced a new method of estimating the state entropy in this model. This estimator is based on the number of values that produce the same value after k iterations. Its computation is expensive for large numbers of iterations; however, for a value of k up to a few hundred, it is much more precise than the upper bound given by the number of image points.

In this model, we have also examined the two collision attacks proposed in [HK05] which are based on the entropy loss in the state. We pointed out that the first attack improves the space complexity at the cost of significantly increasing the query complexity. We proved that the complexity of the second attack is of the same magnitude as a collision search directly in the starting values. In addition we discussed a new variant of this attack, using distinguished points, which reduces the space complexity but leaves the query complexity constant.

The use of a random function in a stream cipher introduces the problem of entropy loss. However, the studied attacks based on this weakness are less effective than expected. Thus, the argument alone that a stream cipher uses a random function is not enough to threaten it due to a collision attack based on the entropy loss.

References

[BD05]   Babbage, S., Dodd, M.: The stream cipher MICKEY (version 1). eS-TREAM, ECRYPT Stream Cipher Project, Report 2005/015 (2005), http://www.ecrypt.eu.org/stream

[DQ88]   Delescaille, J.-P., Quisquater, J.-J.: Other cycling tests for DES (abstract). In: Pomerance, C. (ed.) CRYPTO 1987. LNCS, vol. 293, pp. 255–256. Springer, Heidelberg (1988)

[FO90a] Flajolet, P., Odlyzko, A.M.: Random mapping statistics. In: Advances in Cryptology, Proc. Eurocrypt 1998, vol. 434, pp. 329–354 (1990)

[FO90b] Flajolet, P., Odlyzko, A.M.: Singularity analysis of generating functions. SIAM J. Discrete Math. 3(2), 216–240 (1990)

[FS96]   Flajolet, P., Sedgewick, R.: An introduction to the analysis of algorithms. Addison-Wesley Longman Publishing Co., Inc. Boston (1996)

[Hel80] Hellman, M.: A cryptanalytic time-memory trade-off. Information Theory, IEEE Transactions on 26(4), 401–406 (1980)

[HK05]   Hong, J., Kim, W.H.: TMD-tradeoff and state entropy loss considerations of streamcipher MICKEY. In: Maitra, S., Veni Madhavan, C.E., Venkatesan, R. (eds.) INDOCRYPT 2005. LNCS, vol. 3797, pp. 169–182. Springer, Heidelberg (2005)

[HS05]   Hong, J., Sarkar, P.: New applications of time memory data tradeoffs. In: Roy, B. (ed.) ASIACRYPT 2005. LNCS, vol. 3788, pp. 353–372. Springer, Heidelberg (2005)

[Kol86] Kolchin, V.F.: Random Mappings. Optimization Software, Inc. (1986)

[SS03]   Seznec, A., Sendrier, N.: HAVEGE: A user-level software heuristic for generating empirically strong random numbers. ACM Trans. Model. Comput. Simul. 13(4), 334–346 (2003)

Analysis of Grain's Initialization Algorithm[*]

Christophe De Cannière[1,2], Özgül Küçük[1], and Bart Preneel[1]

[1] Katholieke Universiteit Leuven, Dept. ESAT/SCD-COSIC, and IBBT
Kasteelpark Arenberg 10, B–3001 Heverlee, Belgium
{christophe.decanniere,ozgul.kucuk}@esat.kuleuven.be
[2] Département d'Informatique École Normale Supérieure,
45, rue d'Ulm, F-75230 Paris cedex 05

Abstract. In this paper, we analyze the initialization algorithm of Grain, one of the eSTREAM candidates which made it to the third phase of the project. We point out the existence of a sliding property in the initialization algorithm of the Grain family, and show that it can be used to reduce by half the cost of exhaustive key search (currently the most efficient attack on both Grain v1 and Grain-128). In the second part of the paper, we analyze the differential properties of the initialization, and mount several attacks, including a differential attack on Grain v1 which recovers one out of 2^9 keys using two related keys and 2^{55} chosen IV pairs.

1 Introduction

Symmetric encryption algorithms are traditionally categorized into two types of schemes: block ciphers and stream ciphers. Stream ciphers distinguish themselves from block ciphers by the fact that they process plaintext symbols (typically bits) as soon as they arrive by applying a very simple but ever changing invertible transformation. As opposed to block ciphers, stream ciphers do not derive their security from the complexity of the encryption transformation, but from the unpredictable way in which this transformation depends on the position in the plaintext stream.

The most common type of stream ciphers are binary additive stream ciphers. The encryption transformation in this type of ciphers just consists of an exclusive or (XOR) with an independent sequence of bits called key stream. The key stream bits are derived from a secret internal state which is initialized using a secret key, and is then continuously updated.

The security of a binary additive stream cipher depends directly on the unpredictability of its key stream. In particular, the same sequence of key stream bits should never be reused to encrypt different plaintexts, and hence, a stream

---

[*] The work described in this paper has been partly supported by the European Commission under contract IST-2002-507932 (ECRYPT), by the Fund for Scientific Research – Flanders (FWO), the Chaire France Telecom pour la sécurité des réseaux de télécommunications, and the IAP Programme P6/26 BCRYPT of the Belgian State (Belgian Science Policy).

cipher should never be reinitialized with the same secret key. However, in order to avoid having to perform an expensive key agreement protocol for every single message, all modern stream ciphers accept during their initialization phase an additional parameter, typically called initialization vector (IV), which allows to generate different key streams from the same secret key.

Although the possibility to reuse the same key for several messages is an indispensable feature in many practical applications, the introduction of initialization vectors in stream ciphers also opens new opportunities for the adversary. Several recent stream cipher proposals [1,2,3,4] have succumbed to attacks exploiting relations between key stream bits generated from the same key but different (known or chosen) IVs. This clearly demonstrates the importance of a carefully designed initialization algorithm.

In this paper, we analyze the initialization algorithm of Grain, a family of hardware-oriented stream ciphers submitted to the eSTREAM Stream Cipher Competition. We will first show that a sliding property of the initialization algorithm, which was already noted in [5] but never formally published, not only results in a very efficient related-key attack, but can also be used more generally to reduce the cost of exhaustive key search. We will then study the differential properties of the initialization, and develop a differential attack on Grain v1 which recovers one out of 2^9 keys, and requires two related keys and 2^{55} chosen IV pairs. We will show that similar attacks apply to Grain-128, and that the requirement for related keys can be dropped if we consider reduced-round variants.

We finally note that we do not consider any of the attacks presented in this paper to be a serious threat in practice. However, they certainly expose some non-ideal behavior of the Grain initialization algorithm.

2 Description of Grain

Grain is a family of stream ciphers, proposed by Hell, Johansson, and Meier in 2005 [6], which was designed to be particularly efficient and compact in hardware. Its two members, Grain v1 and Grain-128, accept 80-bit and 128-bit keys respectively. The original version of the cipher, later referred to as Grain v0, was submitted to the eSTREAM project, but contained a serious flaw, as was demonstrated by several researchers [7,8]. As a response, the initial submission was tweaked and extended to a family of ciphers.

In the next two sections we first describe the building blocks common to all members of the Grain family. Afterwards, we will show how these blocks are instantiated for the specific ciphers Grain v1 and Grain-128

2.1 Keystream Generation

All Grain members consist of three building blocks: an n-bit nonlinear feedback shift register (NFSR), an n-bit linear feedback shift register (LFSR), and a

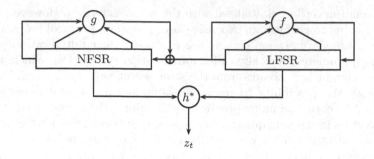

Fig. 1. Grain during the keystream generation phase

nonlinear filtering function. If we denote the content of the NFSR and the LFSR at any time t by $B_t = (b_t, b_{t+1}, \ldots, b_{t+n})$ and $S_t = (s_t, s_{t+1}, \ldots, s_{t+n})$, then the keystream generation process is defined as

$$s_{t+n} = f(S_t),$$
$$b_{t+n} = g(B_t) + s_t,$$
$$z_t = h^*(B_t, S_t),$$

where g and f are the update functions of the NFSR and LFSR respectively, and h^* is the filtering function (see Fig. 1).

2.2   Key and IV Initialization

The initial state of the shift registers is derived from the key and the IV by running an initialization process, which uses the same building blocks as for key stream generation, and will be the main subject of this paper. First, the key and the IV are loaded into the NFSR and LFSR respectively, and the remaining last bits of the LFSR are filled with ones. The cipher is then clocked for as many times as there are state bits. This is done in the same way as before, except that the output of the filtering function is fed back to the shift registers, as shown in Fig. 2 and in the equations below.

$$r_t = h^*(B_t, S_t) + s_t,$$
$$s_{t+n} = f(r_t, s_{t+1}, \ldots, s_{t+n-1}),$$
$$b_{t+n} = g(B_t) + r_t.$$

2.3   Grain v1

Grain v1 is an 80-bit stream cipher which accepts 64-bit IVs. The NFSR and the LFSR are both 80 bits long, and therefore, as explained above, the initialization

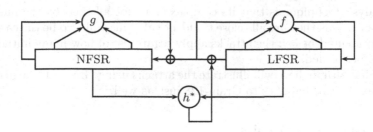

Fig. 2. Grain during the initialization phase

takes 160 cycles. The different functions are instantiated as follows:

$$f(S_t) = s_t + s_{t+13} + s_{t+23} + s_{t+38} + s_{t+51} + s_{t+62},$$
$$g(B_t) = b_t + b_{t+14} + b_{t+62}$$
$$+ g'(b_{t+9}, b_{t+15}, b_{t+21}, b_{t+28}, b_{t+33}, b_{t+37}, b_{t+45}, b_{t+52}, b_{t+60}, b_{t+63}),$$
$$h^*(B_t, S_t) = \sum_{i \in \mathcal{A}} b_{t+i} + h(s_{t+3}, s_{t+25}, s_{t+46}, s_{t+64}, b_{t+63}),$$

with $\mathcal{A} = \{1, 2, 4, 10, 31, 43, 56\}$, g' a function of degree 6, and h a function of degree 3. The exact definitions of these functions can be found in [6].

2.4 Grain-128

Grain-128 is the 128-bit member of the Grain family. The IV size is increased to 96 bits, and the shift registers are now both 128 bits long. The initialization takes 256 cycles, and the functions are defined as follows:

$$f(S_t) = s_t + s_{t+7} + s_{t+38} + s_{t+70} + s_{t+81} + s_{t+96},$$
$$g(B_t) = b_t + b_{t+26} + b_{t+56} + b_{t+91} + b_{t+96}$$
$$+ b_{t+3}b_{t+67} + b_{t+11}b_{t+13} + b_{t+17}b_{t+18}$$
$$+ b_{t+27}b_{t+59} + b_{t+40}b_{t+48} + b_{t+61}b_{t+65} + b_{t+68}b_{t+84},$$
$$h^*(B_t, S_t) = \sum_{i \in \mathcal{A}} b_{t+i} + h(s_{t+8}, s_{t+13}, s_{t+20}, s_{t+42}, s_{t+60}, s_{t+79}, s_{t+95}, b_{t+12}, b_{t+95}).$$

In the equations above, $\mathcal{A} = \{2, 15, 36, 45, 64, 73, 89\}$, and h is a very sparse function of degree 3. Again, we refer to the specifications [9] for the exact definition.

3 Slide Attacks

In this section we discuss a first class of attacks on Grain's initialization phase, which are based on a particular sliding property of the algorithm. Slide attacks have been introduced by Biryukov and Wagner [10] in 1999, and have since then mainly been used to attack block ciphers. A rather unique property of this cryptanalysis technique is that its complexity is not affected by the number of rounds, as long as they are all (close to) identical. This will also be the case in the attacks presented below: the attacks apply regardless of how many initialization steps are performed.

Note that although we will illustrate the attacks using Grain v1, the discussion in the next sections applies to Grain-128 just as well.

3.1   Related (K, IV) Pairs

The sliding property exploited in the next sections is a consequence of the similarity of the operations performed in Grain at any time t, both during initialization and key generation, as well as of the particular way in which the key and IV bits are loaded. More specifically, let us consider a secret key $K = (k_0, \ldots, k_{79})$, used in combination with an initialization vector $\mathrm{IV} = (v_0, \ldots, v_{63})$. During the first 161 cycles (160 initialization steps and 1 key generation step), the registers will contain the following values:

$$
\begin{array}{ll}
\left.\begin{array}{l}
B_0 = (k_0, \ldots \ldots, k_{78}, k_{79}) \\
B_1 = (k_1, \ldots \ldots, k_{79}, b_{80}) \\
\qquad \vdots \\
B_{160} = (b_{160}, \ldots, b_{238}, b_{239}) \\
B_{161} = (b_{161}, \ldots, b_{239}, b_{240})
\end{array}\right\} \text{init. phase}
&
\begin{array}{l}
S_0 = (v_0, \ldots, v_{62}, v_{63}, 1, \ldots, 1, \quad 1) \\
S_1 = (v_1, \ldots, v_{63}, \quad 1, 1, \ldots, 1, s_{80}) \\
\qquad \vdots \\
S_{160} = (s_{160}, \ldots \ldots \ldots \ldots, s_{238}, s_{239}) \\
S_{161} = (s_{161}, \ldots \ldots \ldots \ldots, s_{239}, s_{240})
\end{array}
\end{array}
$$

Let us now assume that $s_{80} = 1$. Note that if this is not the case, it suffices to flip v_{13} for the assumption to hold. We then consider a second key $K^* = (k_1, \ldots, k_{79}, b_{80})$ together with the initialization vector $\mathrm{IV}^* = (v_1, \ldots, v_{63}, 1)$. After loading this pair into the registers, we obtain:

$$
B_0^* = (k_1, \ldots \ldots, k_{79}, b_{80}) \qquad S_0^* = (v_1, \ldots, v_{63}, 1, 1, \ldots, 1, 1)
$$

This, however, is identical to the content of B_1, and since the operations during the initialization are identical as well, the equality $B_t^* = B_{t+1}$ is preserved until step 159, as shown below.

$$
\begin{array}{ll}
\left.\begin{array}{l}
B_0^* = (k_1, \ldots \ldots, k_{79}, b_{80}) \\
\qquad \vdots \\
B_{159}^* = (b_{160}, \ldots, b_{238}, b_{239}) \\
B_{160}^* = (b_{161}, \ldots, b_{239}, b_{239}^*) \\
B_{161}^* = (b_{162}, \ldots, b_{239}^*, b_{240}^*)
\end{array}\right\} \text{init. phase}
&
\begin{array}{l}
S_0^* = (v_1, \ldots, v_{63}, 1, 1, \ldots, 1, 1) \\
\qquad \vdots \\
S_{159}^* = (s_{160}, \ldots \ldots, s_{238}, s_{239}) \\
S_{160}^* = (s_{161}, \ldots \ldots, s_{239}, s_{239}^*) \\
S_{161}^* = (s_{162}, \ldots \ldots, s_{239}^*, s_{240}^*)
\end{array}
\end{array}
$$

In step 160, b_{239}^* and s_{239}^* are not necessarily equal to b_{240} and s_{240}, since the former are computed in initialization mode, whereas the latter are computed in key stream generation mode. Nevertheless, and owing to the tap positions of Grain v1, the equality will still be detectable in the first 15 keystream bits.

Moreover, if $h^*(B_{159}^*, S_{159}^*) = h^*(B_{160}, S_{160}) = 0$ (this happens with probability $1/2$), then both modes of Grain are equivalent, and hence the equality is preserved in the last step as well. After this point, (B_t^*, S_t^*) and (B_{t+1}, S_{t+1}) are both updated in key stream generation mode; their values will therefore stay equal till the end, leading to identical but shifted key streams.

With an appropriate choice of IVs, similar sliding behaviors can also be observed by sliding the keys over more bit positions. In general, we have the following property for $1 \leq n \leq 16$:

Property 1. For a fraction $2^{-2 \cdot n}$ of pairs (K, IV), there exists a related pair (K^*, IV^*) which produces an identical but n-bit shifted key stream.

Note that the existence of different (K, IV) pairs which produce identical but shifted key streams is in itself not so uncommon in stream ciphers. When a stream cipher is iterated, its internal state typically follows a huge predefined cycle, and the role of the initialization algorithm is to assign a different starting position for each (K, IV) pair. Obviously, if the total length of the cycle(s) is smaller than the number of possible (K, IV) pairs multiplied by the maximum allowed key stream length, then some overlap between the key stream sequences generated by different (K, IV) pairs is unavoidable. This is the case in many stream ciphers, including Grain. However, what is shown by the property above, is that the initialization algorithm of Grain has the particularity that it tends to cluster different starting positions together, instead of distributing them evenly over the cycle(s).

3.2   A Related-Key Slide Attack

A first straightforward application of the property described in the previous section is a related-key attack. Suppose that the attacker somehow suspects that two (K, IV) pairs are related in the way explained earlier. In that case, he knows that the corresponding key stream sequences will be shifted over one bit with probability $1/4$, and if that happens, he can conclude that $s_{80} = 1$. This allows him to derive a simple equation in the secret key bits. Note that if the (K, IV) pairs are shifted over $n > 1$ positions, then with probability $2^{-2 \cdot n}$ the attacker will be able to obtain n equations.

As is the case for all related key attacks, the simple attack just described is admittedly based on a rather strong supposition. In principle, however, one could imagine practical situations where different session keys are derived from a single master key in a funny way, making this sort of related keys more likely to occur, or where the attacker has some means to transform the keys before they are used (for example by causing synchronization errors).

3.3   Speeding Up Exhaustive Key Search

A second application, which is definitely of more practical relevance, is to use the sliding property of the initialization algorithm to speed up exhaustive key search by a factor two.

The straightforward way to run an exhaustive search on Grain v1 is described by the pseudo-code below:

for $K = 0$ to $2^{80} - 1$ **do**
    perform 160 initialization steps;
    generate first few key stream bits $(z_0, \ldots, z_t)$;
    check if $(z_0, \ldots, z_t)$ matches given key stream;
end for

Let us now analyze the special case where the given key stream sequence was generated using an IV which equals $I = (1, \ldots, 1)$. In this case, which can easily be enforced if we assume a chosen-IV scenario, the algorithm above can be improved by exploiting the sliding property. In order to see this, it suffices to analyze the contents of the registers during the initialization:

$$
\begin{array}{l}
\text{init. phase}
\begin{cases}
B_0 = (k_0, \ldots \ldots, k_{78}, k_{79}) \\
B_1 = (k_1, \ldots \ldots, k_{79}, b_{80}) \\
B_2 = (k_2, \ldots \ldots, b_{80}, b_{81}) \\
\quad \vdots \\
B_{160} = (b_{160}, \ldots, b_{238}, b_{239}) \\
B_{161} = (b_{161}, \ldots, b_{239}, b_{240})
\end{cases}
\end{array}
\qquad
\begin{array}{l}
S_0 = (1, \ldots, 1, 1, \ldots \ldots, 1 \quad 1) \\
S_1 = (1, \ldots, 1, 1, \ldots \ldots, 1 \; s_{80}) \\
S_2 = (1, \ldots, 1, 1, \ldots, s_{80}, s_{81}) \\
\quad \vdots \\
S_{160} = (s_{160}, \ldots \ldots, s_{238}, s_{239}) \\
S_{161} = (s_{161}, \ldots \ldots, s_{239}, s_{240})
\end{array}
$$

The improvement is based on the observation that if $s_{80} = 1$, we can check two keys without having to recalculate the initialization. If $s_{81} = 1$ as well, then we can simultaneously verify three keys, and so on. In order to use this property to cover the key space in an efficient way, we need to change the order in which the keys are searched, though. This is done in the pseudo-code below:

$K = 0$;
repeat
    perform 160 initialization steps;
    generate first 16 key stream bits $(z_0, \ldots, z_{15})$;
    for $t = 0$ to [largest $n < 16$ for which $S_n = I$] **do**
        check if $(z_t, \ldots, z_{15})$ matches given key stream;
        $K = B_{t+1}$;
    end for
until $K = 0$

Since K is updated in an invertible way, we know that the code will eventually reach $K = 0$ again. At this point, the code will only have checked a cycle of keys with an expected length of 2^{79}. This, however, is done by performing only 2^{78} initializations, making it twice as fast as the standard exhaustive search algorithm. If we are unlucky, and the secret key is not found, then the algorithm can simply be repeated with a different starting key in order to cover a different cycle.

Finally note that the algorithm retains a number of useful properties of the regular exhaustive search algorithm: it can be used to attack several keys simultaneously, and it can easily be parallelized (using distinguished points to limit overlap). One limitation however, is that it can only be applied if the attacker can get hold of a keystream sequence corresponding to $IV = (1, \ldots, 1)$, or of a set of keystream sequences corresponding to a number of related IVs. Depending on how the IVs are picked, the latter might be easier to obtain.

3.4 Avoiding the Sliding Property

As discussed earlier, the existence of related (K, IV) pairs in Grain cannot be avoided without increasing the state size. However, in order to avoid the particular sliding behavior of the initialization algorithm, one could try to act on the two factors that lead to this property: the similarity of the computations performed at different times t, and the self-similarity of the constant loaded into the last bits of the LFSR. The similarity of computations could be destroyed by involving a counter in each step. This would effectively increase the size of the state, but one could argue that this counter needs to be stored anyway to decide when the initialization algorithm finishes. An easier modification, however, would be to eliminate the self-similarity of the initialization constant. If the last 16 bits of the LFSR would for example have been initialized with $(0, \ldots, 0, 1)$, then this would already have significantly reduced the probability of the sliding property.

4 Differential Attacks

In this second part, we will analyze the differential properties of the initialization algorithm. We will first show how to generate (truncated) differential characteristics with useful properties, and will then discuss some additional techniques to efficiently use these characteristics in a key recovery attack.

4.1 Truncated Differential Characteristics

The main idea of differential cryptanalysis [11] is to apply differences at the input of a cryptographic function, and to search for non-random properties in the distribution of the corresponding differences at the output. An important tool to find such non-random properties are differential characteristics, which describe possible ways in which differences propagate throughout the internal structure of a function. If the probability that such a characteristic is followed from input to output is sufficiently high, then it will be possible to detect this in the distribution of the output differences.

In the case of stream ciphers, the inputs and outputs that are assumed to be accessible to the adversary are the IV and the keystream sequence respectively. In the attacks described below, we will only consider the difference in a single keystream bit, and ignore all other outputs. In the case of block ciphers, this technique is referred to as truncated differential cryptanalysis.

Finding high probability differential characteristics often boils down to finding sparse characteristics. In the case of Grain such characteristics are relatively easy to find by starting from a difference in a single bit in the state at some step t, and analyzing how this difference propagates, both backwards and forwards. Each time a difference enters the non-linear functions g' or h, we need to make a choice, since there will typically be several possible output differences. A good approach consists in choosing the difference which introduces as few additional differences in the next step as possible, in particular in the NFSR, since bits in this register are more often used as inputs to the non-linear functions than bits in the LFSR. An example of a characteristic found this way is depicted in Fig. 3.

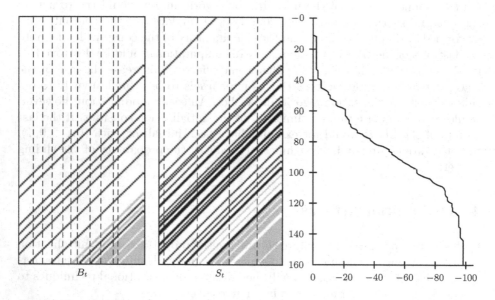

Fig. 3. A truncated differential path in Grain v1 and its probability on a $\log_2$ scale. Differences are denoted by black pixels; bits which do not affect the output are gray. The bit positions which affect either g' or h are marked with dashed lines.

The probability of the truncated characteristic in Fig. 3 is almost 2^{-100}. In order to detect a significant bias in the difference of the single output bit which is predicted by this characteristic, we would in principle need to analyze this difference for at least $2^{2 \cdot 100}$ different IV pairs, which can obviously never be achieved with a 64-bit IV. However, if we allow the attacker to apply changes to the key as well (assuming a related-key scenario), then this hypothetical number can be significantly reduced, as shown in Fig. 4. In this case, the number of required IV pairs is about $2^{2 \cdot 47}$. This is still considerably higher than 2^{63}, the total number of possible IV pairs, but in the next section we will introduce a technique to further reduce this requirement.

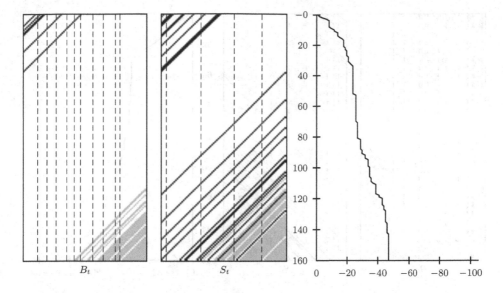

Fig. 4. A related-key truncated differential path in Grain v1 and its probability

4.2   Partitioning the Key and IV Space

In order to the reduce the number of IV pairs needed to detect the bias in the difference of the output bit, we will exploit the fact that the propagation of differences in the first few steps only depends on a rather limited number of key and IV bits (or combinations of them). Hence, if we would guess these key bits, we would be able to get rid of the probabilistic behavior of the first part of the initialization.

Instead of trying to determine exactly which combination of key or IV bits affect the propagation of the differences up to a given step t (which would in fact be relatively easy to do in the case of Grain-128), we will use an alternative technique which allows us to consider the internal operations of the cipher as a black box. To this end, we introduce the function $F_t(K, \text{IV})$ which returns 1 or 0 depending on whether or not the characteristic is followed in the first t steps when the algorithm is initialized with values (K, IV) and (K', IV') satisfying the input difference. The idea now is to partition the key and IV spaces into classes $\{\mathcal{K}_1, \mathcal{K}_2, \ldots\}$ and $\{\mathcal{IV}_1, \mathcal{IV}_2, \ldots\}$ according to the following equivalence relation:

Definition 1. *Two keys K_1 and K_2 are t-equivalent if $F_t(K_1, \text{IV}) = F_t(K_2, \text{IV})$ for all IVs. Similarly, two initialization vectors IV_1 and IV_2 are t-equivalent if $F_t(K, \text{IV}_1) = F_t(K, \text{IV}_2)$ for all keys K.*

In order to check for t-equivalence in practice, we can write F_t as a product $f_1 \cdot f_2 \cdots f_t$, where f_i indicates whether the desired difference at the input of round i propagates to the desired difference at the output of the round. If we observe that $f_i(K_1, \text{IV}) = f_i(K_2, \text{IV})$ for all i and for a sufficient number of random IVs, then we conclude that K_1 and K_2 are most likely t-equivalent.

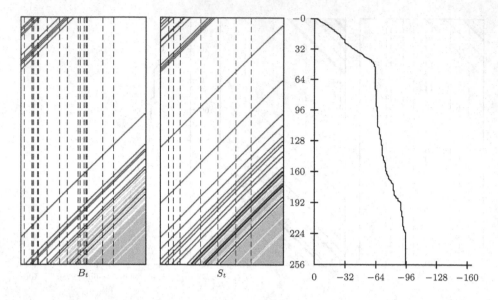

Fig. 5. A related-key truncated differential path in Grain-128 and its probability

Before proceeding with the description of the proposed attack, we introduce some additional notation:

p_1: the total probability of the characteristic in the first t steps.

p_2: the probability of the characteristic in the remaining steps.

p_K: the fraction of keys for which $F_t(K, \cdot) \neq 0$ (weak keys).

p_{IV}: the fraction of IVs for which $F_t(\cdot, \mathrm{IV}) \neq 0$ (weak IVs).

n_K / n_{IV}: the number of key/IV bits.

$N_{\mathcal{K}} / N_{\mathcal{IV}}$: the number of weak equivalence classes.

The attack itself consists of two phases:

1. Initialize the stream cipher with a pair of unknown but related keys (K, K') using N different pairs of related weak IVs $(\mathrm{IV}_i, \mathrm{IV}'_i)$. For each IV_i, compute in which class $\mathcal{IV}_i$ it resides, and depending on the difference in the keystream bit, increment the counter $c^0_{\mathcal{IV}_i}$ or $c^1_{\mathcal{IV}_i}$.

2. For all $N_{\mathcal{K}}$ weak key classes $\mathcal{K}_i$, compute the counters $c^0_{\mathcal{K}_i}$ and $c^1_{\mathcal{K}_i}$, with

$$c^0_{\mathcal{K}_i} = \sum_{F_t(\mathcal{K}_i, \mathcal{IV}_j)=1} c^0_{\mathcal{IV}_j} \quad \text{and} \quad c^1_{\mathcal{K}_i} = \sum_{F_t(\mathcal{K}_i, \mathcal{IV}_j)=1} c^1_{\mathcal{IV}_j} .$$

If N is sufficiently large, and assuming that the unknown key K was indeed weak, we expect the counters above to be biased for the correct key class. In order to get a rough estimate of the minimal value of N, we note that $F_t(\mathcal{K}_i, \mathcal{IV}_j)$ equals 1 with probability $p_1 \cdot p_K^{-1} \cdot p_{\mathrm{IV}}^{-1}$, and hence the expected value of $c^0_{\mathcal{K}_i} + c^1_{\mathcal{K}_i}$ (i.e., the number of IV pairs satisfying the characteristic up to step t, assuming a weak key $K \in \mathcal{K}_i$), is $N \cdot p_1 \cdot p_K^{-1} \cdot p_{\mathrm{IV}}^{-1}$. In order to be able to detect a bias between

Table 1. Summary of the attacks

Cipher	Grain v1	Grain v1	Grain-128	Grain-128	Grain-128
Rounds	160	112	256	224	192
Related keys	yes	no	yes	no	no
# Weak keys	2^{71}	2^{80}	2^{87}	2^{126}	2^{126}
# Weak IVs	2^{57}	2^{63}	2^{84}	2^{93}	2^{93}
# Chosen IV pairs	2^{55}	(2^{72})	2^{73}	(2^{96})	2^{35}
t	33	28	75	78	76
p_1	2^{-23}	2^{-3}	2^{-64}	2^{-6}	2^{-6}
p_2	2^{-24}	2^{-35}	2^{-31}	2^{-47}	2^{-17}
$N_\mathcal{K}$	2^{22}	8	2^{27}	72	72
$N_\mathcal{IV}$	2^{21}	8	2^{32}	64	64

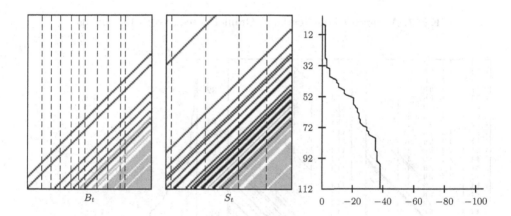

Fig. 6. A truncated differential in Grain v1 reduced to 112 rounds

$c_{\mathcal{K}_i}^0$ and $c_{\mathcal{K}_i}^1$, we need this number to be at least p_2^{-2}, resulting in the following bound:

$$N > \frac{p_K \cdot p_{IV}}{p_1 \cdot p_2^2}.$$

When we apply this idea to Grain v1, we obtain an attack which can successfully recover one key out of 2^9 and requires at least 2^{55} chosen IVs (see Table 1). In the case of Grain-128, we can recover one key out of 2^{41} using 2^{73} chosen IVs.

4.3   Attacks on Reduced Versions

In Sect. 4.1, we were forced to introduce related keys in order to increase the probability of the differential characteristics in Grain v1 and Grain-128. This is not necessary anymore if we consider reduced-round variants of the ciphers, though. Table 1 summarizes the complexity of a number of regular (not related-key) attacks for different reduced versions.

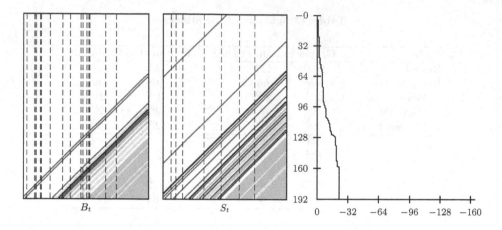

Fig. 7. A truncated differential in Grain-128 reduced to 192 rounds

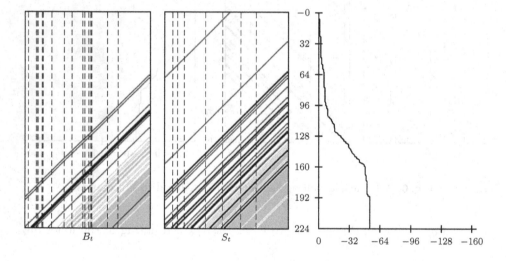

Fig. 8. A truncated differential in Grain-128 reduced to 224 rounds

5   Conclusions

In this paper, we have analyzed the initialization algorithm of Grain taking two very different approaches. First we have studied a sliding property in the initialization algorithm, and shown that it can be used to reduce by half the cost of exhaustive key search. While this might not be significant for Grain-128, it could have some impact on Grain v1, given its relatively short 80-bit key. Moreover, we have shown that this attack could be avoided by making a minor change in the constant used during the initialization.

In the second part of the paper, we have analyzed the differential properties of the initialization. We have constructed truncated differential characteristics for Grain v1, and have shown that by considering a specific partitioning of the key and the IV space, these characteristics can be used to mount a differential attack requiring two related keys and 2^{55} chosen IV pairs, which recovers one key out of 2^9. A similar attack also applies to Grain-128. As is the case for all related-key attacks, the practical impact of this result is debatable, but regardless of this, it can certainly be considered as a non-ideal behavior of the initialization algorithm.

References

1. Wu, H., Preneel, B.: Differential-linear attacks against the stream cipher Phelix. In: Biryukov, A. (ed.) FSE 2007. LNCS, vol. 4593, pp. 87–100. Springer, Heidelberg (2007)
2. Wu, H., Preneel, B.: Differential cryptanalysis of the stream ciphers Py, Py6 and Pypy. In: Naor, M. (ed.) EUROCRYPT 2007. LNCS, vol. 4515, pp. 276–290. Springer, Heidelberg (2007)
3. Joux, A., Reinhard, J.R.: Overtaking Vest. In: Biryukov, A. (ed.) FSE 2007. LNCS, vol. 4593, pp. 58–72. Springer, Heidelberg (2007)
4. Wu, H., Preneel, B.: Resynchronization attacks on WG and LEX. In: Robshaw, M.J.B. (ed.) FSE 2006. LNCS, vol. 4047, pp. 422–432. Springer, Heidelberg (2006)
5. Küçük, Ö.: Slide resynchronization attack on the initialization of Grain 1.0. eSTREAM, ECRYPT Stream Cipher Project, Report 2006/044 (2006), http://www.ecrypt.eu.org/stream
6. Hell, M., Johansson, T., Meier, W.: Grain – A Stream Cipher for Constrained Environments. eSTREAM, ECRYPT Stream Cipher Project, Report 2005/010 (2005) http://www.ecrypt.eu.org/stream
7. Berbain, C., Gilbert, H., Maximov, A.: Cryptanalysis of Grain. In: Robshaw, M.J.B. (ed.) FSE 2006. LNCS, vol. 4047, pp. 15–29. Springer, Heidelberg (2006)
8. Khazaei, S., Kiaei, M.H.M.: Distinguishing attack on Grain. eSTREAM, ECRYPT Stream Cipher Project, Report 2006/071 (2005), http://www.ecrypt.eu.org/stream.
9. Hell, M., Johansson, T., Meier, W.: A Stream Cipher Proposal: Grain-128. eSTREAM, ECRYPT Stream Cipher Project (2006), http://www.ecrypt.eu.org/stream
10. Biryukov, A., Wagner, D.: Slide attacks. In: Knudsen, L.R. (ed.) FSE 1999. LNCS, vol. 1636, pp. 245–259. Springer, Heidelberg (1999)
11. Biham, E., Shamir, A.: Differential Cryptanalysis of the Data Encryption Standard. Springer, Heidelberg (1993)

Password Recovery on Challenge and Response: Impossible Differential Attack on Hash Function

Yu Sasaki[1], Lei Wang[2], Kazuo Ohta[2], and Noboru Kunihiro[2,*]

[1] NTT Information Sharing Platform Laboratories, NTT Corporation
3-9-11 Midorichou, Musashino-shi, Tokyo, 180-8585, Japan
sasaki.yu@lab.ntt.co.jp
[2] The University of Electro-Communications
1-5-1 Chofugaoka, Chofu-shi, Tokyo, 182-8585, Japan

Abstract. We propose practical password recovery attacks against two challenge-response authentication protocols using MD4. When a response is computed as MD4(Password||Challenge), passwords up to 12 characters are practically recovered. To recover up to 8 characters, we need 16 times the amount of eavesdropping and 16 times the number of queries, and the off-line complexity is less than 2^{35} MD4 computations. To recover up to 12 characters, we need 2^{10} times the amount of eavesdropping and 2^{10} times the number of queries, and the off-line complexity is less than 2^{40} MD4 computations. When a response is computed as MD4(Password||Challenge||Password), passwords up to 8 characters are practically recovered by 2^8 times the amount of eavesdropping and 2^8 times the number of queries, and the off-line complexity is less than 2^{39} MD4 computations. Our approach is similar to the "Impossible differential attack", which was originally proposed for recovering the block cipher key. Good impossible differentials for hash functions are achieved by using local collision. This indicates that the presence of one practical local collision can damage the security of protocols.

Keywords: Challenge and Response, Prefix, Hybrid, Impossible Differential Attack, Local Collision, Hash Function, MD4.

1   Introduction

Authentication protocols have recently taken an important role and thus the security of authentication protocols must be carefully considered.

There are many authentication protocols that use hash functions. The security of hash functions is therefore critical for authentication protocols. For hash function H, there is an important property called collision resistance: it must be computationally hard to find a pair of (x, x') such that $H(x) = H(x'), x \neq x'$.

The hash functions MD5 and SHA-1 are widely used, and their designs are based on MD4 [7,8]. In 2005, the collision resistance of such hash functions

---

* Currently with the University of Tokyo.

S. Vaudenay (Ed.): AFRICACRYPT 2008, LNCS 5023, pp. 290–307, 2008.
© Springer-Verlag Berlin Heidelberg 2008

was broken [16,17,18,19]. As a result, reevaluating the security of authentication protocols with widely used hash functions is important.

This paper analyzes challenge-response password authentication protocols. These protocols are classified as Prefix, Suffix, and Hybrid approaches, and their security has been discussed by [14].

Definitions: Prefix, Suffix, and Hybrid

Let C be a challenge, P be a password, and H be a hash function. The responses of the Prefix, Suffix, and Hybrid approaches are computed as $H(P||C)$, $H(C||P)$, and $H(P||C||P)$, respectively. (Two Ps in the hybrid approach are the same password.)

Challenge Handshake Authentication Protocol (CHAP) [13] is an example of practical use of the prefix approach and Authenticated Post Office Protocol (APOP) [9] is that for the suffix approach. The hybrid approach was proposed by Tsudik [14].

Previous Works

Researchers have shown the reevaluation of authentication protocols. Preneel and van Oorschot showed how to attack suffix and hybrid[1] approaches by using collisions [10]. Therefore, the security of the suffix and hybrid approaches has the same level as that of the collision resistance of the hash function used. Preneel and van Oorschot generate collisions by using the birthday paradox, which is too complex for the widely used hash function for practical computation.

In 2006, attacks against NMAC and HMAC were proposed by Contini and Yin [3], Rechberger and Rijmen [11], and Fouque et al. [4]. Among these attacks, the inner key recovery of NMAC-MD4 and HMAC-MD4 by Contini and Yin [3] is important for our work. Contini and Yin recovered the inner key by using the MD4 collision. However, using this collision requires an impractical number of queries. For example, Contini and Yin's method needs 2^{63} queries and so their attack is not practical. Their attack is applicable to password recovery against the prefix- and hybrid-MD4.

In 2007, Leurent and Sasaki et al. independently proposed practical password recovery attacks against APOP using suffix-MD5 [6,12]. These attacks recovered a password by using a property by which many MD5 collisions could be generated in practical time.

Wang et al. improved the password recovery attack against the prefix-MD4 [15]. They used two-round-collisions of MD4 to attack a protocol. As a result, attack complexity is reduced to 2^{37} queries. However, asking 2^{37} queries is completely impractical for real protocols.

A summary of previous works is shown in Table 1.

From Table 1, practical attacks against the prefix and hybrid approaches using practical hash functions such as MD4 or MD5 is appealing.

---

[1] Their attack target was Envelop MAC. This technique can be applied to suffix and hybrid approaches.

Table 1. List of attacks against challenge-response authentication protocols

	Prefix	Suffix	Hybrid
Theoretical attack with general hash function		[10]	[10]
Theoretical attack with MD4 or MD5	[3] [15]	[6] [12]	[3]
Practical attack with MD4 or MD5	Our result	[6] [12]	Our result

Our Contribution

This paper proposes the following two attacks (New results are shown in Table 1.):

- **Practical password recovery attack against prefix-MD4**:
  We propose a password recovery attack against the $MD4(P\|C)$ approach that recovers up to 12 characters. We have experimentally confirmed that up to 8 characters are recovered using this approach. Up to 8 characters are recovered with 16 times the amount of eavesdropping, 16 times the number of queries, and less than 2^{35} off-line MD4 computations. In the case of up to 12 characters, we need 2^{10} times the amount of eavesdropping, 2^{10} times the number of queries, and less than 2^{40} off-line MD4 computations.
- **Practical password recovery attack against hybrid-MD4**:
  We propose a password recovery attack against the $MD4(P\|C\|P)$ approach. This attack is similar to the attack against the prefix-MD4. We have experimentally confirmed up to 8 characters are recovered using this approach. This attack needs 2^8 times the amount of eavesdropping, 2^8 times the number of queries, and less than 2^{39} off-line MD4 computations.

Our attack has a unique technique for recovering a password with an approach that is similar to the "Impossible differential attack", which was named by Biham et al. [1,2] and was originally proposed for recovering a block cipher key. We generate challenges C and C' so that particular differences never occur in the computation of responses R and R'. Then, from R and R', we inversely compute a hash function by guessing part of the passwords. If inverse computation reaches the impossible differentials, we can determine whether our guess was wrong.

In our work we identified the input differences and impossible differentials that make a long impossible differential path. Such a path can be constructed by using the local collision of the hash function. We focused on the characteristic that if two messages collide in an intermediate step, the following several steps never have differences until the next message differences are inserted. Due to this characteristic, the effective long impossible differential path is achieved.

Our password recovery attacks require a small number of queries, while previous works need an impractical number of queries. The most effective improvement is the use of a short local collision. Since probability of forming a local collision is much higher than that of a collision, attack complexity becomes practical. Use of a local collision also enlarges the range of hash functions used for attacking protocols. This concept is shown in Figure 1.

(1): Set of hash functions s.t. "collision" can quickly be found.

(2): Set of hash functions s.t. "local collision" can quickly be found.

Fig. 1. Our improvement: using local collision

Only the hash functions that belong to (1) in Figure 1 are considered in most previous work, whereas, our attack can use both (1) and (2). The use of a local collision for attacking protocols has a large impact since the presence of one practical local collision can damage the security of protocols.

This paper is organized as follows. Section 2 describes related works. Section 3 describes the password recovery attack against the prefix approach. Section 4 describes the password recovery attack against the hybrid approach. Section 5 discusses countermeasures against our attack and the possibility of replacing MD4 with MD5. Finally, we conclude this paper.

2 Related Works

2.1 Description of MD4

The MD4 [7] input is an arbitrary length message M, having 128-bit data $H(M)$. The MD4 input has a Merkle-Damgård structure. First, the input message is padded to be a multiple of 512 bits.

In the padding procedure, bit '1' is added to the tail of the message, then, bit '0's are added until the length of the padded message becomes 448 on modulo 512. Finally, the message length before padding is added to the last 64 bits.

The padded message M^* is divided into 512-bit strings $M_0, \ldots, M_{n-1}$. The initial value (IV) for the hash value is set to
$H_0 = (\text{0x67452301}, \text{0xefcdab89}, \text{0x98badcfe}, \text{0x10325476})$.
Finally, H_n, which is the hash value of M is computed by using compression function h as follows:
$$H_1 \leftarrow h(H_0, M_0), \ H_2 \leftarrow h(H_1, M_1), \ \ldots, \ H_n \leftarrow h(H_{n-1}, M_{n-1}).$$

Compression function of MD4

Basic computations in the compression function are 32-bit. We omit the notation of "mod 2^{32}". The input to the compression function is a 512-bit message M_j and a 128-bit value H_j. First, M_j is divided into $(m_0, \ldots m_{15})$, where each m_i is a 32-bit message, and (a_0, b_0, c_0, d_0) are set to be IV. The compression function consists of 48 steps. Steps 1–16 are called the first round (1R). Steps 17–32 and

Table 2. Message index for each step in MD4

1R (steps 1–16)	0 1 2 3	4 5 6 7	8 9 10 11	12 13 14 15
2R (steps 17–32)	0 4 8 12	1 5 9 13	2 6 10 14	3 7 11 15
3R (steps 33–48)	0 8 4 12	2 10 6 14	1 9 5 13	3 11 7 15

33–48 are the second and third rounds (2R and 3R). In step i, chaining variables a_i, b_i, c_i, d_i ($1 \leq i \leq 48$) are updated by the following expression.

$$a_i = d_{i-1}, \qquad b_i = (a_{i-1} + f(b_{i-1}, c_{i-1}, d_{i-1}) + m_k + t_i) \lll s_i,$$
$$c_i = b_{i-1}, \qquad d_i = c_{i-1},$$

where f is a bitwise Boolean function defined in each round, m_k is one of $(m_0, \ldots m_{15})$, and index k for each step is shown in Table 2.

t_i is a constant number defined in each round, $\lll s_i$ denotes left rotation by s_i bits, and s_i is defined in each step. Details of f and t_i are as follows.

$$1R : t_i = \text{0x00000000}, f(X, Y, Z) = (X \wedge Y) \vee (\neg X \wedge Z),$$
$$2R : t_i = \text{0x5a827999}, f(X, Y, Z) = (X \wedge Y) \vee (Y \wedge Z) \vee (X \wedge Z),$$
$$3R : t_i = \text{0x6ed9eba1}, f(X, Y, Z) = X \oplus Y \oplus Z.$$

After 48 steps are computed, H_{j+1} is calculated as follows.

$$aa_0 \leftarrow a_{48} + a_0, \quad bb_0 \leftarrow b_{48} + b_0, \quad cc_0 \leftarrow c_{48} + c_0, \quad dd_0 \leftarrow d_{48} + d_0,$$
$$H_{j+1} \leftarrow (aa_0, bb_0, cc_0, dd_0).$$

2.2   Key Recovery Attack Against Envelop MAC

The attack against Envelop MAC proposed by Preneel and van Oorschot [10] recovers secret information located behind messages. Therefore, this attack can be applied to the suffix approach and the latter part of hybrid approach. The main idea is that message length is arranged so that part of the secret is located in the second last block and the other part is located in the last block. Then, by using collisions that are generated with the complexity of the birthday paradox, the secret located in the second last block is recovered. Since the complexity of the birthday paradox is practically out of reach, this attack is impractical.

2.3   Practical Password Recovery Attacks Against APOP

The authentication protocol APOP takes the suffix-MD5 approach. Leurent and Sasaki et al. independently proposed attacks against APOP [6,12]. The attack approach is similar to the attack against Envelop MAC [10]. However, since the MD5 collision can be practically generated, Leurent and Sasaki et al. succeeded in practically breaking suffix-MD5.

Different from [10], Leurent and Sasaki et al.'s attack cannot be applied to the hybrid approach. Practically generating collision requires the knowledge of intermediate value of the response computation. However, in the hybrid approach, the attacker cannot determine the intermediate value due to the password before the challenge.

2.4   Key Recovery Attacks Against NMAC-MD4 and HMAC-MD4

NMAC and HMAC have many common structures. In this paper we only describe NMAC. We denote the MD4 digest for message M and the initial value IV by MD4(IV, M). NMAC-MD4 for message M is denoted by MD4(sk2, MD4(sk1, M)), where sk1 and sk2 are the secret keys.

Contini and Yin succeeded in recovering sk1 [3]. First, they determined the differential path that MD4(sk1, M), and MD4(sk1, M') become a collision pair for randomly fixed sk1 with a probability of 2^{-62}. Therefore, they obtain a collision by 2^{63} times the number of queries. After that, they recover the values of intermediate chaining variables by little modification of the collision message pair.

The NMAC-MD4 situation is similar to those for the prefix and hybrid approaches. In NMAC-MD4, all messages can be chosen by the attacker; however, IV is secret, and this makes all intermediate chaining variables unknown. In the prefix and hybrid approaches, IV is public information; however, the first part of the message is secret, and this makes all intermediate chaining variables unknown. Since Contini and Yin's method recovers the value of intermediate chaining variables, their attack can recover a password for the prefix and hybrid approaches. However, their attack needs 2^{63} times the number of queries, which is impractical.

2.5   Password Recovery Attack Against Prefix-MD4

Wang et al. improved Contini and Yin's attack [15]. Their main improvement is in reducing complexity by using two-round-collisions, whereas Contini and Yin used full-collisions. Since the complexity of generating two-round-collisions is smaller than that of generating full-collisions, the number of queries is reduced to 2^{37} MD4. After they obtained a two-round-collision, they recovered the values of intermediate chaining variables in the same way used by Contini and Yin.

2.6   Summary of Related Works

Preneel and van Oorschot proposed a general attack against the suffix and hybrid approaches. Since this attack needs too many queries, the attack is not practical. Leurent and Sasaki et al. proposed practical password recovery attacks against suffix-MD5; however, their method does not work for the prefix and hybrid approaches. Contini and Yin's attack can be applied to the prefix and hybrid approaches, but it needs an impractical number of queries. Wang et al. improved the attack against the prefix-MD4; however, complexity is still impractical. In the end, determining how to recover the passwords for the prefix and hybrid approaches in practical time remains a problem.

3   Password Recovery for Prefix-MD4

This section explains how to recover a password using the MD4($P||C$) approach. In our attack, up to 8 characters are recovered by 16 times the amount of eavesdropping, 16 times the number of queries, and less than 2^{35} off-line MD4 computations. In the case of 9 to 12 characters, we need 2^{10} times the amount of eavesdropping, 2^{10} times the number of queries, and less than 2^{40} off-line MD4 computations.

This attack takes a similar approach as that of the impossible differential attack, so we use a pair of challenges (C, C') that has specific differences. In the prefix approach, the bit position of C and C' after concatenated with a password depends on the password length. Therefore, to insert differences in a desired position, we need to know the password length in advance. In this section, we assume that the password length is already recovered by the password length recovery attack proposed by [15][2].

Sections 3.1 and 3.2 discuss the overall strategy of our attack against the prefix-MD4. Sections 3.3 to 3.6 explain how to recover up to 8 password characters. Finally, in section 3.7, we extend our attack to recover up to 12 password characters.

3.1   Analysis of Problems of Related Works

So far, the best attack against the prefix-MD4 was created by Wang et al. [15]. This method has two main problems.

1. The complexity of the yielding collision up to the second round (2^{35}) is impractical.
2. The complexity of analyzing a password by recovering internal chaining variables (2^{35}) is impractical.

To solve these problems, we use a different approach when analyzing a password. In this attack, we apply an impossible differential attack to recover the password of the prefix-MD4. This enables us to use a short local collision instead of a full-collision, and thus, the complexity of the attack becomes practical.

3.2   Overall Strategy

The high-level overview of the attack is as follows.

1. **Determining ΔC** (Section 3.3):
   Determine differences of challenges ΔC such that a local collision occurs in the first round. If a local collision occurs, differences in the latter step of the second round are fixed or very limited.
2. Generate (C, C') such that $C' - C = \Delta C$. Obtain their responses R and R'.
3. **Backward difference tracing** (Section 3.4):
   Inversely compute the differences of the intermediate chaining variables from R and R'. Here, we exhaustively guess m_1, which is a part of the password.
4. **Matching decision** (Section 3.5):
   If the guess is accurate, the inversely computed differences will match the differences fixed in step 1, and we can thus determine the accuracy of the guess for m_1.

---

[2] Password length recovery attack procedure is described in Appendix A.

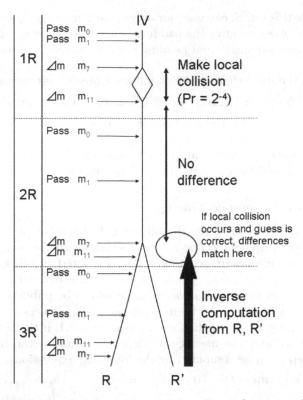

- In the prefix-MD4, an 8-character password is allocated in m_0 and m_1.
- Δm_7 and Δm_{11} make a local collision in 1R with the probability of 2^{-4}. If a local collision occurs, there is no difference until m_7 appears in 2R. On the other hand, intermediate values are inversely computed from R and R'. We exhaustively guess m_1. For each guess, inverse computation is carried out until an unknown m_0 appears in 3R. Finally, by comparing these differences, we determine the accuracy of the guess.

Fig. 2. Overall strategy: impossible differential attack on prefix-MD4

This concept is also described in Figure 2. In Figure 2, we use the fact that message differences are located in m_7 and m_{11} for the sake of simplicity. A detailed strategy for identifying message differences is introduced in section 3.3

3.3   Determining ΔC

When we determine message differences, the following two characteristics are considered.

- A local collision occurs in the first round with a high probability.
- A no-difference state will continue until the latter step of the second round.
  (See Table 2 in order to check the message order in the second round.)

Moreover, since this attack can work for only 1-block messages[3], we need to make certain that the message after the padding is a 1 block message; 512 bits. Considering the password length and padding rules, differences can only be present in m_2 - m_{13}.

Considering the above characteristics, following are the best message differences.

$$\Delta m_7 = \pm 2^j, \Delta m_{11} = \mp 2^{j+19}, \Delta m_i = 0 \text{ for other } i$$

The above differences give the local collision in the first round a probability of 2^{-4}. The value of j is flexible. If both j and $j+19$ are not MSB in each byte, we can construct challenges with only ASCII characters. In the end, we can expect to obtain a local collision by trying 16 challenge pairs.

3.4   Backward Difference Tracing

Backward difference tracing inversely computes the values of intermediate chaining variables from responses R and R'. Let the hash value be (aa_0, bb_0, cc_0, dd_0). Output chaining variables in step 48 are then computed as follows.
$a_{48} = aa_0 - a_0$, $b_{48} = bb_0 - b_0$, $c_{48} = cc_0 - c_0$, $d_{48} = dd_0 - d_0$.
To perform these expressions, (a_0, b_0, c_0, d_0) needs to be public IV. This attack therefore succeeds if and only if the message is a 1-block message.

From the step updating function shown in section 2.1, if the values of output chaining variables and the message in step i are known, the input chaining variables in step i can be computed by the following expressions.

$a_i = (b_{i+1} \ggg s_i) - m_k - t_i - f(c_{i+1}, d_{i+1}, a_{i+1})$, $b_i = c_{i+1}$,
$c_i = d_{i+1}$, $d_i = a_{i+1}$.

In MD4, unknown passwords m_0 and m_1 are used in steps 33 and 41, respectively, in the third round. Here, we exhaustively guess the value of m_1. Since m_1 is 32-bit, the number of guesses is at most 2^{32}. For each guess of m_1, backward tracing is carried out until step 34. By applying this process to R and R', we compute the values of $(a_{33}, b_{33}, c_{33}, d_{33})$ and $(a'_{33}, b'_{33}, c'_{33}, d'_{33})$. Moreover, considering that m_0 has no difference, we can inversely compute $\Delta a_{32} = a_{32} - a'_{32}$ even though we cannot determine the values of a_{32} and a'_{32}. Since $c_{33} = b_{32}, d_{33} = c_{32} = b_{31}, a_{33} = d_{32} = c_{31} = b_{30}$, the values we can obtain by backward difference tracing are $b_{32}, b'_{32}, b_{31}, b'_{31}, b_{30}, b'_{30}$, and Δb_{29}[4]. ($\Delta b_{32}, \Delta b_{31}, and \Delta b_{30}$ can also be computed.)

3.5   Matching Decision

The matching decision determines whether the differences computed by backward difference tracing will follow the differential path where a local collision occurred. The message differences are $(\Delta m_7 = 2^j, \Delta m_{11} = -2^{j+19})$, which are used in steps 30 and 31, respectively, in the second round. Therefore, if a local collision in the first round is archived, the following expression is guaranteed.

---

[3] This restriction is caused by backward difference tracing. See section 3.4 for details.
[4] If a guess of m_1 changes, the result of backward difference tracing also completely changes.

$$\Delta a_{29} = \Delta b_{29} = \Delta c_{29} = \Delta d_{29} = 0,$$
$$\Delta b_{30} = (\Delta a_{29} + \Delta f(b_{29}, c_{29}, d_{29}) + \Delta m_7) \lll 5 = (0 + 0 + 2^j) \lll 5 = 2^{j+5}.$$

If Δb_{29} and Δb_{30} obtained by backward difference tracing are 0 and 2^{j+5}, respectively, we can determine the guess of m_1 is correct, so m_1 is recovered.

Proof
The probability of generating a local collision is 2^{-4} and of making a correct guess is 2^{-32}. Therefore, with a probability of 2^{-36}, we succeed in a matching decision with a correct guess. If the guess is wrong, the matching decision succeeds with a probability of only 2^{-64}. Due to this huge gap, we can recover the correct password.

After m_1 is recovered using this method, we recover m_0 by an exhaustive search.

3.6   Algorithm for Recovering Eight Characters

We summarize the algorithm of password recovery as follows.

1. Eavesdrop a pair of C and R.
2. Generate $C' \leftarrow C + \Delta C$.
3. Send C', then obtain corresponding R'.
4. `for` $(guess_{m1} = 0$ to $\text{0xffffffff})$ {
5.   From R and R', compute Δb_{29} and Δb_{30} by backward difference tracing.
6.   `if` $(\Delta b_{29} = 0 \wedge \Delta b_{30} = 2^{j+5})$ {
7.   Password for m_1 is $guess_{m1}$. `goto` line 10.
8.   }
9. }
10. `if` (m_1 is recovered?) {
11.   Exhaustively search m_0. Then, halt this algorithm.
12. } `else` {
13.   Local collision did not occur. `goto` line 1; repeat this algorithm.
14. }

In this attack, since the local collision occurs with a probability of 2^{-4}, we need to try 16 pairs of C and C'. Therefore, 16 times the amount of eavesdropping and 16 times the number of queries are necessary[5]. For each R and R', the dominant off-line complexity is 2^{32} computation of the backward difference tracing from steps 41 to 33. Considering MD4 consists of 48 steps and we try sixteen R and sixteen R', the total complexity is $32 \times 2^{32} \times 9/48 = $ less than 2^{35} MD4 computations.

---

[5] If eavesdropped C has an inappropriate length for the attack, for example, C is longer than 1 block, we need to make both C and C'. In this case, 32 queries are necessary.

3.7 Extension to Recovering Twelve Characters

When a password is 12 characters, m_2 is fixed as well as m_0 and m_1. The attack procedure is similar to that used in the case of 8 characters.

Determining ΔC: We use the same ΔC as those in the case of 8 characters.

Backward difference tracing: m_0, m_1 and m_2 are used in steps 33, 41 and 37 respectively in the third round. Therefore, by exhaustively guessing the value of m_1, backward tracing can be performed until step 37. As a result, the values of $b_{36}, b'_{36}, b_{35}, b'_{35}, b_{34}, b'_{34}$, and Δb_{33} are computed from R and R'.

Matching decision: As explained in section 3.5, if a local collision occurs, $\Delta a_{29}, \Delta b_{29}, \Delta c_{29}$ and Δd_{29} are certain to be 0, and Δb_{30} is 2^{j+5}. This significantly limits the number of possible differences in the following few steps. As a result of our analysis, we found that when the value of j is fixed to 12, Δb_{33} and Δb_{34} have the following difference with a probability of 2^{-6}. Details are shown in Appendix B. Here, * means that the signs '+' and '-' do not have any influence.

$$\Delta b_{33} = *2^{j+8} * 2^{j+31}, \qquad \Delta b_{34} = *2^{j+5} * 2^{j+8} * 2^{j+14} * 2^{j+17}.$$

If both Δb_{33} and Δb_{34} computed by backward difference tracing have the above differences, we can determine that the guess of m_1 is correct.

Proof
The probability of generating a local collision in the first round is 2^{-4} and of making a correct guess is 2^{-32}. If both occur with a probability of 2^{-36}, Δb_{33} and Δb_{34} have the above differences with a probability of 2^{-6}. Overall, with a probability of 2^{-42}, we succeed in obtaining a matching decision with a correct guess. If the guess is wrong, the probability that both Δb_{33} and Δb_{34} have the above differences is 2^{-64}. Due to the gap of 2^{-42} and 2^{-64}, we can say that if the matching decision succeeds, differences propagate in the same way as we expected, so a local collision occurs and the guess of m_1 is correct. Consequently, password m_1 is recovered.

Complexity analysis: Since the local collision occurs with a probability of 2^{-4} and both Δb_{33} and Δb_{34} have expected differences with a probability of 2^{-6}, we need 2^{10} times the amount of eavesdropping and 2^{10} times the number of queries. For each R and R', the dominant off-line complexity is 2^{32} computation of the backward difference tracing from steps 37 to 33. Therefore, the total complexity is $2 \times 2^{10} \times 2^{32} \times 5/48 =$ less than 2^{40} MD4 computations.

4 Password Recovery for MD4($P||C||P$)

This section explains the password recovery attack against the hybrid-MD4. We assumed that the same password is concatenated both before and after the challenge.

Table 3. Challenge length and password position for hybrid-MD4

Composition of message							
P_1 P_2	Challenge			P_1 P_2	Padding		
↓ ↓				↓ ↓			
m_0 m_1 m_2 m_3	m_4 m_5 m_6 m_7			m_8 m_9 m_{10} m_{11}	m_{12} m_{13} m_{14} m_{15}		

Positions of password in 3R		
P_1 P_1	P_2 P_2	
↓ ↓	↓ ↓	
m_0 m_8 m_4 m_{12} m_2 m_{10} m_6 m_{14}	m_1 m_9 m_5 m_{13} m_3 m_{11} m_7 m_{15}	

The recovering procedure is basically the same as that for the prefix-MD4. To recover the password, we need to know the password length in advance. However, in the hybrid approach, there is no efficient method for recovering the password length. As a result, we first run an exhaustive search for up to 4 characters. If the password is not recovered, we run the procedure for recovering 5 characters. If we fail, we increase the password length by 1 character and run the recovery procedure. This section explains the recovering procedures for 8 characters.

4.1   Determining Challenge Length and Message Differences

Let an 8-characters password be P, the first 4 characters of P be P_1, and the latter 4 characters of P be P_2, so $P = (P_1 \| P_2)$. In the hybrid-MD4, P before the challenge makes $m_0 = P_1$ and $m_1 = P_2$. However, the location of P after the challenge is not fixed. Its location depends on the challenge length. We determine the challenge length so that backward difference tracing can go back as many steps as possible. With this strategy, challenge length is determined to be $m_2 - m_7$, as shown in Table 3.

Since the challenge string finishes at m_7, message differences can be present only for m_2 - m_7. We therefore choose message difference ($\Delta m_3 = 2^j$, $\Delta m_7 = -2^{j+19}$), which generate a local collision in the first round with a probability of 2^{-4}.

4.2   Backward Difference Tracing and Matching Decision

The backward difference tracing procedure is the same as that for the prefix-MD4. We exhaustively guess the 32-bit value P_2, and for each guess, inversely compute values of intermediate chaining variables until step 34. As a result, we obtain the value of $b_{33}, b'_{33}, b_{32}, b'_{32}, b_{31}, b'_{31}$, and Δb_{30}.

On the other hand, if a local collision occurs, $\Delta m_3 = 2^j$ in step 29 gives $\Delta b_{29} = 2^{j+3}$ a probability of 1, and the possible form of Δb_{30} and Δb_{31} is very limited. As a result of our analysis, we found that when the value of j is fixed to 12, Δb_{30} and Δb_{31} have the following difference with a probability of 2^{-4}. Details are shown in Appendix C.

$$\Delta b_{30} = -2^{j+24}, \qquad \Delta b_{31} = 0$$

Table 4. Results of our attacks

Attack target	Number of eavesdropping	Number of queries	Off-line complexity (Unit: MD4 computation)
Prefix-MD4 8 characters	16	16	Less than 2^{35}
Prefix-MD4 12 characters	2^{10}	2^{10}	Less than 2^{40}
Hybrid-MD4 8 characters	2^8	2^8	Less than 2^{39}

If both Δb_{30} and Δb_{31} computed by backward difference tracing have the above differences, we can determine the guess of P_2 is correct. (The proof is the same as that for the prefix-MD4 so it is omitted here.)

After P_2 is recovered, we exhaustively guess the value of P_1. Finally, 8 password characters of the hybrid-MD4 are recovered.

4.3   Complexity Analysis

Since the local collision occurs with a probability of 2^{-4} and both Δb_{30} and Δb_{31} have expected differences with a probability of 2^{-4}, we need 2^8 times the amount of eavesdropping and 2^8 times the number of queries. For each R and R', the dominant off-line complexity is 2^{32} computation of the backward difference trace from steps 42 to 34. Overall, the complexity is $2 \times 2^8 \times 2^{32} \times 9/48 =$ less than 2^{39} MD4 computations.

5   Discussion and Conclusion

Summary of proposed attacks
This paper proposed password recovery attacks against the prefix- and hybrid-MD4. The attack against the prefix-MD4 recovers up to 12 password characters and the attack against the hybrid-MD4 recovers up to 8 password characters. Their complexity is summarized in Table 4.

The critical idea behind our attack is applying an impossible differential attack to recover the password of the prefix-MD4. This enables us to use a short local collision that occurs with a probability of 2^{-4}, and thus the complexity becomes practical.

Countermeasures and consideration of hash function design
First, we propose countermeasures for our attack.

- The simple solution is replacing MD4 with strong hash functions.
- Since our attack works if and only if a message is a 1-block message, only allowing challenges that are longer than 1 block is effective.
- If the password length is longer than the recoverable limit, our attack cannot recover even one character. Therefore, using passwords longer than 12 characters for the prefix-MD4 and 8 characters for the hybrid-MD4 can prevent our attack.

Table 5. Message index for each step in MD5

1R (steps 1-16)	0	1	2	3	4	5	6	7	8	9	10	11	12	13	14	15
2R (steps 17-32)	1	6	11	0	5	10	15	4	9	14	3	8	13	2	7	12
3R (steps 33-48)	5	8	11	14	1	4	7	10	13	0	3	6	9	12	15	2
4R (steps 49-64)	0	7	14	5	12	3	10	1	8	15	6	13	4	11	2	9

From our attacks, we have developed ideas regarding good hash function design.

- Our attack does not use full-collision, but instead uses local collision. Therefore, considering only the collision resistance of the hash function is not enough to discuss the security of protocols. Construction that prevents a practical local collision is therefore important.
- Our attack uses a characteristics by which the differential path up to the intermediate step can be controlled with a high probability. (We did not attempt controlling the differential path for all the steps. Actually, our attack disregards what differences exist in R and R'.) Considering this, the step updating function must avoid the partial differential path that occurs with a high probability. (A local collision can be used to achieve a long partial differential path.)
- Backward difference tracing can reduce the security of hash functions. In authentication protocols using a password, the bit length of a password tends to be short, and this makes the attack against the protocol easier. To avoid backward difference tracing, message expansion is important. Message expansion should expand short secret bits to various parts of messages.

Possibility of attacks against prefix-MD5 and hybrid-MD5
We conclude this paper by discussing attacks against the prefix- and hybrid-MD5. Analysis on the prefix-MD5 is important since it is practically used, for example [13]. MD5 consists of 64 steps. Table 5 shows the message index of MD5.

Prefix-MD5: An interesting observation is that m_0 in the fourth round is located in the initial step of the fourth round. Therefore, similar to MD4, backward difference tracing is effectively applied by exhaustively guessing m_1.

Hybrid-MD5: Another interesting observation is that the MD5 message order is also suitable for attacking the hybrid-MD5. If a challenge will locate in m_2-m_6, the password before the challenge becomes (m_0, m_1), and the password after the challenge becomes (m_7, m_8). Therefore, by exhaustively guessing $m_1(= m_8)$, backward tracing is effectively performed.

In MD5, the presence of an effective short local collision is not yet known; however, it may be found if the analysis technique is improved. Therefore, use of the prefix- and hybrid-MD5 approaches need to be carefully considered.

References

1. Biham, E., Biryukov, A., Dunkelman, O., Richardson, E., Shamir, A.: Initial Observations on Skipjack: Cryptanalysis of Skipjack-3XOR. In: Tavares, S., Meijer, H. (eds.) SAC 1998. LNCS, vol. 1556, pp. 362–376. Springer, Heidelberg (1999)
2. Biham, E., Biryukov, A., Shamir, A.: Cryptanalysis of Skipjack Reduced to 31 Rounds using Impossible Differentials, Technical Report CS0947, Technion - Computer Science Department (1998), http://www.cs.technion. ac.il/~biham/Reports/SkipJack.txt
3. Contini, S., Yin, Y.L.: Forgery and partial key-recovery attacks on HMAC and NMAC using hash collisions. In: Lai, X., Chen, K. (eds.) ASIACRYPT 2006. LNCS, vol. 4284, pp. 37–53. Springer, Heidelberg (2006)
4. Fouque, P.-A., Leurent, G., Nguyen, P.: Full Key-Recovery Attacks on HMAC/NMAC-MD4 and NMAC-MD5. In: Menezes, A. (ed.) CRYPTO 2007. LNCS, vol. 4622, pp. 15–30. Springer, Heidelberg (2007)
5. Kaliski Jr, B.S., Robshaw, M.J.B.: Message authentication with MD5. Crypto-Bytes 1(1), 5–8 (1995)
6. Leurent, G.: Message Freedom in MD4 and MD5 Collisions: Application to APOP. In: Biryukov, A. (ed.) FSE 2007. LNCS, vol. 4593, pp. 309–328. Springer, Heidelberg (2007)
7. Rivest, R.L.: The MD4 Message-Digest Algorithm, RFC 1320 (April 1992), http://www.ietf.org/rfc/rfc1320.txt
8. Rivest, R.L.: The MD4 Message Digest Algorithm. In: Menezes, A., Vanstone, S.A. (eds.) CRYPTO 1990. LNCS, vol. 537, pp. 303–311. Springer, Heidelberg (1991)
9. Myers, J., Rose, M.: Post Office Protocol - Version 3, RFC 1939, (Standard). Updated by RFCs 1957, 2449. (May 1996), http://www.ietf.org/rfc/rfc1939.txt
10. Preneel, B., van Oorschot, P.C.: On the Security of Two MAC Algorithms. In: Maurer, U.M. (ed.) EUROCRYPT 1996. LNCS, vol. 1070, pp. 19–32. Springer, Heidelberg (1996)
11. Rechberger, C., Rijmen, V.: On Authentication with HMAC and Non-Random Properties, Cryptology ePrint Archive, Report 2006/290, http://eprint.iacr. org/2006/290.pdf
12. Sasaki, Y., Yamamoto, G., Aoki, K.: Practical Password Recovery on an MD5 Challenge and Response. Cryptology ePrint Archive, Report 2007/101, http://eprint.iacr.org/2007/101.pdf
13. Simpson, W.: PPP Challenge Handshake Authentication Protocol (CHAP), RFC 1994, Updated by RFC 2484, (August 1996), http://www.ietf.org/rfc/ rfc1994.txt
14. Tsudik, G.: Message Authentication with One-Way Hash Functions. ACM Computer Communication Review 22(5), 29–38 (1992)
15. Wang, L., Ohta, K., Kunihiro, N.: Password Recovery Attack on Authentication Protocol MD4(Password||Challenge). In: ASIACCS 2008 (to appear, 2008)
16. Wang, X., Lai, X., Feng, D., Chen, H., Yu, X.: Cryptanalysis of the Hash Functions MD4 and RIPEMD. In: Cramer, R.J.F. (ed.) EUROCRYPT 2005. LNCS, vol. 3494, pp. 1–18. Springer, Heidelberg (2005)
17. Wang, X., Yu, H.: How to Break MD5 and Other Hash Functions. In: Cramer, R.J.F. (ed.) EUROCRYPT 2005. LNCS, vol. 3494, pp. 19–25. Springer, Heidelberg (2005)
18. Wang, X., Yu, H., Yin, Y.L.: Efficient Collision Search Attacks on SHA-0. In: Shoup, V. (ed.) CRYPTO 2005. LNCS, vol. 3621, pp. 1–16. Springer, Heidelberg (2005)
19. Wang, X., Yin, Y.L., Yu, H.: Finding Collisions in the Full SHA-1. In: Shoup, V. (ed.) CRYPTO 2005. LNCS, vol. 3621, pp. 17–36. Springer, Heidelberg (2005)

A   Password Length Recovery for Prefix Approach

Wang et al. showed that the password length of the prefix approach can be recovered with a very small number of queries [15]. This attack works against hash functions that attach message length as padding string and compute the hash value by iteratively updating initial values, for example, MD5, SHA-1, and SHA-2. At first, the attacker eavesdrops challenge C and response R. Then, the attacker makes a guess at the password length. Here, the attacker can check the accuracy of the guess by only one query.

To recover the password length, the padding part must be carefully considered. An important characteristic is that the padding string is dependent on only the message length, not the message itself.

The recovering procedure is shown in Table 6. Here, we denote a computation for a hash value of M by using hash function H and initial value A by $H(A, M)$. We also denote the computation of compression function h for processed message m and initial value A by $h(A, m)$.

For line 5 of the above procedure, if the guess is correct, $(\mathsf{Pass}||C||\mathsf{Pad}_1)$ becomes the end of the block, and X becomes another block. Since the output of the compression function for $(\mathsf{Pass}||C||\mathsf{Pad}_1)$ is exactly R, R' should be equal to $h(R, (X||\mathsf{Pad}_2))$. Figure 3 shows the behavior of the hash computation when the guess is correct.

Finally, the password length is recovered. To check the accuracy of n guesses, one times the amount of eavesdropping and n times the number of queries is needed.

This attack takes a similar approach as that used in the "Extension Attack" that was mentioned in [5] by Kaliski and Robshaw. The extension attack is used for forging MAC. On the other hand, the password length recovery attack focuses on the property in which the padding string is dependent on only message length, and the password length is recovered by the chosen message attack.

Remarks

This method can be applied to the prefix approach, but not to the suffix and

Table 6. Algorithm for recovering password length for prefix approach

---

1. Eavesdrop a pair of C and $R = H(IV, (\mathsf{Pass}||C))$.
2. Determine L, which is a guess at the password length.
3. Based on L, identify the padding string for $(\mathsf{Pass}||C)$. Let this be Pad_1.
4. Generate C' so that $C' = (C||\mathsf{Pad}_1||X)$ where X is any string except for $\mathtt{Null}$.
5. Send C' and obtain $R' = H(IV, (\mathsf{Pass}||C')) = H(IV, (\mathsf{Pass}||C||\mathsf{Pad}_1||X))$.
6. Compute the padding string for $(\mathsf{Pass}||C||\mathsf{Pad}_1||X)$. Let this be (Pad_2).
7. Locally computes $h(R, (X||\mathsf{Pad}_2))$, and check if it matches with R'.
8. If they are matched, the guess is right, and halt this procedure. Otherwise change L and goto line 3.

Table 7. Differential path of matching part for prefix-MD4

Step 30	$(\Delta a_{29}, \Delta b_{29}, \Delta c_{29}, \Delta d_{29})$	$(0,0,0,0)$
	$\Delta MAJ(b_{29}, c_{29}, d_{29})$	0
	Δm_7	2^j
	s_{30}	5
Step 31	$(\Delta a_{30}, \Delta b_{30}, \Delta c_{30}, \Delta d_{30})$	$(0, 2^{j+5}, 0, 0)$
	$\Delta MAJ(b_{30}, c_{30}, d_{30})$	0
	Δm_{11}	-2^{j+19}
	s_{31}	9
Step 32	$(\Delta a_{31}, \Delta b_{31}, \Delta c_{31}, \Delta d_{31})$	$(0, -2^{j+28}, 2^{j+5}, 0)$
	$\Delta MAJ(b_{31}, c_{31}, d_{31})$	0
	Δm_{15}	0
	s_{32}	13
Step 33	$(\Delta a_{32}, \Delta b_{32}, \Delta c_{32}, \Delta d_{32})$	$(0, 0, -2^{j+28}, 2^{j+5})$
	$\Delta XOR(b_{32}, c_{32}, d_{32})$	$*2^{j+5} * 2^{j+28}$
	Δm_0	0
	s_{33}	3
Step 34	$(\Delta a_{33}, \Delta b_{33}, \Delta c_{33}, \Delta d_{33})$	$(2^{j+5}, *2^{j+8} * 2^{j+31}, 0, -2^{j+28})$
	$\Delta XOR(b_{33}, c_{33}, d_{33})$	$*2^{j+8} * 2^{j+28} * 2^{j+31}$
	Δm_8	0
	s_{34}	9
Step 35	$(\Delta a_{34}, \Delta b_{34}, \Delta c_{34}, \Delta d_{34})$	$(-2^{j+28}, *2^{j+5} * 2^{j+8} + 2^{j+14} * 2^{j+17},$ $*2^{j+8} * 2^{j+31}, 0)$

hybrid approaches. In line 4 of the above procedure, string X is added to the tails of messages. However, if we add X in the suffix and hybrid approaches, the password is moved to before X, and we thus cannot keep the message before X unchanged.

B   Differential Path of Matching Part for Prefix-MD4

Table 7 shows the differential path that $\Delta b_{33} = *2^{j+8} * 2^{j+31}$, and $\Delta b_{34} = *2^{j+5} * 2^{j+8} + 2^{j+14} * 2^{j+17}$. The probability that this differential path holds is 2^{-4}. The analysis is as follows.

1. $\Delta b_{30} = 2^{j+5}$ must not have carry. This succeeds with a probability of $1/2$.

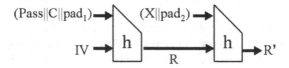

Fig. 3. Behavior when guess is correct

2. In step 31, ΔMAJ must be 0. This succeeds with a probability of $1/2$.
3. $\Delta b_{31} = -2^{j+28}$ must not have carry. This succeeds with a probability of $1/2$.
4. In step 32, ΔMAJ must be 0. This succeeds with a probability of $1/4$.
5. $\Delta b_{33} = *2^{j+8} * 2^{j+31}$ must not have carry. This succeeds with a probability of $1/4$.

If we choose $j = 12$, $\Delta b_{31} = (-2^{j+19}) << 9 = -2^{j+28}$ never has carry since -2^{j+19} becomes MSB. Therefore, the total probability of this differential path is 2^{-6}.

C Differential Path of Matching Part for Hybrid-MD4

Table 8 shows the differential path that $\Delta b_{30} = -2^{j+24}$, and $\Delta b_{31} = 0$. The probability that this differential path holds is 2^{-4}. The analysis is as follows.

1. $\Delta b_{29} = 2^{j+3}$ must not have carry. This succeeds with a probability of $1/2$.
2. In step 30, ΔMAJ must be 0. This succeeds with a probability of $1/2$.
3. $\Delta b_{30} = -2^{j+24}$ must not have carry. This succeeds with a probability of $1/2$.
4. In step 31, ΔMAJ must be 0. This succeeds with a probability of $1/4$.

If we choose $j = 12$, $\Delta b_{30} = (-2^{j+19}) << 5 = -2^{j+24}$ never has carry since -2^{j+19} becomes MSB. Therefore, the total probability of this differential path is 2^{-4}.

Table 8. Differential path of matching part for hybrid-MD4

Step 29	$(\Delta a_{28}, \Delta b_{28}, \Delta c_{28}, \Delta d_{28})$	$(0,0,0,0)$
	$\Delta MAJ(b_{28}, c_{28}, d_{28})$	0
	Δm_3	2^j
	s_{29}	3
Step 30	$(\Delta a_{29}, \Delta b_{29}, \Delta c_{29}, \Delta d_{29})$	$(0, 2^{j+3}, 0, 0)$
	$\Delta MAJ(b_{29}, c_{29}, d_{29})$	0
	Δm_7	-2^{j+19}
	s_{30}	5
Step 31	$(\Delta a_{30}, \Delta b_{30}, \Delta c_{30}, \Delta d_{30})$	$(0, -2^{j+24}, 2^{j+3}, 0)$
	$\Delta MAJ(b_{30}, c_{30}, d_{30})$	0
	Δm_{11}	0
	s_{31}	9
Step 32	$(\Delta a_{31}, \Delta b_{31}, \Delta c_{31}, \Delta d_{31})$	$(0, 0, -2^{j+24}, 2^{j+3})$

How (Not) to Efficiently Dither Blockcipher-Based Hash Functions?

Jean-Philippe Aumasson[1],* and Raphael C.-W. Phan[2],**

[1] FHNW, 5210 Windisch, Switzerland
[2] Loughborough Uni, LE11 3TU Leics, UK

Abstract. In the context of iterated hash functions, "dithering" designates the technique of adding an iteration-dependent input to the compression function in order to defeat certain generic attacks. The purpose of this paper is to identify methods for dithering blockcipher-based hash functions that provide security bounds and efficiency, contrary to the previous proposals. We considered 56 different constructions, based on the 12 secure PGV schemes. Proofs are given in the blackbox model that 12 of them preserve the bounds on collision and inversion resistance given by Black et al. These 12 schemes avoid the need for short dither values, induce negligible extra-computation, and achieve security independent of the dither sequence used. We also identify 8 schemes that lead to strong compression functions but potentially insecure hash functions. Application of our results can be considered to popular hash functions like SHA-1 or Whirlpool.

1 Introduction

The idea of making hash functions out of blockciphers goes back to 1978, when Rabin [40] proposed to hash $(m_1, \ldots, m_\ell)$ as $\mathsf{DES}_{m_\ell}(\ldots(\mathsf{DES}_{m_1}(IV)\ldots)$. Subsequent works devised less straightforward schemes, with either one or two calls to the blockcipher within a compression function [30, 33, 37, 39, 28]. In 1993 research went a step further when Preneel *et al.* [38] conducted a systematic analysis of all 64 compression functions of the form $f(h, m) = E_K(P) \oplus F$, for $K, P, F \in \{m, h, m \oplus h, v\}$, where v is a constant. They showed that only 4 of these schemes resist all considered vulnerabilities, and 8 others just have the non-critical attribute of easily found fixed-points. A decade later, Black et al. [9] proved the security of hash functions based on these 12 PGV schemes in the blackbox model.

Like a majority of hash functions blockcipher-based hash functions follow the Merkle-Damgård (MD) paradigm [16,31]. Recent generic attacks [17,22,24,23,20] that exploit its structure led to proposals to extend the basic MD construction. These include the idea of *dithering*, i.e. adding an input (the *dither*) to the

---

* Supported by the Swiss National Science Foundation under project number 113329.
** Work done while the author was with the Security & Cryptography Lab (LASEC), EPFL, Lausanne, Switzerland.

S. Vaudenay (Ed.): AFRICACRYPT 2008, LNCS 5023, pp. 308–324, 2008.

compression function, whose value depends on the iteration count. The goal is to defeat attacks based on message block repetitions (like [17, 24]).

Proposals of dither sequences came from Kelsey and Schneier [24] (using a counter), from Biham and Dunkelman [5, 6] (as part of the HAIFA framework, using the number of bits hashed so far), and from Rivest [41] (using an abelian square-free sequence). However, the method proposed [6, 41] for integrating the dither value into concrete hash functions is inefficient, in the sense that it increases the number of calls to the compression function. This method indeed consists in reducing the effective size of a message block to make way for the dither, i.e. filling the dither into the space freed up. This motivated Rivest's proposal to use short dithers (2-byte) encoding particular patterns over a small alphabet. Another drawback of this method is that system parameters have to be modified such that message chunks become, for example, 448 bits long instead of 512 with a 64-bit counter, or 496 with Rivest's method. It thus seems valuable to explore generic dithering methods that preserve efficiency and system parameters, and that are still simple to apply.

1.1 Contribution

We will be concerned with the problem of constructing dithered compression functions from blockcipher-based schemes, grounding our work on the 12 secure PGV schemes. We first introduce 56 dithered variants, along with security definitions adapted for dithered functions. Our blackbox analysis singles out 12 dithered schemes leading to hash functions as secure as the original (undithered) ones, as far as collision and inversion resistance are concerned. The bounds given are *independent* of the dither sequence use, contrary to 32 other constructions. A counter-intuitive fact is proven, that 8 out of 56 dithered schemes lead to hash functions which are *not collision resistant* when the dithering method of HAIFA [6] is used, despite having a collision-resistant compression function. This re-opens the suitability issue of the Merkle-Damgård theorem for dithered hash functions, and suggests that a careful revisit is required.

We emphasize that our results say nothing on the resistance to generic second-preimage attacks as [17, 24] that dithering aims at preventing. The resistance to these attacks depends on the dither sequence used, whereas our point is to show that previously known security bounds can hold as well when dithering is used, independently of the sequence chosen.

Apart from our formal security analysis, the interest of our constructions is twofold: Firstly, they are efficient, because the number of calls to the compression function is no longer increased by dithering; secondly, they allow dither values of arbitrary length (up to the size of the key of the blockcipher), with no performance penalty. As a result, a counter supporting large messages can now be used. More generally, it avoids the need for short dither value with non-trivial patterns like [41], which in addition provides fewer security guarantees than a counter (see [12]).

1.2 Related Work

After a very calm period during the 90s, the results of [9] seem to have triggered a regain of interest for blockcipher-based hashing: In 2005 Black et al. [8] proved that a compression function of the form $f_2(h_{i-1}, m_i, E_K(f_1(h_{i-1}, m_i)))$ cannot be provably secure with respect to E_K. This result has been recently extended by Rogaway and Steinberger [43], who proved generic upper bounds on the security of permutation-based hash functions.

Along the same lines, combinations of fixed permutations were previously studied by Shrimpton and Stam [45]. Another impossibility result is due to Boneh and Boyen [11] for hash functions combiners, later generalized by Pietrzak [35]. In [29], Lee et al. extend the [9] results to 22 other constructions, using similar blackbox proofs, and in [46] Stam simplifies the [9] proofs. In [26], Knudsen and Rijmen study known-key distinguishers for blockciphers; though unrealistic for attacking encryption primitives, this scenario can be relevant for blockcipher-based hashing (they show near-collisions for Matyas-Meyer-Oseas, see Appendix B).

More concretely, the recent hash functions Maelstrom [19] and Grindahl [27] are based on AES, and blockcipher-based designs remain a promising alternative for several researchers (e.g. [25]). The NIST hash competition may also mark a revival of hash functions built on blockciphers.

Stream-cipher-based hash function attracted less attention. They offer a less confortable framework because (1) they are generally not defined to operate over "blocks", (2) until now they have been less reliable than blockciphers, and (3) they often have a slow initialization. A counter-example is Bernstein's compression function Rumba [3] is based on the stream cipher Salsa20. We can also cite [14], based on RC4.

Fewer works have been produced about dithering. We can cite Shoup's construction [44] for universal one-way hash functions, which can be seen as a kind of dithering (a sequence of values called the "schedule" is input through iterations, see also [32]). More recently, Bouillaguet et al. [12] presented another generic second-preimage attack, slower than [24] in general, but performing slightly better when certain dither sequences are used. For etymological issues, see Appendix C.

1.3 Notations

We adopt the notations of [9], with only minor changes: A *blockcipher* is a map $E : \{0,1\}^\kappa \times \{0,1\}^\mu \mapsto \{0,1\}^\mu$, such that $E_k(\cdot) = E(k, \cdot)$ is a permutation on $\{0,1\}^\mu$ for all $k \in \{0,1\}^\kappa$, and its inverse is written E^{-1}. The set of all blockciphers with κ-bit key and μ-bit messages is denoted $\mathsf{Bloc}(\kappa, \mu)$. A *blockcipher-based hash function* is a map $H : \mathsf{Bloc}(\kappa, \mu) \times D \mapsto R$, where $D \subseteq \{0,1\}^\star$ and $R = \{0,1\}^n$, defined iteratively by a compression function $f : \mathsf{Bloc}(\kappa, \mu) \times \{0,1\}^{n_1} \times \{0,1\}^{n_2} \mapsto \{0,1\}^{n_2}$, where n_1 is the size of a message block, and n_2 the size of chaining values. In the remainder of the paper, we assume $\mu = n_1 = n_2 = n$. We write f_E (resp. H_E) to denote the compression (resp. hash) function instantiated with a particular E. Eventually, an *adversary* is an algorithm with oracle-access to E and E^{-1}—working within this setting is also known as analysis in the blackbox model, along with the assumption that each E_k is a random permutation. Furthermore, we set $h^+ = f(h, m)$ the output of a compression function. In the context of iterated hashing, blocks and chaining values are indexed as follows: h_0 is the IV, $\bar{m} = (m_1, \ldots, m_\ell)$ is the (encoded) message, $h_1 = f(h_0, m_1)$, and so on until $h_\ell = f(h_{\ell-1}, m_\ell) = H(\bar{m})$.

2   Dithering the PGV Schemes

The 12 PGV schemes $f_1, \ldots, f_{12}$ are depicted in Fig. 1 and 2: f_1 is the Matyas-Meyer-Oseas [30] construction (MMO), and one of the simplest schemes; f_3 is the Miyaguchi-Preneel construction, notably employed in Whirlpool [2], with as blockcipher a variant of Rijndael; f_5 is the Davies-Meyer construction, somehow the dual of MMO: Its structure is similar, except that the inputs of h and m play reversed roles. However, it has the undesirable attribute of easily found fixed-points; indeed, for an arbitrary m, choosing $h = E^{-1}{}_m(0)$ implies $f_5(h, m) = h$. The Davies-Meyer construction is used by the hash function Maelstrom-0 [19] (a variant of Whirlpool), and implicitly by some dedicated hash functions like MD5 and SHA-1.

We consider dithered versions obtained with a single input of the dither value d through an xor operation (in practice, it might be replaced by any easy-to-invert mapping which is a permutation for one of its inputs fixed). We suppose d non-null, and of convenient length (that is, not larger than its input slots in the blockcipher). The dithered PGV (dPGV) schemes are then classified into five subsets, describing the possible points for the dither d, see Table 1.

Table 1. Subsets of dithered PGV schemes.

Subset	Input point	Modification
$\mathcal{C}_1$	chaining value	$h \leftarrow h \oplus d$
$\mathcal{C}_2$	message block	$m \leftarrow m \oplus d$
$\mathcal{C}_3$	output	$h^+ \leftarrow h^+ \oplus d$
$\mathcal{C}_4$	key	$E_k \leftarrow E_{k \oplus d}$
$\mathcal{C}_5$	plaintext	$E_k(\cdot) \leftarrow E_k(\cdot \oplus d)$

We write $f_{i,j}$ for the dithered scheme obtained by applying the j-th transform to f_i; thus $f_{i,j} \in \mathcal{C}_j$, for all $(i, j) \in \{1, \ldots, 12\} \times \{1, \ldots, 5\}$. Clearly, $|\mathcal{C}_i| = 12$ for $i = 1, \ldots, 5$, but there are only 56 distinct dithered schemes, rather than 60, because $f_{1,1} \equiv f_{1,4}$, $f_{5,2} \equiv f_{5,4}$, $f_{9,1} \equiv f_{9,4}$, and $f_{10,2} \equiv f_{10,5}$.

A crucial observation is that almost all schemes of $\mathcal{C}_4$ and $\mathcal{C}_5$ are formally equivalent to a $\mathcal{C}_3$ scheme, up to variable renaming, e.g.

$$f_{4,4}(h, m, d) = E_{h \oplus d}(h \oplus m) \oplus m = E_{h'}(h' \oplus m') \oplus m' \oplus d = f_{4,3}(h', m', d),$$
$$f_{9,5}(h, m, d) = E_{h \oplus m}(m \oplus d) \oplus m = E_{h' \oplus m'}(m') \oplus m' \oplus d = f_{9,3}(h', m', d),$$

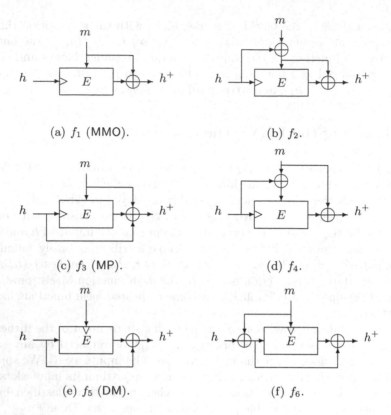

(a) f_1 (MMO).

(b) f_2.

(c) f_3 (MP).

(d) f_4.

(e) f_5 (DM).

(f) f_6.

Fig. 1. PGV schemes f_1 to f_6, where a hatch marks the key input (we assume keys and message blocks of same size, cf. §1.3)

for $h' = h \oplus d$, $m' = m \oplus d$. We denote $\mathcal{C}_3^+$ the set of schemes that can be expressed in $\mathcal{C}_3$ form. Only four members of $\mathcal{C}_4 \cup \mathcal{C}_5$ do not admit such rewriting, namely the ones equivalent to a scheme of $\mathcal{C}_1$ or $\mathcal{C}_2$. We thus have

$$\mathcal{C}_3^+ = (\mathcal{C}_3 \cup \mathcal{C}_4 \cup \mathcal{C}_5) \setminus \{f_{1,4}, f_{5,4}, f_{9,4}, f_{10,5}\}.$$

This set is used later for simplifying security proofs (we shall exploit the $\mathcal{C}_3$ structure for proving security bounds on $\mathcal{C}_3^+$ schemes). To summarize, we have $|\mathcal{C}_1 \cup \mathcal{C}_2| = 24$, $|\mathcal{C}_3^+| = 32$, and $(\mathcal{C}_1 \cup \mathcal{C}_2) \cap \mathcal{C}_3^+ = \emptyset$.

Note that if f_i admits easily found fixed-points (as do 8 of the 12 PGV schemes), then any dithered variant also possesses the property. For instance, $f_{5,5}$ admits the fixed-point $E^{-1}{}_m(0) \oplus d$, for any choice of m. It follows that exactly 37 among the 56 dithered schemes have trivial fixed-points.

To illustrate our constructions, Figure 3 depicts the dithered variants of MMO (f_1): For a given d, dMMO$_1$ is similar to MMO up to a reordering of the permutation indexes (more precisely, the h-th permutation takes $(h \oplus d)$ as new index). dMMO$_2$ simulates the undithered MMO for a blockcipher $E'(\cdot) = E(\cdot \oplus d)$, while dMMO$_3$ has simply the output of MMO xored with d, as in Shoup's method [44].

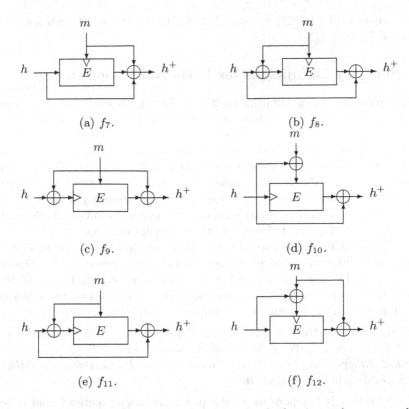

(a) f_7.

(b) f_8.

(c) f_9.

(d) f_{10}.

(e) f_{11}.

(f) f_{12}.

Fig. 2. PGV schemes f_7 to f_{10}, where a hatch marks the key input (we assume keys and message blocks of same size, cf. §1.3)

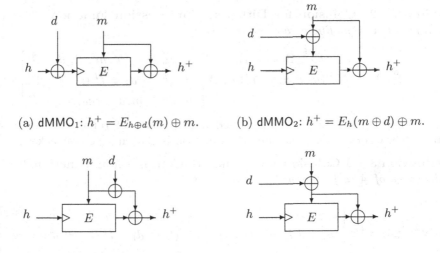

(a) dMMO$_1$: $h^+ = E_{h \oplus d}(m) \oplus m$.

(b) dMMO$_2$: $h^+ = E_h(m \oplus d) \oplus m$.

(c) dMMO$_3$: $h^+ = E_h(m) \oplus m \oplus d$.

(d) dMMO$_4$: $h^+ = E_h(m \oplus d) \oplus m \oplus d$.

Fig. 3. Dithered versions of the Matyas-Meyer-Oseas scheme

The structure of dMMO_4 is somewhat similar to the RMX transform for randomized hashing [21].

3   Security Definitions for Dithered Functions

We build on the formal definitions of [9] (recalled in Appendix A), extending them to the case of dithered functions: A collision for dithered compression functions where dithers are distinct is termed a Δ-collision—in essence, this is somewhat analogous to a free-start collision for hash functions, in the sense that here the dithers are distinct and public. Such a collision does not trivially translate into a collision for the derived hash function, mainly due to the MD-strengthening padding. We reserve the term *collision* to the case where a pair of inputs map to the same image with *same dither values*. This is in order to maintain the usefulness of the MD paradigm ported over to dithered hash functions.

In the following definitions, $\mathcal{A}$ is an adversary that has access to E and E^{-1}, and f is a dithered blockcipher-based compression function, $f : \mathsf{Bloc}(\kappa, n) \times \{0,1\}^n \times \{0,1\}^n \times \{0,1\}^k \mapsto \{0,1\}^n$, for some fixed $k > 0$. The IV of the hash function is an arbitrary constant h_0, introduced for considering collisions with the empty string. Furthermore, we introduce the following definition:

Definition 1 (Dither sequence). *A dither sequence is defined by a triplet* $(\mathcal{I}, \mathcal{D}, d)$, *where* $\mathcal{I} \subseteq \mathbb{N}$ *is the set of iteration indexes,* $\mathcal{D} \subseteq \{0,1\}^k$ *is the set of valid dither values, and d is a function* $\mathcal{I} \mapsto \mathcal{D}$ *returning the dither value corresponding to an iteration index.*

This definition is independent of the particular input method, and is relevant for any iterated hash function. In the remainder, we let $\delta = |\mathcal{D}| \leq 2^k$, and write d for a dither value[1].

Definition 2 (Collision for Dithered Compression Function). *The advantage of* $\mathcal{A}$ *in finding a collision in f is*

$$\mathbf{Adv}_f^{\mathsf{col}}(\mathcal{A}) = \Pr \left[\begin{array}{c} E \overset{\$}{\leftarrow} \mathsf{Bloc}(\kappa, \mu), \\ (h, m, d, h', m') \overset{\$}{\leftarrow} \mathcal{A} \end{array} \middle| \begin{array}{c} (h, m, d) \neq (h', m', d),\ d \in \mathcal{D}, \\ [f^E(h, m, d) = f^E(h', m', d) \\ or\ f^E(h, m, d) = h_0] \end{array} \right].$$

This notion of collision can be viewed as a variant of target-collision resistance [34], where the key indexing the function is chosen by the attacker.

Definition 3 (Δ-Collision for Dithered Compression Functions). *The advantage of* $\mathcal{A}$ *in finding a* Δ-collision in f is

$$\mathbf{Adv}_f^{\Delta\mathsf{col}}(\mathcal{A}) = \Pr \left[\begin{array}{c} E \overset{\$}{\leftarrow} \mathsf{Bloc}(\kappa, \mu), \\ (h, m, d, h', m', d') \overset{\$}{\leftarrow} \mathcal{A} \end{array} \middle| \begin{array}{c} (h, m, d) \neq (h', m', d'),\ (d, d') \in \mathcal{D}^2, \\ [f^E(h, m, d) = f^E(h', m', d') \\ or\ f^E(h, m, d) = h_0] \end{array} \right].$$

---

[1] The notation $\Pr[\alpha|\beta]$ stands here for the probability of the event β *after* the experiment α. This should not be confused with the notation of conditional probabilities.

The notion capturing one-wayness is termed as "inversion" rather than "preimage", merely because of the different sampling rule for the challenge image (see [9, Ap. B] for a discussion).

Definition 4 (Inversion for Dithered Compression Function). *The advantage of $\mathcal{A}$ in inverting f is*

$$\mathbf{Adv}_f^{\text{inv}}(\mathcal{A}) = \Pr \left[\begin{array}{c} E \xleftarrow{\$} \text{Bloc}(\kappa, \mu), \ h^+ \xleftarrow{\$} \text{Range}(f^E), \\ (h, m, d) \xleftarrow{\$} \mathcal{A} \end{array} \middle| \begin{array}{c} d \in \mathcal{D}, \\ f(h, m, d) = h^+ \end{array} \right].$$

Let **Adv** be any of the advantages defined above. For $q \geq 0$, we write $\mathbf{Adv}(q) = \max_{\mathcal{A}} (\mathbf{Adv}(\mathcal{A}))$, where the maximum is taken over all adversaries making at most q oracle queries. The definitions for hash functions apply as well for dithered hash functions, where a random blockcipher is used, and a given dither sequence is considered.

4   Collision Resistance

4.1   Blackbox Bounds

Theorem 1 (Collision Resistance of dPGV Hash Functions). *Let H be a hash function built on a dithered PGV scheme $f \notin \mathcal{C}_2$, where MD-strengthening is applied. Then the best advantage for a q-bounded adversary in finding collisions is*

$$\mathbf{Adv}_H^{\text{col}}(q) \leq \frac{q(q+1)}{2^n}, \ \text{for } f \in \mathcal{C}_1,$$

$$\mathbf{Adv}_H^{\text{col}}(q) \leq \frac{(\delta^2 + \delta)(q^2 + q)}{2^{n+1}}, \ \text{for } f \in \mathcal{C}_3^+,$$

where $\delta = |\mathcal{D}|$ is the number of valid dither values.

This gives for $\mathcal{C}_1$ schemes a bound on collision resistance independent on the dither sequence used. But for $\mathcal{C}_3$ schemes the bound depends on the size of the dither domain $\mathcal{D}$: Clearly, when $\mathcal{D}$ is large (e.g. when $\delta = |\mathcal{D}| = 2^n$) this bound is not relevant. However, it makes sense for example for Rivest's dithering proposal, for which $\delta \leq 2^{15}$.

We prove Theorem 1 by first upper bounding $\mathbf{Adv}_H^{\text{col}}$ by a collision-finding advantage for the compression function (see Lemma 1), then bounding this advantage in the blackbox model (Propositions 1 and 2).

Lemma 1 (Dithered Extension of MD Theorem). *Let H be a hash function built on a dithered PGV scheme $f \in \mathcal{C}_1 \cup \mathcal{C}_3^+$ and using MD-strengthening. Then $\mathbf{Adv}_H^{\text{col}}(q) \leq \mathbf{Adv}_f^{\Delta\text{col}}(q)$. Furthermore, if $f \in \mathcal{C}_1$, then $\mathbf{Adv}_H^{\text{col}}(q) \leq \mathbf{Adv}_f^{\text{col}}(q)$.*

This lemma states that the security of the hash function built on a dPGV scheme can be reduced to the security of its compression function, except for $\mathcal{C}_2$ functions. We show later a counter-example of $\mathcal{C}_2$-based hash functions which are not collision resistant, despite having a collision-resistant compression function.

Proof. Assume given an arbitrary colliding pair $(\bar{m}, \bar{m}')$ for a H^E with random $E \in \mathsf{Bloc}(n, n)$, and set $\ell = |\bar{m}|$, $\ell' = |\bar{m}'|$ (in blocks). We distinguish two cases:

1. $\ell = \ell'$: If $m_\ell \neq m'_\ell$ or $h_{\ell-1} \neq h'_{\ell-1}$, then we get a collision on f with the distinct tuples $(h_{\ell-1}, m_\ell, d_\ell)$ and $(h'_{\ell-1}, m'_\ell, d_\ell)$; otherwise, $h_{\ell-1} = h'_{\ell-1}$ and $m_\ell = m'_\ell$; we then work inductively with the same argument backwards until a collision is found, which necessarily exists, because $\bar{m} \neq \bar{m}'$ by hypothesis. Therefore, $\mathbf{Adv}_H^{\mathsf{col}}(q) \leq \mathbf{Adv}_f^{\mathsf{col}}(q)$ for messages of same length.

2. $\ell \neq \ell'$: Since MD-strengthening is applied, we have $m_\ell \neq m'_{\ell'}$, that necessarily leads to distinct Δ-colliding tuples for f with distinct message block, thus $\mathbf{Adv}_H^{\mathsf{col}}(q) \leq \mathbf{Adv}_f^{\Delta\mathsf{col}}(q)$. If $d_\ell = d_{\ell'}$, we even get a collision on f (with same dither value). Furthermore, for $\mathcal{C}_1$ functions, the pairs $(h_{\ell-1} \oplus d_\ell, m_\ell)$ and $(h_{\ell'-1} \oplus d_{\ell'}, m_{\ell'})$ form a collision for the original (undithered) scheme, hence $\mathbf{Adv}_H^{\mathsf{col}}(q) \leq \mathbf{Adv}_f^{\mathsf{col}}(q)$ in this case.

This covers all possible cases, showing reductions to the security of the dithered compression function f, which completes the proof. □

For functions of $\mathcal{C}_2$, the advantage cannot be bounded by $\mathbf{Adv}_f^{\mathsf{col}}(q)$ since the case $m_i \oplus d_i = m'_j \oplus d'_j$ may occur, for $d_i \neq d'_j$, which does not necessarily lead to a collision on the original undithered scheme. To prove such inequality, one should add the assumption that the dither and the message length padded in the last block do not overlap; e.g. consider the dither coded on the $n/2$ first bits of the blocks, while at most $n/2$ bits are dedicated to encoding the message length.

Proposition 1 (Collision Resistance of dPGV Schemes). *Let f be a dithered PGV scheme. Then the best advantage of a q-bounded adversary in finding collisions in f is $\mathbf{Adv}_f^{\mathsf{col}}(q) \leq q(q+1)/2^n$.*

Proof. For ease of exposition, consider the dithered MMO schemes, instantiated with a random E: From arbitrary colliding inputs (h, m, d) and (h', m', d) with image h^+, we can construct colliding inputs $(h_\star, m_\star)$ and $(h'_\star, m'_\star)$ for the original undithered scheme as follows:

	$h_\star$	$h'_\star$	$m_\star$	$m'_\star$	$h_\star^+$
dMMO_1	$h \oplus d$	$h' \oplus d$	m	m'	h^+
dMMO_2	h	h'	$m \oplus d$	$m' \oplus d$	$h^+ \oplus d$
dMMO_3	h	h'	m	m'	$h^+ \oplus d$
dMMO_4	h	h'	$m \oplus d$	$m' \oplus d$	h^+

A similar method applies for all dithered PGV schemes. The proposition now follows from the bound $q(q+1)/2$ given in Lemma 3.3 of [9]. □

Proposition 2 (Δ-Collision Resistance of dPGV Schemes). *Let f be a dithered PGV scheme. If $f \in \mathcal{C}_1 \cup \mathcal{C}_2$, then the best advantage of a q-bounded adversary in finding Δ-collisions in f is $\mathbf{Adv}_f^{\Delta\mathsf{col}}(q) = 1$. If $f \in \mathcal{C}_3^+$, then $\mathbf{Adv}_f^{\Delta\mathsf{col}}(q) \leq (\delta^2 + \delta)(q^2 + q)/2^{n+1}$.*

The idea of the proof of Proposition 2 for $f \in C_3^+$ is similar to the one of the proof of [9, Lemma 3.3]. In short, we first show that any collision for a C_3^+ scheme can be used to find values $(x_r, k_r, E_{k_r}(x_r))$ and $(x_s, k_s, E_{k_s}(x_s))$ satisfying a particular relationship, then we bound the cost of finding such values. The proof strategy is fairly standard. The simulator used for E and E^{-1} is described in [9, Fig. 4].

Proof. For $f \in C_1 \cup C_2$, we simply show how to construct a collision: For $f \in C_1$, pick an arbitrary triplet (h, m, d) such that $d \in \mathcal{D}$. Then construct (h', m', d') by choosing an arbitrary $d' \in \mathcal{D}$ distinct from d, and setting $h' = h \oplus d \oplus d'$, $m' = m$. For $f \in C_2$, a similar method can be applied with $h' = h$, and $m' = m \oplus d \oplus d'$. In both cases the constructed pairs map to the same image.

For $f \in C_3^+$, we just give the proof for MMO dithered variants (a similar one can easily be derived for any C_3^+ scheme): First, observe that dMMO$_2$ ($\in C_5$) and dMMO$_3$ ($\in C_3$) are in C_3^+, while dMMO$_1$, dMMO$_4$, dMMO$_5$ are not in C_3^+. Hence, the proof considers only dMMO$_2$ and dMMO$_3$.

Then, observe that for both dMMO$_2$ and dMMO$_3$ finding a Δ-Collision is *equivalent* to finding a tuple $(h, h', m, m', \tilde{d})$ such that $(E_h(m) \oplus E_{h'}(m') \oplus m \oplus m') \in \mathcal{D}_\oplus$, for

$$\mathcal{D}_\oplus = \left\{ \tilde{d} = d \oplus d', (d, d') \in \mathcal{D}^2 \right\}, \text{ and } \delta_\oplus = |\mathcal{D}_\oplus|.$$

Indeed, for such a tuple $(h, h', m, m', \tilde{d} = d \oplus d')$, we have that

- (h, m, d) and (h', m', d') form a collision for dMMO$_3$, because

$$E_h(m) \oplus m \oplus d = E_{h'}(m') \oplus m \oplus d'$$

- $(h, m \oplus d, d)$ and $(h', m' \oplus d', d')$ form a collision for dMMO$_2$, because

$$E_h((m \oplus d) \oplus d) \oplus (m \oplus d) = E_{h'}((m' \oplus d') \oplus d') \oplus (m' \oplus d')$$

We have thus shown that for any Δ-collision for dMMO$_2$ (or dMMO$_3$), one can return two triplets (x_r, k_r, y_r) and (x_s, k_s, y_s) such that $x_r \oplus x_s \oplus y_r \oplus y_s \in \mathcal{D}_\oplus$ and $y_r = E_{k_r}(x_r)$, $y_s = E_{k_s}(x_s)$. Using arguments similar to [9, Lemma 3.3 proof], we will show that this event is unlikely.

As preliminaries, consider an adversary $\mathcal{A}$ making q queries (to E or E^{-1}) and who gets q triplets (x_i, k_i, y_i), such that $y_i = E_{k_i}(x_i)$, $i = 1, \ldots, q$. These triplets are constructed by the simulator described in [9, Fig. 4]. Following these notations, $\mathcal{A}$ succeeds only if there exists distinct r, s such that $(x_r \oplus x_s \oplus y_r \oplus y_s) \in \mathcal{D}_\oplus$, or $x_r \oplus y_r = h_0$.

Now, in the process of simulating E (and E^{-1}) we let C_i stand for the event "$x_i \oplus y_i = h_0$ or there exists $j < i$ such that $(x_i \oplus x_j \oplus y_i \oplus y_j) \in \mathcal{D}_\oplus$"; in other words, this is the event "$\mathcal{A}$ succeeds".

The probabilistic argument is that, depending on the oracle queried, either y_i or x_i was (uniformly) randomly selected from a set of size $\geq 2^n - (i - 1)$ (see the definition of the simulator in [9, Fig. 4]). Hence $\Pr[C_i] \leq i \cdot \delta_\oplus / (2^n - (i - 1))$, because there are $\delta_\oplus = |\mathcal{D}_\oplus|$ values of $(x_i \oplus x_j \oplus y_i \oplus y_j)$ for which $\mathcal{A}$ succeeds.

It follows that for a number of queries $q \leq 2^{n-1}$,

$$\mathbf{Adv}_f^{\Delta \text{col}}(q) \leq \Pr[C_1 \vee \cdots \vee C_q] \leq \sum_{0 < i \leq q} \frac{i \cdot \delta_\oplus}{(2^n - i + 1)} \leq \frac{\delta_\oplus}{2^n - 2^{n-1}} \sum_{0 < i \leq q} i,$$

that is, $\mathbf{Adv}_f^{\Delta \text{col}}(q) \leq \delta_\oplus \cdot q(q+1)/2^n \leq (\delta^2 + \delta)(q^2 + q)/2^{n+1}$.

We have proven the bound when f is $\mathsf{dMMO_2}$ or $\mathsf{dMMO_3}$. As suggested in §2, a similar proof can be given for any $f \in \mathcal{C}_3^+$: the only difference will be in the conversion of the tuple $(h, h', m, m', \tilde{d} = d \oplus d')$ for which $(E_h(m) \oplus E_{h'}(m') \oplus m \oplus m') \in \mathcal{D}_\oplus$ to a collision for f. For instance, consider $f_{2,5}(h, m, d) = E_h(m \oplus h \oplus d) \oplus m \oplus h$; from the tuple above we can construct the collision

$$f_{2,5}(h, m \oplus h \oplus d) = E_h(m) \oplus m \oplus d = f_{2,5}(h', m' \oplus h' \oplus d', d').$$

Similar conversion can be given for the other $\mathcal{C}_3^+$ schemes. The rest of the proof is then independent of the scheme considered, hence apply as well to any $f \in \mathcal{C}_3^+$. □

Proof (Theorem 1). The result follows directly from Lemma 1 and the bounds given in Propositions 1 and 2. □

4.2   Finding Collisions for $\mathcal{C}_2$ Hash Functions

We describe an attack for 8 of the 12 schemes of $\mathcal{C}_2$ (namely $f_{5,2}, \ldots, f_{12,2}$), when the dither sequence scheme is the one of HAIFA [6], i.e. where d_i is the number of message bits hashed so far, and when MD-strengthening is applied. The attack exploits the structure of the compression function, and computes a pair of message colliding for any choice of a blockcipher.

The method is inspired from slide attacks on blockciphers [7]: Consider an arbitrary message $\bar{m} = (m_1, \ldots, m_\ell)$, split into ℓ blocks, with $\{d_i\}_{0 < i \leq \ell}$ the dither sequence. Compute the fixed-point $h_0 = h_1$ corresponding to m_1, and construct the message $\bar{m}' = (m_1', \ldots, m_{\ell-1}')$ by setting $m_i' = m_{i+1} \oplus d_{i+1} \oplus d_i'$, for $i = 1, \ldots, \ell - 2$. The last message blocks m_ℓ and $m_{\ell-1}'$ have to follow the MD-strengthening rule, that is, having the number of bits of the message coded in their least significant bits. Since we consider a dither sequence coding the number of message bits hashed so far, the padded values will be equal to d_ℓ and $d_{\ell-1}$, respectively. Therefore $m_{\ell-1}' = m_\ell \oplus d_\ell \oplus d_{\ell-1}$ is a valid last block, and we end up with $h_\ell = h_{\ell-1}'$, giving a collision with $\bar{m}$. Note that no call to the compression is needed, nor to the blockcipher.

When MD-strengthening is not used, this technique can be applied for any dither sequence, for the 8 schemes $f_{5,2}, \ldots, f_{12,2}$. This concerns for instance Rivest's dithering, that uses a special dither for the last block instead of MD-strengthening. The fact that a secure compression function (i.e. a provably secure PGV scheme) can lead to a weak hash function contrasts with the result of Black et al. where certain weak compression functions (namely non-preimage-resistant) are shown to provide collision-resistant hash functions.

5   Inversion Resistance

Theorem 2 below holds for inverting a random range point, rather than the image of a random domain point (the latter problem being refered as "preimage"). Quoting [9, Ap. B], though these measures *"can, in general, be far apart, it is natural to guess that they coincide for 'reasonable' hash functions"*.

Theorem 2 (Inversion Resistance of dPGV Hash Functions). *Let H be a hash function built on a dithered PGV scheme f, where MD-strengthening is applied. Then the best advantage of a q-bounded adversary in inverting H is*

$$\mathbf{Adv}_H^{\mathrm{inv}}(q) \leq \frac{q}{2^{n-1}}, \ for \ f \in \mathcal{C}_1 \cup \mathcal{C}_2,$$

$$\mathbf{Adv}_H^{\mathrm{inv}}(q) \leq \frac{\delta \cdot q}{2^{n-1}}, \ for \ f \in \mathcal{C}_3^+.$$

Proposition 3 (Inversion Resistance of dPGV Schemes). *Let f be a dithered PGV scheme. Then the best advantage of a q-bounded adversary in inverting f is* $\mathbf{Adv}_f^{\mathrm{inv}}(q) \leq \delta \cdot q/2^{n-1}$. *Furthermore, if* $f \in \mathcal{C}_1 \cup \mathcal{C}_2$, *then* $\mathbf{Adv}_f^{\mathrm{inv}}(q) \leq q/2^{n-1}$.

Proof. For $\mathcal{C}_1$ and $\mathcal{C}_2$, just observe that a preimage oracle for the $\mathcal{C}_1$ or $\mathcal{C}_2$ version of a PGV scheme can be used to solve the preimage problem for the original scheme, whose bound is $q/2^{n-1}$, from [9].

For $\mathcal{C}_3^+$, the problem is equivalent to finding h and m such that $(F(h, m) \oplus h^+) \in \mathcal{D}$, with F the original (undithered) scheme, for a fixed h^+. This equation is satisfied for a random permutation and arbitrary h, m, h^+ with probability $\delta/2^n$. We can use the same strategy as for proving Proposition 1: e.g. for dMMO$_3 \in \mathcal{C}_3$, let C_i be the event "the i-th query (x_i, k_i, y_i) satisfies $(y_i \oplus x_i \oplus h^+) \in \mathcal{D}$", $i \in \{1, \ldots, q\}$; then we have $\Pr[\mathsf{C}_i] \leq \delta/(2^n - (i-1))$. By the union bound, we get

$$\mathbf{Adv}_f^{\mathrm{inv}}(q) \leq \Pr[\mathsf{C}_1 \vee \cdots \vee \mathsf{C}_q] \leq \frac{\delta \cdot q}{2^{n-1}}.$$

$\square$

Proof (Theorem 2). An oracle inverting H can be trivially used for inverting its dithered compression function. The result of the theorem then follows from Proposition 3. $\square$

6   Conclusions

Among the 56 dPGV schemes studied,

- 12 inherit the bounds on collision and inversion resistance of the the original (undithered) constructions, independently of the dither sequence considered (these are of the form $f_{i,1}$, $i \in [1, 12]$)

- 37 have the "fixed-point" attribute ($f_{i,j}$, $i \in [5, 12]$)
- 8 lead to weak hash functions for HAIFA's dithering ($f_{i,2}$, $i \in [5, 12]$)

It appears that the most reliable schemes have the dither value simply xored with the initial chaining value h (subset $\mathcal{C}_1$). Nevertheless, the schemes of $\mathcal{C}_2$ fail to achieve similar security just because the overlap of dither and padding might allow collisions, for particular dither sequence. This problem can be easily avoided in practice, e.g. by encoding the dither in big-endian, and the message-dependent padding in little-endian, such that the two values do not overlap. In this case, $\mathcal{C}_2$ becomes as secure as $\mathcal{C}_1$, with the added benefit that it requires no change in the implementation of the hash function (whereas all other $\mathcal{C}_i$'s do).

Another desirable property concerns all 56 schemes considered: The fact that efficiency is no longer affected by the length of dither values allows to use a large counter, which provides better protection against attacks as [17,24,20] than schemes with short dithers [12]. An additional feature which can be derived from our constructions is *randomized hashing*, e.g. by choosing a random starting point for the counter. This would avoid extra changes to the compression function, like the RMX transform [21].

Eventually, we stress that dithering not only protects against generic short-cut attacks for second-preimage—which we might live with, since they remain much slower than collision search—but also provides a safety net against more elaborate attacks, and is expected to complicate some existing dedicated attacks.

Further work may consider the existence of generic second-preimage attacks for dPGV schemes instantiated with particular dither sequences, as well as refinement of our proofs at the light of Stam's recent improvements [46].

Acknowledgments

Particular thanks to Africacrypt anonymous referee #4 for his/her constructive criticism, and to referees #1, #5 and #6 for their pointing out several ways to improve readability.

References

1. Andreeva, E., Bouillaguet, C., Fouque, P.-A., Hoch, J., Kelsey, J., Shamir, A., Zimmer, S.: Second preimage attacks on dithered hash functions. In: Smart, N. (ed.) EUROCRYPT 2008. LNCS, vol. 4965. Springer, Heidelberg (2008)
2. Barreto, P., Rijmen, V.: The Whirlpool hashing function. First Open NESSIE Workshop (2000)
3. Bernstein, D.J.: The Rumba20 compression function. In: Function introduced in [4], http://cr.yp.to/rumba20.html
4. Bernstein, D.J.: What output size resists collisions in a xor of independent expansions? In: ECRYPT Workshop on Hash Functions (2007) see, http://cr.yp.to/rumba20.html#expandxor
5. Biham, E.: Recent advances in hash functions - the way to go. In: ECRYPT Hash Function Workshop (2005)

6. Biham, E., Dunkelman, O.: A framework for iterative hash functions - HAIFA. In: Cryptology ePrint Archive, Report 2007/278 (2007); Previously presented at the second NIST Hash Function Workshop (2006)
7. Biryukov, A., Wagner, D.: Slide attacks. In: Knudsen, L.R. (ed.) FSE 1999. LNCS, vol. 1636, pp. 245–259. Springer, Heidelberg (1999)
8. Black, J., Cochran, M., Shrimpton, T.: On the impossibility of highly-efficient blockcipher-based hash functions. In: Cramer [15], pp. 526–541
9. Black, J., Rogaway, P., Shrimpton, T.: Black-box analysis of the block-cipher-based hash-function constructions from PGV. Cryptology ePrint Archive, Report 2002/066, Full version of [10] (2002)
10. Black, J., Rogaway, P., Shrimpton, T.: Black-box analysis of the block-cipher-based hash-function constructions from PGV. In: Yung, M. (ed.) CRYPTO 2002. LNCS, vol. 2442, pp. 330–335. Springer, Heidelberg (2002)
11. Boneh, D., Boyen, X.: On the impossibility of efficiently combining collision resistant hash functions. In: Dwork [18], pp. 570–583
12. Bouillaguet, C., Fouque, P.-A., Shamir, A., Zimmer, S.: Second preimage attacks on dithered hash functions. Cryptology ePrint Archive, Report 2007/395. See also [1].
13. Brassard, G. (ed.): CRYPTO 1989. LNCS, vol. 435. Springer, Heidelberg (1990)
14. Chang, D., Gupta, K.C., Nandi, M.: A new hash function based on RC4. In: Barua, R., Lange, T. (eds.) INDOCRYPT 2006. LNCS, vol. 4329, pp. 80–94. Springer, Heidelberg (2006)
15. Cramer, R.J.F. (ed.): EUROCRYPT 2005. LNCS, vol. 3494. Springer, Heidelberg (2005)
16. Damgård, I.: A design principle for hash functions. In: Brassard [13], pp. 416–427.
17. Dean, R.D.: Formal Aspects of Mobile Code Security. PhD thesis, Princeton University (1999)
18. Dwork, C. (ed.): CRYPTO 2006. LNCS, vol. 4117. Springer, Heidelberg (2006)
19. Filho, D.G., Barreto, P., Rijmen, V.: The Maelstrom-0 hash function. In: 6th Brazilian Symposium on Information and Computer Security (2006)
20. Gauravaram, P., Kelsey, J.: Cryptanalysis of a class of cryptographic hash functions. In: Cryptology ePrint Archive, Report 2007/277 (2007)
21. Halevi, S., Krawczyk, H.: Strengthening digital signatures via randomized hashing. In: Dwork [18], pp. 41–59
22. Joux, A.: Multicollisions in iterated hash functions. application to cascaded constructions. In: Franklin, M. (ed.) CRYPTO 2004. LNCS, vol. 3152, pp. 306–316. Springer, Heidelberg (2004)
23. Kelsey, J., Kohno, T.: Herding hash functions and the Nostradamus attack. In: First NIST Cryptographic Hash Function Workshop (2005)
24. Kelsey, J., Schneier, B.: Second preimages on n-bit hash functions for much less than 2^n work. In: Cramer [15], pp. 474–490
25. Knudsen, L.: Hash functions and SHA-3. In: FSE 2008 (2008)
26. Knudsen, L., Rijmen, V.: Known-key distinguishers for some block ciphers. In: Kurosawa, K. (ed.) ASIACRYPT 2007. LNCS, vol. 4833, pp. 315–324. Springer, Heidelberg (2007)
27. Knudsen, L.R., Rechberger, C., Thomsen, S.S.: The Grindahl hash functions. In: Biryukov, A. (ed.) FSE 2007. LNCS, vol. 4593, pp. 39–57. Springer, Heidelberg (2007)
28. Lai, X., Massey, J.: Hash function based on block ciphers. In: Rueppel, R.A. (ed.) EUROCRYPT 1992. LNCS, vol. 658, pp. 55–70. Springer, Heidelberg (1993)

29. Lee, W., Nandi, M., Sarkar, P., Chang, D., Lee, S., Sakurai, K.: PGV-style block-cipher-based hash families and black-box analysis. IEICE Transactions 88-A(1), 39–48 (2005)

30. Matyas, S., Meyer, C., Oseas, J.: Generating strong one-way functions with cryptographic algorithm. IBM Technical Disclosure Bulletin 27(10A), 5658–5659 (1985)

31. Merkle, R.C.: One way hash functions and DES. In: Brassard [13], pp. 428–446

32. Mironov, I.: Hash functions: From Merkle-Damgård to Shoup. In: Pfitzmann, B. (ed.) EUROCRYPT 2001. LNCS, vol. 2045, pp. 166–181. Springer, Heidelberg (2001)

33. Miyaguchi, S., Ohta, K., Iwata, M.: New 128-bit hash function. In: 4th International Joint Workshop on Computer Communications, pp. 279–288 (1989)

34. Naor, M., Yung, M.: Universal one-way hash functions and their cryptographic applications. In: STOC, pp. 33–43. ACM, New York (1989)

35. Pietrzak, K.: Non-trivial black-box combiners for collision-resistant hash-functions don't exist. In: Naor, M. (ed.) EUROCRYPT 2007. LNCS, vol. 4515, pp. 23–33. Springer, Heidelberg (2007)

36. Pohlmann, K.: Principles of Digital Audio, 4th edn. McGraw-Hill, New York (2005)

37. Preneel, B., Bosselaers, A., Govaerts, R., Vandewalle, J.: Collision-free hash functions based on block cipher algorithms. In: Carnahan Conference on Security Technology, pp. 203–210 (1989)

38. Preneel, B., Govaerts, R., Vandewalle, J.: Hash functions based on block ciphers: A synthetic approach. In: Stinson, D.R. (ed.) CRYPTO 1993. LNCS, vol. 773, pp. 368–378. Springer, Heidelberg (1994)

39. Quisquater, J.-J., Girault, M.: 2n-bit hash-functions using n-bit symmetric block cipher algorithms. In: Quisquater, J.-J., Vandewalle, J. (eds.) EUROCRYPT 1989. LNCS, vol. 434, pp. 102–109. Springer, Heidelberg (1990)

40. Rabin, M.: Digitalized signatures. In: Lipton, R., DeMillo, R. (eds.) Foundations of Secure Computation, pp. 155–166. Academic Press, London (1978)

41. Rivest, R.: Abelian square-free dithering for iterated hash functions. In: ECRYPT Workshop on Hash Functions, Also presented in [42] (2005)

42. Rivest, R.: Abelian square-free dithering for iterated hash functions. In: NIST Hash Function Workshop (2005)

43. Rogaway, P., Steinberger, J.: Security/efficiency tradeoffs for permutation-based hashing. In: Smart, N. (ed.) EUROCRYPT 2008. LNCS, Springer, Heidelberg (to appear, 2008)

44. Shoup, V.: A composition theorem for universal one-way hash functions. In: Preneel, B. (ed.) EUROCRYPT 2000. LNCS, vol. 1807, pp. 445–452. Springer, Heidelberg (2000)

45. Shrimpton, T., Stam, M.: Building a collision-resistant compression function from non-compressing primitives. In: Cryptology ePrint Archive, Report 2007/409 (2007)

46. Stam, M.: Another glance at blockcipher based hashing. Cryptology ePrint Archive, Report 2008/071 (2008)

47. Wikipedia. Dither — Wikipedia, The Free Encyclopedia, Accessed (November 22, 2007)

A Definitions

Definition 5 (Collision for Hash Functions). *Let H be a blockcipher-based hash function. The advantage of $\mathcal{A}$ in finding collisions in H is*

$$\mathbf{Adv}_H^{\mathrm{col}}(\mathcal{A}) = \Pr\left[\begin{array}{c|c} E \stackrel{\$}{\leftarrow} \mathsf{Bloc}(\kappa, n), & \bar{m} \neq \bar{m}', \\ (\bar{m}, \bar{m}') \stackrel{\$}{\leftarrow} \mathcal{A} & H^E(\bar{m}) = H^E(\bar{m}') \end{array} \right].$$

Definition 6 (Collision for Compression Functions). *Let f be a blockcipher-based compression function. The advantage of $\mathcal{A}$ in finding collisions in f is*

$$\mathbf{Adv}_f^{\mathrm{col}}(\mathcal{A}) = \Pr\left[\begin{array}{c|c} E \stackrel{\$}{\leftarrow} \mathsf{Bloc}(\kappa, \mu), & (h, m) \neq (h', m'), \\ (h, m, h', m') \stackrel{\$}{\leftarrow} \mathcal{A} & [f^E(h, m) = f^E(h', m') \\ & or\ f^E(h, m) = h_0] \end{array} \right].$$

B Near-Collisions for dPGV and PGV Schemes

For the 32 schemes of C_3^+, which can be rewritten as $f(h, m, d) = F(h, m) \oplus d$, near-collisions might be easily found, depending on the structure of $\mathcal{D}$: suppose there exists $d, d' \in \mathcal{D}$ such that $d \oplus d'$ has weight w. Then, $f(h, m, d) \oplus f(h, m, d') = d \oplus d'$ and has weight w as well. This trivial property seems not to imply any weakness on the hash functions, since an adversary has no freedom on choosing the dither value for a given iteration count.

In the "known-key" scenario for blockciphers, Knudsen and Rijmen [26] presents a distinguisher for 7-round Feistel blockciphers based on the finding messages m, m' such that $E_k(m) \oplus m \oplus E_k(m') \oplus m' = 0 \ldots 0 \| x$, where x is a random $n/2$-bit value. As they observe, it can be applied to find "half-collisions" on MMO (f_1) instantiated with a similar blockcipher; indeed, MMO sets $f(h, m) = E_h(m) \oplus m$, thus one can choose a h which shall play the role of the "known-key" (note that in [26] the key cannot be chosen). We observe that a similar method can be applied to the other PGV schemes $f_2, \ldots, f_8$ (for some of them, by conveniently choosing the null value for h or m).

C Origins of Dithering

The use of the term "dither" in the context of hash functions finds its origin in signal processing, which itself borrowed it from engineers, who adapted the ancient word "didder" to a mechanical problem. The three quotes below give a bit more details about this story.

Quoting Rivest [41]: *"The word 'dithering' derives from image-processing, where a variety of gray or colored values can be represented by mixing together pixels of a small number of basic shades or colors; this is done in a random or pseudo-random manner to prevent simple visual patterns from being visible.*

We adapt the term dithering here to refer to the process of adding an additional 'dithering' input to a sequence of processing steps, to prevent an adversary from causing and exploiting simple repetitive patterns in the input."

Quoting Wikipedia [47]: *"Dither is an intentionally applied form of noise, used to randomize quantization error, thereby preventing large-scale patterns such as contouring that are more objectionable than uncorrelated noise. (...) Dither most often surfaces in the fields of digital audio and video, where it is applied to rate conversions and (usually optionally) to bit-depth transitions; it is utilized in many different fields where digital processing and analysis is used— especially waveform analysis."*

Quoting Pohlman [36]: *"one of the earliest [applications] of dither came in World War II. Airplane bombers used mechanical computers to perform navigation and bomb trajectory calculations. Curiously, these computers (boxes filled with hundreds of gears and cogs) performed more accurately when flying on board the aircraft, and less well on ground. Engineers realized that the vibration from the aircraft reduced the error from sticky moving parts. Instead of moving in short jerks, they moved more continuously. Small vibrating motors were built into the computers, and their vibration was called 'dither' from the Middle English verb 'didderen,' meaning 'to tremble.' Today, when you tap a mechanical meter to increase its accuracy, you are applying dither, and modern dictionaries define 'dither' as 'a highly nervous, confused, or agitated state.' In minute quantities, dither successfully makes a digitization system a little more analog in the good sense of the word."*

Attribute-Based Broadcast Encryption Scheme Made Efficient

David Lubicz[1,2] and Thomas Sirvent[1,2]

[1] DGA-CELAR, Bruz, France
[2] IRMAR, Université de Rennes 1, France
david.lubicz@univ-rennes1.fr, thomas.sirvent@m4x.org

Abstract. In this paper, we describe a new broadcast encryption scheme for stateless receivers. The main difference between our scheme and the classical ones derived from the complete subtree paradigm is that the group of privileged users is described by attributes. Actually, some real applications have been described where the use of a more adaptable access structure brings more efficiency and ease of deployment. On the other side, the decryption algorithm in so far existing attribute-based encryption schemes adapted for broadcast applications is time-consuming for the receiver, since it entails the computation of a large number of pairings. This is a real drawback for broadcast applications where most of the technological constraints are on the receiver side.

Our scheme can be viewed as a way to benefit at the same time from the performance of decryption of the classical broadcast schemes and the management easiness provided by the use of a more adaptable data structure based on attributes. More precisely, our scheme allows one to select or revoke users by sending ciphertexts of linear size with respect to the number of attributes, which is in general far less than the number of users. We prove that our scheme is fully collusion secure in the generic model of groups with pairing.

Keywords: Public-key broadcast encryption, Attribute-based encryption, Generic model of groups with pairing.

1 Introduction

A broadcast encryption scheme [FN93] is used whenever an emitter wants to send messages to several recipients using an unsecured channel. Such a scheme actually allows the broadcaster to choose dynamically a subset of privileged users inside the set of all possible recipients and to send a ciphertext, readable only by the privileged users. This kind of schemes is helpful in numerous commercial applications such as the broadcast of multimedia content or pay-per-view television.

Many schemes have been suggested to solve this problem regarding two main settings. The first one deals with almost fixed sets of privileged users. In this case the encryption is efficient but modifying the set of privileged users entails the sending of a long message. The second setting is aimed at the management

S. Vaudenay (Ed.): AFRICACRYPT 2008, LNCS 5023, pp. 325–342, 2008.
© Springer-Verlag Berlin Heidelberg 2008

of very large or very small sets of privileged users. Schemes designed for that purpose allow one to change at no cost the set of privileged users but the size of the encryption grows linearly with the size of the set of revoked users.

In this paper, we consider the real application where an emitter produces different kinds of content for different categories of users. This is a natural problem to deal with for a broadcaster which proposes to its customers several subscription packages, or for different broadcasters using the same asymmetrical broadcast encryption scheme. In this case, it is very possible that the set of privileged users has to be changed dramatically along with the type of content. As this set can not be considered as being particularly small or large, this situation is not covered by usual broadcast encryption schemes.

Recently, a notion of attribute-based encryption has been introduced in [SW05]. This notion seems to address that kind of problem. In [GPSW06], the authors present a declination of these ideas with applications in "targeted" broadcast encryption. In ciphertext-policy schemes, which is our concern here, each user is associated with a set of attributes and its decryption key depends on this set. A ciphertext contains an access policy based on these attributes: only users satisfying this policy may obtain the plaintext, and even a collusion of other users can not obtain it. In broadcast applications, the main drawback of this family of schemes is that the decryption may require large computations which cannot be quickly achieved by low-cost decoders.

Our Contribution. In this paper, we propose a broadcast encryption scheme, with attribute-based mechanisms: it allows the broadcaster to select or to revoke not only single users, but groups of users defined by their attributes. This scheme can be seen as an attribute-based encryption scheme, with efficient decryption and restriction of access policy: the restriction of access policy (using AND and NOT functions) is enough to provide broadcast encryption since the OR function can be simulated using concatenation, exactly like in the Subset-Cover framework.

The idea behind this scheme is the ability to compute a specific greatest common divisor of polynomials. Each receiver is associated with a polynomial (with roots depending on its attributes), and a ciphertext is associated with another polynomial (with roots depending on required attributes and revoked attributes). A receiver in the access policy defined by a ciphertext computes the greatest common divisor of its polynomial and of the polynomial associated with the ciphertext: this divisor is the same for all receivers in the access policy. A receiver not in this access policy can not compute this specific polynomial.

In this scheme, the size of the decryption key given to a receiver is linear in the number of attributes associated with this receiver. The size of a ciphertext is linear in the number of attributes used in the access policy. The public encryption key is quite long: its size is linear in the total number of attributes used in the scheme. This is not a real drawback for realistic situations where anyway the broadcaster must have a database containing the list of users together with their attributes. Moreover, a broadcaster which intends to use only a small set of attributes requires only an encryption key with a size linear in the size of this small set.

This scheme has a new design, since it is not based on secret sharing like previous attribute-based schemes. This design allows the decryption algorithm to use only a fixed number (3) of pairing computations. As a broadcast encryption scheme, it uses the Subset-Cover framework suggested in [NNL01]. We prove the security of this scheme against full collusions in the generic model of groups with pairings. Another interesting feature in this scheme is that new decryption keys can be built without any modification of previously distributed decryption keys: adding new decryption keys requires only to extend the public key to take new attributes into account.

1.1   Related Work

Stateful Broadcast Schemes. The first broadcast schemes were based upon stateful receivers, which means that the receivers have a memory that can store some information about the past messages. Such receivers have the possibility to refresh their decryption key using information given in broadcasted messages. This is the case of "Logical Key hierarchy" (LKH) presented independently in [WGL98] and in [WHA99]: users have assigned positions as leaves in a tree, and have keys corresponding to nodes on the path from user's leaf to the root. The key corresponding to the root is used to encrypt messages to users. When users are revoked or when a new user joins, a rekey occurs, using keys corresponding to internal nodes. These techniques have been later improved in [CGI$^+$99, CMN99, PST01].

These schemes are aimed at practical applications where the set of privileged users is updated rarely and in a marginal way. The ciphertexts are very short and are computed from a key known by all current users. In return, changing the set of privileged users (add or exclude a user) is bandwidth-consuming and must be done on a per user basis: each change entails the distribution of a new global key to privileged users. Moreover, this can only be done if all users are on-line which is a strong limitation in some applications. The frequent and important changes in the set of privileged users make these schemes inappropriate for the previously mentioned applications.

Stateless Broadcast Schemes. A different kind of broadcast schemes have been introduced later on: the goal is to avoid frequent rekeys. In [KRS99, GSW00], users have different decryption keys, and each decryption key is known by a well-chosen set of users. When the broadcaster wants to exclude a given set of users, it builds ciphertexts corresponding to decryption keys that these specific users do not know. Rekey occurs only after large permanent modifications of the privileged set of users. The ciphertexts are longer than with the LKH schemes mentioned in the previous paragraph.

Stateless receivers extend this last case: in [NNL01], the broadcaster can choose any set of privileged users without any rekey, i.e. the receivers can keep the same decryption keys during the whole life of the broadcast system. These

schemes, called Complete Subtree (CS) and Subset Difference (SD) are based
on a binary tree structure, where users are placed in the leaves. They have subsequently been improved in [HS02, GST04], and an efficient extension to the
public-key case based on hierarchical identity-based encryption has been proposed in [DF02]. This extension has been confirmed in [BBG05] with the first
hierarchical identity-based encryption with constant-size ciphertexts.

The efficiency of these schemes are only proved when few users are revoked,
but the binary tree structure presented in [NNL01] and its following improvements may be used to characterize groups of users by attributes: for example, the
left subtrees of the internal nodes at a given level may correspond to users with
a given attribute, and the right subtrees to users with this attribute missing.
This seems doable, even if the tree structure constrains the organization of the
attributes (the binary tree must be balanced to keep a good efficiency, so every
attribute must concern about half of the users). The Figure 1.1 shows that the
selection of users with a given attribute, or the revocation of users without this
attribute, is efficient if the attribute corresponds to a high level in the tree, but
very inefficient when the attribute is near the leafs. As a consequence, the use
of these schemes for selection or revocation of users regarding to their attributes
i s not practical, since the size of ciphertexts may be linear in the number of
revoked users.

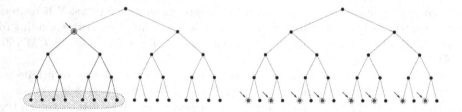

Fig. 1. Selection with CS/SD scheme: first attribute versus last attribute

New public-key broadcast schemes with constant-size ciphertexts have been
proposed in [BGW05] (scheme 1) and in [DPP07] (scheme 2). In these schemes, a
receiver needs however the exact knowledge of the set of privileged users, which
means the transmission of an information with non-constant size, which is not
mentioned in the ciphertexts.

These schemes require moreover decryption keys of size linear in the number
of users (this is clearly stated in [DPP07]; in [BGW05], a receiver has a constant-
size private key, but needs the encryption key to perform a decryption). This
storage may be excessive for low-cost devices.

Broadcast Scheme from HIBE with Wildcards. Management of attributes
can be performed by the combination of the scheme given in [DF02] with a
hierarchical identity-based encryption scheme with wildcards, like presented in
[ACD+06, BDNS06]. The resulting scheme would allow the selection of users
with given attributes, i.e. build ciphertexts addressed to intersections of groups.

The revocation of all users with a fixed attribute from the SD technique is however unclear, and its use is not efficient since the size of the ciphertexts is not constant in the hierarchical identity based encryption (see [BBG05]).

Attribute-Based Encryption. Attribute-based encryption has been suggested in [SW05], and later developed in [GPSW06]. In a first version (later called key-policy attribute-based encryption), the goal is to define access policies, and to allow a user to obtain some information if the access policy associated with this user is valid for this content. In this way, the decryption key given to a user depends on an access policy, and the encryption of a content relies on attributes, which are used in the evaluation of an access policy. Even a collusion of users with invalid access policies for a given ciphertext should not be able to obtain the corresponding plaintext.

Later, in [BSW07], a new scheme is proposed, but with an inversion: the access policy is defined with the content, and attributes are used to build decryption keys given to users. These ciphertext-policy attribute-based encryption schemes have direct applications for broadcast: the access policy defines a set of privileged users. With a relevant distribution of attributes, any set of privileged users may be described by an access policy.

In these schemes, an access policy is build using secret sharing techniques, like Shamir's one based on polynomials. An access policy is defined by a tree, where leaves correspond to the presence of an attribute (the evaluation of a leaf is true if the corresponding attribute is used) and internal nodes are threshold functions (in particular, these nodes may be AND, or OR functions). With such structure, revocation is quite difficult, since adding attributes can only provide a larger access to the content.

This problem is solved in [OSW07], where the access policy may be non-monotonic: the use of NOT functions becomes possible. Combining results from [BSW07, OSW07] gives rise to a ciphertext-policy attribute-based encryption which can be used for broadcast applications. The design of these schemes requires however a receiver to perform a large number of pairing computations (linear in the number of attributes used in the access policy). A low-cost receiver may not be able to compute so much pairings in complex access policies.

Our scheme has a completely different design, and it allows only very specific access policies. An access policy in this scheme is a disjunction (OR function, using the Subset-Cover framework) of conjunctions (AND functions) of attributes and of negations of attributes. Such access policy is more restrictive, but it is enough for practical broadcast applications. In return, a receiver performs only 3 pairing computations whatever the access policy is.

Dynamic Broadcast Encryption Scheme. The notion of dynamic schemes has been defined in [DPP07]. In such schemes, new users can be added without modification of previously distributed decryption keys. The encryption key has only to be slightly extended. This feature seems to be very useful in practical applications. The dynamic schemes suggested in [DPP07] requires ciphertexts of size linear in the number of revoked users. This feature is quite rare in broadcast schemes, but common in attribute-based encryption schemes.

1.2   Organization

The paper is organized as follows. In Section 2, we give a formal definition of groups of users, and an associated definition of attribute-based broadcast encryption schemes. In Section 3, we describe our scheme and prove its correctness. In Section 4, we prove the security of this scheme.

2   Preliminaries

We give a formal definition of groups of users and an associated definition of attribute-based broadcast encryption schemes deduced from the definition given in [BGW05]. We present then the security model. The last part explains how to define groups of users in concrete applications.

2.1   Groups of Users

In our applications, we have a large number of users, and a large number of groups (in practice, we need for each user a group containing this single user). Each user belongs to a few groups of users. We choose a description which takes advantage of this fact.

Let $\mathcal{U}$ be the set of all users. We represent an element of $\mathcal{U}$ by an integer in $\{1, \ldots, n\}$. A group of users is a subset $\mathcal{G}$ of $\mathcal{U}$. From the inverse point of view, for a fixed number l of groups of users, we can associate with a user $u \in \mathcal{U}$ the set of groups he belongs to: $\mathcal{B}(u) = \{i \in \{1, \ldots, l\} \, / \, u \in \mathcal{G}_i\} \subset \{1, \ldots, l\}$.

2.2   Attribute-Based Broadcast Encryption Schemes

In this part, we give a formal definition of an attribute-based broadcast encryption scheme. This model does not take into account the fact that the scheme could be dynamic, like in [DPP07], even if our scheme seems dynamic. The following definitions are just a slight adaptation of [BGW05, BSW07] to deal with groups of users.

A public-key attribute-based broadcast encryption scheme with security parameter λ is a tuple of three randomized algorithms:

- **Setup**$(\lambda, n, (\mathcal{B}(u))_{1 \leq u \leq n})$: takes as input the security parameter λ, the number of users n, and groups of users. It outputs an encryption key EK, and n decryption keys $(\mathrm{dk}_u)_{1 \leq u \leq n}$.
- **Encrypt**$(\mathrm{EK}, \mathcal{B}^N, \mathcal{B}^R)$: takes as input the encryption key EK and two sets of groups $\mathcal{B}^N$ and $\mathcal{B}^R$. It outputs a header hdr and a message encryption key $K \in \mathcal{K}$, where $\mathcal{K}$ is a finite set of message encryption keys.
- **Decrypt**$(\mathrm{dk}_u, \mathrm{hdr})$: takes as input a decryption key given to a user u and a header hdr. If the header hdr comes from an encryption using $(\mathcal{B}^N, \mathcal{B}^R)$ such that $\mathcal{B}^N \subset \mathcal{B}(u)$ and $\mathcal{B}(u) \cap \mathcal{B}^R = \emptyset$, then it outputs a message encryption key $K \in \mathcal{K}$. In the other case, it outputs $\perp$.

In the encryption process, a message M is encrypted with a key K and the resulting ciphertext C is sent together with the header hdr. Users in all groups mentioned in $\mathcal{B}^N$ (needed groups) and outside all groups mentioned in $\mathcal{B}^R$ (revoked groups) can compute K from the header hdr and their decryption key dk_u. Using the key K, a user recovers M from C.

Note that in these definitions, the decryption key and the header are the only elements that a user needs in the computation of the key K. The encryption key and the knowledge of the set of privileged users are not necessary for decryption. The header corresponds then exactly to the cost of the broadcast scheme in terms of transmission. In fact, in our scheme, the knowledge of the set of privileged users is implicitly included in the header, encoded in the attributes corresponding to the required and revoked groups.

In this description, we do not allow an encryption for an arbitrary set of privileged users, which is the usual definition of a broadcast encryption scheme. Any set of privileged users can however be represented by a union of sets used in this "basic encryption" for well-chosen groups of users (in fact, it is enough that each user belongs to a group containing only this single user). Different basic encryptions are then used to encrypt a common key, instead of a message. The full message can then be sent, using this common key.

2.3   Security Model

We consider semantic security of attribute-based broadcast encryption schemes. The adversary is assumed static, as in previous models: the only difference with standard definitions is that the groups of users are given to the adversary before the beginning of the game played by the challenger and the adversary $\mathcal{A}$:

- The challenger and the adversary are given l fixed groups of users, defined by $(\mathcal{B}(u))_{1 \leq u \leq n}$.
- The adversary $\mathcal{A}$ outputs two sets of groups $\mathcal{B}^N$ and $\mathcal{B}^R$ corresponding to a configuration it intends to attack.
- The challenger runs $Setup(\lambda, n, (\mathcal{B}(u))_{1 < u \leq n})$ and gives to $\mathcal{A}$ the encryption key EK, and the decryption keys dk_u corresponding to users that the adversary may control, i.e. such that $\mathcal{B}^N \cap \mathcal{B}(u) \neq \mathcal{B}^N$ or $\mathcal{B}^R \cap \mathcal{B}(u) \neq \emptyset$.
- The challenger runs $Encrypt(EK, \mathcal{B}^N, \mathcal{B}^R)$, and obtains a header hdr and a key $K \in \mathcal{K}$. Next, the challenger draws a random bit b, sets $K_b = K$, picks up randomly K_{1-b} in $\mathcal{K}$, and gives (hdr, K_0, K_1) to the adversary $\mathcal{A}$.
- The adversary $\mathcal{A}$ outputs a bit b'.

The adversary $\mathcal{A}$ wins the previous game when $b' = b$. The advantage of $\mathcal{A}$ in this game, with parameters $(\lambda, n, (\mathcal{B}(u))_{1 \leq u \leq n})$, is $|2\Pr[b' = b] - 1|$, where the probability is taken over the choices of b and all the random bits used in the simulation of the $Setup$ and $Encrypt$ algorithms:

$$\mathrm{Adv}^{\mathrm{ind}}(\lambda, n, (\mathcal{B}(u)), \mathcal{A}) = |2\Pr[b' = b] - 1|.$$

An attribute-based broadcast encryption scheme is semantically secure against full static collusions if for all randomized polynomial-time (in λ) adversary $\mathcal{A}$ and for all groups of users $(\mathcal{B}(u))_{1 \leq u \leq n}$ with at most l groups, $\mathrm{Adv}^{\mathrm{ind}}(\lambda, n, (\mathcal{B}(u)), \mathcal{A})$ is a negligible function in λ when n and l are at most polynomials in λ.

From such semantically secure schemes, we can build schemes secure in a stronger model: the use of generic transformations, like the ones presented in [FO99a, FO99b, OP01] has a negligible cost, and we obtain chosen-ciphertext security in the random oracle model. This explains why our security model is limited to chosen-plaintexts attacks.

2.4   Well-Chosen Groups of Users

In real broadcast applications, one has often to deal with obvious groups of users, because users are classified for instance by subscription package or subscription period. These groups are easily managed by an attribute-based broadcast encryption scheme, by simply using one attribute for each obvious group of users.

In some circumstances, it may happen that the group of privileged users does not fit easily with a description based on these obvious groups of users. Even if rare, it is preferable to be able to deal with such situations.

A solution consists in adding some extra attributes to the set of attributes corresponding to obvious groups. These new attributes describe a binary tree structure over the users, and allows the same management of users as in the SD-scheme. More precisely, we place users in the leaves of a binary tree, each node corresponds to a new attribute and each user receives the attributes of its parent nodes. At most $2n$ new attributes are added, and a user belongs to at most $\lceil \log_2(n) \rceil + 1$ new groups.

With this setting, there is an attribute for each user and this simple fact guarantees that any subset of users can be described by attributes. Moreover, basic encryption with privileged users corresponding to members of one group, excluding members of another group give at least the same sets as in the SD-method presented in [NNL01]. The efficiency of the attribute-based broadcast encryption scheme is then at least as good as in the SD-method, for any set of privileged users.

3   Construction

In this section, we first present bilinear maps. We describe next the *Setup, Encrypt* and *Decrypt* algorithms of a public-key attribute-based broadcast encryption scheme based on groups with a bilinear map. The correctness can then be verified.

3.1   Bilinear Maps

In the following definitions, we consider the symmetric setting of bilinear maps, like in [Jou00, BF01]. Let $\mathbb{G}_1$ and $\mathbb{G}_2$ be two cyclic groups of prime order p. The group laws in $\mathbb{G}_1$ and $\mathbb{G}_2$ are noted additively. Let g_1 be a generator of $\mathbb{G}_1$. Let $e : \mathbb{G}_1 \times \mathbb{G}_1 \to \mathbb{G}_2$ be a non-degenerate pairing:

- for all $a, b \in (\mathbb{Z}/p\mathbb{Z})$, $e(a\,g_1, b\,g_1) = ab.e(g_1, g_1)$,
- let $g_2 = e(g_1, g_1)$, g_2 is a generator of $\mathbb{G}_2$.

We make the assumption that the group laws in $\mathbb{G}_1$ and $\mathbb{G}_2$, and the bilinear map e can be computed efficiently.

3.2 Setup Algorithm

From the security parameter λ, the first step of the setup consists in constructing a tuple $(\mathbb{G}_1, \mathbb{G}_2, g_1, g_2, e, p)$, where:

- p is a prime, the length of which is λ,
- $\mathbb{G}_1$ and $\mathbb{G}_2$ are two cyclic groups of prime order p,
- e is a non-degenerate pairing from $\mathbb{G}_1 \times \mathbb{G}_1$ into $\mathbb{G}_2$,
- g_1 is a generator of $\mathbb{G}_1$ and $g_2 = e(g_1, g_1)$.

Four elements (α, β, γ and δ) are randomly chosen in $(\mathbb{Z}/p\mathbb{Z})^*$. Each group of users $\mathcal{G}_i$, mentioned in $(\mathcal{B}(u))_{1 \leq u \leq n}$ is then associated with an attribute μ_i randomly chosen in $(\mathbb{Z}/p\mathbb{Z})$, such that all these attributes are pairwise different and different from α. Another attribute μ_0 is chosen with the same constraints, corresponding to a virtual group containing no users. The encryption key is:

$$\text{EK} = \left(g_1, \beta\gamma\delta\,g_1, (\mu_i)_{0 \leq i \leq l}, (\alpha^i g_1)_{0 \leq i \leq l}, (\alpha^i \gamma\,g_1)_{0 \leq i \leq l}, (\alpha^i \delta\,g_1)_{0 \leq i \leq l} \right).$$

For each user $u \in \mathcal{U}$, s_u is randomly chosen in $(\mathbb{Z}/p\mathbb{Z})^*$. Let $\Omega(u)$ be the set of attributes corresponding to the groups he belongs to: $\Omega(u) = \{\mu_i \in (\mathbb{Z}/p\mathbb{Z}) \,/\, i \in \mathcal{B}(u)\}$. Let $l(u)$ be the size of $\Omega(u)$, i.e. the number of groups containing u. Let $\Pi(u) = \prod_{\mu \in \Omega(u)}(\alpha - \mu)$. The decryption key of u is:

$$\text{dku} = \left(\Omega(u), (\beta + s_u)\,\delta\,g_1, \gamma\,s_u\,\Pi(u)\,g_1, (\alpha^i \gamma\,\delta\,s_u\,g_1)_{0 \leq i < l(u)} \right).$$

3.3 Encryption Algorithm

If $\mathcal{B}^N \cap \mathcal{B}^R \neq \emptyset$, the encryption algorithm aborts and returns $\bot$, since a user can not be simultaneously inside and outside a given group of users. Otherwise, let $\Omega^N = \{\mu_i \,/\, i \in \mathcal{B}^N\}$ and $\Omega^R = \{\mu_i \,/\, i \in \mathcal{B}^R\}$. Let $l^N = |\mathcal{B}^N|$ be the number of required groups and $l^R = |\mathcal{B}^R|$ be the number of revoked groups[1]. Let $\Pi^N = \prod_{\mu \in \Omega^N}(\alpha - \mu)$, let $\Pi^R = \prod_{\mu \in \Omega^R}(\alpha - \mu)$ and let $\Pi^{NR} = \Pi^N \Pi^R$. Let z be randomly chosen in $(\mathbb{Z}/p\mathbb{Z})^*$. The result of the encryption is:

$$\text{hdr} = \left(\Omega^N, \Omega^R, z\,\Pi^{NR}\,g_1, \gamma\,z\,\Pi^N\,g_1, (\alpha^i \delta\,z\,g_1)_{0 \leq i < l^R} \right), \quad K = \beta\gamma\delta\,z\,\Pi^N\,g_2.$$

All these elements can be computed using only the encryption key EK.

---

[1] A slight modification occurs when $\mathcal{B}^R$ is empty: in such case, the encryption considers that the virtual group containing no users is revoked and then $\Omega^R = \{\mu_0\}$, $l^R = 1$.

3.4 Decryption Algorithm

We consider here the decryption of a header hdr with a decryption key dk_u:

$$
\begin{cases}
dk_u = \left(\Omega(u), dk_1, dk_2, dk_{3,0}, \ldots, dk_{3,l(u)-1} \right), \\
hdr = \left(\Omega^N, \Omega^R, hdr_1, hdr_2, hdr_{3,0}, \ldots, hdr_{3,l^R-1} \right).
\end{cases}
$$

The receiver u is valid for this header if $\Omega(u)$ contains Ω^N and if the intersection between Ω^R and $\Omega(u)$ is empty. To decrypt the header, the valid receiver u uses the extended Euclidean algorithm over the polynomials $\prod_{\mu \in (\Omega^N \cup \Omega^R)} (X - \mu)$ and $\prod_{\mu \in \Omega(u)} (X - \mu)$. It obtains two unitary polynomials, $V(X) = \sum_{0 \le i < l(u)} v_i X^i$ and $W(X) = \sum_{0 \le i < l^R} w_i X^i$, in $(\mathbb{Z}/p\mathbb{Z})[X]$, such that:

$$
V(X) \prod_{\mu \in (\Omega^N \cup \Omega^R)} (X - \mu) + W(X) \prod_{\mu \in \Omega(u)} (X - \mu) = \prod_{\mu \in \Omega^N} (X - \mu).
$$

From these polynomials, the receiver computes the key:

$$
K(dk_u, hdr) = e(dk_1, hdr_2) - e \left(\sum_{i=0}^{l(u)-1} v_i \, dk_{3,i} \, , \, hdr_1 \right) - e \left(dk_2, \sum_{i=0}^{l^R-1} w_i \, hdr_{3,i} \right).
$$

3.5 Proof of Correctness

If dk_u is the valid decryption key given to a user u, if hdr is a header built using the encryption and if u is a valid user for hdr, then the decryption gives:

$$
K(dk_u, hdr) = (\beta + s_u) \, \gamma \, \delta \, z \, \Pi^N \, g_2 - \gamma \, \delta \, z \, s_u \, V(\alpha) \, \Pi^{NR} \, g_2 - \gamma \, \delta \, z \, s_u \, W(\alpha) \, \Pi(u) \, g_2.
$$

By definition of the two polynomials V and W, we have the following relation: $V(\alpha) \, \Pi^{NR} + W(\alpha) \Pi(u) = \Pi^N$. The computed key is then exactly the key associated with the header in the encryption:

$$
K(dk_u, hdr) = (\beta + s_u) \, \gamma \, \delta \, z \, \Pi^N \, g_2 - \gamma \, \delta \, z \, s_u \, \Pi^N \, g_2 = \beta \, \gamma \, \delta \, z \, \Pi^N \, g_2.
$$

4 Security of the Scheme

The previous scheme can be proved in different ways. The usual strategy is first to define some security assumption and to prove this assumption in the generic model of groups with pairing. The reduction of the security of the scheme to this assumption concludes the proof. Following this strategy, we need a new security assumption which is an extension of the decisional version of the General Diffie-Hellman Exponent (GDHE) problem, precisely studied in the full version of [BBG05]. For the sake of simplicity, we prefer here a more direct proof in the generic model of groups with pairing.

In this section, we define the decisional problem upon which our broadcast encryption mechanism is built. We assess its security in the framework of the generic model of groups with pairing.

4.1   A Decisional Problem

Let $\mathbb{G}_1$ and $\mathbb{G}_2$ be two cyclic groups of prime order p and e be a non-degenerate pairing from $\mathbb{G}_1 \times \mathbb{G}_1$ into $\mathbb{G}_2$. Let g_1 be a generator of $\mathbb{G}_1$ and $g_2 = e(g_1, g_1)$. Let α, β, γ, δ, z be elements of $(\mathbb{Z}/p\mathbb{Z})^*$. For all $i \in \{0, \ldots, l\}$, let μ_i be an element of $(\mathbb{Z}/p\mathbb{Z})$ different from α and from μ_j where $j < i$.

The encryption key is:

$$\mathrm{EK} = \left(g_1, \ \beta\gamma\delta g_1, \ (\mu_i)_{0 \le i \le l}, \ (\alpha^i g_1)_{0 \le i \le l}, \ (\alpha^i \gamma g_1)_{0 \le i \le l}, \ (\alpha^i \delta g_1)_{0 \le i \le l} \right).$$

For each user $u \in \mathcal{U}$, $\Omega(u)$ is a subset of $\{\mu_1, \ldots, \mu_l\}$. Let $l(u) = |\Omega(u)|$ and let $\Pi(u) = \prod_{\mu \in \Omega(u)} (\alpha - \mu)$. The decryption key dk_u of the user u is:

$$\mathrm{dk}_u = \left(\Omega(u), (\beta + s_u)\delta g_1, \ \gamma s_u \Pi(u) g_1, \ (\alpha^i \gamma \delta s_u g_1)_{0 \le i < l(u)} \right).$$

Let Ω^N be a subset of $\{\mu_1, \ldots, \mu_l\}$, let Ω^R be a non-empty subset of $\{\mu_0, \ldots, \mu_l\}$ such that $\Omega^N \cap \Omega^R = \emptyset$, let $l^R = |\Omega^R|$. Let $\mathcal{R}$ be the set of revoked users for these sets:

$$\mathcal{R} = \left\{ u \in \mathcal{U} \ / \ \Omega(u) \cap \Omega^N \ne \Omega^N \ \text{ or } \ \Omega(u) \cap \Omega^R \ne \emptyset \right\}.$$

Let $\Pi^N = \prod_{\mu \in \Omega^N} (\alpha - \mu)$, let $\Pi^R = \prod_{\mu \in \Omega^R} (\alpha - \mu)$ and let $\Pi^{NR} = \Pi^N \Pi^R$. The header hdr and the key K are defined by:

$$\mathrm{hdr} = \left(\Omega^N, \Omega^R, z \Pi^{NR} g_1, \ \gamma z \Pi^N g_1, \ (\alpha^i \delta z g_1)_{0 \le i < l^R} \right), \quad K = \beta\gamma\delta z \Pi^N g_2.$$

Let b be a bit, let K_{1-b} be an element of $(\mathbb{Z}/p\mathbb{Z})^*$, let $K_b = K$. The decisional problem is the following: guess b from the knowledge of EK, hdr, K_0, K_1 and all the dk_u, where $u \in \mathcal{R}$.

4.2   Interpretation in the Generic Model

In this section, we use the notations of the full version of [BBG05] in order to assess the difficulty of the preceding decisional problem in the generic model of groups with pairing model. This extends the classical model of generic groups presented in [Nec93, Sho97].

The first part of the proof consists in showing that there exists no formula giving the key from the header, the encryption key, and the decryption keys corresponding to revoked users. The second part details why an adversary can not distinguish the key from a random element in the generic model of groups with pairing.

No Formula. Let $\mathcal{P}$ be the ring of polynomials over the variables A, B, C, D, Z and $\{S_u, u \in \mathcal{R}\}$. Each of these variables represent an element picked at random in the decisional problem and not explicitly unveiled: A is used for α, B for β, C for γ, D for δ, Z for z and for all $u \in \mathcal{U}$, S_u is used for s_u.

Let $\mathcal{D}$ be the tuple of elements in $\mathcal{P}$, corresponding to the discrete logarithms of elements in $\mathbb{G}_1$ given to an adversary in the problem. The tuple $\mathcal{D}$ contains 1, $B\,C\,D$, $Z\,\Pi^{NR}(A)$, $C\,Z\,\Pi^{N}(A)$ and the following polynomials:

- A^i, $A^i\,C$ and $A^i\,D$ for all $i \in \{0,\dots,l\}$,
- $(B + S_u)\,D$ and $C\,S_u\,\Pi_u(A)$, for all $u \in \mathcal{R}$,
- $A^i\,C\,D\,S_u$, for all $u \in \mathcal{R}$ and $i \in \{0,\dots,l(u)-1\}$,
- $A^i\,D\,Z$ for all $i \in \{0,\dots,l^R-1\}$,

where

$$\Pi^N(A) = \prod_{\mu \in \Omega^N} (A - \mu), \quad \Pi^R(A) = \prod_{\mu \in \Omega^R} (A - \mu),$$

$$\Pi_u(A) = \prod_{\mu \in \Omega(u)} (A - \mu), \quad \Pi^{NR}(A) = \Pi^N(A)\,\Pi^R(A).$$

Lemma 1. *Let $\mathcal{M}$ be the sub-$\mathbb{Z}$-module of $\mathcal{P}$ generated by all products of elements of $\mathcal{D}$. If $l^R \leq \sqrt{p}/2$ and for all u, $l(u) \leq \sqrt{p}/2$, the element $B\,C\,D\,Z\,\Pi^N(A)$ is an element of $\mathcal{M}$ with probability less than $1/\sqrt{p}$, this last probability being taken over all possible choices of the attributes μ_i in $(\mathbb{Z}/p\mathbb{Z})$.*

Proof. This lemma is proved in appendix A.1. $\qquad\qquad$

Indistinguishability in the Generic Model. In the generic model of groups with pairing, we consider two injective maps ξ_1 and ξ_2 from $(\mathbb{Z}/p\mathbb{Z})$ into $\{0,1\}^*$, also known as encoding functions. The additive law on $(\mathbb{Z}/p\mathbb{Z})$ induces a group law over $\xi_1(\mathbb{Z}/p\mathbb{Z})$ and $\xi_2(\mathbb{Z}/p\mathbb{Z})$, and the sets $\xi_1(\mathbb{Z}/p\mathbb{Z})$ and $\xi_2(\mathbb{Z}/p\mathbb{Z})$ together with these group laws are respectively denoted by $\mathbb{G}_1$ and $\mathbb{G}_2$. Oracles corresponding to the group law and the inverse law of each group are provided. A new law, corresponding to the pairing, is also given as an oracle: for all $x, y \in \mathbb{G}_1, e(x,y) = \xi_2(\xi_1^{-1}(x) \times \xi_1^{-1}(y)) \in \mathbb{G}_2$. An algorithm computing in this model has only access to these 5 oracles, and has no information about ξ_1 and ξ_2: its computations are based on queries to these oracles.

In our case, this model means that a challenger will use randomly chosen encoding functions from $(\mathbb{Z}/p\mathbb{Z})$ into a set of p binary strings. The challenger randomly chooses α, β, γ, δ, z, $(\mu_i)_{0 \leq i \leq l}$, $(s_u)_{u \in \mathcal{U}}$ following their constraints, and gives to the adversary all values $\xi_1\left(f(\alpha, \beta, \gamma, \delta, z, s_1, \dots, s_n)\right)$, where f is in the tuple $\mathcal{D}$. The adversary receives moreover $\xi_2(\kappa_0)$ and $\xi_2(\kappa_1)$, where κ_{1-b} is chosen randomly in $(\mathbb{Z}/p\mathbb{Z})^*$ and $\kappa_b = \beta\,\gamma\,\delta\,z\Pi^N$. The adversary makes then queries to oracles and finally outputs its guess b'.

We use the following theorem, proposed and proved in the full version of [BBG05] (Theorem A.2):

Theorem 1. *Let $\mathcal{D}$ be a subset of $\mathcal{P}$ of size k and suppose that for all $f \in \mathcal{D}$, $\deg(f) \leq d$. Let ϕ be an element of $\mathcal{P}$ such that ϕ is not is the sub-$\mathbb{Z}$-module spanned by the products of any two elements of $\mathcal{D}$. We consider an adversary which receives the set $\{\xi_1\left(f(\alpha, \beta, \gamma, \delta, z, s_1, \dots, s_n)\right) / f \in \mathcal{D}\}$, $\xi_2(\kappa_0)$ and $\xi_2(\kappa_1)$,* *where κ_{1-b} is chosen randomly in $(\mathbb{Z}/p\mathbb{Z})^*$ and $\kappa_b = \phi(\alpha, \beta, \gamma, \delta, z, s_1, \ldots, s_n)$. All such adversary which is allowed to issue at most q queries to the oracles can not guess the bit b with a probability significantly better than $1/2$:*

$$\left| Pr[b' = b] - \frac{1}{2} \right| \leq \frac{\max(2d, \deg(\phi))\,(q + 2k + 2)^2}{2p}.$$

In our context, the set $\mathcal{D}$ contains at most $nl + 3(n+l) + 7$ elements. Moreover these elements have degree less than $l+2$ and the degree of $\phi = B\,C\,D\,R\,\Pi^N(A)$ is less than $l + 4$. If ϕ is not in the span generated by the products of any two elements of $\mathcal{D}$, this lemma implies:

$$\left| Pr[b' = b] - \frac{1}{2} \right| \leq \frac{(l+2)\,(q + 2nl + 6n + 6l + 14)^2}{p}.$$

The results of Lemma 1 and Theorem 1 give the following theorem:

Theorem 2. *In the generic model of groups with pairing, the advantage of an adversary for the problem defined in Part 2.3 of the attribute-based broadcast encryption scheme presented in Section 3, issuing at most q queries to the oracles is bounded by:*

$$\frac{(l+2)\,(q + 2nl + 6n + 6l + 14)^2}{p - \sqrt{p}},$$

where n is the number of users and l is the number of groups of users.

Proof. We only have to divide the maximum probability obtained by the Theorem 1 by the factor $1 - 1/\sqrt{p}$ which is a lower bound for the probability that the polynomial ϕ is not in the sub-$\mathbb{Z}$-module generated by products of elements of $\mathcal{D}$ which is a consequence of the Lemma 1. The condition on the degrees in the Lemma 1 is verified, l being polynomial in the security parameter λ whereas p is exponential in this same parameter.

The arguments that n, q and l are at most polynomials in the security parameter λ, whereas p is exponential in λ, yield moreover that the given bound is a negligible function of the security parameter. This concludes the proof of security of our attribute-based broadcast encryption scheme.

5   Conclusion

In this paper, we have built a new public-key broadcast encryption scheme especially interesting when dealing with groups of users defined by the conjunction and exclusion of some attributes. We have described a practical application where none of previously existing broadcast or attribute-based encryption schemes behave in a suitable manner.

We have given a generic way to use attributes in order to manage groups of users in an efficient way. Finally, we have proved that our scheme is semantically secure against full static collusions in the generic model of groups with pairing.

It would be interesting to investigate the possibility to improve the access structure of our scheme by implementing efficiently the OR, or a threshold functionality. We also believe that the underlying problem of our scheme, based upon the reconstruction of the greatest common divisor of polynomials, may have some other interesting applications.

Acknowledgments. The authors would like to thank Cécile Delerablée for helpful comments on earlier drafts of this paper.

References

[ACD⁺06] Abdalla, M., Catalano, D., Dent, A.W., Malone-Lee, J., Neven, G., Smart, N.P.: Identity-based encryption gone wild. In: Bugliesi, M., Preneel, B., Sassone, V., Wegener, I. (eds.) ICALP 2006. LNCS, vol. 4052, pp. 300–311. Springer, Heidelberg (2006)

[BBG05] Boneh, D., Boyen, X., Goh, E.-J.: Hierarchical identity based encryption with constant size ciphertext. In: Cramer, R.J.F. (ed.) EUROCRYPT 2005. LNCS, vol. 3494, pp. 440–456. Springer, Heidelberg (2005)

[BDNS06] Birkett, J., Dent, A.W., Neven, G., Schuldt, J.: Efficient chosen-ciphertext secure identity-based encryption with wildcards. Technical Report 2006/377, Cryptology ePrint Archive (2006)

[BF01] Boneh, D., Franklin, M.: Identity-based encryption from the Weil pairing. In: Kilian, J. (ed.) CRYPTO 2001. LNCS, vol. 2139, pp. 213–229. Springer, Heidelberg (2001)

[BGW05] Boneh, D., Gentry, C., Waters, B.: Collusion resistant broadcast encryption with short ciphertexts and private keys. In: Shoup, V. (ed.) CRYPTO 2005. LNCS, vol. 3621, pp. 258–275. Springer, Heidelberg (2005)

[BSW07] Bethencourt, J., Sahai, A., Waters, B.: Ciphertext-policy attribute-based encryption. In: Proc. of IEEE Symposium on Security and Privacy, pp. 321–334 (2007)

[CGI⁺99] Canetti, R., Garay, J., Itkis, G., Micciancio, D., Naor, M., Pinkas, B.: Multicast security: A taxonomy and efficient constructions. In: IEEE Infocom 1999, vol. 2, pp. 708–716 (1999)

[CMN99] Canetti, R., Malkin, T., Nissim, K.: Efficient communication-storage tradeoffs for multicast encryption. In: Stern, J. (ed.) EUROCRYPT 1999. LNCS, vol. 1592, pp. 459–474. Springer, Heidelberg (1999)

[DF02] Dodis, Y., Fazio, N.: Public key broadcast encryption for stateless receivers. In: Feigenbaum, J. (ed.) DRM 2002. LNCS, vol. 2696, pp. 61–80. Springer, Heidelberg (2003)

[DPP07] Delerablee, C., Paillier, P., Pointcheval, D.: Fully collusion secure dynamic broadcast encryption with constant-size ciphertexts and decryption keys. Technical report, Prepublication accepted in Pairing 2007 (2007)

[FN93] Fiat, A., Naor, M.: Broadcast encryption. In: Stinson, D.R. (ed.) CRYPTO 1993. LNCS, vol. 773, pp. 480–491. Springer, Heidelberg (1994)

[FO99a] Fujisaki, E., Okamoto, T.: How to enhance the security of public-key encryption at minimum cost. In: Imai, H., Zheng, Y. (eds.) PKC 1999. LNCS, vol. 1560, pp. 53–68. Springer, Heidelberg (1999)

[FO99b] Fujisaki, E., Okamoto, T.: Secure integration of asymmetric and symmetric encryption schemes. In: Wiener, M.J. (ed.) CRYPTO 1999. LNCS, vol. 1666, pp. 537–554. Springer, Heidelberg (1999)

[GPSW06]   Goyal, V., Pandey, O., Sahai, A., Waters, B.: Attribute-based encryption for fine-grained access control of encrypted data. In: Proc. of ACM-CCS 2006, pp. 89–98 (2006)

[GST04]   Goodrich, M.T., Sun, J.Z., Tamassia, R.: Efficient tree-based revocation in groups of low-state devices. In: Franklin, M. (ed.) CRYPTO 2004. LNCS, vol. 3152, pp. 511–527. Springer, Heidelberg (2004)

[GSW00]   Garay, J.A., Staddon, J., Wool, A.: Long-lived broadcast encryption. In: Bellare, M. (ed.) CRYPTO 2000. LNCS, vol. 1880, pp. 333–352. Springer, Heidelberg (2000)

[HS02]   Halevy, D., Shamir, A.: The LSD broadcast encryption scheme. In: Yung, M. (ed.) CRYPTO 2002. LNCS, vol. 2442, pp. 47–60. Springer, Heidelberg (2002)

[Jou00]   Joux, A.: A one round protocol for tripartite Diffie-Hellman. In: Bosma, W. (ed.) ANTS 2000. LNCS, vol. 1838, pp. 385–393. Springer, Heidelberg (2000)

[KRS99]   Kumar, R., Rajagopalan, S., Sahai, A.: Coding constructions for blacklisting problems without computational assumptions. In: Wiener, M.J. (ed.) CRYPTO 1999. LNCS, vol. 1666, pp. 609–623. Springer, Heidelberg (1999)

[Nec93]   Nechaev, V.I.: Complexity of a determinate algorithm for the discrete logarithm. Mathematicheskie Zametki 55(2), 91–101 (1993)

[NNL01]   Naor, M., Naor, D., Lotspiech, J.: Revocation and tracing schemes for stateless receivers. In: Kilian, J. (ed.) CRYPTO 2001. LNCS, vol. 2139, pp. 41–62. Springer, Heidelberg (2001)

[OP01]   Okamoto, T., Pointcheval, D.: React: Rapid enhanced-security asymmetric cryptosystem transform. In: Naccache, D. (ed.) CT-RSA 2001. LNCS, vol. 2020, pp. 159–175. Springer, Heidelberg (2001)

[OSW07]   Ostrovsky, R., Sahai, A., Waters, B.: Attribute-based encryption with non-monotonic access structures. In: Proc. of ACM-CCS 2007, pp. 195–203 (2007)

[PST01]   Perrig, A., Song, D., Tygar, J.D.: Elk, a new protocol for efficient large-group key distribution. In: Proc. of IEEE Symposium on Security and Privacy, pp. 247–262 (2001)

[Sch80]   Schwartz, J.T.: Fast probabilistic algorithms for verification of polynomial identities. J. Assoc. Comput. Mach. 27(4), 701–717 (1980)

[Sho97]   Shoup, V.: Lower bounds for discrete logarithms and related problems. In: Fumy, W. (ed.) EUROCRYPT 1997. LNCS, vol. 1233, pp. 256–266. Springer, Heidelberg (1997)

[SW05]   Sahai, A., Waters, B.: Fuzzy identity-based encryption. In: Cramer, R.J.F. (ed.) EUROCRYPT 2005. LNCS, vol. 3494, pp. 457–473. Springer, Heidelberg (2005)

[WGL98]   Wong, C.K., Gouda, M., Lam, S.S.: Secure group communications using key graphs. In: Proc. of ACM-SIGCOMM 1998, pp. 68–79 (1998)

[WHA99]   Wallner, D.M., Harder, E.J., Agee, R.C.: Key management for multicast: Issues and architectures. RFC 2627 (1999)

A   Proof of Lemma 1

A.1   Proof of Lemma 1

Let $\mathcal{D}'$ be the set of elements of $\mathcal{P}$ which are products of pairs of elements of $\mathcal{D}$. By definition, $\mathcal{D}'$ generates $\mathcal{M}$. Suppose that $B\,C\,D\,Z\,\Pi^N(A) \in \mathcal{M}$. Then it is a linear combination with coefficients in $\mathbb{Z}$ of elements of $\mathcal{D}'$. Considering the elements of $\mathcal{D}'$ as polynomials with respect to the variable C, we see that $B\,C\,D\,Z\,\Pi^N(A)$ can only be obtained as a linear combination of terms linear in the variable C. In the same way, it can only be obtained as a linear combination of terms linear in the variables D and Z.

All elements in $\mathcal{P}$ are homogeneous of degree 0 or 1 in the set of variables $\{B\} \cup \{S_u, u \in \mathcal{R}\}$. Elements in $\mathcal{D}'$ are then homogeneous of degree 0, 1 or 2 in the same set of variables, and the polynomial $B\,C\,D\,Z\,\Pi^N(A)$ can only be obtained as a linear combination of homogeneous terms of degree 1.

These terms of $\mathcal{D}'$ which are simultaneously linear in the variables C, D and Z, and homogeneous of degree 1 in the set of variables $\{B\} \cup \{S_u, u \in \mathcal{R}\}$ are listed in the four sets below:

$$\mathcal{D}'_1 = \left\{ B\,C\,D\,Z\,\Pi^{NR}(A) \right\},$$

$$\mathcal{D}'_2 = \left\{ (B + S_u)\,C\,D\,Z\,\Pi^N(A)\,/\,u \in \mathcal{R} \right\},$$

$$\mathcal{D}'_3 = \left\{ A^i\,C\,D\,Z\,S_u\,\Pi_u(A)\,/\,u \in \mathcal{R}, i \in \{0, \ldots, l^R - 1\} \right\},$$

$$\mathcal{D}'_4 = \left\{ A^i\,C\,D\,Z\,S_u\,\Pi^{NR}(A)\,/\,u \in \mathcal{R}, i \in \{0, \ldots, l(u) - 1\} \right\}.$$

The polynomial in $\mathcal{D}'_1$ is not $B\,C\,D\,Z\,\Pi^N(A)$, since $\Omega^R \neq \emptyset$. As B only appears in polynomials in $\mathcal{D}'_1$ and $\mathcal{D}'_2$, at least one polynomial in $\mathcal{D}'_2$ must be used in the linear combination of elements of $\mathcal{D}'$ which is equal to $B\,C\,D\,Z\,\Pi^N(A)$.

We have to cancel linearly independent terms of the form $S_u\,C\,D\,Z\,\Pi^N(A)$ appearing in the elements of $\mathcal{D}'_2$ used in the linear combination. By considering only linear terms in this specific S_u in the sets $\mathcal{D}'_3$ and $\mathcal{D}'_4$, one can see that it is necessary to build a relation of the form

$$\Pi^N(A) = \left(\sum_{i=0}^{l^R-1} \lambda_i A^i \right) \Pi_u(A) + \left(\sum_{i=0}^{l(u)-1} \lambda'_i A^i \right) \Pi^{NR}(A). \tag{1}$$

By hypothesis, the user u is revoked. We have two cases:

– Either u is in a revoked group, and $\Omega(u) \cap \Omega^R \neq \emptyset$. We consider an attribute μ in this intersection: the polynomial $A - \mu$ divides $\Pi_u(A)$ and $\Pi^{NR}(A)$, and thus it divides the right part of the equation. Since $\Omega^N \cap \Omega^R$ is empty, $A - \mu$ does not divide $\Pi^N(A)$, and the relation (1) can not exist.

– Either u is not in an imposed group, and Ω^N is not included in $\Omega(u)$. So $\Pi^N(A)$ does not divide $\Pi_u(A)$. As $\Pi^N(A)$ divides $\Pi^{NR}(A)$, it divides $\left(\sum_{j=0}^{l^R-1} \lambda'_j A^j \right) \Pi_u(A)$ as well. It means that we have:

$$\left(\sum_{i=0}^{l^R-1} \lambda_i A^i \right) \Pi_u(A) = \Pi^N(A)\,Q(A)\,\pi_u(A),$$

where $Q(A)$ is a strict divisor of $\sum_{i=0}^{l^R-1} \lambda_i A^i$ and $\pi_u(A)$ is divisor of $\Pi_u(A)$. So Equation (1) is equivalent to the following equation:

$$Q(A)\,\pi_u(A) + \left(\sum_{i=0}^{l(u)-1} \lambda_i' A^i\right) \Pi^R(A) = 1, \quad \text{with}\ \deg(Q) < \deg(\Pi^R) - 1.$$

According to Lemma 2 given in next section of this appendix, such a relation does happen with probability less than $1/\sqrt{p}$.

In one case the relation (1) does not exist, in the other case such a relation exists with a probability less than $1/\sqrt{p}$. So with probability greater than $1 - 1/\sqrt{p}$ there is a contradiction with the hypothesis that $B\,C\,D\,Z\,\Pi^N(A)$ is an element of $\mathcal{M}$.

A.2   Lemma 2

Consider P_1 and P_2 two unitary polynomials of the ring $(\mathbb{Z}/p\mathbb{Z})[X]$ with $\deg P_1 = d_1$ and $\deg P_2 = d_2$. We suppose that P_1 and P_2 are relatively prime. By Bezout's Theorem, there exists V_1, V_2 in $(\mathbb{Z}/p\mathbb{Z})[X]$ unitary such that

$$V_1 P_1 + V_2 P_2 = 1, \quad \text{with}\ \deg V_1 < d_2\ \text{and}\ \deg V_2 < d_1. \tag{2}$$

The condition over the degrees determines uniquely V_1 and V_2. We are interested here in computing the probability that $\deg V_1 < d_2 - 1$. We have the following lemma:

Lemma 2. *For all $(d_1, d_2) \in (\mathbb{N}^*)^2$, for all prime p such that $p \geq (d_1 + d_2)^2$, the probability taken over all the pairs of relatively prime unitary polynomials (P_1, P_2) in $(\mathbb{Z}/p\mathbb{Z})[X]$ with degree d_1 and d_2 that the pair (V_1, V_2) of unitary polynomials defined uniquely by the relation (2) satisfies $\deg V_1 < d_2 - 1$ is upper bounded by $1/\sqrt{p}$.*

Proof. Let $P_1 = X^{d_1} + \sum_{k=0}^{d_1-1} \nu_k X^k$ and $P_2 = X^{d_2} + \sum_{k=0}^{d_2-1} \nu_k' X^k$ be two unitary polynomials of $(\mathbb{Z}/p\mathbb{Z})[X]$, with degrees $d_1 \in \mathbb{N}^*$ and $d_2 \in \mathbb{N}^*$. These two polynomials are relatively non primes if and only if the Sylvester determinant of dimension $d_1 + d_2$ cancels:

$$\det \begin{pmatrix} \nu_0 & 0 & \cdots & \cdots & 0 & \nu_0' & 0 & \cdots & 0 \\ \nu_1 & \nu_0 & \ddots & & \vdots & \nu_1' & \nu_0' & \ddots & \vdots \\ \vdots & \nu_1 & \ddots & \ddots & \vdots & \vdots & \nu_1' & \ddots & 0 \\ \nu_{d_1-1} & \vdots & \ddots & \ddots & 0 & \vdots & \vdots & \ddots & \nu_0' \\ 1 & \nu_{d_1-1} & \ddots & \nu_0 & \nu_{d_2-1}' & \vdots & & & \nu_1' \\ 0 & 1 & \ddots & \nu_1 & 1 & \nu_{d_2-1}' & & & \vdots \\ \vdots & \ddots & \ddots & \vdots & 0 & 1 & \ddots & & \vdots \\ \vdots & & \ddots & \ddots & \nu_{d_1-1} & \vdots & \ddots & \ddots & \nu_{d_2-1}' \\ 0 & \cdots & \cdots & 0 & 1 & 0 & \cdots & 0 & 1 \end{pmatrix} = 0.$$

Expanding this determinant, one obtains a polynomial of degree $d_1 + d_2 - 1$ in the variables $\nu_0, \ldots, \nu_{d_1-1}, \nu'_0, \ldots, \nu'_{d_2-1}$ over $(\mathbb{Z}/p\mathbb{Z})$. By Lemma 1 of [Sch80], the probability that this polynomial cancels is bounded by $(d_1 + d_2 - 1)/p$, where the probability is taken over the values of $\nu_0, \ldots, \nu_{d_1-1}, \nu'_0, \ldots, \nu'_{d_2-1}$. As a consequence, there is at least $(p+1-d_1-d_2)\, p^{d_1+d_2-1}$ pairs of relatively prime unitary polynomials of degree d_1 and d_2.

From now on, we suppose that P_1 and P_2 are relatively prime unitary polynomials. Let (V_1, V_2) be defined by the relation (2), we suppose that $\deg V_1 < d_2 - 1$. We have immediately that $\deg V_2 < d_1 - 1$. The relation (2) with these degree conditions in the $(\mathbb{Z}/p\mathbb{Z})$ vector space $(\mathbb{Z}/p\mathbb{Z})[X]$ implies that the following family is non free:

$$\left(1, \{P_1(X)\, X^k \,/\, k \in \{0, \ldots, d_2 - 2\}\}, \{P_2(X)\, X^k \,/\, k \in \{0, \ldots d_1 - 2\}\}\right).$$

This property is captured by the cancellation of the following determinant of dimension $d_1 + d_2 - 1$ depending on the coefficients of P_1 and P_2:

$$\det \begin{pmatrix}
1 & \nu_0 & 0 & \cdots & \cdots & 0 & \nu'_0 & 0 & \cdots & 0 \\
0 & \nu_1 & \nu_0 & \ddots & & \vdots & \nu'_1 & \nu'_0 & \ddots & \vdots \\
\vdots & \vdots & \nu_1 & \ddots & \ddots & \vdots & \vdots & \nu'_1 & \ddots & 0 \\
\vdots & \nu_{d_1-1} & \vdots & \ddots & \ddots & 0 & \vdots & \vdots & \ddots & \nu'_0 \\
\vdots & 1 & \nu_{d_1-1} & \ddots & \nu_0 & \nu'_{d_2-1} & \vdots & & & \nu'_1 \\
\vdots & 0 & 1 & \ddots & \nu_1 & 1 & \nu'_{d_2-1} & & & \vdots \\
\vdots & \vdots & \ddots & \ddots & \vdots & 0 & 1 & \ddots & & \vdots \\
\vdots & \vdots & & \ddots & \ddots \nu_{d_1-1} & \vdots & & \ddots & \ddots & \nu'_{d_2-1} \\
0 & 0 & \cdots & \cdots & 0 & 1 & 0 & \cdots & 0 & 1
\end{pmatrix} = 0.$$

Expanding this determinant, one obtains a polynomial of degree $d_1 + d_2 - 3$ in the variables $\nu_0, \ldots, \nu_{d_1-1}, \nu'_0, \ldots, \nu'_{d_2-1}$ over $(\mathbb{Z}/p\mathbb{Z})$. Again by Lemma 1 of [Sch80], the probability that this polynomial cancels is bounded by $(d_1+d_2-3)/p$, where the probability is taken over the values of $\nu_0, \ldots, \nu_{d_1-1}, \nu'_0, \ldots, \nu'_{d_2-1}$. As a consequence, there exists at most $(d_1 + d_2 - 3)\, p^{d_1+d_2-1}$ pairs of relatively prime unitary polynomials of degree d_1 and d_2 such that Bezout's equation returns a unitary polynomial V_1 of degree strictly less than $d_2 - 1$.

We just have to compute the quotient of the sizes of the two aforementioned sets in order to bound the probability that a pair of relatively prime unitary polynomials verifies Bezout's equation (2) with $\deg(V_1) < d_2 - 1$:

$$\frac{d_1 + d_2 - 3}{p + 1 - d_1 - d_2}.$$

If $d_1 + d_2 \leq \sqrt{p}$, this probability is bounded by $1/\sqrt{p}$.

Lower Bounds for Subset Cover Based Broadcast Encryption

Per Austrin* and Gunnar Kreitz

KTH – Royal Institute of Technology, Stockholm, Sweden
{austrin,gkreitz}@kth.se

Abstract. In this paper, we prove lower bounds for a large class of Subset Cover schemes (including all existing schemes based on pseudo-random sequence generators). In particular, we show that
- For small r, bandwidth is $\Omega(r)$
- For some r, bandwidth is $\Omega(n/\log(s))$
- For large r, bandwidth is $n - r$

where n is the number of users, r is the number of revoked users, and s is the space required per user.

These bounds are all tight in the sense that they match known constructions up to small constants.

Keywords: Broadcast Encryption, Subset Cover, key revocation, lower bounds.

1 Introduction

A Broadcast Encryption scheme is a cryptographic construction allowing a trusted sender to efficiently and securely broadcast information to a dynamically changing group of users over an untrusted network. The area is well studied and there are numerous applications, such as pay-per-view TV, CD/DVD content protection, and secure group communication. For instance, the new Advanced Access Content System (AACS) standard, which is used for content protection with next-generation video disks, employs Broadcast Encryption.

A Broadcast Encryption scheme begins with an initialization phase where every user is given a set of secrets. Depending on the application, a "user" in the scheme could be an individual, a subscriber module for a cable TV receiver, or a model of HD-DVD players. When the initialization is complete, the sender can transmit messages. For each message it wants to transmit, it selects a subset of users to receive the message. We will refer to this subset of intended recipients as *members* (another common name is the privileged set). It then encrypts and broadcasts the message, using the secrets of the members, in such a way that only the members can decrypt the broadcast. Even if all the non-members (or *revoked* users) collude, they should not be able to decrypt the broadcast. The term key revocation scheme is also used for Broadcast Encryption schemes.

---

* Research funded by Swedish Research Council Project Number 50394001.

S. Vaudenay (Ed.): AFRICACRYPT 2008, LNCS 5023, pp. 343–356, 2008.

The performance of a Broadcast Encryption scheme is generally measured in three parameters: bandwidth, space and time. Bandwidth is the size of the transmission *overhead* incurred by the scheme, space is the amount of storage for each user, and time is a measurement of the computation time needed for users to decrypt a message. In this paper, we will focus on the tradeoff between bandwidth and space.

In general, Broadcast Encryption schemes work by distributing a fresh *message key*, so that only the current members can recover the message key. The actual message is then encrypted under the message key and broadcast. This construction means that the bandwidth, i.e., the overhead incurred by the scheme, does not depend on the sizes of the messages the sender wants to transmit.

The problem of Broadcast Encryption was first described by Berkovits in [5], and later Fiat and Naor started a more formal study of the subject [7].

There are two naive schemes solving the broadcast encryption problem. In the first naive scheme, we give each user her own secret key shared with the sender. With this scheme, the space is 1 and the bandwidth is m, where m the number of members. In the second naive scheme, we assign a key to every possible subset of users, and give all users belonging to a subset access to the key for that subset. In this case the space is 2^{n-1}, where n is the number of users, and the bandwidth is 1.

In 2001, the *Subset Cover* framework was introduced by Naor *et al.* [16], along with two schemes, Complete Subtree and Subset Difference. In Subset Cover based schemes, there is a family of subsets of users, where each subset is associated with a key. When the sender wishes to make a broadcast, she finds a *cover* of the members using the subsets and encrypts the message key with each of the subset keys used in the cover. Both naive schemes can be seen as Subset Cover schemes; in the scheme with constant space, the family consists only of singleton subsets, and in the scheme with constant bandwidth, the family consists of all subsets of users. The Subset Cover principle is very general, and most published schemes are Subset Cover schemes.

In most Subset Cover schemes, each user is a member of a large number of subsets, so storing the key for each subset would be expensive in terms of memory. To solve this, the keys are chosen in such a way that users can compute the keys they should have using some smaller set of secrets and a *key derivation algorithm*. Schemes where keys are unrelated are called *information-theoretic*.

The most common key derivation algorithm is a straightforward application of a Pseudo-Random Sequence Generator (PRSG). The first protocol of this type was Subset Difference which has a bandwidth of $\min(2r - 1, n - r)$ with a user space $s = \mathcal{O}(\log^2 n)$ where r is the number of revoked users. Many more schemes [11,4,10,13,12] have been proposed using the same kind of key derivation. They all have a bandwidth of $\mathcal{O}(r)$, the same as Subset Difference, and their improvements lie in that some of them have a space of $\mathcal{O}(\log n)$, some offer increased flexibility, and some improve the bandwidth to $c \cdot r$ for $c < 1$ (as opposed to $c = 2$ in the original scheme).

Other forms of key derivation, such as RSA accumulators [2,3,8] and bilinear pairings [6], have also been studied for Subset Cover based broadcast encryption.

There have been attempts to reduce bandwidth by modifying the problem, for instance by allowing some *free-riders* (non-members who can still decrypt the broadcast) [1] or relaxing the security requirements [14].

There has been some analysis of lower bounds for Broadcast Encryption schemes. In 1998 Luby and Staddon [15] showed $s \geq \left(\frac{\binom{n}{r}^{1/b}}{b} - 1 \right) /r$ for Broadcast Encryption without key derivation, using the Sunflower lemma. This bound was sharpened in 2006 by Gentry *et al.* [9] to $s \geq (\binom{n}{r}^{1/b} - 1)/r$. We remark that schemes using key derivation beat these bounds.

1.1 Our Contribution

Table 1. Upper and lower bounds for Subset Cover schemes

Key derivation	Lower bound	Assumption	Upper bound	Space
None	$r\frac{\log(n/r)}{\log(rs)}$	—	$r\log(n/r)$	$s = \mathcal{O}(\log n)$
PRSG, small r	$\Omega(r)$ (**our**)	$s \leq \text{poly}(n)$	$\mathcal{O}(r)$	$s = \mathcal{O}(\log n)$
PRSG, worst r	$\Omega\left(\frac{n}{\log s}\right)$ (**our**)	—	$\mathcal{O}\left(\frac{n}{\log s}\right)$	—
PRSG, large r	$n - r$ (**our**)	$r \geq n - \frac{n}{6s}$	$n - r$	$s = \mathcal{O}(1)$

We present lower bounds for a large class of Subset Cover schemes, including existing schemes based on PRSGs. These lower bounds match known constructions up to a small constant, showing that current PRSG-based schemes are essentially optimal. Table 1 gives a summary of our results.

Our bounds on the bandwidth usage are strong, and show that the early Subset Difference scheme is in fact very close to being optimal. For instance, our bound for small r shows that improving the bandwidth to $o(r)$ would require super-polynomial space, which is unreasonable. In fact, depending on the application, space is generally considered reasonable if it is at most logarithmic (or possibly polylogarithmic) in n.

Our second result implies that, in order to get constant bandwidth b, the space required is exponential. It also implies that, using polylogarithmic space, the worst case bandwidth will be almost linear, $n/\log\log n$.

The third result says that, for a small number of members, the first naive scheme is optimal. With polylogarithmic amount of space, this holds even if the number of members is almost linear, $n/\text{poly}\log n$.

Also, in most current schemes, the decryption time for members is limited to be polylogarithmic in n. Our proofs do not make use of any such restrictions, so allowing longer decryption time than current schemes cannot lower the bandwidth requirements.

1.2   Organization of This Paper

In Section 2 we discuss the structure of Subset Cover based Broadcast Encryption schemes and define the class of schemes, Unique Predecessor schemes, for which we prove lower bounds. In Section 3 we give a proof showing that with polynomial memory in clients, the bandwidth consumption is $\Omega(r)$ for "small" r. In Section 4, we prove a bound for generic r, and in particular show that for $r \approx \frac{n}{e}$, the bandwidth is at least $\frac{n}{1.89 \log s}$. Section 5 shows that for a large number of revoked users, the worst case bandwidth is $n - r$, i.e., the same as for the naive scheme where every user has a single key.

2   Preliminiaries

In this section, we review some preliminaries. The concepts of Broadcast Encryption and Subset Cover schemes are described, and notation will be introduced. We also define a class of Broadcast Encryption schemes called Unique Predecessor (UP) schemes to which our lower bounds apply.

2.1   Broadcast Encryption

In Broadcast Encryption, we have a trusted sender, and a set of users. After some initialization, the sender can securely broadcast messages to some subset of the users in a way which is efficient for both the sender and users. We will refer to the users who are targeted by a broadcast as *members* and the users who are not as *revoked* users. As the name Broadcast Encryption implies, we assume there is a single broadcast medium, so all users see the messages transmitted by the sender.

When evaluating the efficiency, three parameters are measured: bandwidth, space, and time (for decryption). Most Broadcast Encryption schemes transmit encrypted keys, so we will measure the bandwidth in terms of the number of encrypted keys to be transmitted. The sender uses the broadcast encryption to distribute a message key K_m and then encrypts the actual message under K_m, so the bandwidth overhead incurred does not depend on the size of the actual message.

The bandwidth required for a scheme with n users out of which r are revoked can, and generally will, vary, depending on which r users are revoked. We define the bandwidth $b = f(n, r)$ of the Subset Cover scheme as being that of the maximum bandwidth over the choice of the set of revoked users $R \subseteq [n]$ such that $|R| = r$. Thus, when we say that the bandwidth is at least $c_1 r$ for $r \leq n^{c_2}$, we mean that for every such r, there is at least one choice of r revoked users which requires bandwidth at least $c_1 r$.

Similarly, we measure the space as the number of keys, seeds, or other secrets that a user must store to be able to correctly decrypt transmissions she should be able to decrypt. It need not be the case that all users have to store the same amount of secrets, so we let the space of a scheme be the size of the largest amount of secrets any one user must store.

We remark that, in general, the keys and secrets may vary in length, so that our convention of simply counting the number of keys may not measure the exact bandwidth or space. However, such differences are generally small and not taking them into account costs us at most a small constant factor.

In this paper, we will not concern ourselves with the computational time of the clients. Our only assumption will be the very natural (and necessary) assumption that users cannot derive keys which they should not have access to.

Broadcast Encryption schemes can be classified as either stateful or stateless. In a stateful scheme, a transmission from the sender may update the set of secrets a user uses to decrypt future broadcasts, whereas in the stateless case, the secrets are given to the user at initialization and then remain constant. We focus on the largest family of Broadcast Encryption schemes, Subset Cover schemes, and such schemes are stateless.

2.2   Notation

Throughout the paper, we will use the following notation. We let n denote the total number of users, and identify the set of users with $[n] = \{1, \ldots, n\}$. We let m denote the number of members and r the number of revoked users (so $n = r + m$). The space of a scheme is denoted by s and the bandwidth by b. Note that we are generally interested in the bandwidth as a function of r (or equivalently, of m).

2.3   Subset Cover Schemes

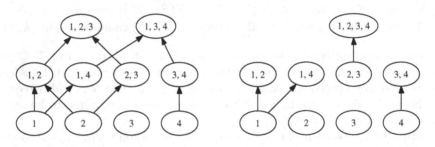

(a) Example of a Subset Cover scheme

(b) Example of a Subset Cover scheme $\mathcal{S}$ with indegree 1

Fig. 1. Illustration of Subset Cover schemes

In this paper, we consider a family of Broadcast Encryption schemes known as Subset Cover schemes, introduced in [16]. In a Subset Cover scheme, the sender starts by creating a family of subsets of users. Each such subset is associated with a key. To make an encrypted broadcast, the sender first computes a *cover* of the current members. A cover is a choice of subsets from the family, so that all members belong to at least one chosen subset, and no revoked user belong to any chosen subset. The message broadcasted will then contain, for each subset in the cover, the message key encrypted under that subset's key.

Without key derivation, each user would have to store the key for each subset of which she is a member. However, when using key derivation, keys of subsets are related in a way that allows a user to derive keys of subsets by applying a suitable function, typically a one way function, to her set of secrets. Thus the space decreases, as one secret can be used to derive multiple keys.

Example 1. Figure 1(a) shows an example of a Subset Cover scheme on $n = 4$ users. In the example, the family of subsets consists of all four singleton subsets, four subsets of size 2, and two subsets of size 3. An edge from S_i to S_j indicates that the secrets used to derive the key for S_i can also be used to derive the key for S_j. Thus, the secret used by user 2 to derive the key for her singleton set $\{2\}$ can also be used to derive the keys for nodes $\{1, 2\}$, $\{2, 3\}$, and $\{1, 2, 3\}$. Without key derivation, she would have had to store four keys, but now she only needs to store one secret.

More formally, a Subset Cover scheme consists of a family of subsets $\mathcal{F} = \{S\} \subseteq 2^{[n]}$ with the property that for every selection of members $M \subseteq [n]$ there is a cover $T \subseteq \mathcal{F}$ such that $\cup_{S \in T} S = M$. There is a set of "secrets" $\mathcal{K}$, and each user $i \in [n]$ is given a subset $P(i)$ of these secrets. Additionally, there is, for each $S \in \mathcal{F}$, a set $K(S) \subseteq \mathcal{K}$ of secrets and a secret key $k(S)$, with the following properties:

- Any user with access to a secret in $K(S)$ can compute $k(S)$.
- For every $S \in \mathcal{F}$ and user $i \in [n]$, $P(i) \cap K(S) \neq \emptyset$ if and only if $i \in S$.
- An adversary with access to all secrets in $\mathcal{K} \backslash K(S)$ cannot compute $k(S)$

To send a message key to the set $M \subseteq [n]$ of members, the cover $T \subseteq \mathcal{F}$ of subsets is chosen in such a way that $\cup_{S \in T} S = M$. The server then broadcasts the message key encrypted using $k(S)$ for each $S \in T$. The bandwidth required for this is $|T|$. We remark that a Subset Cover scheme is required to be able to cover any member set $M \subseteq [n]$.

Naturally, a Subset Cover scheme should also include efficient ways of computing $k(S)$ and the cover T, but as we are interested in lower bounds on the tradeoff between space and bandwidth, these computational issues are not relevant to us.

We denote by $B(\mathcal{F})$ the partially ordered set on the elements of $\mathcal{F}$ in which $S_1 \leq S_2$ if $K(S_1) \subseteq K(S_2)$, i.e., if any secret that can be used to deduce $k(S_1)$ can also be used to deduce $k(S_2)$. Note that $S_1 \leq S_2$ implies $S_1 \subseteq S_2$ (since any user $u \in S_1$ will be able to compute $k(S_2)$ and thus has to be an element of S_2). From now on, we will ignore the set of secrets and the keys, and only study the poset $B(\mathcal{F})$, since it captures all information that we need for our lower bounds. In Figure 1 we show Hasse diagrams of $B(\mathcal{F})$ for two toy example Subset Cover schemes.

The number of secrets a user u needs to store, i.e., the space s, is precisely the number of elements S of $B(\mathcal{F})$ such that u occurs in S, but not in any of the predecessors of S.

Lemma 1. *Any Subset Cover scheme will have at least one singleton node for each user.*

Proof. If there is a user which does not occur in a singleton node, the Broadcast Encryption scheme would fail when the sender attempts to broadcast only to that user. □

2.4 Key Derivation Based on a PRSG

The most common type of key derivation uses a Pseudo-Random Sequence Generator (PRSG), or equivalently, a family of hash functions. This type of key derivation was first used in the context of Broadcast Encryption in the Subset Difference scheme [16]. In [4] it is called Sequential Key Derivation Pattern. The key derivation described here is the intuitive way to do key derivation using a PRSG, and all Subset Cover schemes that the authors are aware of that use a PRSG (or a family of hash functions) do have this form of key derivation.

Let ℓ be a security parameter and let $H(x)$ be a pseudo-random sequence generator taking as seed a string x of length ℓ. Let $H_0(x)$ denote the first ℓ bits of output when running $H(x)$, let $H_1(x)$ denote the next ℓ bits, and so on.

Each subset S in the scheme will be assigned a seed $p(S)$ and a key $k(S)$. The key $k(S)$ will be computed as $k(S) = H_0(p(S))$, so from the seed for a subset, one can always compute the key for that subset. All secrets given to users will be seeds, no user is ever given a key directly. The reason for this is that it gives an almost immediate proof of the security of the scheme by giving the keys the property of *key indistinguishability*, which was proved in [16] to be sufficient for the scheme to be secure in a model also defined in [16].

Consider an edge $e = (S_i, S_j)$ in the Hasse diagram of $B(\mathcal{F})$. The edge means that someone with access to the secrets to deduce $k(S_i)$, i.e. $p(S_i)$ should also be able to deduce $p(S_j)$. If we let $p(S_j) = H_c(p(S_i))$ for some $c \geq 1$, anyone with $p(S_i)$ can derive $p(S_j)$. For a node S_i with edges to $S_{j_1}, S_{j_2}, \ldots, S_{j_k}$ we let $p(S_{j_1}) = H_1(p(S_i)), p(S_{j_2}) = H_2(p(S_i)), \ldots p(S_{j_k}) = H_k(p(S_i))$.

This construction cannot support nodes with indegree greater than 1, since that would require $p(S_j) = H_{c_1}(p(S_{i_1})) = H_{c_2}(p(S_{i_2}))$, which, in general, we cannot hope to achieve. This means that the Hasse diagram will be a forest, since all nodes have an indegree of either 0 or 1.

2.5 UP-Schemes

When the Hasse diagram of $B(\mathcal{F})$ is a forest, we say that the Subset Cover scheme is a *Unique Predecessor scheme* (UP-scheme). Schemes using key derivation as described in Subsection 2.4 will always be UP-schemes. Schemes not using any key derivation (there are no edges in $B(\mathcal{F})$) are also UP-schemes, this class of schemes is sometimes referred to as information-theoretic.

Example 2. The scheme in Figure 1(a) is not a UP-scheme, since there are several sets which have multiple incoming edges, for instance the set $\{1, 2\}$. However, the scheme S in Figure 1(b) is a UP-scheme. In this case, user 1 would have to store two secrets, one for her singleton node, and one for the node $\{1, 2, 3, 4\}$. The keys for nodes $\{1, 2\}$ and $\{1, 4\}$ can be derived from the same secret used to derive the key for her singleton node.

We view a UP-scheme as a rooted forest S, in which each node $V \subseteq [n]$ is labelled with the set of users which are in V, but are not in the parent node. The number of node labels in which a user occurs is the same as the number of secrets that a user will need to store. Thus, when we say that a scheme S has space s we mean that every user can be used in a label at most s times.

Lemma 2. *Any Unique Predecessor scheme will have at most ns distinct subsets.*

Proof. Adding a new node to a Unique Predecessor scheme means increasing the space for at least one member. Starting from an "empty" scheme, this can be done at most ns times. □

2.6   Normalized UP-Schemes

To simplify the proofs, we will work with *normalized* UP-schemes. We will show that we can perform a simple normalization of a UP-scheme which gives a new scheme with the same set of users, no more space, and at most the same bandwidth. This normalization is similar to the construction of the Flexible SD scheme in [4].

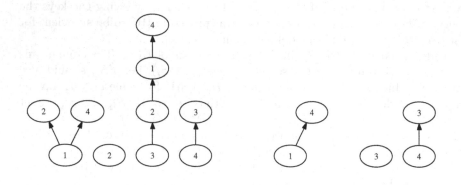

(a) Normalization S' of the UP-scheme S in Figure 1(b)   (b) The subscheme $S'(\{1, 3, 4\})$

Fig. 2. Normalization of UP-schemes

Definition 1. *A UP-scheme S is normalized when every node of S is labelled with exactly one user and S has exactly n trees.*

Example 3. The scheme S from Figure 1(b) is a UP-scheme, but it is not normalized. Two nodes violate the normalization criteria. First, the key for $\{1,2,3,4\}$ can be directly derived from the secrets used for $\{2,3\}$, which adds two new users at the same time. Second, the node $\{2,3\}$ also adds two users at once. In Figure 2(a) shows a normalized scheme S' which is essentially equivalent with S. The key for $\{2,3\}$ can now be derived from the secret for $\{3\}$, and an extra node $\{1,2,3\}$ was inserted between $\{2,3\}$ and $\{1,2,3,4\}$.

Lemma 3. *Let S be an arbitrary UP-scheme (on n users) with space s and bandwidth b. Then there exists a normalized UP-scheme S' (on n users) with space $s' \leq s$ and bandwidth $b' \leq b$.*

Proof. The proof consists of two steps. First we ensure that each node is labelled with exactly one user. Second, we merge identical nodes, which will ensure that S' has exactly n trees.

Consider a node labelled with a set $U = \{u_1, \ldots, u_k\}$ of users with $k > 1$. Now, split this node into a chain of k nodes, adding one user at a time (in arbitrary order) rather than all k at once. Call the resulting forest S_0. Note that, strictly speaking, it is possible that S_0 is not a UP-scheme as we have defined it, since there may be several nodes representing the same subset S of users. However, it still makes sense to speak of the space and bandwidth of S_0, and we note that the space of S_0 is the same as that of S, as each user occurs in the same number of labels in both. Furthermore, the bandwidth of S_0 is no more than that of S, since all subsets of users present in S are also present in S_0, and thus, any cover in S is also valid in S_0.

Next, we describe how to merge nodes representing the same set $S \subseteq [n]$. Given two such nodes v_1 and v_2, attach the children of v_2 as children to v_1, and remove v_2 from the scheme. Note that this operation does not change the bandwidth of the scheme, since only a single node is removed, and this node represents a subset which is still present in the resulting scheme. Also, the space for the resulting scheme will be no larger than the space for the original scheme. The user which was the label for v_2 will now need to store one secret less, whereas the space will be the same for all other users. Let S' be the result of applying this merging until every set is represented by at most one node.

It remains to show that S' has exactly n trees. By Lemma 1, there must be at least n trees in S'. Because of the first step every root of a tree will represent a singleton set, and because of the second step every singleton set can be present at most once, implying that there are at most n trees. □

We will, without loss of generality, from now on assume UP-schemes we deal with are normalized. See Figure 2(a) for an example of a normalized UP-scheme. We would like to remark that while normalization can only improve bandwidth and space, it does so at the cost of time. Thus, when applied to improve the performance of practical schemes, one has to take into account the computation time of users, as discussed in [4].

We remark that, in general, normalization will introduce key derivation, even if the original UP-scheme had completely independent keys.

Definition 2. *Given a UP-scheme S and a set $X \subseteq [n]$ of users, the subscheme of S induced by X, denoted $S(X)$, is defined as follows: for every user $y \notin X$, we remove all nodes of S labelled with y, and their subtrees.*

In other words, $S(X)$ contains the nodes (and thus subsets) which are still usable when $[n] \setminus X$ have been revoked. See Figure 2(b) for an example.

3   Few Revoked Users

We prove that when the number of revoked users r is small, any UP-scheme using at most polynomial space will require bandwidth $\Omega(r)$.

As noted in the introduction, the requirement that the space is polynomial is *very* generous. Anything beyond polylogarithmic space per user is generally considered impractical.

Theorem 1. *Let $c \geq 0$ and $0 \leq \delta < 1$. Then, any UP-scheme with n users and space $s \leq n^c$ will, when the number of revoked users $r \leq n^\delta$, require bandwidth*

$$b \geq \frac{1 - \delta}{c + 1} \cdot r \tag{1}$$

Proof. Let S be an arbitrary UP-scheme with $s \leq n^c$ and let $r \leq n^\delta$. An upper bound on the number t of sets of users that can be handled using bandwidth at most b is given by the number of sets of nodes of S of cardinality at most b. Since S contains at most ns nodes, this is upper-bounded by

$$\sum_{i=1}^{b} \binom{ns}{i} \leq (ns)^b \leq n^{(c+1)b} \tag{2}$$

In order for S to be able to handle every set of revoked users of size r, we need t to be at least $\binom{n}{r}$, giving

$$n^{(c+1)b} \geq \binom{n}{r} \geq (n/r)^r \geq n^{(1-\delta)r} \tag{3}$$

and the theorem follows.   □

Theorem 1 comes very close to matching many of the previous works, for instance Subset Difference [16] with $s = \mathcal{O}(\log^2 n)$ and $b = \min(2r-1, n-r)$. For $r \leq \sqrt{n}$, our bound gives $b \geq \frac{r}{2(1+c)}$ which is within a factor $4 + o(1)$.

As mentioned in the introduction, [9] has shown a stronger bound, roughly $r \frac{\log(n/r)}{\log(rs)}$, using the Sunflower lemma. However, their bound applies only to Subset Cover schemes without key derivation, and is in fact stronger than existing schemes using key derivation – e.g. the Subset Difference scheme mentioned above for $r < n^{1/3}$.

4 Arbitrarily Many Revoked Users

In this section we show that, for a certain choice of r, any UP-scheme has to use bandwidth at least $\frac{n}{1.89 \log s}$. We start with Theorem 2, which gives a lower bound on the bandwidth as a function of m/n. Plugging in a suitable value of m/n in Corollary 1 will then give the desired result.

Theorem 2. *Let* $\delta \in (0,1]$ *and* $\epsilon > 0$. *Then for every UP-scheme* S *with* $n > \frac{2\delta(1-\delta)}{\epsilon^2}$ *there exists a set of users* M *of size* $\delta - 3\epsilon \leq |M|/n \leq \delta + \epsilon$ *which requires bandwidth* $b \geq |M| \frac{\log(1/\delta)}{\log(s/\epsilon)}$

Proof. Pick $M_0 \subseteq [n]$ randomly where every element is chosen with probability δ, independently.

Set $d = \log_\delta(\epsilon/s)$ and let X be the set of users which occur at depth exactly d in $S(M_0)$ (where the roots are considered to be at depth 1). Let $M = M_0 \setminus X$. Since each node can cover at most d users of M, the bandwidth required for M is at least

$$\frac{|M|}{d} = |M| \frac{\log(1/\delta)}{\log(s/\epsilon)}$$

It remains to show that there is a positive probability (over the random choice of M_0) that M ends up having the required size, as this implies that such an M exists.

The probability that a node at depth d of S remains in $S(M_0)$ is $\delta^d = \epsilon/s$. The total number of nodes at depth d in S is upper-bounded by ns, and thus, the expected number of nodes at depth d in $S(M_0)$, i.e. the expected size of X, is at most $\delta^d ns = \epsilon n$. By Chebyshev's inequality, we have $\Pr\left[\left|\frac{|M_0|}{n} - \delta\right| \geq \epsilon\right] \leq \frac{\delta(1-\delta)}{n\epsilon^2} < 1/2$. By Markov's inequality, we have $\Pr\left[\frac{|X|}{n} \geq 2\epsilon\right] \leq 1/2$. The union bound then gives that $\Pr[\delta - 3\epsilon \leq |M|/n \leq \delta + \epsilon] > 0$. Thus, there exists some choice of M_0 such that $|M|$ falls within this range. $\square$

As a corollary, we have:

Corollary 1. *For any* $\epsilon > 0$ *there exist* n_0 *and* s_0 *such that any UP-scheme* S *with* $n \geq n_0$ *and* $s \geq s_0$ *uses bandwidth at least*

$$\frac{n}{(e \ln(2) + \epsilon) \log_2(s)} \approx \frac{n}{1.89 \log_2(s)} \tag{4}$$

Proof. Let $\delta = 1/e$. Invoking Theorem 2 with parameters δ and ϵ' (the value of which will be addressed momentarily), we get a set M of size at least $(\delta - 3\epsilon')n$ requiring bandwidth at least

$$n \frac{\delta - 3\epsilon'}{\ln(s/\epsilon')} = n \frac{1 - 3e\epsilon'}{e \ln(2) \log_2(s) + e \ln(1/\epsilon')} \tag{5}$$

Pick ϵ' small enough so that

$$\frac{e \ln(2)}{1 - 3e\epsilon'} \leq e \ln(2) + \epsilon/2.$$

Then Equation (5) is lower-bounded by Equation (4) for any s satisfying

$$\frac{e\ln(1/\epsilon')}{1-3e\epsilon'} \leq \frac{\epsilon}{2}\log_2(s)$$

$$\log_2(s) \geq \frac{2e\ln(1/\epsilon')}{\epsilon(1-3e\epsilon')},$$

and we are done.   □

We remark that Corollary 1 is tight up to the small constant 1.89, as seen by the following theorem.

Theorem 3. *There exists a UP-scheme $\mathcal{S}$ using bandwidth at most $\left\lceil \frac{n}{\log_2(s)} \right\rceil$.*

Proof. Partition the users into $\lceil n/\log_2(s)\rceil$ blocks of size $\leq \log_2(s)$. Then, in each block, use the naive scheme with exponential space and bandwidth 1, independently of the other blocks.   □

5   Bandwidth is $n-r$ for Large r

We show that when the number of revoked users gets very large, all UP-schemes will have a bandwith of $n-r$, e.g. one encryption per member. Exactly how large r has to be for this bound to apply depends on s. This is the same bandwidth as is achieved by the naive solution of just giving each user her own private key.

Theorem 4. *For any UP-scheme $\mathcal{S}$ and $m \leq \frac{n}{6s}$, there is a member set M of size $|M| = m$ requiring bandwidth $b = |M|$.*

Proof. We will build a sequence $M_0 \subseteq M_1 \subseteq M_2 \subseteq \ldots \subseteq M_m$ of sets of members with the properties that $|M_i| = i$, and that the bandwidth required for M_i is i.

The initial set M_0 is the empty set. To construct M_{i+1} from M_i, we pick a user $u \notin M_i$ satisfying:

- There is no $v \in M_i$ such that some node labelled with u occurs as the parent of some node labelled with v
- There is no $v \in M_i$ such that the root node labelled with v occurs as the parent of some node labelled with u
- The root node labelled u has outdegree $\leq 2s$

We then set $M_{i+1} = M_i \cup \{u\}$. Clearly $|M_i| = i$, so there are two claims which remain to be proved. First, that the required bandwidth of M_i is i. Second, that the process can be repeated at least m times.

To compute the bandwidth of M_i, we prove that the only way to cover M_i is to pick the singleton sets of every $u \in M_i$. To see this, assume for contradiction that there exists some set S with $|S| > 1$ that can be used when constructing a cover. This corresponds to a node x at depth $|S|$ of some tree, and S is given by the labels of all nodes from x up to the root. In order for us to be able to use

S when constructing a cover, all these nodes need to belong to M_i. However, the first two criterions in the selection of u above guarantee that not all of these nodes can belong to M_i. The first criterion states that, once we have added a node, we can never add its parent. The second criterion states that, once we have added a root, we can never add any of its children. This shows that there can be no such S.

To see how many steps the process can be repeated, let r_i be the total number of nodes which are "disqualified" after having constructed M_i. Then, M_{i+1} can be constructed if and only if $r_i < n$. First, r_0 equals the number of roots which have degree $> 2s$. Since the total number of nodes is at most ns, this number is at most $r_0 \leq n/2$. When going from M_i to M_{i+1}, the total number of new disqualified nodes can be at most $3s$ – the node added, the parents of the at most $s - 1$ non-root occurrences of u, and the at most $2s$ children of the root labelled with u. Thus, we have that $r_i \leq n/2 + 3si$, which is less then n if $i < \frac{n}{6s}$. $\qquad \square$

The lower bound of Theorem 4 is tight up to a small constant in the following sense.

Theorem 5. *There exists a UP-scheme S such that for any set M of $|M| > \lceil \frac{n}{s} \rceil$ members, the bandwidth is $b < |M|$.*

Proof. Partition the set of users into $B = \lceil \frac{n}{s} \rceil$ blocks of size $\leq s$, and let each user share a key with each of the $s - 1$ other users in her block. Then, given a set M of size $|M| > B$, there must be two users $i, j \in M$ belonging to the same block. Using the key shared by i and j to cover both them both, we see that the bandwidth of M is at most $b \leq |M| - 1$. $\qquad \square$

6   Conclusion

In this paper, we have shown lower bounds for a large class of Subset Cover based Broadcast Encryption schemes. This type of scheme is probably the most explored class of schemes today, with many constructions. Our proofs are in a model with very relaxed constraints compared to what is considered practical, so it would not help to simply relax requirements slightly (e.g. allowing more space or time). The lower bounds shown in this paper match known constructions very well.

In particular, our bounds show that it will be impossible to get a bandwidth of $o(r)$ without increasing the space requirements to unreasonable levels or using some new form of key derivation. We do not have any lower bounds on the memory needed for $\mathcal{O}(r)$ bandwidth, an open question is thus if it is possible to get $\mathcal{O}(r)$ bandwidth with space $o(\log n)$.

Acknowledgements. The authors are grateful to Johan Håstad for many useful comments and discussions.

References

1. Adelsbach, A., Greveler, U.: A broadcast encryption scheme with free-riders but unconditional security. In: Safavi-Naini, R., Yung, M. (eds.) DRMTICS 2005. LNCS, vol. 3919, pp. 246–257. Springer, Heidelberg (2006)
2. Asano, T.: A revocation scheme with minimal storage at receivers. In: Zheng, Y. (ed.) ASIACRYPT 2002. LNCS, vol. 2501, pp. 433–450. Springer, Heidelberg (2002)
3. Asano, T.: Reducing Storage at Receivers in SD and LSD Broadcast Encryption Schemes. In: Chae, K.-J., Yung, M. (eds.) WISA 2003. LNCS, vol. 2908, pp. 317–332. Springer, Heidelberg (2004)
4. Attrapadung, N., Kobara, K., Imai, H.: Sequential key derivation patterns for broadcast encryption and key predistribution schemes. In: Laih, C.S. (ed.) ASIACRYPT 2003. LNCS, vol. 2894, pp. 374–391. Springer, Heidelberg (2003)
5. Berkovits, S.: How to broadcast a secret. In: Davies, D.W. (ed.) EUROCRYPT 1991. LNCS, vol. 547, pp. 535–541. Springer, Heidelberg (1991)
6. Boneh, D., Gentry, C., Waters, B.: Collusion resistant broadcast encryption with short ciphertexts and private keys. In: Shoup, V. (ed.) CRYPTO 2005. LNCS, vol. 3621, pp. 258–275. Springer, Heidelberg (2005)
7. Fiat, A., Naor, M.: Broadcast encryption. In: Stinson, D.R. (ed.) CRYPTO 1993. LNCS, vol. 773, pp. 480–491. Springer, Heidelberg (1994)
8. Gentry, C., Ramzan, Z.: RSA accumulator based broadcast encryption. In: Zhang, K., Zheng, Y. (eds.) ISC 2004. LNCS, vol. 3225, pp. 73–86. Springer, Heidelberg (2004)
9. Gentry, C., Ramzan, Z., Woodruff, D.P.: Explicit exclusive set systems with applications to broadcast encryption. In: Proceedings of the 47th Annual IEEE Symposium on Foundations of Computer Science (FOCS 2006), pp. 27–38. IEEE Computer Society, Washington (2006)
10. Goodrich, M.T., Sun, J.Z., Tamassia, R.: Efficient tree-based revocation in groups of low-state devices. In: Franklin, M. (ed.) CRYPTO 2004. LNCS, vol. 3152, pp. 511–527. Springer, Heidelberg (2004)
11. Halevy, D., Shamir, A.: The LSD broadcast encryption scheme. In: Yung, M. (ed.) CRYPTO 2002. LNCS, vol. 2442, pp. 47–60. Springer, Heidelberg (2002)
12. Hwang, J.Y., Lee, D.H., Lim, J.: Generic transformation for scalable broadcast encryption schemes. In: Shoup, V. (ed.) CRYPTO 2005. LNCS, vol. 3621, pp. 276–292. Springer, Heidelberg (2005)
13. Jho, N.S., Hwang, J.Y., Cheon, J.H., Kim, M.H., Lee, D.H., Yoo, E.S.: One-way chain based broadcast encryption schemes. In: Cramer, R.J.F. (ed.) EUROCRYPT 2005. LNCS, vol. 3494, pp. 559–574. Springer, Heidelberg (2005)
14. Johansson, M., Kreitz, G., Lindholm, F.: Stateful subset cover. In: Zhou, J., Yung, M., Bao, F. (eds.) ACNS 2006. LNCS, vol. 3989, pp. 178–193. Springer, Heidelberg (2006)
15. Luby, M., Staddon, J.: Combinatorial bounds for broadcast encryption. In: Nyberg, K. (ed.) EUROCRYPT 1998. LNCS, vol. 1403, pp. 512–526. Springer, Heidelberg (1998)
16. Naor, D., Naor, M., Lotspiech, J.: Revocation and tracing schemes for stateless receivers. In: Kilian, J. (ed.) CRYPTO 2001. LNCS, vol. 2139, pp. 41–62. Springer, Heidelberg (2001)

A Brief History of Provably-Secure Public-Key Encryption

Alexander W. Dent

Royal Holloway, University of London
Egham, Surrey, TW20 0EX, UK
a.dent@rhul.ac.uk

Abstract. Public-key encryption schemes are a useful and interesting field of cryptographic study. The ultimate goal for the cryptographer in the field of public-key encryption would be the production of a very efficient encryption scheme with a proof of security in a strong security model using a weak and reasonable computational assumption. This ultimate goal has yet to be reached. In this invited paper, we survey the major results that have been achieved in the quest to find such a scheme.

1 Introduction

The most popular field of study within public-key cryptography is that of public-key encryption, and the ultimate goal of public-key encryption is the production of a simple and efficient encryption scheme that is provably secure in a strong security model under a weak and reasonable computational assumption. The cryptographic community has had a lot of successes in this area, but these successes tend to fall into two categories: the production of very efficient encryption schemes with security proofs in idealised models, and the production of less-efficient encryption schemes with full proofs of security in strong models. The ultimate prize has yet to be claimed.

However, we are getting closer to that important break-through. Schemes with full security proofs are getting more efficient and the efficient schemes are getting stronger security guarantees. This paper aims to briefly discuss some of the history behind the production of standard-model-secure encryption schemes and to give a personal interpretation of some of the major results.

The first attempt to prove the security of a public-key encryption scheme was by Rabin [25] in 1979, who described an encryption scheme for which recovering the message was as intractable as factoring an RSA modulus. Later, Goldwasser and Micali [21] described a scheme which they could prove hid all information about the plaintext. However, it wasn't until the early 1990s that researchers began to establish reliable and easy to use formal models for the security of an encryption scheme and that the cryptographic community began to think about constructing practical and efficient provably-secure public-key encryption schemes.

S. Vaudenay (Ed.): AFRICACRYPT 2008, LNCS 5023, pp. 357–370, 2008.
© Springer-Verlag Berlin Heidelberg 2008

1.1   Notation

We will use standard notation. For a natural number $k \in \mathbb{N}$, we let $\{0,1\}^k$ denote the set of k-bit strings and $\{0,1\}^*$ denote the set of bit strings of finite length. We let 1^k denote a string of k ones.

We let $\leftarrow$ denote assignment; hence, $y \leftarrow x$ denotes the assignment to y of the value x. For a set S, we let $x \xleftarrow{\$} S$ denote the assignment to x of a uniformly random element of S. If $\mathcal{A}$ is a randomised algorithm, then $y \xleftarrow{\$} \mathcal{A}(x)$ denotes the assignment to y of the output of $\mathcal{A}$ when run on input x with a fresh set of random coins. If we wish to execute $\mathcal{A}$ using a particular set of random coins R, then we write $y \leftarrow \mathcal{A}(x; R)$, and if $\mathcal{A}$ is deterministic, then we write $y \leftarrow \mathcal{A}(x)$.

1.2   The IND-CCA2 Security Model

A public-key encryption scheme is formally defined as a triple of probabilistic, polynomial-time algorithms $(\mathcal{G}, \mathcal{E}, \mathcal{D})$. The key generation algorithm $\mathcal{G}$ takes as input a security parameter 1^k and outputs a public/private key pair (pk, sk). The public key implicitly defines a message space $\mathcal{M}$ and a ciphertext space $\mathcal{C}$. The encryption algorithm takes as input the public key pk and a message $m \in \mathcal{M}$, and outputs a ciphertext $C \in \mathcal{C}$. The decryption algorithm takes as input the private key sk and a ciphertext $C \in \mathcal{C}$, and outputs either a message $m \in \mathcal{M}$ or the error symbol $\perp$. We demand that the encryption scheme is sound in the sense that if $C \xleftarrow{\$} \mathcal{E}(pk, m)$, then $m \leftarrow \mathcal{D}(sk, C)$, for all keys $(pk, sk) \xleftarrow{\$} \mathcal{G}(1^k)$ and $m \in \mathcal{M}$.

If we are going to prove that an encryption scheme is secure, then we need to have some formal notion of confidentiality. The commonly accepted "correct" definition is that of indistinguishability under adaptive chosen ciphertext attack (IND-CCA2) was proposed by Rackoff and Simon [26]. It built on the weaker notion of IND-CCA1 security proposed by Naor and Yung [24].

Definition 1. *An attacker $\mathcal{A}$ against the IND-CCA2 security of an encryption scheme $(\mathcal{G}, \mathcal{E}, \mathcal{D})$ is a pair of probabilistic polynomial-time algorithms $(\mathcal{A}_1, \mathcal{A}_2)$. The success of the attacker is defined via the IND-CCA2 game:*

$$(pk, sk) \xleftarrow{\$} \mathcal{G}(1^k)$$
$$(m_0, m_1, state) \xleftarrow{\$} \mathcal{A}_1^{\mathcal{D}}(pk)$$
$$b \xleftarrow{\$} \{0,1\}$$
$$C^* \xleftarrow{\$} \mathcal{E}(pk, m_b)$$
$$b' \xleftarrow{\$} \mathcal{A}_2^{\mathcal{D}}(C^*, state)$$

The attacker may query a decryption oracle with a ciphertext C at any point during its execution, with the exception that $\mathcal{A}_2$ may not query the decryption oracle on C^*. The decryption oracle returns $m \leftarrow \mathcal{D}(sk, C)$. The attacker wins the game if $b = b'$. An attacker's advantage is defined to be

$$Adv_{\mathcal{A}}^{IND}(k) = |Pr[b = b'] - 1/2|. \tag{1}$$

We require that a "reasonable" attacker's advantage is "small". This can either be phrased by saying that every polynomial-time attacker must have negligible advantage under the assumption that it is hard to solve some underlying problem (asymptotic security) or by relating the advantage ϵ of an attacker that runs in time t to the success probability ϵ' that an algorithm that runs in time t' has in breaking some underlying hard problem (concrete security). Much is often made of the difference in these two approaches, but in practice they are very similar – they both require that the proof demonstrate a tight reduction from the encryption scheme to the underlying problem. This issue is discussed in more detail in a previous paper [17].

It is sometimes convenient to work with a slightly different definition for advantage. If the IND-CCA2 game encrypts a message defined by the bit b and the attacker outputs the bit b', then

$$Adv_{\mathcal{A}}^{\text{IND}*}(k) = |Pr[b' = 0|b = 0] - Pr[b' = 0|b = 1]|\,. \tag{2}$$

It can easily be shown that

$$Adv_{\mathcal{A}}^{\text{IND}*}(k) = 2 \cdot Adv_{\mathcal{A}}^{\text{IND}}(k)\,. \tag{3}$$

Hence, it is sufficient to bound $Adv_{\mathcal{A}}^{\text{IND}*}$ in order to prove security.

A scheme that is secure against attackers that can only make decryption oracle queries before receiving the challenge ciphertext C^* is said to be IND-CCA1 secure. A scheme that is secure against attackers that do not make any decryption oracle queries at all is said to be IND-CPA or passively secure.

2   The Random Oracle Methodology

No paper on the history of secure encryption schemes would be complete without a mention of the random oracle methodology. In the early 1990s, after the development of the IND-CCA2 security model, researchers turned to the random oracle methodology [4] in order to provide proofs of security for practical public key encryption schemes. The intuition is simple: secure hash functions would share many properties with random functions. Hence, it made sense to model a secure hash function as a completely random function in a security analysis.

This greatly simplifies the process of proving the security of a cryptographic scheme. By modelling the hash function as a random function, we know that the hash function will output completely random and independently generated values on different inputs. Knowledge of the hash values for several different inputs gives absolutely no information about the hash value for any other input and therefore the only way that an attacker can compute the hash value for a given input is to query the hash function oracle on that input. This means that the attacker's behaviour is no longer completely black-box – we may now observe the attacker's behaviour during the attack process (in some limited way). We may even construct the responses that the hash function oracle gives in ways that help prove the security of the cryptosystem (subject to the restriction that they appear to the attacker to be chosen at random).

Of course, schemes proven secure using the random oracle methodology are not necessarily secure when the hash function is instantiated with a given fixed hash function. There is always the possibility that the particular hash function will interact badly with the mathematics of the encryption scheme, and that the resulting system will be insecure. It was, however, hoped that the number of hash functions that "interacted badly" would be small and that a scheme proven secure using the random oracle methodology would be secure when the random oracle was replaced with almost any hash function.

This turned out not to be true. In an amazing paper by Canetti, Goldreich and Halevi [12], it was shown that it was possible to construct an encryption scheme that was provably secure using the random oracle methodology, but was insecure when the random oracle was instantiated with *any* hash function. The paper notes that in the standard model (i.e. when we are not using the random oracle methodology) the attacker has an extra piece of information not available to the attacker in the random oracle model: the attacker has a description of the hash function. The paper gives a scheme for which an attacker can use this description like a password – the attacker submits the description of the hash function to the decryption oracle as a ciphertext and the decryption oracle helpfully returns the private key of the encryption scheme.

It is clear that the encryption scheme of Canetti, Goldreich and Halevi is completely artificial – no real encryption scheme would make use of a decryption algorithm that would output the private key if it were given a ciphertext of a particular (checkable) form. However, it does act as a proof of concept: it is possible to construct a scheme that is secure in the random oracle model, but insecure in the standard model. We therefore cannot completely trust schemes that are only proven secure in the random oracle model. A lot of effort has been expended by cryptographers attempting to find a non-artificial scheme which is secure in the random oracle model, but insecure in practice, but so far no such scheme has been found.

Personally, I still think the random oracle model is a useful tool in cryptography. I believe that it provides trustworthy security guarantees for the vast majority of practical cryptosystems. Furthermore, I don't think I know of a single industrial company or standardisation body that would reject an efficient cryptosystem because it "only" had a proof of security in the random oracle model.

3   Double-and-Add Schemes

We now turn our attention to schemes that can be proven secure in the standard model. The approaches to constructing encryption schemes secure in the standard model tend to fall several categories. The first approach is to use a "double-and-add" technique, in which a message is encrypted twice (using two weak encryption schemes) and a checksum value is added to the ciphertext.

3.1   The NIZK Schemes

The first attempt to prove the security of a scheme against chosen ciphertext attacks was given by Naor and Yung [24]. Their approach was to encrypt a message twice using two independent IND-CPA secure encryption schemes, and then to provide a non-interactive zero-knowledge (NIZK) proof that the two ciphertexts were encryptions of the same message. The Naor-Yung result only produced an encryption scheme that was IND-CCA1 secure. Their approach was extended by Sahai [27] to cover IND-CCA2 attacks by using a slightly more powerful NIZK proof system.

It is not going to be possible, due to space constraints, to fully explain the technical details of this scheme. However, we will give an overview of the scheme. Suppose $(\mathcal{G}, \mathcal{E}, \mathcal{D})$ is an IND-CPA secure encryption scheme. The Sahai encryption scheme works as follows:

- **Key generation.** Generate two independent key pairs $(pk_1, sk_1) \xleftarrow{\$} \mathcal{G}(1^k)$ and $(pk_2, sk_2) \xleftarrow{\$} \mathcal{G}(1^k)$, and a random string σ (for use by the NIZK proof). The public key is $pk = (pk_1, pk_2, \sigma)$ and the private key is $sk = (sk_1, sk_2)$.
- **Encryption.** To encrypt a message m, compute $C_1 \xleftarrow{\$} \mathcal{E}(pk_1, m)$ and $C_2 \xleftarrow{\$} \mathcal{E}(pk_2, m)$, and give a NIZK proof π that C_1 and C_2 are encryptions of the same message (using the random string σ). The ciphertext is (C_1, C_2, π).
- **Decryption.** To decrypt a message, first check the proof π. If the proof fails, then output $\perp$. Otherwise, output $m \leftarrow \mathcal{D}(sk_1, C_1)$.

Of course, as the NIZK proof π proves that C_1 and C_2 are the encryption of the same message, we could have equivalently computed $m \leftarrow \mathcal{D}(sk_2, C_2)$ in the decryption algorithm.

The key to understanding the security of this scheme is in understanding the security properties of the NIZK proof system. We require two properties from the NIZK proof system:

- **Zero knowledge.** It should be possible to choose the random string σ in such a way that the NIZK proof system has a trapdoor τ that allows an entity in possession of the trapdoor to produce false proofs – i.e. it should be possible to "prove" that any pair of ciphertexts (C_1, C_2) are the encryption of the same message using the trapdoor τ, even if (C_1, C_2) are encryptions of different messages. Furthermore, it should be impossible for the attacker (who only knows the string σ and not the trapdoor τ) to be able to distinguish false proofs from real ones.
- **Simulation Sound.** It should be impossible for the attacker to produce a proof π that two ciphertexts (C_1, C_2) are encryptions of the same message unless the ciphertexts actually are the encryptions of the same message. Furthermore, this property should hold even if the attacker is given a false proof π which is computed using the trapdoor τ.

The ideas behind the proof become very simple to understand if one considers bounding $Adv_{\mathcal{A}}^{\text{IND}*}$ rather than $Adv_{\mathcal{A}}^{\text{IND}}$. In the IND* security model, we observe the difference in the attacker's behaviour when the challenge encryption C^* is an encryption of m_0 and when the challenge encryption C^* is an encryption of m_1. Recall that this C^* is of the form (C_1^*, C_2^*, π^*) where $C_i^* \xleftarrow{\$} \mathcal{E}(pk_i, m_b)$. First, since the NIZK proof system is zero knowledge, we may assume that the challenger has

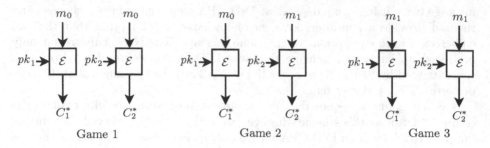

Fig. 1. The games used in security proof of the Sahai construction (with the NIZK proof omitted)

chosen a random string with a trapdoor, and that the NIZK proof π^* is produced using the trapdoor τ, rather than by using the normal proof algorithm.

We use a simple game-hopping argument (as illustrated in Figure 1). Let Game 1 be the game in which the challenge ciphertext C^* is computed as an encryption of m_0. In other words,

$$C_1^* \xleftarrow{\$} \mathcal{E}(pk_1, m_0) \qquad C_2^* \xleftarrow{\$} \mathcal{E}(pk_2, m_0)$$

and π^* is a proof that (C_1^*, C_2^*) are encryptions of the same message. Let Game 2 be the game in which the challenge ciphertext C^* is computed as

$$C_1^* \xleftarrow{\$} \mathcal{E}(pk_1, m_0) \qquad C_2'^* \xleftarrow{\$} \mathcal{E}(pk_2, m_1)$$

and π^* is a false proof that (C_1^*, C_2^*) are encryptions of the same message computed using the trapdoor τ. We claim that any attacker that can distinguish between Game 1 and Game 2 can also break the IND-CPA security of the second encryption scheme. The reduction makes use of the fact that we may decrypt a valid ciphertext using the secret key for the first encryption scheme – this allows us to simulate the decryption oracle.

Similarly, let Game 3 be the game in which the challenge ciphertext C^* is computed as

$$C_1^* \xleftarrow{\$} \mathcal{E}(pk_1, m_1) \qquad C_2^* \xleftarrow{\$} \mathcal{E}(pk_2, m_1)$$

and π^* is a proof that (C_1^*, C_2^*) are encryptions of the same message. If the attacker can distinguish between Game 2 and Game 3, then the attacker can break the IND-CPA security of the first scheme. This time the reduction makes use of the fact that we may decrypt a valid ciphertext using the secret key for the second encryption scheme.

The beauty of this construction is that it allows us to prove that secure public-key encryption schemes exist assuming only the existence of trapdoor one-way permutations. Sahai [27] notes that passively secure encryption schemes exist under the assumption that trapdoor one-way functions exist [20] and builds suitable NIZK proof systems using the results of Feige, Lapidot and Shamir [19] and Bellare and Yung [6]. This is a wonderful theoretical result, but, due to the theoretical nature of the NIZK proof system used in the construction, the construction is not practical.

3.2 The Cramer-Shoup Encryption Scheme

The first practical public-key encryption scheme that was proven secure in the standard model was the Cramer-Shoup scheme [14]. Although not explicitly presented as an extension of the Sahai construction, it can be thought of as building on these ideas. Suppose $\mathbb{G}$ is a cyclic group of prime order p that is generated by g and that $Hash : \mathbb{G}^3 \to \mathbb{Z}_p$ is a (target collision resistant) hash function. The Cramer-Shoup encryption scheme can be written as[1]:

$\mathcal{G}(1^k)$	$\mathcal{E}(pk, m)$	$\mathcal{D}(sk, C)$
$\hat{g} \xleftarrow{\$} \mathbb{G}$	$r \xleftarrow{\$} \mathbb{Z}_p$	Parse C as $(a, \hat{a}, c, d)$
$x_1, x_2, y_1, y_2, z \xleftarrow{\$} \mathbb{Z}_p$	$a \leftarrow g^r$	$v \leftarrow Hash(a, \hat{a}, c)$
$h \leftarrow g^z$	$\hat{a} \leftarrow \hat{g}^r$	If $d \neq a^{x_1 + y_1 v} \hat{a}^{x_2 + y_2 v}$
$e \leftarrow g^{x_1} \hat{g}^{x_2}$	$c \leftarrow h^r m$	Output $\bot$
$f \leftarrow g^{y_1} \hat{g}^{y_2}$	$v \leftarrow Hash(a, \hat{a}, c)$	$m \leftarrow c/a^z$
$pk \leftarrow (g, \hat{g}, h, e, f)$	$d \leftarrow e^r f^{rv}$	Output m
$sk \leftarrow (x_1, x_2, y_1, y_2, z)$	Output $(a, \hat{a}, c, d)$	

This scheme is proven secure under the assumption that the DDH problem is hard to solve in $\mathbb{G}$ and the hash function is target collision resistant.

On first glance, this scheme does not appear to have much in common with Sahai's double-and-add scheme. However, consider a variant of the ElGamal encryption scheme [18] in the group $\mathbb{G}$ generated by an element h:

$\mathcal{G}(1^k)$	$\mathcal{E}(pk, m)$	$\mathcal{D}(sk, C)$
$z \xleftarrow{\$} \mathbb{Z}_p^*$	$r \xleftarrow{\$} \mathbb{Z}_p$	Parse C as (a, c)
$g \leftarrow h^{1/z}$	$a \leftarrow g^r$	$m \leftarrow c/a^z$
$pk \leftarrow g$	$c \leftarrow h^r m$	Output m
$sk \leftarrow z$	Output (a, c)	

This scheme is known to be IND-CPA secure under the DDH assumption. In order to use this scheme with Sahai's construction, we would need to encrypt the same message twice using separate random values for each encryption. However, Bellare, Boldyreva and Staddon [1] show that the ElGamal scheme remains secure when it is used to encrypt the same message under multiple public keys *even if the same random value r is used in all the encryptions*. Hence, we may think of $(a, \hat{a}, c)$ as a double encryption of the same message under two separate public keys.

To complete the analogy, we must show that d acts in a manner similar to the NIZK proof π in the Sahai construction. Therefore, d would have to have properties similar to simulation soundness and zero knowledge. In the Cramer-Shoup scheme $(a, \hat{a}, c)$ is a valid double encryption of the same message providing that there exists a value r such that $a = g^r$ and $\hat{a} = \hat{g}^r$ – i.e. providing that $(g, \hat{g}, a, \hat{a})$ form a DDH triple. An examination of the security proof for the Cramer-Shoup

---

[1] Technically, this is the CS1a scheme.

scheme shows that a large portion of that proof is devoted to showing that we can reject ciphertexts submitted to the decryption oracle for which $(g, \hat{g}, a, \hat{a})$ is not a DDH triple. This is analogous to simulation soundness. A further examination of the proof shows that it constructs the challenge ciphertext as $a \leftarrow g^r$ and $\hat{a} \leftarrow g^{r'}$ for $r \neq r'$. The "proof" d is falsely constructed from $(a, \hat{a})$ using knowledge of (x_1, x_2, y_1, y_2). This is clearly analogous to the zero knowledge property.

We note that the analogy is not entirely correct. In order to verify the correctness of the "proof" d, it is necessary to know the secret values (x_1, x_2, y_1, y_2). In the analogy, this would be the equivalent to requiring the trapdoor τ to verify the NIZK proof and Sahai's construction does not appear to work if the trapdoor is required to verify proofs. However, the similarities between the Cramer-Shoup encryption scheme and the Sahai construction are striking. Other variants of the Cramer-Shoup scheme, such as the Kurosawa-Desmedt scheme [23], can be viewed similarly, albeit with more complex analyses.

4   Signatures and Identities

The security of the "double-and-add" schemes of the preceding section can be proven because there are two equivalent ways in which a ciphertext can be decrypted. Therefore, if part of the security proof prevents us from using one decryption method, then we may still decrypt ciphertexts correctly using the other decryption method. In this section, we look at a technique which handles decryption in another way.

The elegant technique we will look at was proposed by Canetti, Halevi and Katz [13] and converts a passively secure identity-based encryption scheme into a fully secure public-key encryption scheme using a one-time signature scheme.

The formal security models for identity-based encryption were introduced by Boneh and Franklin [9]. An identity-based encryption scheme is a set of four probabilistic, polynomial-time algorithms $(IGen, Ext, Enc, Dec)$. The $IGen$ algorithm takes as input the security parameter 1^k, and outputs the public parameters of the system mpk and the master private key msk. The key extraction algorithm Ext takes as input an identity ID and the master private key msk, and outputs a decryption key sk_{ID} for that identity. The encryption algorithm Enc takes as input the master public key mpk, an identity ID and a message m, and outputs a ciphertext C. The decryption algorithm takes as input the master public key mpk, a ciphertext C and a decryption key sk_{ID}, and outputs either a message m or an error symbol $\perp$. The security notion in which we are interested is the IND-CPA notion of security for identity-based encryption. The IND-CPA game for an identity-based encryption is:

$$(mpk, msk) \xleftarrow{\$} IGen(1^k)$$
$$(ID^*, m_0, m_1, state) \xleftarrow{\$} \mathcal{A}_1^{Ext}(pk)$$
$$b \xleftarrow{\$} \{0, 1\}$$
$$C^* \xleftarrow{\$} Enc(mpk, ID^*, m_b)$$
$$b' \xleftarrow{\$} \mathcal{A}_2^{Ext}(C^*, state)$$

The attacker may query an extraction oracle Ext with an identity ID and the oracle will return decryption key $Ext(msk, ID)$. The attacker wins the game if $b = b'$ and the attacker never queried the extraction oracle on ID^*. We define the attacker's advantage in the same way as for public-key encryption.

A one-time signature scheme is a triple of probabilistic, polynomial-time algorithms $(SigGen, Sign, Verify)$. The $SigGen$ algorithm takes as input the security parameter 1^k and outputs a public/private key pair (vrk, snk). The signing algorithm $Sign$ takes as input a private signing key snk and a message m, and outputs a signature σ. The verification algorithm $Verify$ takes as input a public verification key vrk, a message m and a signature σ, and outputs either $true$ or $false$. The verification algorithm should verify all signature created using the signing algorithm. Furthermore, the attacker should not be able to forge a new signature on any message after having seen a single message/signature pair.

The complete public-key encryption scheme is as follows:

$$
\begin{array}{lll}
\mathcal{G}(1^k) & \mathcal{E}(pk, m) & \mathcal{D}(sk, C) \\
\quad (mpk, msk) & \quad (vrk, snk) & \quad \text{Parse } C \text{ as } (c, vrk, \sigma) \\
\quad \xleftarrow{\$} IGen(1^k) & \quad \xleftarrow{\$} SigGen(1^k) & \quad \text{If } Verify(vrk, c, \sigma) \neq true \\
\quad pk \leftarrow mpk & \quad ID \leftarrow vrk & \quad \quad \text{Output } \perp \\
\quad sk \leftarrow msk & \quad c \xleftarrow{\$} Enc(mpk, ID, m) & \quad ID \leftarrow vrk \\
& \quad \sigma \xleftarrow{\$} Sign(snk, c) & \quad sk_{ID} \xleftarrow{\$} Ext(msk, ID) \\
& \quad \text{Output } (c, vrk, \sigma) & \quad m \leftarrow Dec(mpk, sk_{ID}, c) \\
& & \quad \text{Output } m
\end{array}
$$

The principle behind the security proof for this elegant construction couldn't be simpler. We know that the identity-based encryption scheme is IND-CPA secure, therefore the public-key encryption scheme is secure if we can find a way to simulate a decryption oracle. Suppose the challenge ciphertext is (c^*, vrk^*, σ^*) and consider a ciphertext (c, vrk, σ) submitted to a decryption oracle. If $vrk \neq vrk^*$ then we may request the decryption key for the identity vrk and decrypt the ciphertext ourselves. If $vrk = vrk^*$ then either the signature σ is invalid or the attacker has broken the unforgeability of the one-time signature scheme. Hence, with overwhelming probability, we may return $\perp$ as the decryption oracle's response.

There are a number of other schemes that prove their security using similar principles [10,11]. In many ways, it is ironic that it was the development of standard-model-secure identity-based encryption schemes (a harder primitive to construct) that produced the next chapter in the development of public-key encryption schemes. However, these schemes are similar to the "double-and-add" schemes in that they convert a passively secure scheme into a fully secure scheme using a cryptographic checksum. This two-stage process is never going to be as efficient as other constructions might be.

5   Extracting Plaintext Awareness

Plaintext awareness is a simple idea with a complicated explanation. An encryption scheme is plaintext aware if it is impossible for a user to create a valid ciphertext without knowing the underlying message. This effectively makes a decryption oracle useless to the attacker – any valid ciphertext he submits to the decryption oracle will return a message that he already knows. If he submits a ciphertext to the decryption oracle for which he does not know the underlying message, then the decryption oracle will return ⊥. This leads to the central theorem of plaintext awareness: that a scheme that is IND-CPA secure and plaintext aware is IND-CCA2 secure.

The difficulty with this idea is formalising what it means to say that a user "knows" an underlying message. The first attempt to produce a formal definition for plaintext awareness was given in the random oracle model [2,5] but had the disadvantage that it could *only* be realised in the random oracle model. It took several years before a definition compatible with the standard model was found.

5.1 Plaintext Awareness Via Key Registration

The first attempt to provide a standard-model definition of plaintext awareness was given by Herzog, Liskov and Micali [22]. In their model, if a sender wishes to send a message to a receiver, then both the sender and the receiver must have a public key. Furthermore, the sender must register their public key with some trusted registration authority in a process that includes a zero-knowledge proof of knowledge for the private key. Now, whenever the sender wants to send a message, it forms two ciphertexts – an encryption of the message using the receiver's public key and an encryption of the message using the sender's own public key – and provides a NIZK proof that the ciphertexts are the encryption of the same message. The receiver decrypts the ciphertext by checking the validity of the NIZK proof and decrypting the component that was encrypted using their public key.

The plaintext awareness of the scheme can be easily shown: since the NIZK proves that the encryptions are identical, we know that both ciphertexts are the encryption of the same message. Furthermore, since the sender has proven knowledge of the private key, we know that the sender can decrypt the component of the ciphertext encrypted using the sender's public key and recover the message. Hence, we can conclude that the sender "knows" the message.

This is an interesting idea, and clearly related to the security of the Sahai construction, but it is never really been adopted to prove the security of practical schemes. The requirement that the sender must have a registered public key creates the need for a huge public-key infrastructure which is unlikely to exist in practice. Furthermore, the scheme still makes use of arbitrary zero-knowledge proofs of knowledge and NIZK proof systems, which are impractical.

5.2 Using Extractors

In 2004, Bellare and Palacio [3] introduced a new standard-model definition for plaintext awareness. Their definition has several advantages over the definition of Herzog, Liskov and Micali. In particular, Bellare and Palacio's definition doesn't require a sender to register a key. It is also compatible with earlier definitions in the random oracle model, in the sense that a scheme proven plaintext aware using the random-oracle-based definition of plaintext awareness is also plaintext aware using the standard-model-based definition of plaintext awareness (although the proof of this fact uses the random oracle model).

The Bellare and Palacio definition of plaintext awareness uses a definition of "knowledge" that is similar to the definition used in zero knowledge. An attacker $\mathcal{A}$ is deemed to "know" a value x if it is possible to alter $\mathcal{A}$ to give a new algorithm $\mathcal{A}^*$ that outputs x.

Let (pk, sk) be a randomly generated key pair for a public-key encryption scheme $(\mathcal{G}, \mathcal{E}, \mathcal{D})$. We consider an attacker $\mathcal{A}$ that takes as input a public key pk and a set of random coins R, and interacts with an "oracle" to which it can submit ciphertexts. The form of the oracle depends upon the game that the attacker is playing. In the REAL game, the oracle is instantiated using the decryption algorithm $\mathcal{D}(sk, \cdot)$. In the FAKE game, the oracle is instantiated by an algorithm $\mathcal{A}^*$ which we call the *plaintext extractor*. This plaintext extractor $\mathcal{A}^*$ is a stateful, probabilistic, polynomial-time algorithm that depends upon $\mathcal{A}$ and initially takes as input the public key pk and the random coins R used by $\mathcal{A}$. Since $\mathcal{A}^*$ has all the inputs of $\mathcal{A}$, one can think of $\mathcal{A}^*$ as observing $\mathcal{A}$'s behaviour as it creates ciphertexts. If the attacker $\mathcal{A}$ submits a ciphertext C to the plaintext extractor $\mathcal{A}^*$, then it is $\mathcal{A}^*$'s task to determine the underlying message from $\mathcal{A}$'s behaviour.

It would be nice if we were done here, but we also need to consider the possibility that the attacker $\mathcal{A}$ can obtain some ciphertexts for which he does not know the underlying message. In the real world, this corresponds to the idea that the attacker might be able to observe ciphertexts created by other people. In the IND security model, this allows for the fact that the attacker is given the challenge ciphertext C^* (for which he does not know the underlying encryption). This possibility is allowed for in the security model for plaintext awareness by giving the attacker access to an encryption oracle that, when queried with some auxiliary information aux, generates a message $m \xleftarrow{\$} \mathcal{P}(aux)$ (using some arbitrary, stateful, probabilistic polynomial-time algorithm $\mathcal{P}$) and returns the ciphertext $C \xleftarrow{\$} \mathcal{E}(pk, m)$. We are forced to give C to the plaintext extractor $\mathcal{A}^*$ so that it may continue to observe $\mathcal{A}$'s behaviour. We forbid $\mathcal{A}$ from asking for the decryption of C. We show the differences between the REAL and FAKE game graphically in Fig. 2.

We say that a scheme is plaintext aware if, for any attacker $\mathcal{A}$, there exists a plaintext extractor $\mathcal{A}^*$ such that, for *any* plaintext creating algorithm $\mathcal{P}$, the output x of $\mathcal{A}$ in the REAL game is indistinguishable from the output x of $\mathcal{A}$ in the FAKE game.

Bellare and Palacio [3] prove that any scheme that is IND-CPA secure and plaintext aware in this model is necessarily IND-CCA2 secure. In an extraordinary paper, Teranishi and Ogata [28] prove that a scheme that is one-way and plaintext aware in this model is necessarily IND-CCA2 secure. There are weaker models for plaintext awareness that are similar to this model, and their relationships to the full security model have been well explored by Bellare and Palacio [3] and by Birkett and Dent [8].

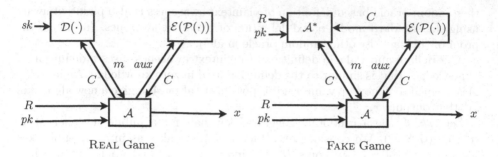

Fig. 2. The REAL and FAKE games for plaintext awareness

The first scheme that was proven fully plaintext aware in the standard model was the Cramer-Shoup encryption scheme [16]. This proof relies heavily on the *Diffie-Hellman Knowledge* assumption first introduced by Damgård [15]. This assumption is meant to capture the intuition that the only way the attacker can compute a Diffie-Hellman tuple (g, h, g^r, h^r) from the pair (g, h) is by generating r and computing (g^r, h^r) directly. The definition states that for every attacker $\mathcal{A}$ that outputs (g^r, h^r), there exists an algorithm $\mathcal{A}^*$ that can output r given the random coins of $\mathcal{A}$. This is known as an extractor assumption, as the algorithm $\mathcal{A}^*$ extracts the random value r by observing the execution of $\mathcal{A}$. Birkett and Dent [7] have shown that other schemes with a similar structures to the Cramer-Shoup [14] and Kurosawa-Desmedt [23] schemes are plaintext aware under similar extractor assumptions.

This highlights the most significant problem with the plaintext awareness approach to proving security: no-one has yet managed to prove the plaintext awareness of an encryption scheme without the use of an extractor assumption. These extractor assumption are poor things on which to base the security of an encryption scheme as it is very difficult to gain any evidence about whether the assumption is true or not. It can be as difficult to prove the assumption is false as it is to prove the assumption is true.

6 Conclusion

The cryptographic community have come a long way in proving the security of public-key encryption schemes. However, the ultimate prize is still yet to be claimed: a proof of security for an ultra-efficient encryption scheme in the standard model. The approaches we have discussed in this paper do make significant advantages in improving the efficiency of schemes with full security proofs. However, none of the approaches seem likely to break the final efficiency barrier. Both the "double-and-add" schemes and the identity-based schemes require separate encryption and checksum operations. Hence, the resulting encryption schemes require two "expensive" calculations. On the other hand, the plaintext awareness approach relies on extractor-based assumptions, which do not engender confidence in the security of the scheme, and still do not seem to be able to prove the security of a scheme that uses less than two "expensive" calculations. It seems as if a new technique has to be developed before this barrier can be broken.

Acknowledgements. I'd like to thank Prof. Serge Vaudenay and Prof. Abdelhak Azhari for extending the invitation to me to give an invited talk at Africacrypt 2008, and I would like to stress that this paper has not been refereed by a peer review process. Thus, any mistakes or inaccuracies in the paper should be considered mine alone and no blame should be attached to the programme committee or any external reviewers. I'd also like to thank James Birkett and Gaven Watson for their comments on the paper.

References

1. Bellare, M., Boldyreva, A., Staddon, J.: Multi-recipient encryption schemes: Security notions and randomness re-use. In: Desmedt, Y.G. (ed.) PKC 2003. LNCS, vol. 2567, pp. 85–99. Springer, Heidelberg (2002)
2. Bellare, M., Desai, A., Pointcheval, D., Rogaway, P.: Relations among notions of security for public-key encryption schemes. In: Krawczyk, H. (ed.) CRYPTO 1998. LNCS, vol. 1462, pp. 26–45. Springer, Heidelberg (1998)
3. Bellare, M., Palacio, A.: Towards plaintext-aware public-key encryption without random oracles. In: Lee, P.J. (ed.) ASIACRYPT 2004. LNCS, vol. 3329, pp. 48–62. Springer, Heidelberg (2004)
4. Bellare, M., Rogaway, P.: Random oracles are practical: A paradigm for designing efficient protocols. In: Proc. of the First ACM Conference on Computer and Communications Security, pp. 62–73 (1993)
5. Bellare, M., Rogaway, P.: Optimal asymmetric encryption. In: De Santis, A. (ed.) EUROCRYPT 1994. LNCS, vol. 950, pp. 92–111. Springer, Heidelberg (1995)
6. Bellare, M., Yung, M.: Certifying permutations: Non-interactive zero-knowledge based on any trapdoor permutation. Journal of Cryptology 9(1), 149–166 (1996)
7. Birkett, J., Dent, A.W.: The generalised Cramer-Shoup and Kurosawa-Desmedt schemes are plaintext aware (unpublished manuscript, 2008)
8. Birkett, J., Dent, A.W.: Relations among notions of plaintext awareness. In: Cramer, R. (ed.) Public Key Cryptography – PKC 2008. LNCS, vol. 4939, pp. 47–64. Springer, Heidelberg (2008)
9. Boneh, D., Franklin, M.: Identity-based encryption from the Weil pairing. In: Kilian, J. (ed.) CRYPTO 2001. LNCS, vol. 2139, pp. 213–229. Springer, Heidelberg (2001)
10. Boneh, D., Katz, J.: Improved efficiency for CCA-secure cryptosystems built using identity-based encryption. In: Menezes, A. (ed.) CT-RSA 2005. LNCS, vol. 3376, pp. 87–103. Springer, Heidelberg (2005)
11. Boyen, X., Mei, Q., Waters, B.: Direct chosen ciphertext security from identity-based techniques. In: Proc. of the 12th ACM Conference on Computer and Communications Security, pp. 320–329 (2005)
12. Canetti, R., Goldreich, O., Halevi, S.: The random oracle model, revisited. In: Proc. of the 30th Annual ACM Symposium on the Theory of Computing – STOC 1998, pp. 209–218 (1998)
13. Canetti, R., Halevi, S., Katz, J.: Chosen-ciphertext security from identity-based encryption. In: Cachin, C., Camenisch, J.L. (eds.) EUROCRYPT 2004. LNCS, vol. 3027, pp. 207–222. Springer, Heidelberg (2004)

14. Cramer, R., Shoup, V.: Design and analysis of practical public-key encryption schemes secure against adaptive chosen ciphertext attack. SIAM Journal on Computing 33(1), 167–226 (2004)
15. Damgård, I.B.: Towards practical public key systems secure against chosen ciphertext attacks. In: Feigenbaum, J. (ed.) CRYPTO 1991. LNCS, vol. 576, pp. 445–456. Springer, Heidelberg (1992)
16. Dent, A.W.: The Cramer-Shoup encryption scheme is plaintext aware in the standard model. In: Vaudenay, S. (ed.) EUROCRYPT 2006. LNCS, vol. 4004, pp. 289–307. Springer, Heidelberg (2006)
17. Dent, A.W.: Fundamental problems in provable security and cryptography. Phil. Trans. R. Soc. A 364(1849), 3215–3230 (1849)
18. ElGamal, T.: A public key cryptosystem and a signature scheme based on discrete logarithms. IEEE Transactions on Information Theory 31, 469–472 (1985)
19. Feige, U., Lapidot, D., Shamir, A.: Multiple noninteractive zero knowledge proofs under general assumptions. SAIM Journal on Computing 29(1), 1–28 (1999)
20. Goldreich, O., Levin, L.A.: A hard-core predicate for all one-way functions. In: Proceedings of the 21st Symposium on Theory of Computer Science – STOC 1989, pp. 25–32. ACM, New York (1989)
21. Goldwasser, S., Micali, S.: Probabilistic encryption. Journal of Computer and System Science 28, 270–299 (1984)
22. Herzog, J., Liskov, M., Micali, S.: Plaintext awareness via key registration. In: Boneh, D. (ed.) CRYPTO 2003. LNCS, vol. 2729, pp. 548–564. Springer, Heidelberg (2003)
23. Kurosawa, K., Desmedt, Y.: A new paradigm of hybrid encryption scheme. In: Franklin, M. (ed.) CRYPTO 2004. LNCS, vol. 3152, pp. 426–442. Springer, Heidelberg (2004)
24. Naor, M., Yung, M.: Public-key cryptosystems provably secure against chosen ciphertext attacks. In: Proc. 22nd Symposium on the Theory of Computing – STOC 1990, pp. 427–437. ACM, New York (1990)
25. Rabin, M.O.: Digitalized signatures and public-key functions as intractable as factorization. Technical Report MIT/LCS/TR-212, MIT Laboratory for Computer Science (1979)
26. Rackoff, C., Simon, D.: Non-interactive zero-knowledge proof of knowledge and chosen ciphertext attack. In: Feigenbaum, J. (ed.) CRYPTO 1991. LNCS, vol. 576, pp. 434–444. Springer, Heidelberg (1992)
27. Sahai, A.: Non-malleable non-interactive zero knowledge and adaptive chosen-ciphertext security. In: 40th Annual Symposium on Foundations of Computer Science, FOCS 1999, pp. 543–553. IEEE Computer Society, Los Alamitos (1999)
28. Teranishi, I., Ogata, W.: Relationship between standard model plaintext awareness and message hiding. In: Lai, X., Chen, K. (eds.) ASIACRYPT 2006. LNCS, vol. 4284, pp. 226–240. Springer, Heidelberg (2006)

On Compressible Pairings and Their Computation

Michael Naehrig[1,*], Paulo S.L.M. Barreto[2,**], and Peter Schwabe[1]

[1] Department of Mathematics and Computer Science
Technische Universiteit Eindhoven, P.O. Box 513, 5600 MB Eindhoven, Netherlands
{michael,peter}@cryptojedi.org
[2] Escola Politécnica, Universidade de São Paulo.
Av. Prof. Luciano Gualberto, tr. 3, n. 158.
BR 05508-900, São Paulo(SP), Brazil
pbarreto@larc.usp.br

Abstract. In this paper we provide explicit formulæ to compute bilinear pairings in compressed form. We indicate families of curves where the proposed compressed computation method can be applied and where particularly generalized versions of the Eta and Ate pairings due to Zhao *et al.* are especially efficient. Our approach introduces more flexibility when trading off computation speed and memory requirement. Furthermore, compressed computation of reduced pairings can be done without any finite field inversions. We also give a performance evaluation and compare the new method with conventional pairing algorithms.

Keywords: pairing-based cryptography, compressible pairings, algebraic tori, Tate pairing, Eta pairing, Ate pairing, twists.

1 Introduction

Cryptographically relevant bilinear maps like the Tate and Weil pairing usually take values over an extension field $\mathbb{F}_{p^k}$ of the base field $\mathbb{F}_p$. Pairing inputs are typically points on an elliptic curve defined over $\mathbb{F}_p$. It has been known for a while (see the work of Scott and Barreto [13] and Granger, Page and Stam [7]) that pairing values can be efficiently represented in compressed form by using either traces over subfields or algebraic tori. The former approach leads to a small loss of functionality: the trace of an exponential, $\mathrm{Tr}(g^x)$, can be computed from the trace $\mathrm{Tr}(g)$ and the exponent x alone, but the trace of a product $\mathrm{Tr}(gh)$ cannot be easily computed from $\mathrm{Tr}(g)$ and $\mathrm{Tr}(h)$. The latter approach does not suffer from this drawback, since torus elements can implicitly be multiplied in the compressed representation. With either approach, pairing values can be

---

* Most of the work presented in this paper was done, while the first author was visiting the Escola Politécnica, Universidade de São Paulo, Brazil. The visit was supported by the German Research Foundation (DFG).

** Supported by the Brazilian National Council for Scientific and Technological Development (CNPq) under grant 312005/2006-7.

S. Vaudenay (Ed.): AFRICACRYPT 2008, LNCS 5023, pp. 371–388, 2008.

efficiently compressed to one half or one third of the original length, depending on the precise setting of the underlying fields and curves.

Our contribution in this paper is to provide explicit formulæ to compute pairings directly in compressed form. Although we do not claim any performance improvement over existing methods, we show that full implementation of arithmetic over $\mathbb{F}_{p^k}$ can be avoided altogether; only operations for manipulating pairing arguments and (compressed) pairing values are needed.

From an implementor's or hardware designer's perspective the contribution of this paper consists of mainly two aspects. Firstly, the explicit formulæ for multiplication and squaring of torus elements give more flexibility in trading off computation speed with memory requirement. The second aspect concerns field inversions during pairing computation. Using projective representation for curve points, inversions can be avoided in the Miller loop. However, a very efficient way to then compute the final exponentiation is to decompose the exponent into three factors and use the Frobenius automorphism to compute powers for two of these factors. This involves an inversion in $\mathbb{F}_{p^k}$, which can be avoided using the compressed representation of pairing values. Hence, we can entirely avoid field inversion during pairing computation and still use fast Frobenius actions in the final exponentiation. From a more theoretical perspective this approach can be seen as a first step to further enhancement of the resulting algorithms, and parallels the case of hyperelliptic curve arithmetic where the introduction of explicit formulæ paved the way to more efficient arithmetic.

We provide timing results for implementations of different pairing algorithms, comparing the newly proposed pairings in compressed form with their conventional counterparts. Additionally, we give examples of curve families amenable to pairing compression where generalized versions of the Eta and Ate pairings due to Zhao et al. are more efficient than the non-generalized versions. We provide examples for the three AES security levels 128, 192 and 256 bits. In this paper we use the notion Eta pairing instead of twisted Ate pairing, because it has originally been used in the non-supersingular case as well.

This paper is organized as follows. In Sections 2 and 3 we review mathematical concepts related to pairings and algebraic tori. In Section 4 we discuss torus-based pairing compression and provide explicit formulæ for pairing computation in compressed form. We describe how to avoid inversions in Section 5. In Section 6 implementation costs are given and we conclude in Section 7.

2   Preliminaries on Pairings

Let E be an elliptic curve defined over a finite field $\mathbb{F}_p$ of characteristic $p \geq 5$. Let r be a prime divisor of the group order $n = \#E(\mathbb{F}_p)$ and let k be the embedding degree of E with respect to r, i.e. k is the smallest integer such that $r \mid p^k - 1$. We assume that $k > 1$.

Let $\mathbb{F}_q$ be an extension of $\mathbb{F}_p$. An elliptic curve E' over $\mathbb{F}_q$ is called a *twist of degree* d if there exists an isomorphism $\psi_d : E' \to E$ defined over $\mathbb{F}_{q^d}$ and d is minimal with this property. There is a nice summary about twists of elliptic curves regarding their existence and the possible group orders of $E'(\mathbb{F}_q)$ given by Hess, Smart and Vercauteren in [8].

We consider an r-torsion point $P \in E(\mathbb{F}_p)[r]$ and an independent r-torsion point $Q \in E(\mathbb{F}_{p^k})[r]$. We fix $G_1 = \langle P \rangle \subseteq E(\mathbb{F}_p)[r]$ and $G_2 = \langle Q \rangle \subseteq E(\mathbb{F}_{p^k})[r]$. If the curve has a twist of order d we may choose the point Q arising as $Q = \psi_d(Q')$, where Q' is an $\mathbb{F}_{p^{k/d}}$-rational point of order r on the twist E', see again [8]. Taking this into account we can represent points in $\langle Q \rangle$ by the points in $\langle Q' \rangle \subseteq E'(\mathbb{F}_{p^{k/d}})[r]$. Let t be the trace of Frobenius on $E/\mathbb{F}_p$ and $\lambda = (t-1)^{k/d} \bmod r$. Notice that λ is a primitive d-th root of unity modulo r.

The i-th Miller function $f_{i,P}$ for P is a function with divisor $(f_{i,P}) = i(P) - ([i]P) - (i-1)(\mathcal{O})$. We use Miller functions to compute pairings. Let the function e_s be defined by

$$e_s : G_1 \times G_2 \to \mu_r, \ (P, Q) \mapsto f_{s,P}(Q)^{(p^k-1)/r}.$$

For certain choices of s this function is a non-degenerate bilinear pairing. For $s = r$ we obtain the reduced Tate pairing τ, $s = \lambda$ yields the reduced Eta pairing η and $s = T = t - 1$ leads to the reduced Ate pairing α by switching the arguments. Altogether we have

- Tate pairing: $\tau(P, Q) = f_{r,P}(Q)^{(p^k-1)/r}$,
- Eta pairing: $\eta(P, Q) = f_{\lambda,P}(Q)^{(p^k-1)/r}$,
- Ate pairing: $\alpha(P, Q) = f_{T,Q}(P)^{(p^k-1)/r}$.

To obtain unique values, all pairings are reduced via the final exponentiation by $(p^k - 1)/r$. The Eta pairing was introduced in the supersingular context by Barreto, Galbraith, Ó' hÉigeartaigh and Scott in [1]. The Ate pairing was introduced by Hess, Smart and Vercauteren [8]. Actually the concept of the Eta pairing can be transferred to ordinary curves as well. Hess, Smart and Vercauteren [8] call it the twisted Ate pairing.

Recently much progress has been made in improving the performance of pairing computation. Main achievements have been made by suggesting variants of the above pairings which shorten the loop length in Miller's algorithm, for example so called generalized pairings [15], optimized pairings [11], the R-Ate pairing [10] as well as optimal pairings [14].

As an example we consider the generalized versions of the Eta and Ate pairings by Zhao, Zhang and Huang [15]:

- generalized Eta pairing: $\eta_c(P, Q) = f_{\lambda^c \bmod r, P}(Q)^{(p^k-1)/r}$, $0 < c < k$,
- generalized Ate pairing: $\alpha_c(P, Q) = f_{T^c \bmod r, Q}(P)^{(p^k-1)/r}$, $0 < c < k$.

For a certain choice of c the loop length of the generalized pairings may turn out shorter than the loop length of the original pairing. Notice that if $T^c \equiv -1 \pmod{r}$ or $\lambda^c \equiv -1 \pmod{r}$ the loop length is $r - 1$ which is the same as for the Tate pairing and does not give any advantage.

For each of the three AES security levels 128, 192 and 256 bits we give examples of elliptic curve families where generalized pairings lead to a shortening of the loop length. The examples all have embedding degree divisible by 6 and a twist of degree 6 such that the compressed pairing computations of Sections 4 and 5 can be applied. We stress that for all example families the generalized Eta pairing is more efficient than the Tate pairing. We emphasize the Eta pairing since this goes along with our compression method, but we note that there are versions of the Ate pairing which have a much shorter loop length than the pairings suggested here. For example the curves in Example 1 can be used for an optimal Ate pairing with loop length $\log_2 r/4$ (see Vercauteren [14]).

Example 1. We consider the family of elliptic curves introduced by Barreto and Naehrig in [2]. Let E be an elliptic curve of the family parameterized by $p = 36u^4 + 36u^3 + 24u^2 + 6u + 1$ and $t = 6u^2 + 1$. From the construction it follows that the curve has prime order, i.e. $r = n$, complex multiplication discriminant $D = -3$ and embedding degree $k = 12$. As shown in [2] E admits a twist E' of degree $d = 6$. This also follows from Lemma 4 in Section 4.2. We consider

$$\lambda = (t-1)^{k/d} = (6u^2)^2 \equiv 36u^4 \pmod{n}.$$

Since $n = 36u^4 + 36u^3 + 18u^2 + 6u + 1$ for positive values of u the length of λ is about the same as n, which means that there is no point in using the eta pairing. But for negative u we obtain $\lambda \equiv -36u^3 - 18u^2 - 6u - 1 \pmod{n}$ which is only 3/4 the size of n. Thus the Eta pairing gets faster than the Tate pairing.

For positive u the generalized version of the Eta pairing suggests to use a different power of λ. For example we could use $\lambda^4 = -\lambda$ since λ is a primitive sixth root of unity. We have $-\lambda \equiv -36u^4 \equiv 36u^3 + 18u^2 + 6u + 1 \pmod{n}$ and the length of $-\lambda$ is as well 3/4 of that of n which yields a faster pairing than the Tate pairing.

Example 2. A family of curves with embedding degree $k = 18$ was found by Kachisa and is described in Example 6.14 of [6]. For those curves we have $r(u) = u^6 + 37u^3 + 343$ and $t(u) = \frac{1}{7}(u^4 + 16u + 7)$. The generalized Ate pairing computing the loop over $T^{12} \equiv u^3 + 18 \pmod{r}$ for positive u and $T^3 \equiv -u^3 - 18 \pmod{r}$ for negative u is more efficient than the standard Ate pairing using $T \equiv \frac{1}{7}(u^4 + 16u)$.

The curves have a sextic twist and can be used for the Eta pairing with a loop over $\lambda = T^3$ which for negative u is as short as the generalized Ate pairing loop. For positive u take T^{12} for the generalized Eta pairing.

Example 3. Recently, Kachisa, Schaefer and Scott [9] found a family of pairing friendly curves with embedding degree $k = 36$. The curves have a sextic twist and lead to shorter loops in pairing computation. The group order is parametrized by a polynomial of degree 12 which we omit for space reasons. The trace of Frobenius is parametrized by the following polynomial of degree 7:

$$\begin{aligned}
t = {} & 125339925335955356307102330222u^7 + 8758840655324856893143016502u^6 \\
& + 2623173607511461889107842784u^5 + 4364504419607578015316190u^4 \\
& + 435706552724399907140970u^3 + 260978358826886222466u^2 \\
& + 868445151522065613u + 1238521382045476.
\end{aligned}$$

Both the generalized Ate and Eta pairings can be computed with a loop over $T_6 = T^6 \bmod r$ with

$$T_6 = 152593042775694370960679733 u^6 + 913997772652077313277994 u^5$$
$$+ 228109987087504077455555 u^4 + 303628259738257192620 u^3$$
$$+ 2273330651802144795 u^2 + 9077823883505034 u + 15103919293237.$$

For details see [9].

3   Preliminaries on Tori

Let $\mathbb{F}_q$ be a finite field and $\mathbb{F}_{q^l} \supseteq \mathbb{F}_q$ a field extension. Then the norm of an element $\alpha \in \mathbb{F}_{q^l}$ with respect to $\mathbb{F}_q$ is defined as the product of all conjugates of α over $\mathbb{F}_q$, namely $N_{\mathbb{F}_{q^l}/\mathbb{F}_q}(\alpha) = \alpha \alpha^q \cdots \alpha^{q^{l-1}} = \alpha^{1+q+\cdots+q^{l-1}} = \alpha^{(q^l-1)/(q-1)}$.

Rubin and Silverberg describe in [12] how algebraic tori can be used in cryptography. We recall the definition of a torus. For a positive integer l define the torus

$$T_l(\mathbb{F}_q) = \bigcap_{\mathbb{F}_q \subseteq F \subsetneq \mathbb{F}_{q^l}} \ker(N_{\mathbb{F}_{q^l}/F}). \tag{1}$$

Thus we have $T_l(\mathbb{F}_q) = \{\alpha \in \mathbb{F}_{q^l} \mid N_{\mathbb{F}_{q^l}/F}(\alpha) = 1, \ \mathbb{F}_q \subseteq F \subsetneq \mathbb{F}_{q^l}\}$. If $\mathbb{F}_q \subseteq F \subsetneq \mathbb{F}_{q^l}$ then $F = \mathbb{F}_{q^d}$ where $d \mid l$ so the relative norm is given as $N_{\mathbb{F}_{q^l}/\mathbb{F}_{q^d}}(\alpha) = \alpha^{(q^l-1)/(q^d-1)}$. The number of elements in the torus is $|T_l(\mathbb{F}_q)| = \Phi_l(q)$, where Φ_l is the l-th cyclotomic polynomial. We know that

$$X^l - 1 = \prod_{d \mid l} \Phi_d(X) = \Phi_l(X) \prod_{d \mid l, d \neq l} \Phi_d(X).$$

Thus the torus $T_l(\mathbb{F}_q)$ is the unique subgroup of order $\Phi_l(q)$ of $\mathbb{F}_{q^l}^*$. Set $\Psi_l(X) = \prod_{d \mid l, d \neq l} \Phi_d(X) = (X^l - 1)/\Phi_l(X)$.

Lemma 1. Let $\alpha \in \mathbb{F}_{q^l}^*$. Then $\alpha^{\Psi_l(q)} \in T_l(\mathbb{F}_q)$.

Proof. Let $\beta = \alpha^{\Psi_l(q)}$, then $\beta^{\Phi_l(q)} = \alpha^{q^l-1} = 1$, thus β has order dividing $\Phi_l(q)$ and therefore lies in $T_l(\mathbb{F}_q)$.   □

Lemma 2. For each divisor $d \mid l$ of l it holds $T_l(\mathbb{F}_q) \subseteq T_{l/d}(\mathbb{F}_{q^d})$.

Proof. Let $\beta \in T_l(\mathbb{F}_q)$. Then $N_{\mathbb{F}_{q^l}/F}(\beta) = 1$ for all fields $\mathbb{F}_q \subseteq F \subsetneq \mathbb{F}_{q^l}$. In particular the norm is 1 for all fields $\mathbb{F}_{q^d} \subseteq F \subsetneq \mathbb{F}_{q^l}$. And so $\beta \in T_{l/d}(\mathbb{F}_{q^d})$.   □

Combining the above two Lemmas shows that the element α raised to the power $\Psi_l(q)$ is an element of each torus $T_{l/d}(\mathbb{F}_{q^d})$ for all divisors $d \mid l, d \neq k$.

Let E be an elliptic curve defined over $\mathbb{F}_p$ with embedding degree k as in the previous section. By the definition of the embedding degree we have $r \nmid \Phi_d(p)$

for all divisors $d \mid k$, $d \neq k$. From that we see that the final exponent can be split up as

$$\frac{p^k - 1}{r} = \Psi_k(p) \frac{\Phi_k(p)}{r}.$$

This means that pairing values lie in the torus $T_k(\mathbb{F}_p)$ und thus by the preceeding Lemmas in each torus $T_{k/d}(\mathbb{F}_{p^d})$ for $d \mid k$, $d \neq k$.

4   Compressed Pairing Computation

Scott and Barreto [13] show how to compress the pairing value before the final exponentiation and how to use traces to compute the result. Also the use of tori has been investigated for the final exponentiation and to save bandwidth.

It is already shown by Granger, Page and Stam [7] how a pairing value in a field extension $\mathbb{F}_{q^6}$ can be compressed to an element in $\mathbb{F}_{q^3}$ plus one bit. We note that the technique of compression that we use here has already been explained in [7] for supersingular curves in characteristic 3. Granger, Page and Stam [7] mention that the technique works as well for curves over large characteristic fields. We describe and use the compression in the case of large characteristic and additionally as a new contribution include the compression into the Miller loop to compress the computation itself. In the following section 4.1 we recapitulate the compression for even embedding degree and show how to use it during pairing computation.

To make the paper as self-contained as possible and to enhance better understanding we derive and prove certain facts which are already known in the literature.

4.1   Compression for Even Embedding Degree

Let k be even and let $p \geq 5$ be a prime. In this section let $q = p^{k/2}$ and thus $\mathbb{F}_q = \mathbb{F}_{p^{k/2}}$ such that $\mathbb{F}_{q^2} = \mathbb{F}_{p^k}$. Choose $\xi \in \mathbb{F}_q$ to be a nonsquare. Then the polynomial $X^2 - \xi$ is irreducible and we may represent $\mathbb{F}_{q^2} = \mathbb{F}_q(\sigma)$ where σ is a root of $X^2 - \xi$.

Lemma 3. *Let $\alpha \in \mathbb{F}_{q^2}$. Then α^{q-1} is an element of $T_2(\mathbb{F}_q)$ and can be represented by a single element in $\mathbb{F}_q$ plus one additional bit. This element can be computed by one inversion in $\mathbb{F}_q$.*

Proof. We compute the q-Frobenius of σ which gives $\pi_q(\sigma) = \sigma^q = -\sigma$. The element α can be written as $\alpha = a_0 + a_1\sigma$ with coefficients $a_0, a_1 \in \mathbb{F}_q$. Raising α to the power of $q - 1$ we obtain

$$(a_0 + a_1\sigma)^{q-1} = \frac{(a_0 + a_1\sigma)^q}{a_0 + a_1\sigma} = \frac{a_0 - a_1\sigma}{a_0 + a_1\sigma}.$$

If $a_1 \neq 0$ we can proceed further by dividing in numerator and denominator by a_1 which gives

$$(a_0 + a_1\sigma)^{q-1} = \frac{a_0/a_1 - \sigma}{a_0/a_1 + \sigma} = \frac{a - \sigma}{a + \sigma}. \tag{2}$$

It is clear that the above fraction is an element of $T_2(\mathbb{F}_q)$. It can be represented by $a \in \mathbb{F}_q$ only. But we need an additional bit to represent 1 in the torus. If $a_1 = 0$ we started with an element of the base field and the exponentiation gives 1. In summary α^{q-1} can be represented by just one value in $\mathbb{F}_q$ plus one bit to describe the unit element 1.   □

The final exponentiation in the reduced pairing algorithm has to be carried out in the large field $\mathbb{F}_{p^k}$. The idea is to do part of the final exponentiation right inside the Miller loop to move elements to the torus $T_2(\mathbb{F}_q)$. Using torus arithmetic we may compute the compressed pairing value by computations in the torus only using less memory than with full extension field arithmetic. The rest of the final exponentiation can be carried out in the end on the compressed pairing value by also using torus arithmetic only.

Now if we have an elliptic curve with embedding degree k, in the final exponentiation we raise the output of the Miller loop to $(p^k - 1)/r = (q^2 - 1)/r$ where r is the order of the used subgroup. Since the embedding degree is k we have that $r \nmid q - 1$. Therefore we may split up the final exponentiation and raise the elements to $q - 1$ right away. This can be done in the above described manner by only one $\mathbb{F}_q$ inversion. Since the pairing value is computed multiplicatively we already exponentiate the line functions in the Miller loop by $q - 1$ and then carry out multiplications in torus arithmetic.

There is no need to have a representation for 1 in the torus during the pairing computation. The remaining part of the final exponentiation $(q + 1)/r$ is even, if q is the power of an odd prime and r is a large prime which thus is also odd. Therefore both values 1 and -1 are mapped to 1 when the final exponentiation is completed. We thus may take the representation for -1 whenever 1 occurs during computation. This will not alter the result of the pairing. Note that the torus element -1 has a regular representation with $a = 0$, since then the fraction (2) assumes the value -1. In this way we can save the bit which is usually needed to represent 1 when working in the torus.

For $\alpha = a_0 + a_1\sigma$ we denote by $\hat{\alpha} \in \mathbb{F}_q$ the torus representation of α^{q-1} for the pairing algorithm, i.e. $\hat{\alpha} = a_0/a_1$ if $a_1 \neq 0$ and $\hat{\alpha} = 0$ if $a_1 = 0$. The latter means we identify 1 and -1. Granger, Page and Stam [7] have demonstrated that arithmetic in the multiplicative group $T_2(\mathbb{F}_q)$ can now be done via

$$\frac{\hat{\alpha} - \sigma}{\hat{\alpha} + \sigma} \cdot \frac{\hat{\beta} - \sigma}{\hat{\beta} + \sigma} = \frac{\widehat{\alpha\beta} - \sigma}{\widehat{\alpha\beta} + \sigma},$$

where

$$\widehat{\alpha\beta} = (\hat{\alpha}\hat{\beta} + \xi)/(\hat{\alpha} + \hat{\beta}) \tag{3}$$

if $\hat{\alpha} \neq -\hat{\beta}$ and $\hat{\alpha} \neq 0$ and $\hat{\beta} \neq 0$. If $\hat{\alpha} = -\hat{\beta}$ the result is simply 1. If one of the values represents 1 we return the other value. For squaring a torus element with $\hat{\alpha} \neq 0$ we compute $\widehat{\alpha^2} = \hat{\alpha}/2 + \xi/(2\hat{\alpha})$.

The representation of the inverse of a torus element given by $\hat{\alpha}$ can be seen to be $-\hat{\alpha}$, since

$$\alpha^{-1} = \left(\frac{\hat{\alpha} - \sigma}{\hat{\alpha} + \sigma}\right)^{-1} = \frac{\hat{\alpha} + \sigma}{\hat{\alpha} - \sigma} = \frac{-\hat{\alpha} - \sigma}{-\hat{\alpha} + \sigma}. \tag{4}$$

We point out that doing inversions in torus representation does not need inversions in a finite field. Instead computation of an inverse only requires negation of a finite field element.

As seen above, we need to compute the result of the Miller loop only up to sign since -1 will be mapped to 1 in the final exponentiation. If we take the negative of a torus element, we obtain

$$-\frac{\hat{\alpha} - \sigma}{\hat{\alpha} + \sigma} = \frac{\sigma^2 - \hat{\alpha}\sigma}{\sigma^2 + \hat{\alpha}\sigma} = \frac{\xi - \hat{\alpha}\sigma}{\xi + \hat{\alpha}\sigma} = \frac{\xi/\hat{\alpha} - \sigma}{\xi/\hat{\alpha} + \sigma},$$

as long as $\hat{\alpha} \neq 0$. If $\hat{\alpha} = 0$ we are dealing with the element -1 and the negative of it is 1. This computation shows that the negative of a torus element $\alpha \neq \pm 1$ represented by $\hat{\alpha}$ is represented by $\xi/\hat{\alpha}$.

There may be potential to even further compress the computation inside the Miller loop. If it is possible to raise elements to $\Psi_k(p)$ in an efficient way, one may use the norm conditions in other tori to deduce equations which allow to achieve even more compact representations for the field elements used in the pairing computation. We will see in section 4.2 how this works in the special case $k \equiv 0 \pmod 6$.

4.2   Curves with a Sextic Twist and $6 \mid k$

From now on we assume that $6 \mid k$, i.e. $k = 6m$, where m is an arbitrary positive integer. In this section we fix $q = p^m$. Then $\mathbb{F}_q = \mathbb{F}_{p^m}$ and $\mathbb{F}_{q^6} = \mathbb{F}_{p^k}$. We have a look at the case where we are dealing with an elliptic curve which has complex multiplication discriminant $D = -3$. Under the above assumptions we give the details of our new method to include compression into the Miller loop. The existence of twists of degree 6 leads to compressed values of line functions which can easily be computed by only a few field operations in $\mathbb{F}_q$.

The description of twists and their orders given by Hess, Smart and Vercauteren in [8] yields the following lemma.

Lemma 4. *Let E be an ordinary elliptic curve with CM discriminant $D = -3$. Let E be defined over $\mathbb{F}_q$ where $q \equiv 1 \pmod 6$ and let r be a divisor of the group order $\#E(\mathbb{F}_q)$. The curve E can be represented as $E : y^2 = x^3 + B$, $B \in \mathbb{F}_q$.*

Then there exists a twist E' of degree $d = 6$ which is defined over $\mathbb{F}_q$ and $E'(\mathbb{F}_q)$ has order divisible by r.

The twist is given by $E' : y^2 = x^3 + B/\xi$, where $\xi \in \mathbb{F}_q^*$ is not a square or a third power. A $\mathbb{F}_{q^6}$-isomorphism is given by

$$\psi_d : E' \to E, \ (x, y) \mapsto (\xi^{1/3}x, \xi^{1/2}y). \tag{5}$$

We can represent the field extensions of $\mathbb{F}_q$ contained in $\mathbb{F}_{q^6}$ as $\mathbb{F}_{q^2} = \mathbb{F}_q(\xi^{1/2})$ and $\mathbb{F}_{q^3} = \mathbb{F}_q(\xi^{1/3})$ respectively. We use the twist to compactly represent the second argument of the pairing. This also implies that elliptic curve arithmetic in the group G_2 can be replaced by arithmetic in $E'(\mathbb{F}_q)$.

The twist also gives rise to further improvements for the compressed pairing computation. We consider terms which arise from line functions inside the Miller loop. Let $l_{U,V}(Q)$ be the line function of the line through the points U and V evaluated at Q. In the Miller loop U and V are points in $E(\mathbb{F}_p)$ and $Q = \psi_d(Q')$ for a point $Q' \in E'(\mathbb{F}_q)$ on the twist. These assumptions can not be made when computing the Ate pairing. Let $U = (x_U, y_U)$, $V = (x_V, y_V)$ and $Q' = (x_{Q'}, y_{Q'})$, and thus $Q = (x_Q, y_Q) = (\tau x_{Q'}, \sigma y_{Q'})$ where $\sigma = \xi^{1/2} \in \mathbb{F}_{q^2}$ and $\tau = \xi^{1/3} \in \mathbb{F}_{q^3}$. Notice that $\sigma^q = -\sigma$ and that $\mathbb{F}_{q^6} = \mathbb{F}_{q^3}(\sigma)$. For $U \neq -V$ the line function then yields

$$l_{U,V}(Q) = \lambda(x_Q - x_U) + (y_U - y_Q),$$

where λ is the slope of the line through U and V, i.e. $\lambda = (y_V - y_U)/(x_V - x_U)$ if $U \neq \pm V$ and $\lambda = (3x_U^2)/(2y_U)$ if $U = V$ respectively. In the case $U = -V$ the line function is $l_{U,-U}(Q) = x_Q - x_U$.

We take advantage of the fact that Q arises as $Q = \psi_d(Q')$ for some point $Q' \in E'(\mathbb{F}_q)$ and obtain

$$l_{U,V}(Q) = \lambda(\tau x_{Q'} - x_U) + (y_U - \sigma y_{Q'})$$
$$= (y_U - \lambda x_U + \lambda x_{Q'}\tau) - y_{Q'}\sigma.$$

For $U = -V$ we have $l_{U,-U}(Q) \in \mathbb{F}_{q^3}$. We thus proved the following lemma.

Lemma 5. *For $U \neq -V$ the torus representation of $(l_{U,V}(Q))^{q^3-1}$ can be computed as $(\lambda x_U - y_U - \lambda x_{Q'}\tau)/y_{Q'} \in \mathbb{F}_{q^3}$.*

Although $(\lambda x_U - y_U - \lambda x_{Q'}\tau)/y_{Q'}$ is an element of $\mathbb{F}_{q^3}$ it is possible to compute it with just a few $\mathbb{F}_q$ computations since λ as well as the coordinates of all involved points are elements of $\mathbb{F}_q$. Note that no exponentiation in $\mathbb{F}_{q^3}$ is required.

Inside the Miller loop we must carry out multiplications and squarings in torus representation. Squarings have to be done with elements represented by full $\mathbb{F}_{q^3}$ elements. But multiplications always include a line function as one factor. Let $\mu = -(y_U - \lambda x_U + \lambda x_{Q'}\tau)$ be the numerator of the representative for the exponentiated line function. If we compute the torus product with $\hat{\alpha}$ an arbitrary $\mathbb{F}_{q^3}$ element and $\hat{\beta} = \mu/y_{Q'}$ we get the following.

$$\widehat{\alpha\beta} = \frac{\hat{\alpha}\hat{\beta} + \xi}{\hat{\alpha} + \hat{\beta}} = \frac{\hat{\alpha}\mu + \xi y_{Q'}}{\hat{\alpha}y_{Q'} + \mu}.$$

There is no need to invert $y_{Q'}$ to compute the corresponding torus representation for $(l_{U,V}(Q))^{q^3-1}$. Instead we directly compute the product representative. Thus there is only one inversion in $\mathbb{F}_{q^3}$ needed to exponentiate the line function and compute the product in the Miller loop.

The final exponentiation is raising to $(q^6 - 1)/r$ in terms of q. We may write this as

$$\frac{q^6 - 1}{r} = (q^3 - 1)(q + 1)\frac{q^2 - q + 1}{r}.$$

What we did up to now is to raise line functions to $q^3 - 1$ in order to already move the elements to $T_2(\mathbb{F}_{q^3})$. But when we now do the exponentiation to $q + 1$ we have raised the element to $\Psi_6(q)$ and therefore end up with an element in $T_6(\mathbb{F}_q)$. This in particular means that our element lies in the kernel of $N_{\mathbb{F}_{q^6}/\mathbb{F}_{q^2}}$. If we use this property we may compress the element $\hat{\alpha}$ to two $\mathbb{F}_q$ elements which also has been demonstrated similarly by Granger, Page and Stam [7], and compute the pairing using this compact representation.

Proposition 1. *Let $p \equiv 1 \pmod 3$ and $\alpha \in \mathbb{F}_{q^6}^*$. Then $\alpha^{\Psi_6(q)}$ can be uniquely represented by a pair (a_0, a_1) of $\mathbb{F}_q$ elements.*

Proof. As seen before we can represent α^{q^3-1} by $\hat{\alpha}$ as $\alpha^{q^3-1} = \frac{\hat{\alpha} - \sigma}{\hat{\alpha} + \sigma}$. Let

$$\beta = \alpha^{\Psi_6(q)} = \left(\frac{\hat{\alpha} - \sigma}{\hat{\alpha} + \sigma}\right)^{q+1}.$$

We represent β by its torus representative $\hat{\beta}$, which can be computed as follows:

$$\beta = \left(\frac{\hat{\alpha} - \sigma}{\hat{\alpha} + \sigma}\right)^q \cdot \frac{\hat{\alpha} - \sigma}{\hat{\alpha} + \sigma} = \frac{\hat{\alpha}^q + \sigma}{\hat{\alpha}^q - \sigma} \cdot \frac{\hat{\alpha} - \sigma}{\hat{\alpha} + \sigma} = \frac{-\hat{\alpha}^q - \sigma}{-\hat{\alpha}^q + \sigma} \cdot \frac{\hat{\alpha} - \sigma}{\hat{\alpha} + \sigma}.$$

If $\hat{\alpha}^q = \hat{\alpha}$ we get $\beta = 1$. Otherwise, using (3) we get $\hat{\beta} = (-\hat{\alpha}^{q+1} + \xi)/(-\hat{\alpha}^q + \hat{\alpha})$. We now make use of the property that α has been raised to $\Psi_6(q)$ and thus lies in the torus $T_6(\mathbb{F}_q)$. We have $N_{\mathbb{F}_{q^6}/\mathbb{F}_{q^2}}(\beta) = 1$, i. e.

$$\left(\frac{\hat{\beta} - \sigma}{\hat{\beta} + \sigma}\right)^{1+q^2+q^4} = 1,$$

which is equivalent to $(\hat{\beta} - \sigma)^{1+q^2+q^4} = (\hat{\beta} + \sigma)^{1+q^2+q^4}$. We write $\hat{\beta} = b_0 + b_1\tau + b_2\tau^2$ with $b_i \in \mathbb{F}_q$ and use the fact that $\tau^q = \zeta^2\tau$ for ζ a primitive third root of unity which lies in $\mathbb{F}_q$ since $q \equiv 1 \pmod 3$. An explicit computation of $(\hat{\beta} \pm \sigma)^{1+q^2+q^4}$ and simplification of the equation $(\hat{\beta} - \sigma)^{1+q^2+q^4} = (\hat{\beta} + \sigma)^{1+q^2+q^4}$ gives the following relation:

$$-3b_1 b_2 \xi + \xi + 3b_0^2 = 0.$$

This equation can be used to recover b_2 from b_0 and b_1 if $b_1 \neq 0$ as

$$b_2 = \frac{3b_0^2 + \xi}{3b_1\xi}. \tag{6}$$

If $b_1 = 0$ we have $\xi = -3b_0^2$. Since $p \equiv 1 \pmod 3$ then -3 is a square modulo p thus ξ is a square which is not true. Therefore b_1 can not be 0 in this case. Summarizing we see that we can represent the element β by b_0 and b_1 only which concludes the proof. $\qquad\square$

We now turn our attention again to the line functions $l_{U,V}(Q)$ used in Miller's algorithm.

Proposition 2. *Let $\zeta \in \mathbb{F}_q$ be a primitive third root of unity such that $\tau^q = \zeta^2 \tau$. Let $\beta = (l_{U,V}(Q))^{\Psi_6(q)}$ where $Q = \psi_d(Q')$. If $\beta \neq 1$ then β can be uniquely represented by*

$$c_0 = \left(\frac{-\zeta}{1 - \zeta^2} y_{Q'}^{-1} \right)(y_U - \lambda x_U), \quad c_1 = \left(\frac{\zeta^2}{1 - \zeta^2} y_{Q'}^{-1} \right) \lambda x_{Q'}. \tag{7}$$

Proof. In the proof of Proposition 1 we have seen how to compute $\hat{\beta} = (-\hat{\alpha}^{q+1} + \xi)/(-\hat{\alpha}^q + \hat{\alpha})$. For the line function we take $\hat{\alpha} = (\lambda x_U - y_U - \lambda x_{Q'}\tau)/y_{Q'}$ from Lemma 5. We thus obtain

$$-\hat{\alpha}^q = \frac{y_U - \lambda x_U + \lambda x_{Q'}\zeta^2 \tau}{y_{Q'}}.$$

Multiplying with $\hat{\alpha}$ yields

$$-\hat{\alpha}^{q+1} = -\frac{1}{y_{Q'}^2}\Big((y_U - \lambda x_U)^2 + (1 + \zeta^2)\lambda x_{Q'}(y_U - \lambda x_U)\tau + \lambda^2 x_{Q'}^2 \zeta^2 \tau^2 \Big).$$

We further have

$$-\hat{\alpha}^q + \hat{\alpha} = \frac{\lambda x_{Q'}(\zeta^2 - 1)\tau}{y_{Q'}}$$

and compute

$$\hat{\beta} = \frac{(1 + \zeta^2)\lambda x_{Q'}(y_U - \lambda x_U)\xi + \lambda^2 x_{Q'}^2 \zeta^2 \xi \tau + ((y_U - \lambda x_U)^2 - \xi y_{Q'}^2)\tau^2}{\lambda(1 - \zeta^2)x_{Q'}y_{Q'}\xi}$$

$$= \frac{1 + \zeta^2}{1 - \zeta^2} \cdot \frac{y_U - \lambda x_U}{y_{Q'}} + \frac{\zeta^2}{1 - \zeta^2} \cdot \frac{\lambda x_{Q'}}{y_{Q'}}\tau + \frac{(y_U - \lambda x_U)^2 - \xi y_{Q'}^2}{\lambda(1 - \zeta^2)x_{Q'}y_{Q'}\xi}\tau^2.$$

Recall that $\tau^3 = \xi$. Taking c_i the coefficient of τ^i we have the property $c_2 = \frac{3c_0^2 + \xi}{3c_1\xi}$ and thus c_2 can be computed from c_0 and c_1. $\square$

The input Q is not changed during one pairing computation. Hence, $y_{Q'}^{-1}$ can be computed at the beginning of the pairing computation and we do not need inversions to compute the values of the exponentiated line functions inside the Miller loop.

For squaring and multiplication in the Miller loop we need formulæ to compute with compressed values. Squaring of an element (a_0, a_1) can be done with the following formulæ which can be derived by computing the square of the corresponding torus elements explicitly and compressing again. Compute

$$r_0 = a_0^5 + \xi(a_0^3 - 2a_0^2 a_1^3) + \xi^2(\tfrac{1}{3}a_0 - a_1^3),$$
$$r_1 = a_0^5 + \xi(2a_0^3 - 2a_0^2 a_1^3) + \xi^2(a_0 - 2a_1^3),$$
$$s_0 = a_0(a_0 r_0 + a_1^6 \xi^2 + \tfrac{1}{27}\xi^3) - \tfrac{1}{3}a_1^3 \xi^3,$$
$$s_1 = a_1(a_0 r_1 + a_1^6 \xi^2 + \tfrac{4}{27}\xi^3),$$
$$s = 2(a_0 r_0 + a_1^6 \xi^2 + \tfrac{1}{27}\xi^3),$$
$$c_0 = \frac{s_0}{s},$$
$$c_1 = \frac{s_1}{s}.$$

Then the square of the $\mathbb{F}_{q^6}$ element represented by (a_0, a_1) is represented by (c_0, c_1). Multiplication can be derived in a similar way. We give formulæ for the computation of the product of two elements given by (a_0, a_1) and (b_0, b_1) in compressed form.

$$r_0 = a_0^2 + \tfrac{1}{3}\xi,$$
$$r_1 = b_0^2 + \tfrac{1}{3}\xi,$$
$$s_0 = \xi(a_1 b_1(a_0 b_0 + \xi) + a_1^2 r_1 + b_1^2 r_0),$$
$$s_1 = a_1 b_1 \xi(a_0 b_1 + a_1 b_0) + r_0 r_1,$$
$$s_2 = a_1^2 b_1^2 \xi + a_0 a_1 r_1 + b_0 b_1 r_0,$$
$$t_0 = a_1 b_1 \xi(a_0 + b_0),$$
$$t_1 = a_1 b_1 \xi(a_1 + b_1),$$
$$t_2 = b_1 r_0 + a_1 r_1,$$
$$u = t_0^3 + t_1^3 \xi + t_2^3 \xi^2 - 3\xi t_0 t_1 t_2,$$
$$u_0 = t_0^2 - t_1 t_2 \xi,$$
$$u_1 = t_2^2 \xi - t_0 t_1,$$
$$u_2 = t_1^2 - t_0 t_2,$$
$$v_0 = s_0 u_0 + s_1 u_2 \xi + s_2 u_1 \xi,$$
$$v_1 = s_0 u_1 + s_1 u_0 + s_2 u_2 \xi,$$
$$c_0 = \frac{v_0}{u},$$
$$c_1 = \frac{v_1}{u}.$$

The product is then represented by (c_0, c_1).

5   Dealing with Field Inversions

In this section we use the assumptions from section 4.2, i. e. $q = p^m$. For compressed computation we need to do inversions during our computations. This is usually unpleasant, because inversions are very expensive. First of all, one can replace inversion of an element a in $\mathbb{F}_{p^m}$ by an inversion in $\mathbb{F}_p$ and at most $\lfloor \lg m \rfloor + 1$ multiplications in $\mathbb{F}_{p^m}$ by

$$\frac{1}{a} = \frac{a^{p+p^2+\cdots+p^{m-1}}}{N_{\mathbb{F}_{p^m}/\mathbb{F}_p}(a)}.$$

The term in the numerator can be computed by addition chain like methods. For a description of this method see section 11.3.4 in [5].

5.1   Avoid Inversions by Storing One More $\mathbb{F}_q$ Element

The above squaring and multiplication formulæ for compressed computation include an inversion in $\mathbb{F}_q$. We may avoid to do the inversions in each step by additionally storing the denominator and homogenizing the formulae. This means we represent compressed elements in a projective space. At the cost of providing memory space for one more $\mathbb{F}_q$ element and some additional multiplications we get rid of all inversions during the Miller loop. For the compressed line functions computed in Proposition 2 this means that we do not store (c_0, c_1) given by equations (7) but instead we store (C_0, C_1, C), where

$$C_0 = \left(\frac{-\zeta}{1 - \zeta^2}\right)(\nu y_U - \mu x_U), \quad C_1 = \left(\frac{\zeta^2}{1 - \zeta^2}\right)\mu x_{Q'}, \quad C = \nu y_{Q'}. \qquad (8)$$

Here $\mu, \nu \in \mathbb{F}_p$ are the numerator and denominator of the slope λ of the line function, i.e. $\lambda = \mu/\nu$. Notice that μ and ν are elements of $\mathbb{F}_p$ since they arise from points in $E(\mathbb{F}_p)$ (when the pairing we compute is the Tate or Eta pairing).

5.2   Storing Only One More $\mathbb{F}_p$ Element

When $m > 1$ we are able to compress further, by using the method described at the beginning of Section 5. The denominator C which has to be stored in a third coordinate can be replaced by a denominator which is an element in $\mathbb{F}_p$, namely the norm $N_{\mathbb{F}_{p^m}/\mathbb{F}_p}(C)$ of the previous denominator in $\mathbb{F}_q$. We only need to multiply the other two coordinates by $C^{p+p^2+\cdots+p^{m-1}}$.

In this way it is possible to avoid inversions during pairing computation. Taking into account that inversion of torus elements can be done by negating the representative, we also do not need finite field inversions for the final exponentiation. Normally an inversion is needed to efficiently implement the exponentiation by using the Frobenius automorphism. Furthermore, the cheap inversion of torus elements makes it possible to use windowing methods for Miller loop computations without any field inversions. This is particularly interesting if the loop scalar can not be chosen to be sparse.

We give an example of the compressed squaring and multiplication formulæ for embedding degree $k = 12$.

Example 4. For embedding degree 12 we have $q = p^2$. Let $\mathbb{F}_{p^2} = \mathbb{F}_p(i)$ and $i^2 = -z$ for some element $z \in \mathbb{F}_p$. Let (A_0, A_1, A) be an element in compressed form, i.e. $A_0, A_1 \in \mathbb{F}_{p^2}$ and $A \in \mathbb{F}_p$. Squarings and Multiplications can be computed using the following formulæ.

Squaring: We can compute the square (C_0, C_1, C) as follows.

$$R_0 = A_0^5 + \xi(A_0^3 A^2 - 2A_0^2 A_1^3) + \xi^2(\tfrac{1}{3}A_0 A^4 - A_1^3 A^2),$$
$$R_1 = A_0^5 + 2\xi(A_0^3 A^2 - A_0^2 A_1^3) + \xi^2(A_0 A^4 - 2A_1^3 A^2),$$
$$S_0 = A_0(A_0 R_0 + A_1^6 \xi^2 + \tfrac{1}{27}A^6 \xi^3) - \tfrac{1}{3}A_1^3 A^4 \xi^3,$$
$$S_1 = A_1(A_0 R_1 + A_1^6 \xi^2 + \tfrac{4}{27}A^6 \xi^3),$$
$$S = 2A(A_0 R_0 + A_1^6 \xi^2 + \tfrac{1}{27}A^6 \xi^3).$$

Write $S = s_0 + i s_1$ with $s_0, s_1 \in \mathbb{F}_p$. Then the square is given by

$$C_0 = S_0(s_0 - i s_1),$$
$$C_1 = S_1(s_0 - i s_1),$$
$$C = s_0^2 + z s_1^2.$$

Multiplication: To multiply two compressed elements (A_0, A_1, A) and (B_0, B_1, B) we have to use the following formulæ.

$$R_0 = A_0^2 + \tfrac{1}{3}A^2 \xi,$$
$$R_1 = B_0^2 + \tfrac{1}{3}B^2 \xi,$$
$$S_0 = \xi(A_1 B_1(A_0 B_0 + \xi A B) + A_1^2 R_1 + B_1^2 R_0),$$
$$S_1 = A_1 B_1 \xi(A_0 B_1 + A_1 B_0) + R_0 R_1,$$
$$S_2 = A_1^2 B_1^2 \xi + A_0 A_1 R_1 + B_0 B_1 R_0,$$
$$T_0 = A_1 B_1 \xi(A_0 B + B_0 A),$$
$$T_1 = A_1 B_1 \xi(A_1 B + B_1 A),$$
$$T_2 = B_1 B R_0 + A_1 A R_1,$$
$$T = T_0^3 + T_1^3 \xi + T_2^3 \xi^2 - 3\xi T_0 T_1 T_2,$$
$$U_0 = T_0^2 - T_1 T_2 \xi,$$
$$U_1 = T_2^2 \xi - T_0 T_1,$$
$$U_2 = T_1^2 - T_0 T_2,$$
$$V_0 = S_0 U_0 + S_1 U_2 \xi + S_2 U_1 \xi,$$
$$V_1 = S_0 U_1 + S_1 U_0 + S_2 U_2 \xi.$$

Write $T = t_0 + i t_1$ where $t_0, t_1 \in \mathbb{F}_p$. Then the product (C_0, C_1, C) of the two elements is given by

$$C_0 = V_0(t_0 - i t_1),$$
$$C_1 = V_1(t_0 - i t_1),$$
$$C = t_0^2 + z t_1^2.$$

For an implementation of a pairing algorithm in compressed form without inversions one can use (8) to compute the evaluated compressed line functions

Table 1. Parameters of the curve used in our implementation

p	8243401665430067972121735350319003883657178181138622892116732241281902949183
n	8243401665430067972121735350319003883628466856429668643011451005255640137769
bitsize	256
t	287113247089542491052812360262628119415
k	12
λ^c	$(t-1)^8 \mod n$

Table 2. Rounded average results of measurements on various CPUs. The upper number describes cycles needed for the Miller loop, the lower number cycles needed for final exponentiation.

	Core 2 Duo	Pentium IV	Athlon XP
Ate	16,750,000	50,400,000	38,000,000
	13,000,000	38,600,000	29,300,000
Generalized Eta	22,370,000	67,400,000	51,700,000
	13,000,000	38,600,000	29,300,000
Tate	30,300,000	90,500,000	69,500,000
	13,000,000	38,600,000	29,300,000
Compressed generalized Eta	31,000,000	107,000,000	84,900,000
	11,700,000	40,300,000	30,900,000
Compressed Tate	41,400,000	146,000,000	115,000,000
	11,700,000	40,300,000	30,900,000

and then use the above formulæ for squaring and multiplication in Miller's algorithm. The remaining part of the exponent for the final exponentiation is $(p^4 - p^2 + 1)/n$. The final pairing value can be computed by use of the Frobenius and a square and multiply algorithm with the above squaring and multiplication formulæ (see Devigili, Scott and Dahab [4]). Pseudocode of the above squaring and multiplication algorithms is given in Appendix A.

6   Performance Evaluation

In order to evaluate the performance of the compressed pairing computation, we implemented several pairing algorithms in C. For all these implementations[1] we used the curve $E : y^2 = x^3 + b$ over $\mathbb{F}_p$ with parameters described in Table 1 which belongs to the family in Example 1. It has been constructed using the method of Barreto and Naehrig described in [2]. This curve has also been used for the performance evaluation of pairing algorithms by Devegili, Scott and Dahab in [4]. For a fair comparison we implemented pairing algorithms with $\mathbb{F}_{p^{12}}$ constructed as a quadratic extension on top of a cubic extension which is again built on top of a quadratic extension, as described in [4] and by Devigili, Scott,

---

[1] The code for our implementation can be found at
http://www.cryptojedi.org/downloads/

Ó' hÉigeartaigh and Dahab in [3]. For Ate, generalized Eta and Tate pairings we thus achieve similar timings as [4]. We do not use windowing methods since the curve parameters are chosen to be sparse. The final exponentiation for the non-compressed pairings uses the decomposition of the exponent $(p^k - 1)/n$ into the factors $(p^6 - 1)$, $(p^2 + 1)$ and $(p^4 - p^2 + 1)/n$.

In the Miller loop we entirely avoided to do field inversions, by computing the elliptic curve operations in Jacobian coordinates and by using the compressed representation and storing denominators separately as described in Subsection 5.2. For multiplication and squaring of torus elements we use the algorithms given in Appendix A. The figures in table 2 indicate that, depending on the machine architecture, compressed pairing computation is about 20-45% slower than standard pairing computation, if both computations are optimized for computation speed rather than memory usage.

Performance was measured on a 2.2 GHz Intel Core 2 Duo (T7500), a 2.4 GHz Intel Pentium IV (Northwood) and an AMD Athlon XP 2600+ running on 1.9 GHz. The CPU cycles required for Miller loop and final exponentiation respectively are given in Table 2.

7   Conclusion

We have described explicit formulæ for pairing computation in compressed form for the Tate and Eta pairings. For different AES security levels we have also indicated families of curves amenable to pairing compression where generalized versions of the Eta and Ate pairings are very efficient. Our implementations and cost measurements show that the pairing algorithms in compressed form are on certain platforms only about 20% slower than the conventional algorithms. The algorithms in compressed form have the advantage that they can be implemented without finite field inversions. This is not only an advantage for pairing computations on restricted devices, but also favors implementation of inversion-free windowing methods for the Miller loop. Furthermore compressed pairing computation gives more flexibility in trading off computation speed versus memory requirement.

Neither the algorithms nor the curve families considered herein are exhaustive; we thus hope that these are the first steps toward further algorithmic enhancements for compressed pairings and towards new, efficient constructions of compressible pairing-friendly curves.

Acknowledgments

We thank the anonymous referees and are grateful to Tanja Lange, Mike Scott, Steven Galbraith and Rob Granger for their valuable comments on earlier versions of this work. The first and third authors did portions of this work at the Institute for Theoretical Information Technology, RWTH Aachen University.

References

1. Barreto, P.S.L.M., Galbraith, S.D., O'hEigeartaigh, C., Scott, M.: Efficient pairing computation on supersingular abelian varieties. Designs, Codes and Cryptography 42(3), 239–271 (2007)
2. Barreto, P.S.L.M., Naehrig, M.: Pairing-friendly elliptic curves of prime order. In: Preneel, B., Tavares, S. (eds.) SAC 2005. LNCS, vol. 3897, pp. 319–331. Springer, Heidelberg (2006)
3. Devegili, A.J., O'hEigeartaigh, C., Scott, M., Dahab, R.: Multiplication and squaring on pairing-friendly fields. Cryptology ePrint Archive, Report 2006/471(2006), http://eprint.iacr.org/
4. Devegili, A.J., Scott, M., Dahab, R.: Implementing cryptographic pairings over barreto-naehrig curves. Cryptology ePrint Archive, Report,2007/390 (2007), http://eprint.iacr.org/2007/390
5. Doche, C.: Finite field arithmetic. In: Cohen, H., Frey, G. (eds.) Handbook of Elliptic and Hyperelliptic Curve Cryptography, ch. 11, pp. 201–238. CRC Press, Boca Raton (2005)
6. Freeman, D., Scott, M., Teske, E.: A taxonomy of pairing-friendly elliptic curves. Cryptology ePrint Archive, Report, 2006/372 (2006), http://eprint.iacr.org/2006/372
7. Granger, R., Page, D., Stam, M.: On small characteristic algebraic tori in pairing based cryptography. LMS Journal of Computation and Mathematics 9, 64–85 (2006)
8. Hess, F., Smart, N., Vercauteren, F.: The eta pairing revisited. IEEE Transactions on Information Theory 52(10), 4595–4602 (2006)
9. Kachisa, E.J., Schaefer, E.F., Scott, M.: Constructing brezing-weng pairing friendly elliptic curves using elements in the cyclotomic field. Cryptology ePrint Archive, Report, 2007/452 (2007), http://eprint.iacr.org/
10. Lee, E., Lee, H., Park, C.: Efficient and generalized pairing computation on abelian varieties. Cryptology ePrint Archive, Report 2008/040 (2008), http://eprint.iacr.org/.
11. Matsuda, S., Kanayama, N., Hess, F., Okamoto, E.: Optimised versions of the ate and twisted ate pairings. In: Galbraith, S.D. (ed.) Cryptography and Coding 2007. LNCS, vol. 4887, pp. 302–312. Springer, Heidelberg (2007)
12. Rubin, K., Silverberg, A.: Torus-based cryptography. In: Boneh, D. (ed.) CRYPTO 2003. LNCS, vol. 2729, pp. 349–365. Springer, Heidelberg (2003)
13. Scott, M., Barreto, P.S.L.M.: Compressed pairings. In: Franklin, M. (ed.) CRYPTO 2004. LNCS, vol. 3152, pp. 140–156. Springer, Heidelberg (2004)
14. Vercauteren, F.: Optimal pairings. Cryptology ePrint Archive, Report 2008/096 (2008), http://eprint.iacr.org/
15. Zhao, C., Zhang, F., Huang, J.: A note on the ate pairing. Cryptology ePrint Archive, Report 2007/247 (2007), http://eprint.iacr.org/2007/247

A   Compressed Multiplication and Squaring Algorithms

Algorithm 1. Squaring of the element (A_0, A_1, A)

Require: $(A_0, A_1, A) \in \mathbb{F}_{p^2} \times \mathbb{F}_{p^2} \setminus \{0\} \times \mathbb{F}_p$

Ensure: $(C_0, C_1, C) = (A_0, A_1, A)^2$

$r_1 \leftarrow A_0^2,\ r_2 \leftarrow A_0 r_1,\ S_0 \leftarrow r_1 r_2,\ t_0 \leftarrow A^2,\ r_4 \leftarrow r_2 t_0,\ r_5 \leftarrow A_1^2,\ r_5 \leftarrow A_1 r_5,$

$r_3 \leftarrow r_1 r_5,\ r_4 \leftarrow r_4 - r_3,\ r_0 \leftarrow r_4 \xi,\ r_0 \leftarrow 2 r_0,\ S_1 \leftarrow S_0 + r_0,\ r_4 \leftarrow r_4 - r_3,\ r_4 \leftarrow r_4 \xi,$

$S_0 \leftarrow S_0 + r_4,\ t_1 \leftarrow t_0^2,\ r_4 \leftarrow t_1 A_0,\ r_0 \leftarrow \frac{1}{3} r_4,\ r_1 \leftarrow r_5 t_0,\ r_0 \leftarrow r_0 - r_1,\ r_1 \leftarrow 2 r_1,$

$r_4 \leftarrow r_4 - r_1,\ r_0 \leftarrow \xi^2 r_0,\ r_4 \leftarrow \xi^2 r_4,\ S_0 \leftarrow S_0 + r_0,\ S_0 \leftarrow S_0 A_0,\ S_1 \leftarrow S_1 + r_4,$

$S_1 \leftarrow S_1 A_0,\ r_2 \leftarrow r_5^2,\ r_2 \leftarrow r_2 \xi^2,\ r_4 \leftarrow t_1 t_0,\ r_4 \leftarrow \frac{1}{27} \xi^3 r_4,\ r_1 \leftarrow r_2 + r_4,\ S_0 \leftarrow S_0 + r_1,$

$S \leftarrow S_0 A,\ S_0 \leftarrow S_0 A_0,\ S \leftarrow 2S,\ r_4 \leftarrow 4 r_4,\ r_1 \leftarrow r_2 + r_4,\ S_1 \leftarrow S_1 + r_1,\ S_1 \leftarrow S_1 A_1,$

$r_1 \leftarrow r_5 t_1,\ r_1 \leftarrow \frac{1}{3} \xi^3 r_1,\ S_0 \leftarrow S_0 - r_1$

Write $S = s_0 + i s_1$

$r_1 \leftarrow (s_0 - i s_1),\ C_0 \leftarrow S_0 r_1,\ C_1 \leftarrow S_1 r_1,\ C \leftarrow S r_1 = s_0^2 + c s_1^2$

return (C_0, C_1, C)

Algorithm 2. Multiplication of elements (A_0, A_1, A) and (B_0, B_1, B)

Require: $(A_0, A_1, A), (B_0, B_1, B) \in \mathbb{F}_{p^2} \times \mathbb{F}_{p^2} \setminus \{0\} \times \mathbb{F}_p$

Ensure: $(C_0, C_1, C) = (A_0, A_1, A) \cdot (B_0, B_1, B)$

$R_0 \leftarrow A_0^2,\ t_1 \leftarrow A^2,\ r_3 \leftarrow \frac{1}{3} \xi t_1,\ R_0 \leftarrow R_0 + r_3,$

$R_1 \leftarrow B_0^2,\ t_1 \leftarrow B^2,\ r_3 \leftarrow \frac{1}{3} \xi t_1,\ R_1 \leftarrow R_1 + r_3$

$r_3 \leftarrow A_1 B_1,\ r_4 \leftarrow A_0 B_0,\ t_1 \leftarrow AB,\ r_5 \leftarrow t_1 \xi,\ r_4 \leftarrow r_4 + r_5,\ S_0 \leftarrow r_3 r_4,\ S_2 \leftarrow r_3^2$

$S_2 \leftarrow S_2 \xi,\ r_4 \leftarrow A_0 B_1,\ r_5 \leftarrow A_1 B_0,\ r_4 \leftarrow r_4 + r_5,\ r_6 \leftarrow r_3 \xi,\ S_1 \leftarrow r_6 r_4,\ r_4 \leftarrow R_0 R_1$

$S_1 \leftarrow S_1 + r_4,\ r_4 \leftarrow A_1 R_1,\ r_5 \leftarrow r_4 A_0,\ S_2 \leftarrow S_2 + r_5,\ T_2 \leftarrow r_4 A,\ r_4 \leftarrow r_4 A_1,$

$S_0 \leftarrow S_0 + r_4,\ r_4 \leftarrow B_1 R_0,\ r_5 \leftarrow r_4 B,\ T_2 \leftarrow T_2 + r_5,\ r_5 \leftarrow r_4 B_0,\ S_2 \leftarrow S_2 + r_5,$

$r_4 \leftarrow r_4 B_1,\ S_0 \leftarrow S_0 + r_4,\ S_0 \leftarrow S_0 \xi,\ T_0 \leftarrow A_0 B,\ r_4 \leftarrow B_0 A,\ T_0 \leftarrow T_0 + r_4,\ T_0 \leftarrow r_6 T_0,$

$T_1 \leftarrow A_1 B,\ r_4 \leftarrow B_1 A,\ T_1 \leftarrow T_1 + r_4,\ T_1 \leftarrow T_1 r_6$

$r_0 \leftarrow T_0^2,\ r_1 \leftarrow T_1^2,\ r_2 \leftarrow T_2^2,\ T \leftarrow r_0 T_0,\ r_3 \leftarrow r_1 T_1,\ r_3 \leftarrow r_3 \xi,\ T \leftarrow T + r_3$

$r_3 \leftarrow r_2 T_2,\ r_3 \leftarrow r_3 \xi^2,\ T \leftarrow T + r_3,\ r_3 \leftarrow T_1 T_2,\ r_3 \leftarrow r_3 \xi,\ U_0 \leftarrow r_0 - r_3,\ r_3 \leftarrow r_3 T_0$

$r_3 \leftarrow 3 r_3,\ T \leftarrow T - r_3,\ r_3 \leftarrow T_0 T_1,\ U_1 \leftarrow r_2 \xi,\ U_1 \leftarrow U_1 - r_3,\ r_3 \leftarrow T_0 T_2,\ U_2 \leftarrow r_1 - r_3$

$V_0 \leftarrow S_0 U_0,\ r_0 \leftarrow S_1 U_2,\ r_1 \leftarrow S_2 U_1,\ r_0 \leftarrow r_0 + r_1,\ r_0 \leftarrow r_0 \xi,\ V_0 \leftarrow V_0 + r_0$

$V_1 \leftarrow S_0 U_1,\ r_0 \leftarrow S_1 U_0,\ V_1 \leftarrow V_1 + r_0,\ r_0 \leftarrow S_2 U_2,\ r_0 \leftarrow r_0 \xi,\ V_1 \leftarrow V_1 r_0$

Write $T = t_0 + i t_1$

$r_1 \leftarrow (t_0 - i t_1),\ C_0 \leftarrow V_0 r_1,\ C_1 \leftarrow V_1 r_1,\ C \leftarrow S r_1 = t_0^2 + c t_1^2$

return (C_0, C_1, C)

Twisted Edwards Curves[*]

Daniel J. Bernstein[1], Peter Birkner[2], Marc Joye[3], Tanja Lange[2],
and Christiane Peters[2]

[1] Department of Mathematics, Statistics, and Computer Science (M/C 249)
University of Illinois at Chicago, Chicago, IL 60607–7045, USA
djb@cr.yp.to
[2] Department of Mathematics and Computer Science
Technische Universiteit Eindhoven, P.O. Box 513, 5600 MB Eindhoven, Netherlands
p.birkner@tue.nl, tanja@hyperelliptic.org, c.p.peters@tue.nl
[3] Thomson R&D France
Technology Group, Corporate Research, Security Laboratory
1 avenue de Belle Fontaine, 35576 Cesson-Sévigné Cedex, France
marc.joye@thomson.net

Abstract. This paper introduces "twisted Edwards curves," a general-ization of the recently introduced Edwards curves; shows that twisted Edwards curves include more curves over finite fields, and in particular every elliptic curve in Montgomery form; shows how to cover even more curves via isogenies; presents fast explicit formulas for twisted Edwards curves in projective and inverted coordinates; and shows that twisted Edwards curves save time for many curves that were already expressible as Edwards curves.

Keywords: Elliptic curves, Edwards curves, twisted Edwards curves, Montgomery curves, isogenies.

1 Introduction

Edwards in [13], generalizing an example from Euler and Gauss, introduced an addition law for the curves $x^2 + y^2 = c^2(1 + x^2y^2)$ over a non-binary field k. Edwards showed that every elliptic curve over k can be expressed in the form $x^2 + y^2 = c^2(1 + x^2y^2)$ if k is algebraically closed. However, over a finite field, only a small fraction of elliptic curves can be expressed in this form.

Bernstein and Lange in [4] presented fast explicit formulas for addition and doubling in coordinates $(X : Y : Z)$ representing $(x, y) = (X/Z, Y/Z)$ on an Edwards curve, and showed that these explicit formulas save time in elliptic-curve cryptography. Bernstein and Lange also generalized the addition law to

[*] Permanent ID of this document: c798703ae3ecfdc375112f19dd0787e4. Date of this document: 2008.03.12. This work has been supported in part by the European Com-mission through the IST Programme under Contract IST–2002–507932 ECRYPT, and in part by the National Science Foundation under grant ITR–0716498. Part of this work was carried out during a visit to INRIA Lorraine (LORIA).

S. Vaudenay (Ed.): AFRICACRYPT 2008, LNCS 5023, pp. 389–405, 2008.

the curves $x^2 + y^2 = c^2(1 + dx^2y^2)$. This shape covers considerably more elliptic curves over a finite field than $x^2 + y^2 = c^2(1 + x^2y^2)$. All curves in the generalized form are isomorphic to curves $x^2 + y^2 = 1 + dx^2y^2$.

In this paper, we further generalize the Edwards addition law to cover all curves $ax^2 + y^2 = 1 + dx^2y^2$. Our explicit formulas for addition and doubling are almost as fast in the general case as they are for the special case $a = 1$. We show that our generalization brings the speed of the Edwards addition law to every Montgomery curve; we also show that, over prime fields $\mathbf{F}_p$ where $p \equiv 1$ (mod 4), many Montgomery curves are not covered by the special case $a = 1$. We further explain how to use isogenies to cover the odd part of every curve whose group order is a multiple of 4; over prime fields $\mathbf{F}_p$ where $p \equiv 3$ (mod 4), the special case $a = 1$ covers all Montgomery curves but does not cover all curves whose group order is a multiple of 4. Our generalization is also of interest for many curves that were already expressible in Edwards form; we explain how the twisting can save time in arithmetic. See [2] for a successful application of twisted Edwards curves to the elliptic-curve method of factorization.

Section 2 reviews Edwards curves, introduces twisted Edwards curves, and shows that each twisted Edwards curve is (as the name would suggest) a twist of an Edwards curve. Section 3 shows that every Montgomery curve can be expressed as a twisted Edwards curve, and vice versa. Section 4 reports the percentages of elliptic curves (over various prime fields) that can be expressed as Edwards curves, twisted Edwards curves, "4 times odd" twisted Edwards curves, etc. Section 5 uses isogenies to cover even more curves: specifically, it shows that every curve with group order a multiple of 4 and with no point of order 4 is 2-isogenous to a twisted Edwards curve. Section 6 generalizes the Edwards addition law, the explicit formulas from [4], and the "inverted" formulas from [5] to handle twisted Edwards curves. Section 7 analyzes the benefits of the generalization for cryptographic applications.

2   Edwards Curves and Twisted Edwards Curves

In this section we briefly review Edwards curves and the Edwards addition law at the level of generality of [4]. We then introduce twisted Edwards curves and discuss their relationship to Edwards curves.

Review of Edwards Curves. Throughout the paper we consider elliptic curves over a non-binary field k, i.e., a field k whose characteristic $\mathrm{char}(k)$ is not 2.

An Edwards curve over k is a curve $E : x^2 + y^2 = 1 + dx^2y^2$ where $d \in k \backslash \{0, 1\}$. The sum of two points (x_1, y_1), (x_2, y_2) on this Edwards curve E is

$$(x_1, y_1) + (x_2, y_2) = \left(\frac{x_1y_2 + y_1x_2}{1 + dx_1x_2y_1y_2}, \frac{y_1y_2 - x_1x_2}{1 - dx_1x_2y_1y_2} \right).$$

The point $(0, 1)$ is the neutral element of the addition law. The point $(0, -1)$ has order 2. The points $(1, 0)$ and $(-1, 0)$ have order 4. The inverse of a point (x_1, y_1) on E is $(-x_1, y_1)$. The addition law is strongly unified: i.e., it can also be used to double a point. The addition law also works for the neutral element and for inverses. If d is a nonsquare in k then, as proven in [4, Theorem 3.3], this addition law is complete: it works for *all* pairs of inputs.

Twisted Edwards Curves. The existence of points of order 4 restricts the number of elliptic curves in Edwards form over k. We embed the set of Edwards curves in a larger set of elliptic curves of a similar shape by introducing twisted Edwards curves.

Definition 2.1 (Twisted Edwards curve). *Fix a field k with* char$(k) \neq 2$. *Fix distinct nonzero elements $a, d \in k$. The* twisted Edwards curve with coefficients a and d *is the curve*

$$\mathrm{E}_{\mathrm{E},a,d} : ax^2 + y^2 = 1 + dx^2y^2.$$

An Edwards curve *is a twisted Edwards curve with $a = 1$.*

In Section 3 we will show that every twisted Edwards curve is birationally equivalent to an elliptic curve in Montgomery form, and vice versa. The elliptic curve has j-invariant $16(a^2 + 14ad + d^2)^3/ad(a - d)^4$.

Twisted Edwards Curves as Twists of Edwards Curves. The twisted Edwards curve $\mathrm{E}_{\mathrm{E},a,d} : ax^2 + y^2 = 1 + dx^2y^2$ is a quadratic twist of the Edwards curve $\mathrm{E}_{\mathrm{E},1,d/a} : \bar{x}^2 + \bar{y}^2 = 1 + (d/a)\bar{x}^2\bar{y}^2$. The map $(\bar{x}, \bar{y}) \mapsto (x, y) = (\bar{x}/\sqrt{a}, \bar{y})$ is an isomorphism from $\mathrm{E}_{\mathrm{E},1,d/a}$ to $\mathrm{E}_{\mathrm{E},a,d}$ over $k(\sqrt{a})$. If a is a square in k then $\mathrm{E}_{\mathrm{E},a,d}$ is isomorphic to $\mathrm{E}_{\mathrm{E},1,d/a}$ over k.

More generally, $\mathrm{E}_{\mathrm{E},a,d}$ is a quadratic twist of $\mathrm{E}_{\mathrm{E},\bar{a},\bar{d}}$ for any $\bar{a}, \bar{d}$ satisfying $\bar{d}/\bar{a} = d/a$. Conversely, every quadratic twist of a twisted Edwards curve is isomorphic to a twisted Edwards curve; i.e., the set of twisted Edwards curves is invariant under quadratic twists.

Furthermore, the twisted Edwards curve $\mathrm{E}_{\mathrm{E},a,d} : ax^2 + y^2 = 1 + dx^2y^2$ is a quadratic twist of (actually is birationally equivalent to) the twisted Edwards curve $\mathrm{E}_{\mathrm{E},d,a} : d\bar{x}^2 + \bar{y}^2 = 1 + a\bar{x}^2\bar{y}^2$. The map $(\bar{x}, \bar{y}) \mapsto (x, y) = (\bar{x}, 1/\bar{y})$ is a birational equivalence from $\mathrm{E}_{\mathrm{E},d,a}$ to $\mathrm{E}_{\mathrm{E},a,d}$. More generally, $\mathrm{E}_{\mathrm{E},a,d}$ is a quadratic twist of $\mathrm{E}_{\mathrm{E},\bar{a},\bar{d}}$ for any $\bar{a}, \bar{d}$ satisfying $\bar{d}/\bar{a} = a/d$. This generalizes the known fact, used in [4, proof of Theorem 2.1], that $\mathrm{E}_{\mathrm{E},1,d}$ is a quadratic twist of $\mathrm{E}_{\mathrm{E},1,1/d}$.

3   Montgomery Curves and Twisted Edwards Curves

Let k be a field with char$(k) \neq 2$. In this section we show that the set of Montgomery curves over k is equivalent to the set of twisted Edwards curves over k. We also analyze the extent to which this is true without twists.

Standard algorithms for transforming a Weierstrass curve into a Montgomery curve if possible (see, e.g., [11, Section 13.2.3.c]) can be combined with our explicit transformation from a Montgomery curve to a twisted Edwards curve.

Definition 3.1 (Montgomery curve). *Fix a field k with* char$(k) \neq 2$. *Fix $A \in k \setminus \{-2, 2\}$ and $B \in k \setminus \{0\}$. The* Montgomery curve with coefficients A and B *is the curve*

$$\mathrm{E}_{\mathrm{M},A,B} : Bv^2 = u^3 + Au^2 + u.$$

Theorem 3.2. *Fix a field k with* $\mathrm{char}(k) \neq 2$.

(i) Every twisted Edwards curve over k is birationally equivalent over k to a Montgomery curve.

Specifically, fix distinct nonzero elements $a, d \in k$. The twisted Edwards curve $E_{E,a,d}$ is birationally equivalent to the Montgomery curve $E_{M,A,B}$, where $A = 2(a+d)/(a-d)$ and $B = 4/(a-d)$. The map $(x,y) \mapsto (u,v) = ((1+y)/(1-y), (1+y)/(1-y)x)$ is a birational equivalence from $E_{E,a,d}$ to $E_{M,A,B}$, with inverse $(u,v) \mapsto (x,y) = (u/v, (u-1)/(u+1))$.

(ii) Conversely, every Montgomery curve over k is birationally equivalent over k to a twisted Edwards curve.

Specifically, fix $A \in k \setminus \{-2, 2\}$ and $B \in k \setminus \{0\}$. The Montgomery curve $E_{M,A,B}$ is birationally equivalent to the twisted Edwards curve $E_{E,a,d}$, where $a = (A+2)/B$ and $d = (A-2)/B$.

Proof. (i) Note that A and B are defined, since $a \neq d$. Note further that $A \in k \setminus \{-2, 2\}$ and $B \in k \setminus \{0\}$: if $A = 2$ then $a + d = a - d$ so $d = 0$, contradiction; if $A = -2$ then $a + d = d - a$ so $a = 0$, contradiction. Thus $E_{M,A,B}$ is a Montgomery curve.

The following script for the Sage computer-algebra system [24] checks that the quantities $u = (1+y)/(1-y)$ and $v = (1+y)/(1-y)x$ satisfy $Bv^2 = u^3 + Au^2 + u$ in the function field of the curve $E_{E,a,d} : ax^2 + y^2 = 1 + dx^2 y^2$:

```
R.<a,d,x,y>=QQ[]
A=2*(a+d)/(a-d)
B=4/(a-d)
S=R.quotient(a*x^2+y^2-(1+d*x^2*y^2))
u=(1+y)/(1-y)
v=(1+y)/((1-y)*x)
0==S((B*v^2-u^3-A*u^2-u).numerator())
```

The exceptional cases $y = 1$ and $x = 0$ occur for only finitely many points (x, y) on $E_{E,a,d}$. Conversely, $x = u/v$ and $y = (u-1)/(u+1)$; the exceptional cases $v = 0$ and $u = -1$ occur for only finitely many points (u, v) on $E_{M,A,B}$.

(ii) Note that a and d are defined, since $B \neq 0$. Note further that $a \neq 0$ since $A \neq -2$; $d \neq 0$ since $A \neq 2$; and $a \neq d$. Thus $E_{E,a,d}$ is a twisted Edwards curve. Furthermore

$$2\frac{a+d}{a-d} = 2\frac{\frac{A+2}{B} + \frac{A-2}{B}}{\frac{A+2}{B} - \frac{A-2}{B}} = A \quad \text{and} \quad \frac{4}{(a-d)} = \frac{4}{\frac{A+2}{B} - \frac{A-2}{B}} = B.$$

Hence $E_{E,a,d}$ is birationally equivalent to $E_{M,A,B}$ by (i). $\qquad \square$

Exceptional Points for the Birational Equivalence. The map $(u,v) \mapsto (u/v, (u-1)/(u+1))$ from $E_{M,A,B}$ to $E_{E,a,d}$ in Theorem 3.2 is undefined at the points of $E_{M,A,B} : Bv^2 = u^3 + Au^2 + u$ with $v = 0$ or $u + 1 = 0$. We investigate these points in more detail:

- The point $(0,0)$ on $E_{M,A,B}$ corresponds to the affine point of order 2 on $E_{E,a,d}$, namely $(0, -1)$. This point and $(0, 1)$ are the only exceptional points of the inverse map $(x, y) \mapsto ((1 + y)/(1 - y), (1 + y)/(1 - y)x)$, where $(0, 1)$ is mapped to the point at infinity.

- If $(A + 2)(A - 2)$ is a square (i.e., if ad is a square) then there are two more points with $v = 0$, namely $((-A \pm \sqrt{(A + 2)(A - 2)})/2, 0)$. These points have order 2. These points correspond to two points of order 2 at infinity on the desingularization of $E_{E,a,d}$.
- If $(A - 2)/B$ is a square (i.e., if d is a square) then there are two points with $u = -1$, namely $(-1, \pm\sqrt{(A - 2)/B})$. These points have order 4. These points correspond to two points of order 4 at infinity on the desingularization of $E_{E,a,d}$.

Eliminating the Twists. Every Montgomery curve $E_{M,A,B}$ is birationally equivalent to a twisted Edwards curve by Theorem 3.2, and therefore to a quadratic twist of an Edwards curve. In other words, there is a quadratic twist of $E_{M,A,B}$ that is birationally equivalent to an Edwards curve.

We now state two situations in which twisting is not necessary. Theorem 3.3 states that every elliptic curve having a point of order 4 is birationally equivalent to an Edwards curve. Theorem 3.4 states that, over a finite field k with $\#k \equiv 3$ (mod 4), every Montgomery curve is birationally equivalent to an Edwards curve.

Some special cases of these results were already known. Bernstein and Lange proved in [4, Theorem 2.1(1)] that every elliptic curve having a point of order 4 is birationally equivalent to a twist of an Edwards curve, and in [4, Theorem 2.1(3)] that, over a finite field, every elliptic curve having a point of order 4 and a unique point of order 2 is birationally equivalent to an Edwards curve. We prove that the twist in [4, Theorem 2.1(1)] is unnecessary, and that the unique point of order 2 in [4, Theorem 2.1(3)] is unnecessary.

Theorem 3.3. *Fix a field k with $\mathrm{char}(k) \neq 2$. Let E be an elliptic curve over k. The group $E(k)$ has an element of order 4 if and only if E is birationally equivalent over k to an Edwards curve.*

Proof. Assume that E is birationally equivalent over k to an Edwards curve $E_{E,1,d}$. The elliptic-curve addition law corresponds to the Edwards addition law; see [4, Theorem 3.2]. The point $(1, 0)$ on $E_{E,1,d}$ has order 4, so E must have a point of order 4.

Conversely, assume that E has a point (u_4, v_4) of order 4. As in [4, Theorem 2.1, proof], observe that $u_4 \neq 0$ and $v_4 \neq 0$; assume without loss of generality that E has the form $v^2 = u^3 + (v_4^2/u_4^2 - 2u_4)u^2 + u_4^2 u$; define $d = 1 - 4u_4^3/v_4^2$; and observe that $d \notin \{0, 1\}$.

The following script for the Sage computer-algebra system checks that the quantities $x = v_4 u/u_4 v$ and $y = (u - u_4)/(u + u_4)$ satisfy $x^2 + y^2 = 1 + dx^2 y^2$ in the function field of E:

```
R.<u,v,u4,v4>=QQ[]
d=1-4*u4^3/v4^2
S=R.quotient((v^2-u^3-(v4^2/u4^2-2*u4)*u^2-u4^2*u).numerator())
x=v4*u/(u4*v)
```

```
y=(u-u4)/(u+u4)
0==S((x^2+y^2-1-d*x^2*y^2).numerator())
```

The exceptional cases $u_4 v = 0$ and $u = -u_4$ occur for only finitely many points (u, v) on E. Conversely, $u = u_4(1 + y)/(1 - y)$ and $v = v_4(1 + y)/(1 - y)x$; the exceptional cases $y = 1$ and $x = 0$ occur for only finitely many points (x, y) on $E_{E,1,d}$.

Therefore the rational map $(u, v) \mapsto (x, y) = (v_4 u/u_4 v, (u-u_4)/(u+u_4))$, with inverse $(x, y) \mapsto (u, v) = (u_4(1 + y)/(1 - y), v_4(1 + y)/(1 - y)x)$, is a birational equivalence from E to the Edwards curve $E_{E,1,d}$.   □

Theorem 3.4. *If k is a finite field with $\#k \equiv 3$ (mod 4) then every Montgomery curve over k is birationally equivalent over k to an Edwards curve.*

Proof. Fix $A \in k \setminus \{-2, 2\}$ and $B \in k \setminus \{0\}$. We will use an idea of Okeya, Kurumatani, and Sakurai [21], building upon the observations credited to Suyama in [20, page 262], to prove that the Montgomery curve $E_{M,A,B}$ has a point of order 4. This fact can be extracted from [21, Theorem 1] when $\#k$ is prime, but to keep this paper self-contained we include a direct proof.

Case 1: $(A+2)/B$ is a square. Then (as mentioned before) $E_{M,A,B}$ has a point $(1, \sqrt{(A + 2)/B})$ of order 4.

Case 2: $(A + 2)/B$ is a nonsquare but $(A - 2)/B$ is a square. Then $E_{M,A,B}$ has a point $(-1, \sqrt{(A - 2)/B})$ of order 4.

Case 3: $(A+2)/B$ and $(A-2)/B$ are nonsquares. Then $(A + 2)(A - 2)$ must be square, since k is finite. The Montgomery curve $E_{M,A,A+2}$ has three points $(0, 0)$, $((-A \pm \sqrt{(A + 2)(A - 2)})/2, 0)$ of order 2, and a point $(1, 1)$ of order 4, so $\#E_{M,A,A+2}(k) \equiv 0$ (mod 8). Furthermore, $E_{M,A,B}$ is a nontrivial quadratic twist of $E_{M,A,A+2}$, so $\#E_{M,A,B}(k) + \#E_{M,A,A+2}(k) = 2\#k + 2 \equiv 0$ (mod 8). Therefore $\#E_{M,A,B}(k) \equiv 0$ (mod 8). The curve $E_{M,A,B}$ cannot have more than three points of order 2, so it must have a point of order 4.

In every case $E_{M,A,B}$ has a point of order 4. By Theorem 3.3, $E_{M,A,B}$ is birationally equivalent to an Edwards curve.   □

This theorem does not generalize to $\#k \equiv 1$ (mod 4). For example, the Montgomery curve $E_{M,9,1}$ over $\mathbf{F}_{17}$ has order 20 and group structure isomorphic to $\mathbf{Z}/2 \times \mathbf{Z}/10$. This curve is birationally equivalent to the twisted Edwards curve $E_{E,11,7}$, but it does not have a point of order 4, so it is not birationally equivalent to an Edwards curve.

Theorem 3.5. *Let k be a finite field with $\#k \equiv 1$ (mod 4). Let $E_{M,A,B}$ be a Montgomery curve so that $(A + 2)(A - 2)$ is a square and let δ be a nonsquare.*

Exactly one of $E_{M,A,B}$ and its nontrivial quadratic twist $E_{M,A,\delta B}$ is birationally equivalent to an Edwards curve.

In particular, $E_{M,A,A+2}$ is birationally equivalent to an Edwards curve.

Proof. Since $(A + 2)(A - 2)$ is a square both $E_{M,A,B}$ and $E_{M,A,\delta B}$ contain a subgroup isomorphic to $\mathbf{Z}/2\mathbf{Z} \times \mathbf{Z}/2\mathbf{Z}$. This subgroup accounts for a factor of 4 in the group order. Since $\#E_{M,A,B}(k) + \#E_{M,A,\delta B}(k) = 2\#k + 2 \equiv 4 \pmod 8$ exactly one of $\#E_{M,A,B}(k)$ and $\#E_{M,A,\delta B}(k)$ is divisible by 4 but not by 8. That curve cannot have a point of order 4 while the other one has a point of order 4. The first statement follows from Theorem 3.3.

The second statement also follows from Theorem 3.3, since the point $(1,1)$ on $E_{M,A,A+2}$ has order 4. □

4   Statistics

It is well known that, when p is a large prime, there are approximately $2p$ isomorphism classes of elliptic curves over the finite field $\mathbf{F}_p$. How many of these elliptic curves are birationally equivalent to twisted Edwards curves $ax^2 + y^2 = 1 + dx^2y^2$? How many are birationally equivalent to Edwards curves $x^2 + y^2 = 1 + dx^2y^2$? How many are birationally equivalent to complete Edwards curves, i.e., Edwards curves with nonsquare d? How many are birationally equivalent to original Edwards curves $x^2 + y^2 = c^2(1 + x^2y^2)$? How do the statistics vary with the number of powers of 2 in the group order?

We computed the answers for various primes p by enumerating all complete Edwards curves, all Edwards curves, all twisted Edwards curves (with a limited set of a's covering all isomorphism classes), and all elliptic curves in Weierstrass form (with similar limitations). We transformed each curve to a birationally equivalent elliptic curve E and then computed $(\#E, j(E))$, where $\#E$ is the number of points on E and $j(E)$ is the j-invariant of E. Recall that $j(E) = j(E')$ if and only if E' is a twist of E, and that twists are distinguished by $\#E$ except for a few isomorphism classes.

Some parts of these experiments have been carried out before. See, e.g., [15]. However, the information in the literature is not sufficient for our comparison of Edwards curves (and complete Edwards curves) to twisted Edwards curves.

Answers for Primes $p \equiv 1 \pmod 4$. For $p = 1009$ we found

- 43 different pairs $(\#E, j(E))$ for original Edwards curves,
- 504 different pairs $(\#E, j(E))$ for complete Edwards curves,
- 673 different pairs $(\#E, j(E))$ for Edwards curves,
- 842 different pairs $(\#E, j(E))$ for twisted Edwards curves,
- 842 different pairs $(\#E, j(E))$ for elliptic curves with group order divisible by 4, and
- 2014 different pairs $(\#E, j(E))$ for elliptic curves.

We looked more closely at the number of powers of 2 dividing $\#E$ and observed the following distribution:

Curves	Total	odd	$2 \cdot$ odd	$4 \cdot$ odd	$8 \cdot$ odd	$16 \cdot$ odd	$32 \cdot$ odd	$64 \cdot$ odd
original Edwards	43	0	0	0	0	23	6	6
complete Edwards	504	0	0	252	130	66	24	16
Edwards	673	0	0	252	195	122	42	30
twisted Edwards	842	0	0	421	195	122	42	30
4 divides group order	842	0	0	421	195	122	42	30
all	2014	676	496	421	195	122	42	30

We observed similar patterns for more than 1000 tested primes $p \equiv 1 \pmod 4$:

Curves	Total	odd	2·odd	4·odd	8·odd
original Edwards	$\approx (1/24)p$	0	0	0	0
complete Edwards	$\approx (1/2)p$	0	0	$\approx (1/4)p$	$\approx (1/8)p$
Edwards	$\approx (2/3)p$	0	0	$\approx (1/4)p$	$\approx (3/16)p$
twisted Edwards	$\approx (5/6)p$	0	0	$\approx (5/12)p$	$\approx (3/16)p$
4 divides group order	$\approx (5/6)p$	0	0	$\approx (5/12)p$	$\approx (3/16)p$
all		$\approx 2p$	$\approx (2/3)p$	$\approx (1/2)p$	$\approx (5/12)p$	$\approx (3/16)p$

We do not claim novelty for statistics regarding the set of Montgomery curves (in other words, the set of twisted Edwards curves) and the set of all elliptic curves; all of these statistics have been observed before, and some of them have been proven. Furthermore, the $(1/2)p$ for complete Edwards curves was pointed out in [4, Abstract]. However, the $(2/3)p$, $(1/4)p$, and $(3/16)p$ for Edwards curves appear to be new observations. We include the old statistics as a basis for comparison.

Answers for Primes $p \equiv 3 \pmod 4$. For primes $p \equiv 3 \pmod 4$ the patterns are different, as one would expect from Theorems 3.4 and 3.5. For example, here is the analogous table for $p = 1019$:

Curves	Total	odd	2·odd	4·odd	8·odd	16·odd	32·odd	64·odd
original Edwards	254	0	0	0	127	68	33	10
complete Edwards	490	0	0	236	127	68	33	10
Edwards	744	0	0	236	254	136	66	20
twisted Edwards	744	0	0	236	254	136	66	20
4 divides group order	822	0	0	314	254	136	66	20
all	2012	680	510	314	254	136	66	20

We observed similar patterns for more than 1000 tested primes $p \equiv 3 \pmod 4$:

Curves	Total	odd	2·odd	4·odd	8·odd
original Edwards	$\approx (1/4)p$	0	0	0	$\approx (1/8)p$
complete Edwards	$\approx (1/2)p$	0	0	$\approx (1/4)p$	$\approx (1/8)p$
Edwards	$\approx (3/4)p$	0	0	$\approx (1/4)p$	$\approx (1/4)p$
twisted Edwards	$\approx (3/4)p$	0	0	$\approx (1/4)p$	$\approx (1/4)p$
4 divides group order	$\approx (5/6)p$	0	0	$\approx (1/3)p$	$\approx (1/4)p$
all		$\approx 2p$	$\approx (2/3)p$	$\approx (1/2)p$	$\approx (1/3)p$	$\approx (1/4)p$

As above, we do not claim novelty for statistics regarding the set of Montgomery curves and the set of all elliptic curves; we include these statistics as a basis for comparison.

Near-Prime Group Orders. We also looked at how often the odd part of $\#E$ was prime and observed the following distribution for $p = 1009$:

Curves	prime	2 · prime	4 · prime	8 · prime	16 · prime	32 · prime
original Edwards	0	0	0	0	8	2
complete Edwards	0	0	64	42	28	8
Edwards	0	0	64	63	50	14
twisted Edwards	0	0	102	63	50	14
4 divides group order	0	0	102	63	50	14
all	189	98	102	63	50	14

Here is the analogous table for $p = 1019$:

Curves	prime	2 · prime	4 · prime	8 · prime	16 · prime	32 · prime
original Edwards	0	0	0	25	22	9
complete Edwards	0	0	48	25	22	9
Edwards	0	0	48	50	44	18
twisted Edwards	0	0	48	50	44	18
4 divides group order	0	0	64	50	44	18
all	148	100	64	50	44	18

Of course, larger primes p have smaller chances of prime $\#E$, smaller chances of prime $\#E/2$, etc.

5   Isogenies: Even More Curves

A curve that is not isomorphic to an Edwards curve, and not even isomorphic to a twisted Edwards curve, might nevertheless be *isogenous* to a twisted Edwards curve. This section shows, in particular, that every curve with three points of order 2 is 2-isogenous to a twisted Edwards curve. This section gives an example of a curve that is not birationally equivalent to a twisted Edwards curve but that is 2-isogenous to a twisted Edwards curve. This section also discusses the use of 2-isogenies for scalar multiplication in an odd-order subgroup of the original curve.

Our use of isogenies to expand the coverage of twisted Edwards curves is analogous to the use of isogenies by Brier and Joye in [9] to expand the coverage of "$a_4 = -3$" Weierstrass curves. We comment that isogenies are also useful for other curve shapes. For example, over fields $\mathbf{F}_p$ with $p \equiv 3 \pmod 4$, every elliptic curve with a point of order 4 is 2-isogenous to a Jacobi-quartic curve $v^2 = u^4 - 2\delta u^2 + 1$; see [6], [12], [16], [17], and [3] for fast explicit formulas to perform computations on curves of this shape.

Theorem 5.1. *Fix a field k with $\mathrm{char}(k) \neq 2$. Every elliptic curve over k having three k-rational points of order 2 is 2-isogenous over k to a twisted Edwards curve.*

Proof. Let E be an elliptic curve over k having three k-rational points of order 2. Write E in Weierstrass form $v^2 = u^3 + a_2 u^2 + a_4 u + a_6$, with points $(u_0, 0)$ and $(u_1, 0)$ and $(u_2, 0)$ of order 2. Assume without loss of generality that $u_0 = 0$; to handle the general case, replace u by $u - u_0$.

The polynomial $u^3 + a_2 u^2 + a_4 u + a_6$ has distinct roots $0, u_1, u_2$ so it factors as $u(u - u_1)(u - u_2)$; i.e., E has the form

$$v^2 = u^3 - (u_1 + u_2)u^2 + (u_1 u_2)u.$$

Therefore E is 2-isogenous to the elliptic curve $\bar{E}$ given by

$$\bar{v}^2 = \bar{u}^3 + 2(u_1 + u_2)\bar{u}^2 + (u_1 - u_2)^2 \bar{u};$$

see, e.g., [22, Chapter III, Example 4.5]. The 2-isogeny from E to $\bar{E}$ is given by

$$\bar{u} = \frac{v^2}{u^2} \qquad \text{and} \qquad \bar{v} = \frac{v(u_1 u_2 - u^2)}{u^2}.$$

The dual 2-isogeny from $\bar{E}$ to E is given by

$$u = \frac{\bar{v}^2}{4\bar{u}^2} \qquad \text{and} \qquad v = \frac{\bar{v}((u_1 - u_2)^2 - \bar{u}^2)}{8\bar{u}^2}.$$

The elliptic curve $\bar{E}$ is isomorphic to $E_{M,2(u_1+u_2)/(u_1-u_2),1/(u_1-u_2)}$, so by Theorem 3.2 it is birationally equivalent to $E_{E,4u_1,4u_2}$. Therefore the original elliptic curve E is 2-isogenous to $E_{E,4u_1,4u_2}$.   □

A Numerical Example. Over fields $\mathbf{F}_p$ with $p \equiv 1 \pmod 4$, every curve with three points of order 2 is already birationally equivalent to a twisted Edwards curve. However, over fields $\mathbf{F}_p$ with $p \equiv 3 \pmod 4$, a curve that has three points of order 2 is not birationally equivalent to a twisted Edwards curve unless it has a point of order 4; see Theorem 3.4. Theorem 5.1 applies whether or not there is a point of order 4.

Consider, for example, the elliptic curve given in [7, Appendix A.1, Example 11]. This is a Weierstrass-form curve $y^2 = x^3 + ax + b$ having n points over a prime field $\mathbf{F}_p$ with $p \equiv 3 \pmod 4$, where

$p =$ 70488450694327127420028164186486186967538228180387437428782357259063646577643090299493711662715469759600817584399431788 7,

$a = 5$,

$b =$ 3866629042208848461581189787552969575881611445812272276326084773948335087614278974368305033461629194634976270793647521 99,

$n/4 =$ 17622112673581781855007041046621546741884557045096859357195584999843883742026613677911440007805459015400711640464440609 71.

There are three roots of $x^3 + ax + b$ modulo p, so this elliptic curve has three points of order 2. It is 2-isogenous to a twisted Edwards curve by Theorem 5.1.

On the other hand, it is not birationally equivalent to a Montgomery curve, or to a twisted Edwards curve; if it were, it would have a point of order 4 by Theorem 3.4, so n would have to be a multiple of 8.

The most important operation in elliptic-curve cryptography is scalar multiplication in a prime-order subgroup of an elliptic curve. Consider a point P in the subgroup of order $n/4$ of the elliptic curve shown above; $n/4$ is prime. To compute $Q = mP$ for any integer m, we do the following:

- compute $P' = \phi(P)$, where ϕ is the 2-isogeny (shown explicitly in the proof of Theorem 5.1) from this elliptic curve to a twisted Edwards curve;
- compute $Q' = ((m/2) \bmod (n/4))P'$ on the twisted Edwards curve; and
- compute $Q = \hat{\phi}(Q')$, where $\hat{\phi}$ is the dual isogeny.

The isogeny and dual isogeny are easy to evaluate, so most of the work consists of the scalar multiplication on the twisted Edwards curve.

6   Arithmetic on Twisted Edwards Curves

Let k be a field with $\mathrm{char}(k) \neq 2$. In this section we present fast explicit formulas for addition and doubling on twisted Edwards curves over k.

The Twisted Edwards Addition Law. Let $(x_1, y_1), (x_2, y_2)$ be points on the twisted Edwards curve $E_{E,a,d} : ax^2 + y^2 = 1 + dx^2y^2$. The sum of these points on $E_{E,a,d}$ is

$$(x_1, y_1) + (x_2, y_2) = \left(\frac{x_1y_2 + y_1x_2}{1 + dx_1x_2y_1y_2}, \frac{y_1y_2 - ax_1x_2}{1 - dx_1x_2y_1y_2} \right).$$

The neutral element is $(0, 1)$, and the negative of (x_1, y_1) is $(-x_1, y_1)$.

For the correctness of the addition law observe that it coincides with the Edwards addition law on $\bar{x}^2 + y^2 = 1 + (d/a)\bar{x}^2y^2$ with $\bar{x} = \sqrt{a}x$ which is proven correct in [4, Section 3].

These formulas also work for doubling. These formulas are complete (i.e., have no exceptional cases) if a is a square in k and d is a nonsquare in k. The latter follows from $E_{E,a,d}$ being isomorphic to $E_{E,1,d/a}$; d/a being a nonsquare in k and from [4, Theorem 3.1] which showed that the Edwards addition law is complete on $E_{E,1,d'}$ if d' is a nonsquare.

Projective Twisted Edwards Coordinates. To avoid inversions we work on the projective twisted Edwards curve

$$(aX^2 + Y^2)Z^2 = Z^4 + dX^2Y^2.$$

For $Z_1 \neq 0$ the homogeneous point $(X_1 : Y_1 : Z_1)$ represents the affine point $(X_1/Z_1, Y_1/Z_1)$ on $E_{E,a,d}$.

We checked the following explicit formulas for addition and doubling with the help of the Sage computer-algebra system, following the approach of the Explicit-Formulas Database [3].

Addition in Projective Twisted Coordinates. The following formulas compute $(X_3 : Y_3 : Z_3) = (X_1 : Y_1 : Z_1) + (X_2 : Y_2 : Z_2)$ in $10\mathbf{M} + 1\mathbf{S} + 2\mathbf{D} + 7\mathbf{add}$, where the $2\mathbf{D}$ are one multiplication by a and one by d:

$$A = Z_1 \cdot Z_2; \; B = A^2; \; C = X_1 \cdot X_2; \; D = Y_1 \cdot Y_2; \; E = dC \cdot D;$$
$$F = B - E; \; G = B + E; \; X_3 = A \cdot F \cdot ((X_1 + Y_1) \cdot (X_2 + Y_2) - C - D);$$
$$Y_3 = A \cdot G \cdot (D - aC); \; Z_3 = F \cdot G.$$

Doubling in Projective Twisted Coordinates. The following formulas compute $(X_3 : Y_3 : Z_3) = 2(X_1 : Y_1 : Z_1)$ in $3\mathbf{M} + 4\mathbf{S} + 1\mathbf{D} + 7\mathbf{add}$, where the $1\mathbf{D}$ is a multiplication by a:

$$B = (X_1 + Y_1)^2; \; C = X_1^2; \; D = Y_1^2; \; E = aC; F := E + D; \; H = Z_1^2;$$
$$J = F - 2H; \; X_3 = (B - C - D) \cdot J; \; Y_3 = F \cdot (E - D); \; Z_3 = F \cdot J.$$

Clearing Denominators in Projective Coordinates. Here is an alternative approach to arithmetic on the twisted Edwards curve $E_{E,a,d}$ when a is a square in k.

The curve $E_{E,a,d} : a\bar{x}^2 + \bar{y}^2 = 1 + d\bar{x}^2\bar{y}^2$ is isomorphic to the Edwards curve $E_{E,1,d/a} : x^2 + y^2 = 1 + (d/a)x^2y^2$ by $x = \sqrt{a}\bar{x}$ and $y = \bar{y}$; see Section 2. The following formulas add on $E_{E,1,d/a}$ using $10\mathbf{M} + 1\mathbf{S} + 3\mathbf{D} + 7\mathbf{add}$, where the $3\mathbf{D}$ are two multiplications by a and one by d:

$$A = Z_1 \cdot Z_2; \; B = aA^2; \; H = aA; \; C = X_1 \cdot X_2; \; D = Y_1 \cdot Y_2; \; E = dC \cdot D;$$
$$F = B - E; \; G = B + E; \; X_3 = H \cdot F \cdot ((X_1 + Y_1) \cdot (X_2 + Y_2) - C - D);$$
$$Y_3 = H \cdot G \cdot (D - C); \; Z_3 = F \cdot G.$$

One can double on $E_{E,1,d/a}$ with $3\mathbf{M} + 4\mathbf{S} + 6\mathbf{add}$, independent of the curve coefficient d/a, using the formulas from [4, Section 4].

Our addition formulas for $E_{E,1,d/a}$ are slower (by 1 multiplication by a) than our addition formulas for $E_{E,a,d}$. On the other hand, doubling for $E_{E,1,d/a}$ is faster (by 1 multiplication by a) than doubling for $E_{E,a,d}$. Some applications (such as batch signature verification) have more additions than doublings, while other applications have more doublings than additions, so all of the formulas are of interest.

Inverted Twisted Edwards Coordinates. Another way to avoid inversions is to let a point $(X_1 : Y_1 : Z_1)$ on the curve

$$(X^2 + aY^2)Z^2 = X^2Y^2 + dZ^4$$

with $X_1 Y_1 Z_1 \neq 0$ correspond to the affine point $(Z_1/X_1, Z_1/Y_1)$ on $E_{E,a,d}$.

Bernstein and Lange introduced these inverted coordinates in [5], for the case $a = 1$, and observed that the coordinates save time in addition. We generalize to arbitrary a.

Addition in Inverted Twisted Coordinates. The following formulas compute $(X_3 : Y_3 : Z_3) = (X_1 : Y_1 : Z_1) + (X_2 : Y_2 : Z_2)$ in $9\mathbf{M} + 1\mathbf{S} + 2\mathbf{D} + 7\mathbf{add}$, where the $2\mathbf{D}$ are one multiplication by a and one by d:

$$A = Z_1 \cdot Z_2; \ B = dA^2; \ C = X_1 \cdot X_2; \ D = Y_1 \cdot Y_2; \ E = C \cdot D;$$
$$H = C - aD; \ I = (X_1 + Y_1) \cdot (X_2 + Y_2) - C - D;$$
$$X_3 = (E + B) \cdot H; \ Y_3 = (E - B) \cdot I; \ Z_3 = A \cdot H \cdot I.$$

Doubling in Inverted Twisted Coordinates. The following formulas compute $(X_3 : Y_3 : Z_3) = 2(X_1 : Y_1 : Z_1)$ in $3\mathbf{M} + 4\mathbf{S} + 2\mathbf{D} + 6\mathbf{add}$, where the $2\mathbf{D}$ are one multiplication by a and one by $2d$:

$$A = X_1^2; \ B = Y_1^2; \ U = aB; \ C = A + U; \ D = A - U;$$
$$E = (X_1 + Y_1)^2 - A - B; \ X_3 = C \cdot D; \ Y_3 = E \cdot (C - 2dZ_1^2); \ Z_3 = D \cdot E.$$

Clearing Denominators in Inverted Coordinates. The following formulas add in inverted coordinates on $E_{E,1,d/a}$ using $9\mathbf{M} + 1\mathbf{S} + 3\mathbf{D} + 7\mathbf{add}$, where the $3\mathbf{D}$ are two multiplications by a and one by d:

$$A = Z_1 \cdot Z_2; \ B = dA^2; \ C = X_1 \cdot X_2; \ D = Y_1 \cdot Y_2; \ E = aC \cdot D;$$
$$H = C - D; \ I = (X_1 + Y_1) \cdot (X_2 + Y_2) - C - D;$$
$$X_3 = (E + B) \cdot H; \ Y_3 = (E - B) \cdot I; \ Z_3 = aA \cdot H \cdot I.$$

The following formulas double in inverted coordinates on $E_{E,1,d/a}$ using $3\mathbf{M} + 4\mathbf{S} + 3\mathbf{D} + 5\mathbf{add}$, where the $3\mathbf{D}$ are two multiplications by a and one by $2d$:

$$A = X_1^2; \ B = Y_1^2; \ C = A + B; \ D = A - B; \ E = (X_1 + Y_1)^2 - C;$$
$$F = aC; \ Z_3 = aD \cdot E; \ X_3 = F \cdot D; \ Y_3 = E \cdot (F - 2dZ_1^2).$$

More Parameters. One could consider the more general curve equation

$$ax^2 + y^2 = c^2(1 + dx^2y^2)$$

with addition law

$$(x_1, y_1) + (x_2, y_2) = \left(\frac{x_1y_2 + y_1x_2}{c(1 + dx_1x_2y_1y_2)}, \frac{y_1y_2 - ax_1x_2}{c(1 - dx_1x_2y_1y_2)} \right).$$

We do not present explicit formulas for this generalization; these curves are always isomorphic to twisted Edwards curves. We comment, however, that there exist curves for which the extra parameter saves a little time.

7 Edwards Versus Twisted Edwards

We introduced twisted Edwards curves as a generalization of Edwards curves. Is this generalization actually useful for cryptographic purposes?

Section 4 showed that, over prime fields $\mathbf{F}_p$ where $p \equiv 1 \pmod 4$, twisted Edwards curves cover considerably more elliptic curves than Edwards curves do. In particular, for "4 times odd" elliptic curves over such prime fields, the coverage of Edwards curves is only about 60% of the coverage of twisted Edwards curves. One can choose a to be very small, making twisted Edwards curves essentially as fast as Edwards curves and thus bringing the speed of the Edwards addition law to a wider variety of elliptic curves.

Even when an elliptic curve *can* be expressed in Edwards form, expressing the same curve in twisted Edwards form often saves time in arithmetic. In this section we review the issues faced by implementors aiming for top speed. We give examples of the impact of twisted Edwards curves for implementors who are faced with externally specified curves, and for implementors who are free to choose their own curves.

How Twisting Can Save Time. The following table summarizes the speeds of addition and doubling in standard (projective) coordinates on Edwards curves, standard coordinates on twisted Edwards curves, inverted coordinates on Edwards curves, and inverted coordinates on twisted Edwards curves:

Coordinates	Source of algorithms	Addition	Doubling
Edwards	[4, §4]	10M+1S+1D (mult by d/a)	3M+4S
Edwards	this paper (clearing denoms)	10M+1S+3D (mult by a, a, d)	3M+4S
Twisted Edwards	this paper	10M+1S+2D (mult by a, d)	3M+4S+1D (mult by a)
Inverted Edwards	[5, §§4–5]	9M+1S+1D (mult by d/a)	3M+4S+1D (mult by d/a)
Inverted Edwards	this paper (clearing denoms)	9M+1S+3D (mult by a, a, d)	3M+4S+3D (mult by a, a, d)
Inverted twisted Edwards	this paper	9M+1S+2D (mult by a, d)	3M+4S+2D (mult by a, d)

If a curve E is expressible as an Edwards curve, is there any reason to consider more general expressions of E as a twisted Edwards curve? One might think, from a glance at the above table, that the answer is no: twisting appears to lose 1D in every coordinate system and for every group operation without gaining anything. However, there are many situations where the answer is yes!

Specifically, instead of performing computations on the Edwards curve $E_{E,1,\bar{d}}$ over k, one can perform computations on the twisted Edwards curve $E_{E,a,d}$ over k for any (a,d) such that $\bar{d} = d/a$ and such that a is a square in k. (It is convenient for computing the isomorphism, but certainly not essential, for a to be the square of a small integer.) In particular, many curves over $\mathbf{F}_p$ have $\bar{d}$ expressible as a ratio d/a where both d and a are small, much smaller than any integer congruent to $\bar{d}$ modulo p. In the non-twisted Edwards case the 1D in the table above is a multiplication by $\bar{d}$ while the 2D in the twisted Edwards case are one multiplication by d and one multiplication by a, often taking *less* time than a multiplication by $\bar{d}$.

Consider, for example, the curve "Curve25519" used in [1] to set speed records for elliptic-curve Diffie-Hellman before the advent of Edwards curves. Curve25519 is a particular elliptic curve over $\mathbf{F}_p$ where $p = 2^{255} - 19$. Bernstein and Lange point out in [4, Section 2] that Curve25519 can be expressed as an Edwards curve $x^2 + y^2 = 1 + (121665/121666)x^2y^2$. We point out that this curve is isomorphic to the twisted Edwards curve $121666x^2 + y^2 = 1 + 121665x^2y^2$, and that the twisted Edwards curve provides faster arithmetic. Each addition on the twisted Edwards curve involves only one multiplication by 121665 and one multiplication by 121666, which together are faster than a multiplication by 20800338683988658368647408995589388737092878452977063003340006470870624536394 $\equiv 121665/121666 \pmod{p}$.

This phenomenon is not an accident. Montgomery curves $E_{M,A,B}$ are normally chosen so that $(A+2)/4$ is a small integer: this speeds up u-coordinate arithmetic, as Montgomery pointed out in [20, page 261, bottom]. The corresponding twisted Edwards curves have d/a equal to $(A - 2)/(A + 2)$, a ratio of small integers, allowing fast arithmetic in twisted Edwards form.

The decision between Edwards curves and twisted Edwards curves interacts with the decision between standard Edwards coordinates and inverted Edwards coordinates. Frequent additions make inverted Edwards coordinates more impressive; large a, d make inverted Edwards coordinates less impressive.

Choosing Twisted Edwards Curves. Often implementors are free to choose their own curves for the best possible speed. To illustrate the benefits of this flexibility we studied "small" twisted Edwards curves modulo several primes of cryptographic size: $2^{160} - 47$, the largest prime below 2^{160}; $2^{192} - 2^{64} - 1$, the prime used for NIST's P-192 elliptic curve; $2^{224} - 2^{96} + 1$, the prime used for NIST's P-224 elliptic curve; and $2^{255} - 19$, the prime used in [1]. Specifically, we enumerated twisted Edwards curves $E_{E,a,d}$ for thousands of small pairs (a, d), and we checked which curves had small cofactors over $\mathbf{F}_p$, i.e., had group orders $h \cdot \text{prime}$ where the cofactor h is small. We give some examples of twisted Edwards curves with small cofactor, tiny a, and tiny d, supporting exceptionally fast arithmetic.

For $p = 2^{192} - 2^{64} - 1$, the twisted Edwards curve $E_{E,102,47} : 102x^2 + y^2 = 1 + 47x^2y^2$ has cofactor 4. Arithmetic on $E_{E,102,47}$ is impressively fast, and the cofactor is minimal. The nontrivial quadratic twist $E_{E,1122,517}$ has cofactor only 28, protecting against the active small-subgroup attacks discussed in (e.g.) [1, Section 3].

For $p = 2^{224} - 2^{96} + 1$, the twisted Edwards curve $E_{E,12,1}$ has cofactor 3456, and its nontrivial quadratic twist $E_{E,132,11}$ has cofactor 20. The coefficients $a = 12$ and $d = 1$ here are spectacularly small. The cofactor 3456 is not minimal but can still be considered for cryptographic purposes.

If active small-subgroup attacks are stopped in other ways then one can find even smaller pairs (a, d). For $p = 2^{160} - 47$ the twisted Edwards curve $E_{E,23,-6}$ has cofactor 4; for comparison, the first Edwards curve we found with small parameter d and with cofactor 4 over the same field was $E_{E,1,268}$. For $p = 2^{255} - 19$ the twisted Edwards curve $E_{E,29,-28}$ has cofactor 4 and the twisted Edwards curve $E_{E,25,2}$ has cofactor 8.

References

1. Bernstein, D.J.: Curve25519: New Diffie-Hellman Speed Records. In: PKC 2006 [25], pp. 207–228. (2006) (Citations in this document: §7, §7, §7), http://cr.yp.to/papers.html#curve25519
2. Bernstein, D.J., Birkner, P., Lange, T., Peters, C.: ECM using Edwards curves (2007) (Citations in this document: §1), http://eprint.iacr.org/2008/016
3. Bernstein, D.J., Lange, T.: Explicit-formulas database (2007) (Citations in this document: §5, §6), http://hyperelliptic.org/EFD
4. Bernstein, D.J., Lange, T.: Faster addition and doubling on elliptic curves. In: Asiacrypt 2007 [19], pp. 29–50 (2007) (Citations in this document: §1, §1, §2, §2, §2, §3, §3, §3, §3, §3, §3, §4, §6, §6, §6, §7, §7), http://cr.yp.to/papers.html#newelliptic
5. Bernstein, D.J., Lange, T.: Inverted Edwards coordinates. In: AAECC 2007 [8], pp. 20–27 (2007) (Citations in this document: §1, §6, §7), http://cr.yp.to/papers.html#inverted
6. Billet, O., Joye, M.: The Jacobi model of an elliptic curve and side-channel analysis. In: AAECC 2003 [14], pp. 34–42 (2003) (Citations in this document: §5), http://eprint.iacr.org/2002/125
7. Blake, I.F., Seroussi, G., Smart, N.P.: Elliptic curves in cryptography. Cambridge University Press, Cambridge (2000) (Citations in this document: §5)
8. Boztaş, S., Lu, H.-F(F.) (eds.): AAECC 2007. LNCS, vol. 4851. Springer, Heidelberg (2007)
9. Brier, É., Joye, M.: Fast point multiplication on elliptic curves through isogenies. In: AAECC 2003 [14], pp. 43–50 (2003) (Citations in this document: §5)
10. Cohen, H., Frey, G. (eds.): Handbook of elliptic and hyperelliptic curve cryptography. CRC Press, Boca Raton (2005) See [11]
11. Doche, C., Lange, T.: Arithmetic of elliptic curves. In: [10] (2005), 267–302. (Citations in this document: §3)
12. Duquesne, S.: Improving the arithmetic of elliptic curves in the Jacobi model. Information Processing Letters 104, 101–105 (2007) (Citations in this document: §5)
13. Edwards, H.M.: A normal form for elliptic curves. Bulletin of the American Mathematical Society 44, 393–422 (2007) (Citations in this document: §1), http://www.ams.org/bull/2007-44-03/S0273-0979-07-01153-6/home.html
14. Fossorier, M.P.C., Høholdt, T., Poli, A. (eds.): AAECC 2003. LNCS, vol. 2643. Springer, Heidelberg (2003) See [6], [9]
15. Galbraith, S.D., McKee, J.: The probability that the number of points on an elliptic curve over a finite field is prime. Journal of the London Mathematical Society 62, 671–684 (2000) (Citations in this document: §4), http://www.isg.rhul.ac.uk/~sdg/pubs.html
16. Hisil, H., Carter, G., Dawson, E.: New formulae for efficient elliptic curve arithmetic. In: INDOCRYPT 2007 [23] (2007) (Citations in this document: §5)
17. Hisil, H., Wong, K., Carter, G., Dawson, E.: Faster group operations on elliptic curves. 25 Feb 2008 version (2008) (Citations in this document: §5), http://eprint.iacr.org/2007/441

18. Imai, H., Zheng, Y. (eds.): PKC 2000. LNCS, vol. 1751. Springer, Heidelberg (2000) see [21]
19. Kurosawa, K. (ed.): ASIACRYPT 2007. LNCS, vol. 4833. Springer, Heidelberg (2007)
20. Montgomery, P.L.: Speeding the Pollard and elliptic curve methods of factorization. Mathematics of Computation 48, 243–264 (1987), (Citations in this document: §3, §7), http://links.jstor.org/sici?sici=0025-5718(198701)48:177243:177243:STPAEC2.0.CO;2-3
21. Okeya, K., Kurumatani, H., Sakurai, K.: Elliptic curves with the Montgomery-form and their cryptographic applications. In: PKC 2000 [18], pp. 238–257 (2000) (Citations in this document: §3 §3)
22. Silverman, J.H.: The arithmetic of elliptic curves. Graduate Texts in Mathematics 106 (1986)
23. Srinathan, K., Rangan, C.P., Yung, M. (eds.): INDOCRYPT 2007. LNCS, vol. 4859. Springer, Heidelberg (2007) See [16]
24. Stein, W. (ed.): Sage Mathematics Software (Version 2.8.12), The Sage Group (2008) (Citations in this document: §3), http://www.sagemath.org
25. Yung, M., Dodis, Y., Kiayias, A., Malkin, T. (eds.): PKC 2006. LNCS, vol. 3958. Springer, Heidelberg (2006) See [1]

Efficient Multiplication in $\mathbb{F}_{3^{\ell m}}$, $m \geq 1$ and $5 \leq \ell \leq 18$

Murat Cenk[1] and Ferruh Özbudak[2]

[1] Department of Mathematics and Computer Science,
Çankaya University, Ankara, Turkey
mcenk@cankaya.edu.tr
[2] Department of Mathematics and Institute of Applied Mathematics,
Middle East Technical University, Ankara, Turkey
ozbudak@metu.edu.tr

Abstract. Using a method based on Chinese Remainder Theorem for polynomial multiplication and suitable reductions, we obtain an efficient multiplication method for finite fields of characteristic 3. Large finite fields of characteristic 3 are important for pairing based cryptography [3]. For $5 \leq \ell \leq 18$, we show that our method gives canonical multiplication formulae over $\mathbb{F}_{3^{\ell m}}$ for any $m \geq 1$ with the best multiplicative complexity improving the bounds in [6]. We give explicit formula in the case $\mathbb{F}_{3^{6 \cdot 97}}$.

Keywords: Chinese Remainder Theorem, finite field multiplication, pairing based cryptography.

1 Introduction

Finite field multiplication plays an important role in public key cryptography and coding theory. Public key cryptographic applications accomplished in very large finite fields. For example, one needs a finite field of at least 2^{160} elements for elliptic curve cryptography. For that reason efficient finite field multiplication has become a crucial part of such applications. A finite field with q^n elements is denoted by $\mathbb{F}_{q^n}$ where q is a prime power and $n \geq 1$. The elements of $\mathbb{F}_{q^n}$ can be represented by n-term polynomials over $\mathbb{F}_q$. Field elements can be multiplied in terms of ordinary multiplication of polynomials and modular reduction of the result product by the defining polynomial of the finite field. The reduction step has no multiplicative complexity [5, p.8]. So finite field multiplication is directly related to the polynomial multiplication.

The finite fields of characteristic three are useful for pairing-based cryptography. Therefore, special attention has been given to $\mathbb{F}_{3^m}$, recently. The elements of $\mathbb{F}_{3^m}$ can be represented by at most $(m-1)$ degree polynomials over $\mathbb{F}_3$. To multiply elements of $\mathbb{F}_{3^m}$ one can use Karatsuba method [4] or Montgomery formulae [6], which are among the main algorithms used in every finite fields. On the other hand, for finite fields of fixed characteristics, there are other methods

S. Vaudenay (Ed.): AFRICACRYPT 2008, LNCS 5023, pp. 406–414, 2008.
© Springer-Verlag Berlin Heidelberg 2008

that give more efficient algorithms for polynomial multiplication than Karatsuba and Montgomery in some cases. Some of those methods are Chinese Remainder Theorem (CRT) method [5] and Discrete Fourier Transform (DFT) method. In [1,2], using DFT method, multiplication formula in [3] for $\mathbb{F}_{3^{6m}}$ is improved.

In this paper, using a method based on CRT for polynomial multiplication over $\mathbb{F}_3$ and suitable reductions, we obtained an efficient multiplication method for finite fields of characteristic 3. For $5 \leq \ell \leq 18$, we show that our method gives canonical multiplication formulae over $\mathbb{F}_{3^{\ell m}}$ for any $m \geq 1$ with the best multiplicative complexity improving the bounds in [6]. Moreover, we give explicit formula in the case $\mathbb{F}_{3^{6 \cdot 97}}$.

The rest of paper is organized as follows. In Section 2, we introduced our method. Applying our method we obtain explicit formulae in Section 3. We also compare our results with the previous results in Section 3. We conclude our paper in Section 4.

2 The Method

Let $\mathbb{F}_q$ be the field with q elements where $q = 3^n$. Unless stated otherwise, all polynomials considered here are in $\mathbb{F}_3[x]$. Let $n \geq 1$ be an integer. A polynomial $A(x)$ of the form

$$A(x) = a_0 + a_1 x + \ldots + a_{n-1} x^{n-1}, \ a_{n-1} \neq 0$$

is called an n-term polynomial. $M(n)$ denotes the minimum number of multiplications needed in $\mathbb{F}_3$ in order to multiply two arbitrary n-term polynomials. We note that $M(n)$ is also called multiplicative complexity of n-term polynomials. Let $n \geq 1$ be an integer, $f(x)$ be an irreducible polynomial and $\ell \geq 1$ be an integer such that

$$\ell \ deg(f(x)) < 2n - 1.$$

Let $A(x)$ and $B(x)$ be arbitrary n-term polynomials, $C(x) = A(x)B(x)$ and $\overline{A}(x), \overline{B}(x), \overline{C}(x)$ be the uniquely determined polynomials of degree strictly less than $\ell \ deg(f(x))$ such that

$$\overline{A}(x) \equiv A(x) \bmod f(x)^{\ell}, \ \overline{B}(x) \equiv B(x) \bmod f(x)^{\ell}, \ \overline{C}(x) \equiv C(x) \bmod f(x)^{\ell}.$$

Notation 1. *Let $M_{f,\ell}(n)$ denote the minimum number of multiplications needed in $\mathbb{F}_q$ in order to obtain $\overline{C}(x)$ from given n-term polynomials $A(x)$ and $B(x)$. Obtaining such $\overline{C}(x)$ from $A(x)$ and $B(x)$ is called multiplication of n-term polynomials modulo $f(x)^{\ell}$.*

Let $1 \leq w \leq 2n - 2$ be an integer and $C(x) = c_0 + c_1 x + \ldots + c_{2n-2}x^{2n-2}$. Obtaining the last w coefficients $c_{2n-2}, c_{2n-3}, \ldots, c_{2n-1-w}$ of $C(x)$ is defined as the multiplication of n-term polynomials modulo $(x - \infty)^w$ [5,7].

Notation 2. *Let $M_{(x-\infty),w}(n)$ denote the minimum number of multiplications needed in $\mathbb{F}_q$ in order to obtain $c_{2n-2}, c_{2n-3}, \ldots, c_{2n-1-w}$ from given n-term polynomials $A(x)$ and $B(x)$.*

CRT method for finite field polynomial multiplication can be summarized as follows. For $1 \le i \le t$, let $m_i(x) = f_i(x)^{\ell_i}$ be the ℓ_i-th power ($\ell_i \ge 1$) of an irreducible polynomial $f_i(x)$ such that $deg(m(x)) \ge 2n - 1$ where $m(x) = \prod_{i=1}^{t} m_i(x)$. Assume that $f_1(x), ..., f_t(x)$ are distinct. Let $w \ge 1$ be an integer which corresponds to multiplication modulo $(x - \infty)^w$ (see [7] and [5, p. 34]). It follows from CRT algorithm that if

$$w + \sum_{i=1}^{t} \ell_i \, deg(f_i(x)) \ge 2n - 1 \tag{1}$$

then

$$M(n) \le M_{(x-\infty),w}(n) + \sum_{i=1}^{t} M_{f,\ell}(n). \tag{2}$$

The value of $M_{f,\ell}(n)$ can be bounded from above by $M(deg(f^\ell)) \le M(\ell \cdot deg(f))$. For example in [7], $M_{f,\ell}(n) \le M(\ell \cdot deg(f))$ is used for binary fields. In [8], we improved the estimate of $M_{f,\ell}(n)$ for the binary field $\mathbb{F}_2$. The same techniques also work for any finite field $\mathbb{F}_q$, in particular for $\mathbb{F}_3$. Before giving the improvement, we give the following definition.

Definition 1. *Let $R = \mathbb{F}_q[x]$ be the ring of polynomials over $\mathbb{F}_q$ in variable x, $\ell \ge 1$ be an integer and*

$$A(Y) = a_0(x) + a_1(x)Y + ... + a_{\ell-1}(x)Y^{\ell-1},$$
$$B(Y) = b_0(x) + b_1(x)Y + ... + b_{\ell-1}(x)Y^{\ell-1}$$

be two ℓ-term polynomials in the polynomial ring $R[Y]$ over R. Let $c_0(x), ..., c_{2\ell-2}(x) \in R$ be given by

$$c_0(x) + c_1(x)Y + ... + c_{2\ell-2}(x)Y^{2\ell-2} = A(Y)B(Y).$$

Let $\lambda(\ell)$ denote the minimum number of multiplications needed in R in order to obtain $c_0(x), c_1(x), ..., c_{\ell-1}(x)$.

For the sake of completeness we prefer to give a full proof of this improvement.

Theorem 1. *Let $f(x)$ be an irreducible polynomial and $\ell \ge 1$ be an integer such that $\ell \, deg(f(x)) < 2n - 1$. We have*

$$M_{f,\ell}(n) \le \lambda(\ell)M(deg(f)). \tag{3}$$

Proof. Let $A(x)$ be an n-term polynomial and $\overline{A}(x)$ be the uniquely determined polynomial of degree strictly less than $\ell \, deg(f(x))$ such that $\overline{A}(x) \equiv A(x) \bmod f(x)^\ell$. Let $a_0(x), a_1(x), ..., a_{\ell-1}(x)$ be uniquely determined polynomials such that

$$\overline{A}(x) = a_0(x) + a_1(x)f(x) + ... + a_{\ell-1}(x)f(x)^{\ell-1} \text{ and } deg(a_i(x)) < deg(f(x))$$

for $0 \le i \le \ell - 1$. Let $\overline{B}(x)$ and $b_0(x), b_1(x), ..., b_{\ell-1}(x)$ be defined similarly. Note that $a_i(x)$ and $b_j(x)$, for $0 \le i \le \ell - 1$, $0 \le j \le \ell - 1$ are obtained without any multiplication. Let $R = \mathbb{F}_3[x]$ and $\widetilde{A}(Y)$ and $\widetilde{B}(Y)$ be polynomials in $R[Y]$ such that

$$\widetilde{A}(Y) = a_0(x) + a_1(x)Y + \ldots + a_{\ell-1}(x)Y^{\ell-1},$$
$$\widetilde{B}(Y) = b_0(x) + b_1(x)Y + \ldots + b_{\ell-1}(x)Y^{\ell-1}.$$

Define $\widetilde{C}(Y) = \widetilde{A}(Y)\widetilde{B}(Y)$ and let $c_0(x), c_1(x), \ldots, c_{\ell-1}(x) \in R$ be the first ℓ coefficients of $\widetilde{C}(Y)$. Since $Y^i \equiv 0 \bmod Y^\ell$ for $i \geq \ell$, $M_{f,\ell}(n)$ refers to computing the first ℓ coefficients of $\widetilde{A}(Y)\widetilde{B}(Y)$. Therefore the first ℓ coefficients $c_0(x), c_1(x), \ldots, c_{\ell-1}(x)$ can be obtained from $A(x)$ and $B(x)$ with at most $\lambda(\ell)$ multiplications of certain coefficients of $\widetilde{A}(Y)$ and $\widetilde{B}(Y)$ in R. Since each coefficient of $\widetilde{A}(Y)$ and $\widetilde{B}(Y)$ is a $deg(f(x))$-term polynomial over $\mathbb{F}_3$, any multiplication can be done with $M(deg(f(x)))$ multiplications over $\mathbb{F}_3$. This completes the proof.

Remark 1. Let $1 \leq w \leq 2n-1$ be an integer. Recall that the notation $M_{(x-\infty),w}(n)$ is given in Notation 2. It is clear that $M(1) = 1$. Using similar methods as in Theorem 1 we also obtain that

$$M_{(x-\infty),w}(n) \leq \lambda(w)M(1) = \lambda(w).$$

Corollary 1. $M_{x,w}(n)$ *corresponds to computing first* w *coefficients* $c_0, c_1, \ldots, c_w$ *of* $c(x)$ *and* $M_{x,w}(n) = M_{(x-\infty),w}(n) \leq \lambda(w)$.

Some effective upper bounds of $\lambda(\ell)$ is given in the following lemma which contributes to improvements on $M_{f,\ell}(n)$.

Proposition 1. $\lambda(3) \leq 5$, $\lambda(4) \leq 8$, $\lambda(5) \leq 11$, $\lambda(6) \leq 15$, $\lambda(7) \leq 19$, $\lambda(8) \leq 24$, *and* $\lambda(9) \leq 29$.

Proof. We use a Karatsuba type method (cf., for example in [9]). Here we present an explicit proof of $\lambda(3) \leq 5$ only. The other statements can be proved similarly (see also [9]). Let $A(x)$ and $B(x)$ be arbitrary n-term polynomials, $C(x) = A(x)B(x)$ and c_0, c_1, c_2 be the first 3 coefficients of $C(x)$. Then

$$c_0 = D_0$$
$$c_1 = D_{01} - D_0 - D_1$$
$$c_2 = D_{02} + D_1 - D_0 - D_2$$

where $D_i = a_i b_i$ and $D_{st} = (a_s + a_t)(b_s + b_t)$. Then

$$\lambda(3) \leq \#\{D_0, D_1, D_2, D_{01}, D_{02}\} = 5.$$

This completes the proof of $\lambda(3) \leq 5$.

In Table 1, we list some improvements on the upper bound on $M_{f,\ell}(n)$. Note that computation of $M_{f,\ell}(n)$ can be done by first computing the polynomial multiplication then reducing the result modulo f^ℓ. Therefore we compare our bounds with bounds in [6]. For the range of indices i and j in Table 1 and Table 2, f_{ij} denotes an irreducible polynomial of degree i over $\mathbb{F}_3$ which are defined as follows: $f_{11} = x$, $f_{12} = x + 1$, $f_{13} = x + 2$, $f_{21} = x^2 + 1$, $f_{22} = x^2 + x + 2$, $f_{23} = x^2 + 2x + 2$, $f_{31} = x^3 + 2x + 1$, $f_{32} = x^3 + 2x + 2$, $f_{33} = x^3 + 2x^2 + 2x + 2$, $f_{34} = x^3 + x^2 + x + 2$, $f_{35} = x^3 + x^2 + 2$, $f_{36} = x^3 + 2x^2 + x + 1$, $f_{37} = x^3 + x^2 + 2x + 1$, $f_{38} = x^3 + 2x^2 + 1$.

Table 1. Upper Bounds for $M_{f,\ell}(n)$

f	l	$M_{f,\ell}(n)[6]$	New $M_{f,\ell}(n)$
f_{11}, f_{12}, f_{13}	3	6	5
f_{11}, f_{12}, f_{13}	4	9	8
f_{11}, f_{12}, f_{13}	5	13	11
f_{11}, f_{12}, f_{13}	6	17	15
f_{11}, f_{12}, f_{13}	7	22	19
f_{11}, f_{12}, f_{13}	8	27	24
f_{11}, f_{12}, f_{13}	9	34	29
f_{21}, f_{22}, f_{23}	3	17	15
f_{21}, f_{22}, f_{23}	4	27	24
f_{21}, f_{22}, f_{23}	5	39	33
$f_{31}, ..., f_{38}$	3	34	30

3   Explicit Formulae and Comparison

In this section, up to our knowledge we give the best known bounds for n-term polynomial multiplication over $\mathbb{F}_3$ for $5 \leq n \leq 18$ and we give an explicit formula for multiplication in $\mathbb{F}_{3^{6m}}$ which is used in id-based cryptography for efficient Tate paring computations. Using Theorem 1, Proposition 1 and (2), the bounds in Table 2 are obtained.

Table 2. Upper Bounds for $M(n)$

n	$M(n)[6]$	New $M(n)$	Modulus polynomials
2	3	3	$(x - \infty), f_{11}, f_{12}$
3	6	6	$(x - \infty), f_{11}^2, f_{12}, f_{13}$
4	9	9	$(x - \infty), f_{11}^2, f_{12}, f_{13}, f_{21}$
5	13	12	$(x - \infty), f_{11}^2, f_{12}, f_{13}, f_{21}, f_{22}$
6	17	15	$(x - \infty)^2, f_{11}, f_{12}, f_{13}, f_{21}, f_{22}, f_{23}$
7	22	19	$(x - \infty)^2, f_{11}^2, f_{12}, f_{13}, f_{21}, f_{22}, f_{23}$
8	27	23	$(x - \infty)^3, f_{11}^3, f_{12}^2, f_{13}, f_{21}, f_{22}, f_{23}$
9	34	27	$(x - \infty)^3, f_{11}^3, f_{12}^2, f_{13}^2, f_{21}, f_{22}, f_{23}$
10	39	31	$(x - \infty)^3, f_{11}^3, f_{12}^2, f_{13}^2, f_{21}, f_{22}, f_{23}, f_{31}$
11	46	35	$(x - \infty)^3, f_{11}^3, f_{12}^3, f_{13}^2, f_{21}, f_{22}, f_{23}, f_{31}$
12	51	39	$(x - \infty)^3, f_{11}^3, f_{12}^3, f_{13}^2, f_{21}, f_{22}, f_{23}, f_{31}, f_{32}$
13	60	43	$(x - \infty)^3, f_{11}^3, f_{12}^3, f_{13}^2, f_{21}, f_{22}, f_{23}, f_{31}, f_{32}, f_{33}$
14	66	47	$(x - \infty)^3, f_{11}^2, f_{12}^3, f_{13}^2, f_{21}, f_{22}, f_{23}, f_{31}, f_{32}, f_{33}, f_{34}$
15	75	51	$(x - \infty)^2, f_{11}^2, f_{12}^3, f_{13}^2, f_{21}, f_{22}, f_{23}, f_{31}, f_{32}, f_{33}, f_{34}, f_{35}$
16	81	55	$(x - \infty)^3, f_{11}^3, f_{12}^2, f_{13}^2, f_{21}, f_{22}, f_{23}, f_{31}, f_{32}, f_{33}, f_{34}, f_{35}$
17	94	59	$(x - \infty)^3, f_{11}^3, f_{12}^3, f_{13}^3, f_{21}, f_{22}, f_{23}, f_{31}, f_{32}, f_{33}, f_{34}, f_{35}$
18	102	63	$(x - \infty)^3, f_{11}^3, f_{12}^3, f_{13}^2, f_{21}, f_{22}, f_{23}, f_{31}, f_{32}, f_{33}, f_{34}, f_{35}, f_{36}$

Note that we can conclude from Table 2

$$M(n) \leq \begin{cases} 3n - 3 & \text{if } 2 \leq n \leq 6 \\ 4n - 9 & \text{if } 7 \leq n \leq 18. \end{cases}$$

The bounds in Table 2 is also valid for any polynomial multiplication over $\mathbb{F}_{3^m}$ because of the following Theorem.

Theorem 2. *The formulae for multiplication of two arbitrary n-term polynomials over $\mathbb{F}_3$ are also valid for multiplication of two arbitrary n-term polynomials over $\mathbb{F}_{3^m}$, where m is any positive integer.*

The proof can be found in [10].

The finite fields of $\mathbb{F}_{3^{6m}}$, where m is prime are used in id-based cryptography for efficient computation of the Tate pairing. In [3], multiplication in $\mathbb{F}_{3^{6m}}$ is used 18 multiplications in $\mathbb{F}_{3^m}$. In [1,2], multiplication in $\mathbb{F}_{3^{6m}}$ is decreased to 15 multiplications in $\mathbb{F}_{3^m}$. In Appendix A, we give a formula for 6 term polynomial multiplication over $\mathbb{F}_3$ which requires 15 multiplications in $\mathbb{F}_3$. Since the formula for multiplication of two arbitrary n-term polynomials over $\mathbb{F}_3$ is also valid for multiplication of two arbitrary n-term polynomials over $\mathbb{F}_{3^m}$, where m is any positive integer, the formula given in the Appendix A can be used for the multiplication in $\mathbb{F}_{3^{6m}}$ with 15 multiplications in $\mathbb{F}_{3^m}$. The following example compares our formula and the formula given in [1,2].

Example 1. We will show that multiplication in $\mathbb{F}_{3^{6 \cdot 97}}$ can be done with 15 multiplications in $\mathbb{F}_{3^{97}}$. Let us construct,

$$\mathbb{F}_{3^{97}} \cong \mathbb{F}_3[x]/(x^{97} + x^{16} + 2),$$
$$\mathbb{F}_{3^{6 \cdot 97}} \cong \mathbb{F}_{3^{97}}[y]/(y^6 + y - 1).$$

Let $\alpha, \beta, \gamma \in \mathbb{F}_{3^{6 \cdot 97}}$ such that $\alpha = \sum_{i=0}^{5} a_i y^i$, $\beta = \sum_{i=0}^{5} b_i y^i$ and $\gamma = \alpha \cdot \beta = \sum_{i=0}^{5} c_i y^i$.

Then the coefficients of γ can be found as follows: First compute the coefficients of $\left(\sum_{i=0}^{5} a_i y^i \right) \left(\sum_{i=0}^{5} b_i y^i \right)$ and then reduce it modulo $y^6 + y - 1$. Therefore, using the formula in Appendix A we get

$c_0 = -m_{15} - m_1 + m_{10} - m_6 - m_5 + m_7 - m_8 - m_9 - m_{12} - m_{11};$
$c_1 = m_{15} + m_2 - m_3 - m_4 + m_5 - m_7 - m_8 + m_{10} - m_{11} + m_{12} + m_{13} + m_{14};$
$c_2 = -m_3 + m_5 + m_4 - m_6 - m_1 - m_2 - m_8 + m_9 - m_{13};$
$c_3 = -m_3 - m_5 + m_7 - m_1 - m_8 - m_9 - m_{13} - m_{15};$
$c_4 = m_6 + m_{13} - m_{12} - m_{11} - m_8 - m_{10} - m_5 - m_7 + m_2 - m_3 - m_4;$
$c_5 = m_{14} - m_8 + m_9 - m_{10} - m_6 + m_{13} - m_1 + m_3 - m_{11} + m_{12};$

where m_i's are given in Appendix A.

The explicit formula for multiplication in $\mathbb{F}_{3^{6 \cdot 97}}$ in [1,2] can be seen in Appendix B. $\mathbb{F}_{3^{6 \cdot 97}}$ is constructed in [1,2] using tower field representation, i.e.

$$\mathbb{F}_{3^{97}} \cong \mathbb{F}_3[x]/(x^{97} + x^{16} + 2),$$
$$\mathbb{F}_{3^{2 \cdot 97}} \cong \mathbb{F}_{3^{97}}[y]/(y^2 + 1),$$
$$\mathbb{F}_{3^{6 \cdot 97}} \cong \mathbb{F}_{3^{2 \cdot 97}}[z]/(z^3 - z - 1).$$

Therefore, the formula in [1,2] contains multiplication by $\mp s$, $\mp(s+1)$ and $\mp(s-1)$, where $s \in \mathbb{F}_{3^{2 \cdot 97}}$ is a root of $y^2 + 1$. For both our proposed formula and the formula in [2], the number of multiplications is 15. The number of additions for our proposed formula is 137. Note that there are multiplications of form $(s \mp 1)m_i$ in the formula in [2]. Here $s \notin \mathbb{F}_3$. In calculation of the number of additions, if we disregard the multiplication by s for the formula in [2], and if we consider the cost of each multiplication of the form $(s \mp 1)m_i$ for the formula in [2] as 1 addition only, then the number of additions for the formula in [2] is still 138. Moreover, in our formula the only nonzero coefficients are ∓ 1 and we do not need to introduce intermediate field extensions like $\mathbb{F}_{3^{2 \cdot 97}}$ containing $s \notin \mathbb{F}_3$. Therefore it seems that our construction would be preferable to the construction in [1,2].

4   Conclusion

For each $5 \le \ell \le 18$ we obtain a canonical multiplication formula in $\mathbb{F}_{3^{\ell m}}$ which is valid for any $m \ge 1$. To the best of our knowledge, these formulae have the best known multiplication complexity in the literature improving the bounds in [6]. Moreover, we give explicit formula in the case $\mathbb{F}_{3^{6 \cdot 97}}$.

Acknowledgments. The authors would like to thank the anonymous referees for their useful suggestions. This work was supported by TÜBİTAK under Grant No. TBAG-107T826.

References

1. Gorla, E., Puttmann, C., Shokrollahi, J.: Explicit formulas for efficient multiplication in $\mathbb{F}_{3^{6m}}$. In: Adams, C., Miri, A., Wiener, M. (eds.) Selected Areas in Cryptography (SAC 2007). LNCS, vol. 4876, pp. 173–183. Springer, Heidelberg (2007), http://www.arxiv.org/PS_cache/arxiv/pdf/0708/0708.3014v1.pdf
2. Shokrollahi, J., Gorla, E., Puttmann, C.: Efficient FPGA-Based Multipliers for $\mathbb{F}_{3^{97}}$ and $\mathbb{F}_{3^{6 \cdot 97}}$. In: Field Programmable Logic and Applications (FPL 2007), http://www.arxiv.org/PS_cache/arxiv/pdf/0708/0708.3022v1.pdf
3. Kerins, T., Marnane, W.P., Popovici, E.M., Barreto, P.S.L.M.: Efficient hardware for the tate pairing calculation in characteristic three. In: Rao, J.R., Sunar, B. (eds.) CHES 2005. LNCS, vol. 3659, pp. 412–426. Springer, Heidelberg (2005)
4. Karatsuba, A., Ofman, Y.: Multiplication of multidigit numbers by automata. Soviet Physics-Doklady 7, 595–596 (1963)
5. Winograd, S.: Arithmetic Complexity of Computations. SIAM, Philadelphia (1980)
6. Montgomery, P.L.: Five, six, and seven-term Karatsuba-like formulae. IEEE Transactions on Computers 54(3), 362–369 (2005)
7. Fan, H., Anwar Hasan, M.: Comments on Five, Six, and Seven-Term Karatsuba-Like Formulae. IEEE Transactions on Computers 56(5), 716–717 (2007)
8. Cenk, M., Özbudak, F.: Improved Polynomial Multiplication Formulae over $\mathbb{F}_2$ Using Chinese Remainder Theorem. IEEE Transactions on Computers (submitted)
9. Weimerskirch, A., Paar, C.: Generalizations of the Karatsuba Algorithm for Polynomial Multiplication. Technical Report, Ruhr-Universität Bochum, Germany (2003), http://www.crypto.ruhr-uni-bochum.de/imperia/md/content/texte/publications/tecreports/kaweb.pdf

10. Wagh, M.D., Morgera, S.D.: A new structured design method for convolutions over finite fields. Part I", IEEE Transactions on Information Theory 29(4), 583–594 (1983)

Appendix A

In Appendix A, we give explicit formula for 6-term polynomial multiplication over $\mathbb{F}_3$. Let $A(x) = \sum_{i=0}^{5} a_i x^i$ and $B(x) = \sum_{i=0}^{5} b_i x^i$ be polynomials over $\mathbb{F}_3$. Let $C(x) = \sum_{i=0}^{10} c_i x^i \in \mathbb{F}_3[x]$ be the polynomial defined by $C(x) = A(x)B(x)$. We obtain the following explicit formula consisting of the 15 multiplications. We first define the multiplications m_i for $1 \leq i \leq 15$ and then we give the formula for obtaining the coefficients of the polynomial $C(x)$ using these multiplications.

$m_1 = (a_0 + a_1 + a_2 + a_3 + a_4 + a_5)(b_0 + b_1 + b_2 + b_3 + b_4 + b_5)$;
$m_2 = (a_0 + a_1)(b_0 + b_1)$;
$m_3 = a_0 b_0$;
$m_4 = a_1 b_1$;
$m_5 = (a_1 - a_3 - a_5 + a_2)(b_1 - b_3 - b_5 + b_2)$;
$m_6 = (a_0 - a_2 - a_4 + a_1 - a_5)(b_0 - b_2 - b_4 + b_1 - b_5)$;
$m_7 = (a_0 - a_2 + a_4 + a_1 - a_3 + a_5)(b_0 - b_2 + b_4 + b_1 - b_3 + b_5)$;
$m_8 = (a_0 - a_2 + a_4)(b_0 - b_2 + b_4)$;
$m_9 = (a_1 - a_3 + a_5)(b_1 - b_3 + b_5)$;
$m_{10} = (a_0 - a_1 + a_2 - a_3 + a_4 - a_5)(b_0 - b_1 + b_2 - b_3 + b_4 - b_5)$;
$m_{11} = (a_0 + a_2 - a_4 - a_3)(b_0 + b_2 - b_4 - b_3)$;
$m_{12} = (a_0 - a_4 + a_3 + a_1 - a_5)(b_0 - b_4 + b_3 + b_1 - b_5)$;
$m_{13} = (a_0 + a_2 - a_4 + a_3)(b_0 + b_2 - b_4 + b_3)$;
$m_{14} = (a_1 - a_3 - a_5 - a_2)(b_1 - b_3 - b_5 - b_2)$;
$m_{15} = a_5 b_5$;
$c_0 = m_3$;
$c_1 = (m_2 - m_3 - m_4)$;
$c_2 = -m_{15} + m_6 - m_{13} - m_{12} + m_{11} - m_{14} - m_8 + m_9 - m_{10} - m_1$;
$c_3 = m_{13} + m_5 + m_{10} - m_{11} - m_{14} - m_1 - m_7 + m_8 + m_9$;
$c_4 = m_{13} - m_5 + m_6 - m_{10} + m_{14} - m_{12} + m_8 - m_9 - m_1$;
$c_5 = -m_1 + m_{10} - m_6 - m_5 + m_7 - m_8 - m_9 - m_{12} - m_{11}$;
$c_6 = -m_6 + m_{13} - m_1 + m_{12} - m_{11} + m_{14} - m_8 + m_9 - m_{10}$;
$c_7 = -m_{13} - m_5 + m_{10} + m_{11} + m_{14} - m_1 - m_7 + m_8 + m_9$;
$c_8 = -m_3 - m_6 - m_{13} + m_5 - m_1 - m_{10} + m_{12} - m_{14} + m_8 - m_9$;
$c_9 = -m_1 - m_2 + m_3 + m_4 + m_5 + m_6 + m_7 - m_8 - m_9 + m_{10} + m_{11} + m_{12}$;
$c_{10} = m_{15}$

Appendix B

In Appendix B, we give the multiplication formula for $\mathbb{F}_{3^{6 \cdot 97}}$ given in [2]. Let $\alpha, \beta \in \mathbb{F}_{3^{6 \cdot 97}}$ be give as:

$$\alpha = a_0 + a_1 s + a_2 r + a_3 rs + a_4 r^2 + a_5 r^2 s,$$

$$\beta = b_0 + b_1 s + b_2 r + b_3 rs + b_4 r^2 + b_5 r^2 s,$$

where $a_0, ..., b_5 \in \mathbb{F}_{3^{97}}$, $s \in \mathbb{F}_{3^{2\cdot97}}$ and $r \in \mathbb{F}_{3^{6\cdot97}}$ are roots of $y^2 + 1$ and $z^3 - z - 1$, respectively. Let $\gamma = \alpha\beta$ be

$$\gamma = c_0 + c_1 s + c_2 r + c_3 rs + c_4 r^2 + c_5 r^2 s.$$

The coefficients $c_0, ..., c_5 \in \mathbb{F}_{3^{97}}$ of the product can be computed as follows:

$m_0 = (a_0 + a_2 + a_4)(b_0 + b_2 + b_4)$

$m_1 = (a_0 + a_1 + a_2 + a_3 + a_4 + a_5)(b_0 + b_1 + b_2 + b_3 + b_4 + b_5)$

$m_2 = (a_1 + a_3 + a_5)(b_1 + b_3 + b_5)$

$m_3 = (a_0 + sa_2 - a_4)(b_0 + sb_2 - b_4)$

$m_4 = (a_0 + a_1 + sa_2 + sa_3 - a_4 - a_5)(b_0 + b_1 + sb_2 + sb_3 - b_4 - b_5)$

$m_5 = (a_1 + sa_3 - a_5)(b_1 + sb_3 - b_5)$

$m_6 = (a_0 - a_2 + a_4)(b_0 - b_2 + b_4)$

$m_7 = (a_0 + a_1 - a_2 - a_3 + a_4 + a_5)(b_0 + b_1 - b_2 - b_3 + b_4 + b_5)$

$m_8 = (a_1 - a_3 + a_5)(b_1 - b_3 + b_5)$

$m_9 = (a_0 - sa_2 - a_4)(b_0 - sb_2 - b_4)$

$m_{10} = (a_0 + a_1 - sa_2 - sa_3 - a_4 - a_5)(b_0 + b_1 - sb_2 - sb_3 - b_4 - b_5)$

$m_{11} = (a_1 - sa_3 - a_5)(b_1 - sb_3 - b_5)$

$m_{12} = a_4 b_4$

$m_{13} = (a_4 + a_5)(b_4 + b_5)$

$m_{14} = a_5 b_5$

$c_0 = -m_0 + m_2 + (s+1)m_3 - (s+1)m_5 - (s-1)m_9 + (s-1)m_{11} - m_{12} + m_{14}$

$c_1 = m_0 - m_1 + m_2 - (s+1)m_3 + (s+1)m_4 - (s+1)m_5 + (s-1)m_9 - (s-1)m_{10} + (s-1)m_{11} - m_{12} - m_{13} + m_{14}$

$c_2 = -m_0 + m_2 + m_6 - m_8 + m_{12} - m_{14}$

$c_3 = m_0 - m_1 + m_2 - m_6 + m_7 - m_8 - m_{12} + m_{13} - m_{14}$

$c_4 = m_0 - m_2 - m_3 + m_5 + m_6 - m_8 - m_9 + m_{11} + m_{12} - m_{14}$

$c_5 = m_0 + m_1 - m_2 + m_3 - m_4 + m_5 - m_6 + m_7 - m_8 + m_9 - m_{10} + m_{11} - m_{12} + m_{13} - m_{14}$

Author Index

Lecture Notes in Computer Science

Sublibrary 4: Security and Cryptology

For information about Vols. 1– 3876
please contact your bookseller or Springer

Vol. 4499: Y.Q. Shi (Ed.), Transactions on Data Hiding and Multimedia Security II. IX, 117 pages. 2007.

Vol. 4464: E. Dawson, D.S. Wong (Eds.), Information Security Practice and Experience. XIII, 361 pages. 2007.

Vol. 4462: D. Sauveron, K. Markantonakis, A. Bilas, J.-J. Quisquater (Eds.), Information Security Theory and Practices. XII, 255 pages. 2007.

Vol. 4450: T. Okamoto, X. Wang (Eds.), Public Key Cryptography – PKC 2007. XIII, 491 pages. 2007.

Vol. 4437: J.L. Camenisch, C.S. Collberg, N.F. Johnson, P. Sallee (Eds.), Information Hiding. VIII, 389 pages. 2007.

Vol. 4392: S.P. Vadhan (Ed.), Theory of Cryptography. XI, 595 pages. 2007.

Vol. 4377: M. Abe (Ed.), Topics in Cryptology – CT-RSA 2007. XI, 403 pages. 2006.

Vol. 4356: E. Biham, A.M. Youssef (Eds.), Selected Areas in Cryptography. XI, 395 pages. 2007.

Vol. 4341: P.Q. Nguyên (Ed.), Progress in Cryptology - VIETCRYPT 2006. XI, 385 pages. 2006.

Vol. 4332: A. Bagchi, V. Atluri (Eds.), Information Systems Security. XV, 382 pages. 2006.

Vol. 4329: R. Barua, T. Lange (Eds.), Progress in Cryptology - INDOCRYPT 2006. X, 454 pages. 2006.

Vol. 4318: H. Lipmaa, M. Yung, D. Lin (Eds.), Information Security and Cryptology. XI, 305 pages. 2006.

Vol. 4307: P. Ning, S. Qing, N. Li (Eds.), Information and Communications Security. XIV, 558 pages. 2006.

Vol. 4301: D. Pointcheval, Y. Mu, K. Chen (Eds.), Cryptology and Network Security. XIII, 381 pages. 2006.

Vol. 4300: Y.Q. Shi (Ed.), Transactions on Data Hiding and Multimedia Security I. IX, 139 pages. 2006.

Vol. 4298: J.K. Lee, O. Yi, M. Yung (Eds.), Information Security Applications. XIV, 406 pages. 2007.

Vol. 4296: M.S. Rhee, B. Lee (Eds.), Information Security and Cryptology – ICISC 2006. XIII, 358 pages. 2006.

Vol. 4284: X. Lai, K. Chen (Eds.), Advances in Cryptology – ASIACRYPT 2006. XIV, 468 pages. 2006.

Vol. 4283: Y.Q. Shi, B. Jeon (Eds.), Digital Watermarking. XII, 474 pages. 2006.

Vol. 4266: H. Yoshiura, K. Sakurai, K. Rannenberg, Y. Murayama, S.-i. Kawamura (Eds.), Advances in Information and Computer Security. XIII, 438 pages. 2006.

Vol. 4258: G. Danezis, P. Golle (Eds.), Privacy Enhancing Technologies. VIII, 431 pages. 2006.

Vol. 4249: L. Goubin, M. Matsui (Eds.), Cryptographic Hardware and Embedded Systems - CHES 2006. XII, 462 pages. 2006.

Vol. 4237: H. Leitold, E.P. Markatos (Eds.), Communications and Multimedia Security. XII, 253 pages. 2006.

Vol. 4236: L. Breveglieri, I. Koren, D. Naccache, J.-P. Seifert (Eds.), Fault Diagnosis and Tolerance in Cryptography. XIII, 253 pages. 2006.

Vol. 4219: D. Zamboni, C. Krügel (Eds.), Recent Advances in Intrusion Detection. XII, 331 pages. 2006.

Vol. 4189: D. Gollmann, J. Meier, A. Sabelfeld (Eds.), Computer Security – ESORICS 2006. XI, 548 pages. 2006.

Vol. 4176: S.K. Katsikas, J. López, M. Backes, S. Gritzalis, B. Preneel (Eds.), Information Security. XIV, 548 pages. 2006.

Vol. 4117: C. Dwork (Ed.), Advances in Cryptology - CRYPTO 2006. XIII, 621 pages. 2006.

Vol. 4116: R. De Prisco, M. Yung (Eds.), Security and Cryptography for Networks. XI, 366 pages. 2006.

Vol. 4107: G. Di Crescenzo, A. Rubin (Eds.), Financial Cryptography and Data Security. XI, 327 pages. 2006.

Vol. 4083: S. Fischer-Hübner, S. Furnell, C. Lambrinoudakis (Eds.), Trust and Privacy in Digital Business. XIII, 243 pages. 2006.

Vol. 4064: R. Büschkes, P. Laskov (Eds.), Detection of Intrusions and Malware & Vulnerability Assessment. X, 195 pages. 2006.

Vol. 4058: L.M. Batten, R. Safavi-Naini (Eds.), Information Security and Privacy. XII, 446 pages. 2006.

Vol. 4047: M.J.B. Robshaw (Ed.), Fast Software Encryption. XI, 434 pages. 2006.

Vol. 4043: A.S. Atzeni, A. Lioy (Eds.), Public Key Infrastructure. XI, 261 pages. 2006.

Vol. 4004: S. Vaudenay (Ed.), Advances in Cryptology - EUROCRYPT 2006. XIV, 613 pages. 2006.

Vol. 3995: G. Müller (Ed.), Emerging Trends in Information and Communication Security. XX, 524 pages. 2006.

Vol. 3989: J. Zhou, M. Yung, F. Bao (Eds.), Applied Cryptography and Network Security. XIV, 488 pages. 2006.

Vol. 3969: Ø. Ytrehus (Ed.), Coding and Cryptography. XI, 443 pages. 2006.

Vol. 3958: M. Yung, Y. Dodis, A. Kiayias, T. Malkin (Eds.), Public Key Cryptography - PKC 2006. XIV, 543 pages. 2006.

Vol. 3957: B. Christianson, B. Crispo, J.A. Malcolm, M. Roe (Eds.), Security Protocols. IX, 325 pages. 2006.

Vol. 3956: G. Barthe, B. Grégoire, M. Huisman, J.-L. Lanet (Eds.), Construction and Analysis of Safe, Secure, and Interoperable Smart Devices. IX, 175 pages. 2006.

Vol. 3935: D.H. Won, S. Kim (Eds.), Information Security and Cryptology - ICISC 2005. XIV, 458 pages. 2006.

Vol. 3934: J.A. Clark, R.F. Paige, F.A.C. Polack, P.J. Brooke (Eds.), Security in Pervasive Computing. X, 243 pages. 2006.

Vol. 3928: J. Domingo-Ferrer, J. Posegga, D. Schreckling (Eds.), Smart Card Research and Advanced Applications. XI, 359 pages. 2006.

Vol. 3919: R. Safavi-Naini, M. Yung (Eds.), Digital Rights Management. XI, 357 pages. 2006.

Vol. 3903: K. Chen, R. Deng, X. Lai, J. Zhou (Eds.), Information Security Practice and Experience. XIV, 392 pages. 2006.

Vol. 3897: B. Preneel, S. Tavares (Eds.), Selected Areas in Cryptography. XI, 371 pages. 2006.